Demand-Side Management and Electricity End-Use Efficiency

NATO ASI Series

Advanced Science Institutes Series

A Series presenting the results of activities sponsored by the NATO Science Committee, which aims at the dissemination of advanced scientific and technological knowledge, with a view to strengthening links between scientific communities.

The Series is published by an international board of publishers in conjunction with the NATO Scientific Affairs Division

A	Life Sciences	Plenum Publishing Corporation
B	Physics	London and New York
C	Mathematical and Physical Sciences	Kluwer Academic Publishers Dordrecht, Boston and London
D	Behavioural and Social Sciences	
E	Applied Sciences	
F	Computer and Systems Sciences	Springer-Verlag
G	Ecological Sciences	Berlin, Heidelberg, New York, London,
H	Cell Biology	Paris and Tokyo

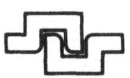

Series E: Applied Sciences - Vol. 149

Demand-Side Management and Electricity End-Use Efficiency

edited by

Anibal T. De Almeida

Departamento de Engenharia Electrotécnica,
University of Coimbra, Coimbra, Portugal

and

Arthur H. Rosenfeld

Center for Building Science,
Lawrence Berkeley Laboratory,
Berkeley, California, U.S.A.

Kluwer Academic Publishers

Dordrecht / Boston / London

Published in cooperation with NATO Scientific Affairs Division

Proceedings of the NATO Advanced Study Institute on
Demand-Side Management and Electricity End-Use Efficiency
Povoa de Varzim, Portugal
July 20–31, 1987

Library of Congress Cataloging in Publication Data

```
NATO Advanced Study Institute on "Demand-Side Management and
  Electricity End-Use Efficiency" (1987 : Póvoa do Varzim, Portugal)
    Demand-side management and electricity end-use efficiency /
  editors, Anibal T. De Almeida, Arthur H. Rosenfeld.
       p.   cm. -- (NATO ASI series. Series E, Applied sciences ; no.
  149)
    "Proceedings of the NATO Advanced Study Institute on 'Demand-Side
  Management and Electricity End-Use Efficiency', Póvoa do Varzim,
  Portugal, July 20-21, 1987"--T.p. verso.
    "Published in cooperation with NATO Scientific Affairs Division."
    Includes index.
    ISBN 9024736986
    1. Electric power--Conservation--Congresses.   I. Almeida, Anibal
  T. de.   II. Rosenfeld, Arthur H., 1926-    . III. North Atlantic
  Treaty Organization. Scientific Affairs Division.  IV. Title.
  V. Series.
  TK4015.N37 1987
  333.79'3216--dc19                                      88-14803
                                                            CIP
```

ISBN-13: 978-94-010-7127-7   e-ISBN-13: 978-94-009-1403-2
DOI: 10.1007/978-94-009-1403-2

CONTENTS

PREFACE

A NATO Advanced Study Institute on "Demand-Side Management and Electricity End-Use Efficiency" was held in order to present and to discuss some of the most recent developments in demand-side electric power management and planning methodologies as well as research progress in relevant end-use technologies.

Electricity is assuming an increasingly important role in buildings and industry, due to its flexibility, efficiency of conversion and cleanliness at the point of use. However the production and transmission of electricity requires huge investments and may have undesirable environmental impacts. The recent nuclear accident in Chernobyl and the damage caused by acid precipitation are creating increasing concerns about the impacts of power plants. Some environmental problems are local or regional, others such as global warming can affect the whole world. Although environmental impacts may be minimized with additional investments, electricity generation will become even more capital intensive.

Energy, and electricity in particular, is not directly consumed by people. To achieve improved standards of living, what is important is the level of production of goods and services. If it is possible to produce the same quantity of goods and services with less electricity and in a cost-effective way, substantial benefits can be gained. By reducing costs, electricity efficiency can raise the standards of living and increase the competitiveness of an economy. Electricity efficiency also leads to reduced requirements in power plant operation, thus leading to reduced consumption of primary energy supplies and a higher quality environment.

Traditionally the planning and management of the electricity sector has been conducted by adjusting the supply to the demand with little consideration of influencing demand. Demand-side management encompasses the technologies and methodologies which influence load behavior in a manner which optimizes the use of existing capacity, leading to benefits both to the electric utilities and the consumers. By managing the demand, the installation of costly new power plants can be deferred. Additionally, the introduction of cost-effective electricity end-use technologies can lead to a substantial decrease in load growth.

This text is organized in seven chapters, covering the technologies, methodologies, programs and policies which can lead to increased electricity productivity. The increasingly relevant role of microelectronics and computers for load management, metering and power conditioning is emphasized. The final chapter includes case studies in electricity conservation in Brazil, Denmark, Japan, Pakistan and in USA. These countries have a per capita electricity consumption ranging from 200 kWh/year in

Pakistan, to 10,000 kWh/year in USA, but in all the cases the potential for electricity savings is impressive.

The articles in this volume have been judged and accepted on their scientific quality, and language corrections may have been sacrificed in order to allow quick dissemination of knowledge to prevail.

This book presents the contents of the major lectures delivered during the Institute in Povoa do Varzim, between July 20 and July 31, 1987. The publication of this text will convey the proceedings of the Institute to a larger audience.

The editors want to thank the NATO Scientific Affairs Division, the University of Coimbra and the Center for Building Science-Lawrence Berkeley Laboratory, for the main financial and logistical support given to this Institute. Additionally the co-sponsorship of the following organizations is acknowledged:
-Junta Nacional de Investigacao Cientifica e Tecnologica
-International Energy Agency
-Instituto Nacional de Investigacao Cientifica
-Fundacao Luso-Americana
-Instituto de Apoio as Pequenas e Medias Empresas
-National Science Foundation
-Secretaria de Estado do Ambiente
-CIMPOR, Cimentos de Portugal
-Foundation Calouste Gulbenkian
-EDP, Electricidade de Portugal
-Electric Power Research Institute

The Editors

Anibal T. De Almeida
Arthur H. Rosenfeld

February 1988

I. Electricity: Trends and the Impact of Efficiency

ELECTRICITY - THE POLITICAL AND ECONOMIC CONTEXT

DAVID JONES

INTERNATIONAL ENERGY AGENCY

May I first say how much the International Energy Agency welcomes the fact that the NATO Advanced Study Institute are holding a conference on electricity demand and how pleased I am to have the chance to speak to you. At first sight this is a somewhat esoteric topic for two weeks of discussion. In fact, policy for electricity - and particularly for electricity use - is today one of the central problems of energy policy in IEA and indeed OECD and NATO countries. Incidentally, I should make it clear at this point that the IEA is an associated body of the OECD and consists of all the Member countries of that Organisation except France, Finland and Iceland.

As I have the privilege of being your opening speaker, I would like to try and do three things. First, I will say briefly why electricity policy is so central to energy policy. Second, I will describe the outlook for electricity consumption and supply in the OECD countries as we see them and define some of the problems to which this outlook gives rise. Third, I will throw out some suggestions as to what needs to be done. Inevitably in the time available I will have to generalise - it is not possible to cover 21 countries individually - but I should make clear at the start that there are important variations between Member countries and that what applies to one does not necessarily apply to another.

The General Energy Situation

But before moving to these three topics I should like to set the context by describing briefly the general energy situation in which the position of the electricity industry needs to be seen. 1986 has been a year of considerable turmoil in the energy markets. The most obvious aspect of this was the sharp decline in world oil prices - after five years of somewhat steadier decline in dollar terms. Prices reached their low point in July last year, when the average cost of crude oil imported into IEA countries was $10.69/barrel. Of course, actual market prices at which oil was traded included some prices well below the average. After wide fluctuations, the average cost of crude imported to the IEA rose again to $14.12/barrel by December, and it has continued on an unsteady path upwards through the early months of this year.

3

A. T. De Almeida and A. H. Rosenfeld (eds.), Demand-Side Management and Electricity End-Use Efficiency, 3–16.
© 1988 by Kluwer Academic Publishers.

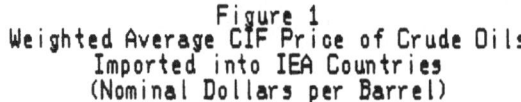

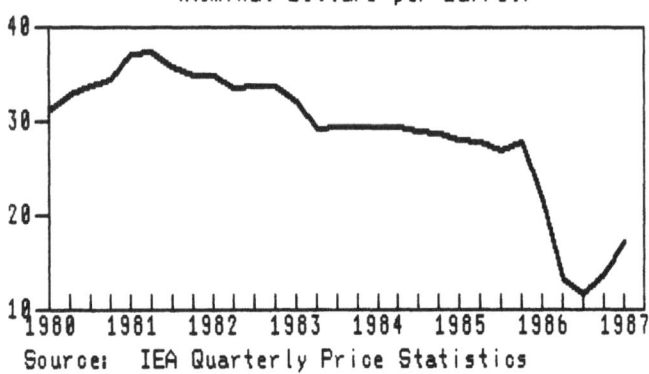

Figure 1
Weighted Average CIF Price of Crude Oils
Imported into IEA Countries
(Nominal Dollars per Barrel)

Source: IEA Quarterly Price Statistics

This fall in prices has brought problems - sometimes severe - to certain countries and energy industries. But it has brought benefits to OECD countries as a whole through lower inflation and faster economic growth than would otherwise have been the case. Any policies developed to deal with the problems of the energy industries must be so designed as not to lose these broad economic advantages.

Forecasting is a hazardous business and energy forecasters have an unenviable record of being proven wrong. There are, however, certain fundamental trends which point to tighter energy markets in the 1990s and to an increasing dependence by OECD countries on oil from the Middle East:

- the improvement in the efficiency with which energy is used which was such a marked feature of the energy scene between 1979 and 1983 is likely to slow down although the gains which have been made will certainly not be lost. As a result total energy consumption in OECD countries is likely to rise, perhaps at about 1.5 per cent a year between 1986 and 2000;

- oil production in the OECD and some other non-OECD countries is likely to fall;

- demand for oil on the world market from the developing countries is likely to rise.

Projections made by the IEA Secretariat - and I emphasize that these are illustrative projections and not a forecast of what is going to happen - suggest that demand for oil from the OPEC countries, which in 1986 was just under 18 million barrels a day, could rise to between 21 and 24 mbd in 1990 and to between 25 and 30 mbd by 2000. Past experience suggests that, when the demand for oil from OPEC countries goes much above 25 mbd, there is likely to be steady upward pressure on oil prices and a risk that even a small interruption in supplies could lead to a sharp and substantial increase in price.

But this development is not inevitable. The situation can be eased by effective energy policy to which policy for electricity is central.

The Importance of Electricity in Energy Policy

A study of Electricity in IEA Countries(1), which we published two years ago, gave the following reasons why electricity was central to energy policy:

- electricity generation consumes far more primary fuel than any other single industry - indeed in 1986 electricity generation accounted for 36 per cent of all the primary energy used in IEA countries;

- electricity generation is the only large-scale means of rendering water power, nuclear energy and some of the renewable energies useful to consumers and it also provides the dominant market for coal;

- electricity can promote energy security by directly displacing oil and in some cases gas; moreover, in the event of a disruption in supplies of primary energy, the electricity industry may be able to help by changing quickly the mix of fuels used in electricity generation;

- the electricity supply industry is normally organised as a public service. Governments can if they wish exercise through the electricity industry a substantial influence on the patterns of energy production and use in their countries.

If we were rewriting that study today we would add the point that electricity is at the centre of the political debate about energy and the environment and in particular the debate on nuclear energy which has followed the Chernobyl accident. It is no exaggeration to say that the development of policies for electricity which command wide public acceptance is essential to the maintenance at reasonable cost of the energy supplies on which the economic and social well being of our countries depend.

---

(1) Electricity in IEA Countries. Issues and Outlook. OECD/IEA Paris 1985.

Future Developments and Problems

(a) The Outlook for Electricity Demand

Since the 1950s, electricity has played an important and rapidly growing role in meeting the energy requirements of OECD countries. The proportion of electricity in total final consumption of energy in OECD countries grew from 9 per cent in 1960 to 16 per cent in 1984. Especially since 1973, electricity has been viewed as a broad substitute for oil in many economic sectors and for this reason has been a primary part of the efforts of many countries to reduce their dependence on imported oil.

There are many difficulties in forecasting the future path of electricity demand. In summary the main issues are:

- Economic growth. In general growth in overall economic activity increases the demand for electricity although the relationship is by no means a simple, or even a stable, one.

- Changes in economic structure and personal lifestyles. Shifts towards more electricity intensive industries and the rapidly increasing penetration of electric home appliances contributed substantially to the electricity demand growth of some 6 per cent per annum in the 1960s and 1970s, considerably in excess of the 4 per cent p.a. GDP growth of that period. Some developments, such as home and business electronics, have become much less electricity intensive and most electricity intensive industries in OECD Member countries are not growing, or not growing as fast, as other industries and services. There has recently been a major restructuring of energy intensive industry in the OECD, for example the steel industry in the United States and aluminium smelting in Japan. However, new ways of using electricity in industry, commerce and the home continually arise. The future balance between these conflicting tendencies is, however, difficult to foresee with precision.

- Substitution of electricity for other fuels. In the 1960s and 1970s, electricity increasingly replaced the use of steam in industry as a source of motive power and in consequence replaced the primary fuels coal, oil and gas used in the production of steam. The use of electricity also grew in place of other fuels in heating and drying where convenience, cleanliness and ease of control are important requirements, and in the steel industry the use of electric arc furnaces has grown in some countries relative to the use of blast furnaces. The prediction of future trends is difficult, partly because of uncertainty over the range of new technical possibilities for substitution arising in the future and the extent to which past substitution possibilities have been exhausted.

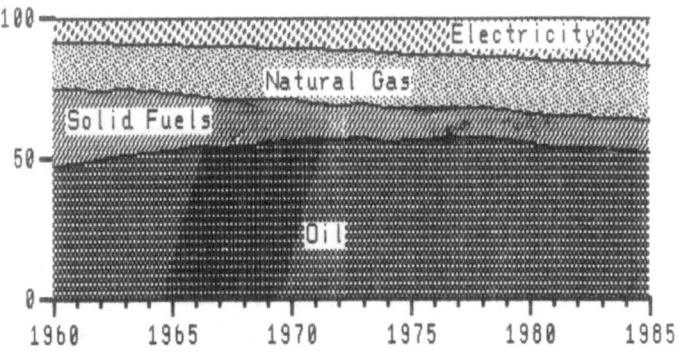

Figure 2
Shares of Total Final Energy Consumption
in IEA Countries

Source: Energy Balances of OECD Countries

— Improved efficiency in electricity use. Although often
   overshadowed by the factors described above, improvements in
   electricity efficiency have been, and will continue to be, an
   important factor affecting future electricity demand levels.
   Improved energy management practices and some technical
   efficiency improvements contributed significantly to the reduced
   electricity demand growth rates experienced in the early 1980s.
   There remains a large economic potential for further energy
   efficiency improvements in many end uses of electricity – a
   point to which I shall return later.

The impact of these factors can be seen in Figure 3, which shows the
development of the ratio of electricity demand to gross domestic product
in the three main OECD areas over the last 25 years. In Europe, there has
been a general upward movement, apart from a flat period between 1979 and
1982, indicating that electricity demand has grown faster than the rate of
economic growth. In North America the upward trend to 1976 was followed
by a relatively flat period to 1985 in which electricity demand kept pace
with economic activity. In the Pacific region, a sharp decline took place
in 1980 followed by a period of lower values.

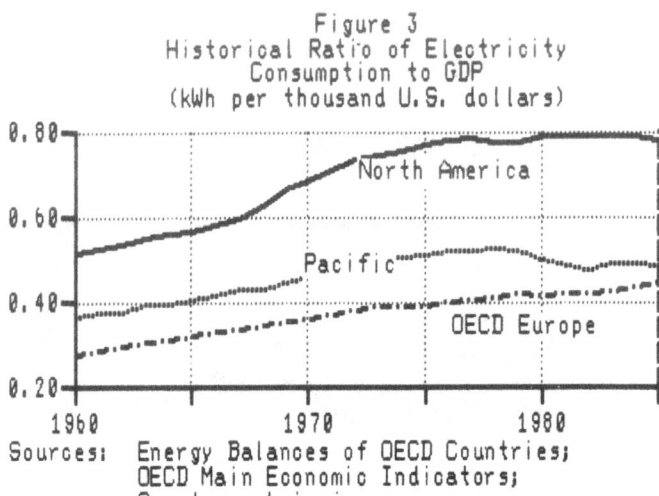

Figure 3
Historical Ratio of Electricity
Consumption to GDP
(kWh per thousand U.S. dollars)

Sources: Energy Balances of OECD Countries;
OECD Main Economic Indicators;
Country submissions.

Turning to the future, the IEA Secretariat has developed an econometric model to examine the dynamics of the demand for and supply of the various forms of energy in a consistent way. Various scenarios have been analysed, including the effects of a range of different economic growth and energy price assumptions. As illustrative working hypotheses, alternative assumptions were made that crude oil prices (in constant 1986 United States dollars):

(i) would range between $15 and $20 per barrel for the next three to five years, accompanied by occasional swings above and below; then that prices would rise gradually to levels of $25 to $35 by 2000;

(ii) would fluctuate around a relatively constant level of $15 to $20 ($17.50 on average) per barrel for the rest of the century.

The rate of growth of GDP in the OECD as a whole was assumed to average around 3 per cent per annum to 1990, and around 2.5 per cent per annum from 1990 to 2000.

Figure 4 shows the outcome of this work. Total primary energy requirements are shown as increasing from just over 3,800 million tons of oil equivalent in 1985 to somewhere between about 4,650 Mtoe and 5,080 Mtoe by the end of the century. The way in which this requirement will be split between the different sources of primary energy depends very much on the assumptions taken. In the constant oil price case the split is much the same as in 1985. On the increasing oil price case there is a considerable displacement of oil by solid fuels and nuclear energy.

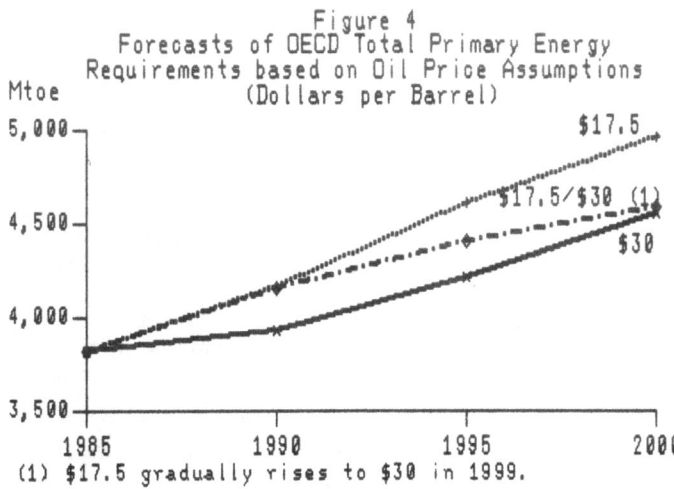

Figure 4
Forecasts of OECD Total Primary Energy
Requirements based on Oil Price Assumptions
(Dollars per Barrel)

(1) $17.5 gradually rises to $30 in 1999.

The outcome of this work also indicates, however, that lower oil prices are not expected to have a significant impact on electricity demand. The effect of crude oil price changes on consumer electricity prices, and therefore on demand, is small (with a few exceptions, such as the Netherlands and Japan where temporary tariff cuts have been implemented), and the potential for direct substitution of gas, coal and oil for electricity is limited (with the possible exception of Sweden). An area of considerable uncertainty regarding electricity demand arises from its possible penetration of the large space-heating market. However, in the OECD as a whole, electricity contributes only a small share of energy requirements in this market, even after ten years of relatively high hydrocarbon prices. If a major increase in this penetration were to occur, it would be necessary for the prices of alternative fuels to rise much higher than implied in all the scenarios examined by the Secretariat. In broad terms, OECD electricity demand may grow on average to the year 2000 by about 2.4 per cent per annum, somewhat lower than GDP.

The conversion of annual demand for electricity into a generating capacity requirement involves a number of assumptions, especially concerning the associated peak level of electricity demand and about adequate levels of reserve capacity to ensure reliable operation of the system. Estimating these parameters for the future is a difficult task, complicated by structural change in the economy, the possible increase in household peak demand and other factors. Our best estimate, which is subject to a wide margin of error, is that to meet an increased demand averaging 2.4 per cent per annum would require an increase by 2000 of about 200 GW in electricity generating capacity in OECD countries in addition to the new capacity which will be needed to replace old capacity which reaches the end of its economic life.

(b) The Options for New Generating Capacity

How is this demand for new generating capacity going to be met? The answer will vary greatly depending on the circumstances of particular utility systems. In determining their plans for meeting growing electricity demand, most utilities already realise they must carefully examine a broad range of options. These are likely to include conventional coal or nuclear powered power plants, decentralised generating capacity, such as cogeneration, or measures to improve the efficiency or level the peaks of end-use demand. Which combination of these options would be least costly to the utilities involved and their customers will vary depending on the cost-characteristics of current generating capacity, the environmental restrictions placed on new generating facilities and the existing potential for conservation, as well as other factors.

Some countries, such as Norway, rely primarily on comparatively cheap sources of electricity, such as hydro. But virtually all of these countries are now being forced to construct more conventional and more expensive power plants to meet future demand. There are other IEA countries, such as Italy, which still rely heavily on more expensive oil-fired generating capacity. As these countries gradually shift to coal or nuclear-powered facilities, their costs may actually decline. Most IEA countries, of course, fall between these two extremes. Although it is difficult to generalise about the cost trends of most utility systems, there seems to be a trend towards higher costs, especially in new facilities.

Another factor affecting utilities' options for meeting increasing demand is the method used for setting tariffs or rates. In many countries, rates are based largely on historical costs, which usually means some type of average between the highest and lowest cost sources of electricity. In other countries, however, utilities have wider freedom to set rates which reflect actual marginal costs.

These differences have important effects on the choice of options to meet future energy demand. For example, utilities that are experiencing rising costs for new generating capacity and which base their tariffs on historical, rather than marginal costs, usually have a strong financial incentive to seek reductions in the rate of demand growth. In such cases, increased energy efficiency that results in lower new capacity requirements will have financial benefits not only for the utility but also for all of its customers. On the other hand, utilities that enjoy stable or declining costs for new capacity or which base their tariffs on long-run marginal costs will not find it in their financial interest to reduce growth rates.

Other important factors that may influence national choices include restrictions on the siting of new generating plants, decisions on the use of nuclear power, the access of independent power producers to the utility system and government tax policies.

But in most countries three factors are likely to be dominant - comparative costs, security of supply and environmental issues. These considerations suggest that nuclear power and coal are the main contenders for new electricity supply. In large new plants they are signficantly cheaper than generation from oil or gas. A return to oil would also increase the risks, to which I have already referred, of much tighter oil markets.

Following the Chernobyl accident, a number of IEA Governments, such as Canada, the Federal Republic of Germany, Japan, the United Kingdom and the United States, have restated their commitment to nuclear power as an attractive and safe long-term option for electricity generation. Another group of countries have confirmed their already-established intention not to use nuclear power, either because they believe that even the residual element of risk in operating nuclear plants is too high to be acceptable or because for economic, scale, and other reasons, its use is inappropriate in their countries. Australia, Austria, Denmark, Ireland, New Zealand and Norway take this approach. Sweden has confirmed its intention to phase out nuclear power from the energy system. A few countries such as Italy, the Netherlands and Switzerland are still considering their position on nuclear power following Chernobyl. The half dozen or so nuclear power stations in these countries which may not see the light of day - and that is not certain - are of course, only a tiny blip on the graph of the world's energy balance.

But let us look at the story of nuclear power construction in the IEA in the last 20 years. Figure 5 shows, above the line, the amount of nuclear capacity which has been newly connected to the grids in IEA Member countries each year since 1965. As you can see, two distinct "cycles" of new connections can be observed, peaking in the mid-1970s and the mid-1980s respectively. Corresponding to these cycles, of course, there were years of peak activity in planning and in the start of construction - seen below the line, in the late 1960s and mid-1970s respectively. Now, the evidence of much more limited construction starts since 1977 is very clear. This virtually guarantees that, for the rest of this century, there will be no rapid growth in the contribution of nuclear power to IEA energy balances on a scale comparable with what we have seen in recent years or with the mid-1970s. Unless, that is, major new ordering of nuclear power plants begins very soon. Outside Japan, there is little evidence that this is likely at the moment. Even were a new cycle of ordering and construction to begin, it could have no impact until around the turn of the century.

12

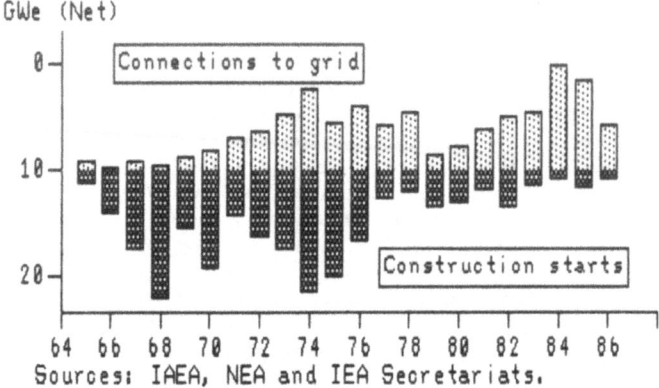

Figure 5
IEA New Nuclear Power Capacity
1965 to 1986

Sources: IAEA, NEA and IEA Secretariats.

The consumption of coal in the electricity sector has grown steadily in recent years. From 406 million tonnes of oil equivalent in 1978, inputs of coal to public power generation had increased to 550 Mtoe by 1985. Our current projections - and they may well prove wrong - are that demand for coal in electricity generation could increase by something under 300 Mtoe by the end of the century. But as I shall show in a moment, this growth in the use of coal is not without problems.

(c) The Problems of Providing Increased Generating Capacity

What are these problems? First, there is uncertainty. Costs and lead times vary between countries. However, a large new nuclear power station, the building of which started today, would probably cost between $1.5 and 3.5 billion and construction would take between six and ten years. The figures for a corresponding amount of coal capacity are between $1 and 1.5 billion and three to six years although, of course, the operating costs of coal-fired are much higher than those of nuclear stations. Premature building would thus involve heavy costs to the utilities and to the electricity consumer and would weaken the competitive position of electricity in relation to other sources of energy. But under building of new generating capacity can also lead to higher costs as well as reduced reliability. Underestimating demand can force the use of old inefficient power plants or the rapid construction of turbine generators that rely on more costly gas or oil. In the worst case, under-building can lead to power cuts.

Second, there is a problem of raising finance for investment on this scale. This is particularly severe in the United States, where tariff regulation which is operated by State utility commissions, has been implemented in a number of States in a way which made obtaining an adequate return on investment very difficult. For example, the recovery of investment costs is normally permitted only when a plant becomes "used and useful", effectively prohibiting the recovery of costs of construction and work in progress before the plant is complete and operating. There have been a growing number of cases in which the costs of new plants have been wholly or partially disallowed for inclusion in the rate base on the grounds that the decision to construct or the way construction was executed were imprudent. But this problem has also arisen in other countries. In Italy for example until two or three years ago, tariffs were held by the Government at a level which denied the nationalised electricity industry the ability to raise the finance needed to undertake necessary new investment.

Third, there is the debate about energy and the environment which I have already mentioned. All forms of electricity generation have environmental and safety consequences. It is vital to the future development of a satisfactory mix of electricity sources that these issues be clearly considered and appropriate responses proposed. The consequences, which can be local or transboundary, are different in kind for fossil fuels and for nuclear energy. In the case of fossil fuels, the most serious problem is that of emissions of sulphur and nitrogen oxides and, perhaps, of carbon dioxide. Emissions can be substantially reduced but not eliminated. Nuclear energy, on the other hand, gives rise to few, if any, environmental problems in normal operation but there is a risk – very remote in OECD countries – of an accident which could lead to widespread deposits of radioactive products. This remote risk can be further reduced by improved safety but cannot be eliminated. The choice of a balanced mix of generating options for each country will need to take account of both relative advantages and relative disadvantages of the alternative options available.

What should be done?

I turn now to what should be done. Energy Ministers of IEA countries met in Paris on 11th May to discuss the long-term energy outlook and whether policies needed to be changed or developed. They paid particular attention to the electricity sector. Let me comment on a few points from their Conclusions.

First, Ministers emphasized that there is important potential for improving the efficiency with which electricity is used, generated and transmitted – the theme of this Conference. It is essential that everything reasonably possible should be done to realise this potential. The acceptability of new generating stations to public opinion may well turn on people being satisfied that the promotion of efficiency is being seriously tackled by utilities and by governments. And a reduction in demand for new capacity as a result of greater efficiency will make it easier to deal with the problems of finance and uncertainty which I have described.

The IEA has reviewed data on the potential for efficiency improvements in a number of key electricity-using technologies including lighting, industrial motors, electric space conditioning and selected major appliances. This preliminary review suggests that the potential for efficiency improvements may be as much as 25 per cent using current best technologies. Because much of this potential can only be achieved by the gradual replacement of existing capital stocks with new, more energy-efficient devices and materials the process of improvement is necessarily slow. The process is underway but current trends and analyses of investment behaviour suggest that a significant proportion of the potential will not be realised. On the basis of a number of studies in the United States - it is not clear how far they can be applied to other IEA countries - there seems to be scope for reducing on an economic basis demand for electricity by between 0.5 and 1.5 percentage points a year below the levels that would otherwise have been realised.

Achievment of this reduction will require strong policy measures. Ministers at the 11th May meeting agreed to promote coherent and forceful strategies to advance efficiency in all the main sectors of energy consumption. To do this, they committed themselves to make a major effort, together with other government and industry leaders, to publicise the advantages of efficient energy use and the ways in which it could be achieved. They will support their efficiency strategies by such measures as wide-ranging information and education activities, fiscal incentives, and the development of innovative methods of private financing of energy conservation investments; voluntary or mandatory energy efficiency standards; the systematic and vigorous pursuit in all public sector activities of efficiency in energy use on an economic basis; and the dissemination of new, proved technologies in accordance with their conclusions on research, development and demonstration. The various organisations in both the public and private sectors concerned with efficient use of energy, particularly the energy-producing and consuming industries, should be actively involved in these activities.

The electric utilities have a key role in promoting the efficient use of electricity. In the United States, with strong encouragement from the Department of Energy, utilities are adopting a comprehensive planning approach which involves regular comparison of the cost of measures to reduce electricity use with the provision of more supply capacity. A utility like the Bonneville Power Administration has a wide range of measures - information programmes, audit schemes, cheap loans and grants. To some extent this is due to the special problems caused by United States' regulations but I have little doubt that there is an example here which could usefully be studied by the European utilities.

It should be emphasized, however, that there are both benefits and costs associated with attempts to meet growing electricity demand by conservation or other demand-side management activities. There is still much uncertainty in the estimates of the incremental effects of specific demand-side management activities and there is concern that conservation efforts may have less impact on peak demand than they do on off-peak consumption. This certainly does not mean that such efforts are too risky for utilities to undertake - utilities face similar risks in many areas - but it does mean that it is especially important to ensure that demand-side management activities are very carefully designed to reduce demand at the appropriate times and continuously evaluated to ensure they are achieving their objective.

Ministers also referred to the need to improve the efficiency with which electricity is generated. This will not reduce the need for new generating capacity. It will reduce the amount of primary energy used and thus reduce emissions. In striking contrast with developments in end-use sectors, there appears to have been little change in the efficiency of electricity generation in the past 20 years. Recent technological developments towards more efficient types of generating systems (e.g. combined cycle plants) seem capable of changing that trend in the longer term. In addition, "non-centralised" technologies, such as cogeneration, could increase generation efficiency considerably. Although these technologies have existed for a long time, their application has often not been thought to be economic. In addition, there are institutional factors which prevent more energy-efficient technologies, especially co-generation, from achieving their full economic potential. The IEA has work under way in this field.

Second, IEA Ministers stressed the importance of maintaining all the non-oil options for electricity generation, but particularly coal and nuclear. This means the environmental and safety problems associated with those forms of electricity generation must be tackled. Ministers emphasized particularly two points:

- new technologies - for example fluidised bed combustion - are being developed which will improve both the competitiveness and the environmental impact of using coal. It is essential that no time be lost in commercialising these new technologies. This is an area where collaboration between the private and public sectors and between countries should bring substantial benefit. The IEA is seeking to promote such collaboration through a series of workshops on the clean use of coal which we hope will lead to new collaborative projects involving the private as well as the public sector;

- the safety issues associated with the production of electricity are of fundamental importance, particularly in the case of nuclear energy. IEA countries have already made substantial progress but Ministers pledged their Governments to continue their efforts and in particular to give full political and technological support to arrangements for international co-operation on nuclear safety.

Third, new and renewable energies. The IEA, with the help of Member governments and outside experts, published a major study on this subject in April(2). This study showed that, except in certain specific situations, renewable energies were unlikely to be competitive with conventional sources of electricity generation. Except for hydropower,

---

(2) Renewable Sources of Energy. OECD/IEA Paris 1987

renewable energies were not therefore likely to make a major contribution
to the energy supplies of IEA countries in the foreseeable future. But
the study also provided a basis for Ministers to conclude that the
development of the renewable sources of energy could provide important new
options in the longer term in relation both to electricity generation and
energy suplies generally and should be actively pursued. The IEA will be
seeking to promote joint research activities in this field.

Fourth, but by no means least, sound pricing policies are
fundamental. If electricity is priced too cheaply demand will rise and
the utilities will face a dilemma between investment which may not earn an
adequate return and a failure to meet demand. If electricity is priced
too dearly, efficiency in use will certainly be encouraged but it will be
at the cost of discouraging the replacement of other forms of energy,
particularly oil, by electricity to the optimal economic extent. There is
a wide measure of agreement among IEA countries that, in principle, this
means relating prices to long-run costs of supply. There are in practice
many obstacles to achieving that result – the practical difficulty of
determining what long-run marginal costs are and the problems which arise
if prices based on long-run marginal costs result in utilities making
either very large profits or losses. In practice, we therefore
concentrate on more modest objectives – ensuring that prices give weight
to current and future as well as past costs, that changes in fuel and
operating costs are reflected promptly in tariffs and that cross subsidies
and the use of electricity prices to promote social, industrial and other
policies in a way inconsistent with energy policy are avoided.

Conclusions

To sum up let me emphasize again three points:

-   electricity is essential to economic and social well-being. So
    demand for electricity is likely to grow;

-   new generating capacity will be needed to meet demand: ways
    must be found of overcoming the obstacles to the provision of
    that capacity;

-   but the obstacles will be easier to overcome if the economic
    potential for efficient use of electricity is more fully
    realised.

Your Conference is one of a growing number of examples of the
importance now being attached to efficiency in the use of electricity. I
wish you every success.

THE SUCCESSES OF CONSERVATION

Arthur H. Rosenfeld and Evan Mills

Center for Building Science, Lawrence Berkeley Laboratory, University of California.
Berkeley, CA 94720 USA.

1. INTRODUCTION

People who have worked in the field of conservation have a lot to be proud of.
In the last 14 years, we have made dramatic improvements in the efficiency with
which we use energy, and have made an impressive head start on weaning ourselves
away from our fossil fuel habit. We'll be looking at how far the U.S. and OECD
have come, and looking ahead a bit to some accomplishments in the not-too-distant
future. We'll talk about conservation in general, but most of our examples will focus
on buildings, the sector we know best and one that accounts for 38% of the $440 Bil-
lion annual U.S. energy bill.

2. CONSERVATION HAS TEMPORARILY OVERWHELMED OPEC

2.1. Savings in the U.S. and within the OECD

The first point to remember is that we have saved a truly staggering amount of
energy through conservation—by which we mean efficiency improvements, not freezing in the dark—since the first oil embargo. We introduce Figure 1 to illustrate
these savings, which have accelerated since the second and more serious oil price
shock in 1979.

Before 1973, energy prices were low and there was little interest in improving
our efficiency. It was conventional wisdom that energy use would grow at least as
fast as GNP. In Figure 1a (for the U.S.), the heavy solid line represents the actual
consumption of total primary energy. The lighter solid line is simply GNP, scaled to
go through the 1973 energy use of 73 quadrillion BTUs (73 "quads"). Backcast to
1965, we see that GNP and energy use tracked nicely, corresponding to frozen
efficiency, but forecast to '85 we see GNP rising 33%, while actual use has leveled off
at 73 quads. Thus we have achieved an astounding 33% increase in efficiency, and a
remarkable annual saving of $150 Billion, but are still left with a $440 Billion annual
energy bill.

In the figure, the broken lines represent oil plus natural gas, which are partially
interchangeable in our economy since many boilers switch from one fuel to the other
depending on the price. Despite the 33% growth in our GNP, our oil & gas use has
declined even faster than our (also declining) domestic production of fossil fuels (indicated by the dotted line). Compared to 1973, we are now annually saving ½ of
OPEC's current capacity of 29 million barrels of oil/day. We believe that if the U.S.
and OECD had not reduced our need for this oil and gas, it could have come only

A. T. De Almeida and A. H. Rosenfeld (eds.), Demand-Side Management and Electricity End-Use Efficiency, 17–61.
© 1988 by Kluwer Academic Publishers.

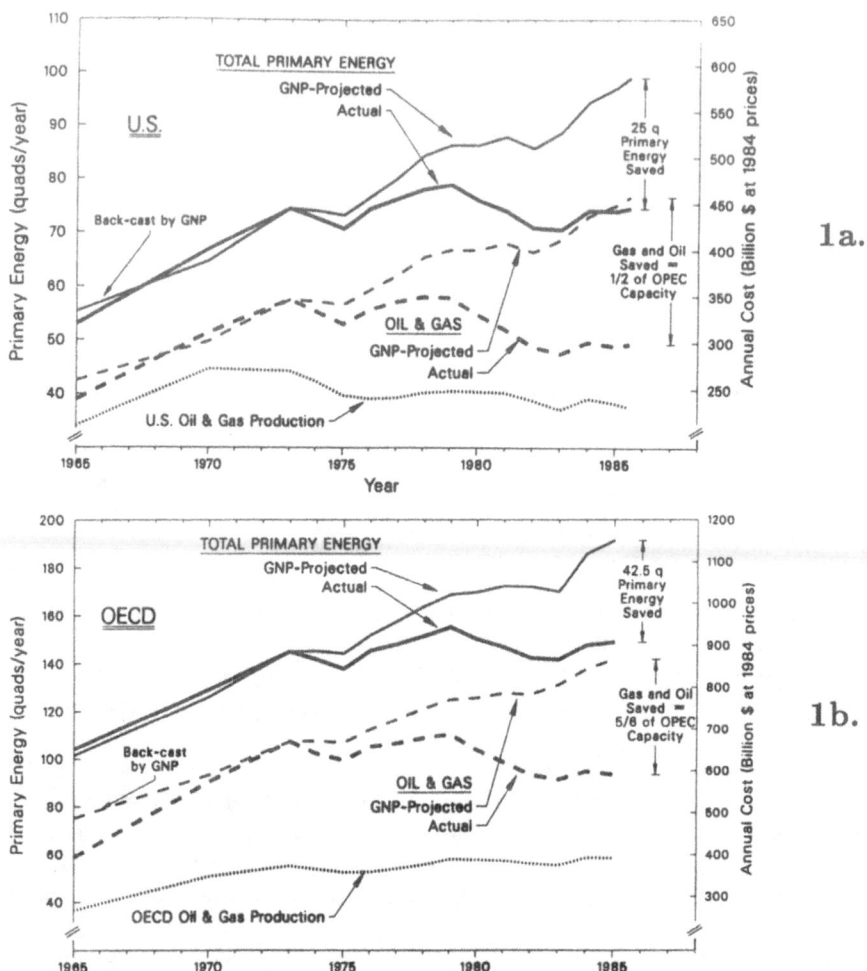

FIGURE 1a and 1b. U.S. and OECD Energy Use: Actual and Projected by GNP. The upper figure is for the U.S. and the lower figure shows comparable data for the entire the Organization for Economic Cooperation and Development (OECD). Projected energy is calculated on a GNP basis in constant dollars, with both forecast and "back-cast" values from 1973. Note that the GNP back-cast generally follows the actual consumption curve before OPEC. The "primary energy" on the left-hand scales includes fuel burned at the power plant, in units of "quads" [quadrillion (10^{15} Btu]. The oil and gas savings were converted from quads to fractions of OPEC capacity using an estimated 1986 total OPEC production capacity of 29 Million barrels per day (58 quads). For the right-hand scales, quads were converted to 1985 dollars using the 1984 U.S. cost of energy (about \$440 billion for 73 quads). Savings for the U.S. in 1985 were one-half of OPEC total capacity. The OECD includes all of North America, Western Europe, Japan, and Australasia, and consumes about twice as much total resource energy as the U.S. alone. Oil and gas savings for the OECD in 1985 were five-sixths of total OPEC capacity.

from imports, since our domestic production is steadily declining.

Figure 1b tells the same story for the OECD, which includes all of North America, Western Europe, and Japan. The OECD annual energy bill is $900 B, but (compared to 1973 efficiency) we are saving $250 B/year. Our oil & gas savings are 5/6 of current OPEC capacity. Because of the North Sea, OECD production of oil and gas is still rising, but nowhere near enough to supply the amount that we have saved. So, again, OECD imports would be nearly 5/6 of OPEC capacity higher.

What would we be paying today for oil and gas if OPEC were at 100% of capacity, and in addition there were still a major shortage of oil? Figure 2, taken from DOE/EIA's International Energy Outlook, hints at the answer—OPEC was able to raise prices in all those years that 80% or more of its capacity was in use. This suggests substantial price increases every year above the $30/barrel which we paid in 1980, disastrous increases of $100, $200, or even $300 Billion in our trade and budget deficits, and a global security problem, compared to which the present problems in the Persian Gulf pale into insignificance.

We conclude that conservation has bought us valuable time, and that we had best continue to support this winning strategy. But how long can we maintain the "glut," i.e., keep OPEC down to 60% of its capacity?

A vigorous government/utility conservation program can continue the flat demand of Figure 1 almost indefinitely, despite a reasonable growth in GNP. But oil production is going to drop, faster and faster for the U.S., and will peak in about 10 years for the North Sea and for the Soviets. Even OPEC, running at full capacity, is good for only about another 40 years.

Figure 1 covers only 20 years, so the decline in production does not appear very steep. Lest the viewer be deceived, we present Figure 3 on U.S. oil production, which goes out past 2020, when our children will still be paying energy bills but living without much domestic oil. The figure comes from *Beyond Oil*, by the Complex Systems Research Center, of the University of New Hampshire. It shows our inexorable decline in oil production. To emphasize this, its authors point out that in the 1950s we discovered 50 barrels of oil for every barrel invested in drilling and pumping. Today the ratio is 5:1, and by about 2000 it will have dropped to 1:1, at which time domestic exploration will become uneconomic.

What is more, the two smooth curves reflect reserves at a time when oil was very inexpensive. We spent $¼ *Trillion* exploring for oil in the 1980s. The bullets to the right of the curves show that this has bought us a mere 8-year delay in the day of reckoning.

Note that buildings generally last for 50 years, so a sub-optimal building constructed today will still be guzzling expensive energy long after American oil and gas have run dry. And today's buildings are very sub-optimal, as can be seen by noting American ideas about acceptable payback times. Builders (including the U.S. government) will not tie up their money in efficiency investments if the payback time is more than 2-3 years; yet, on the supply side, the typical investor will accept a payback time of 25-30 years from a power plant or an oil-and-gas venture. So the playing field is badly tilted in favor of supply. Thus a conservation measure such as thermal storage, which avoids running air conditioners at peak power times, has a payback time of only 2-3 years yet is largely ignored (and completely ignored in new federal buildings). If we persist in ignoring thermal storage until the turn of the

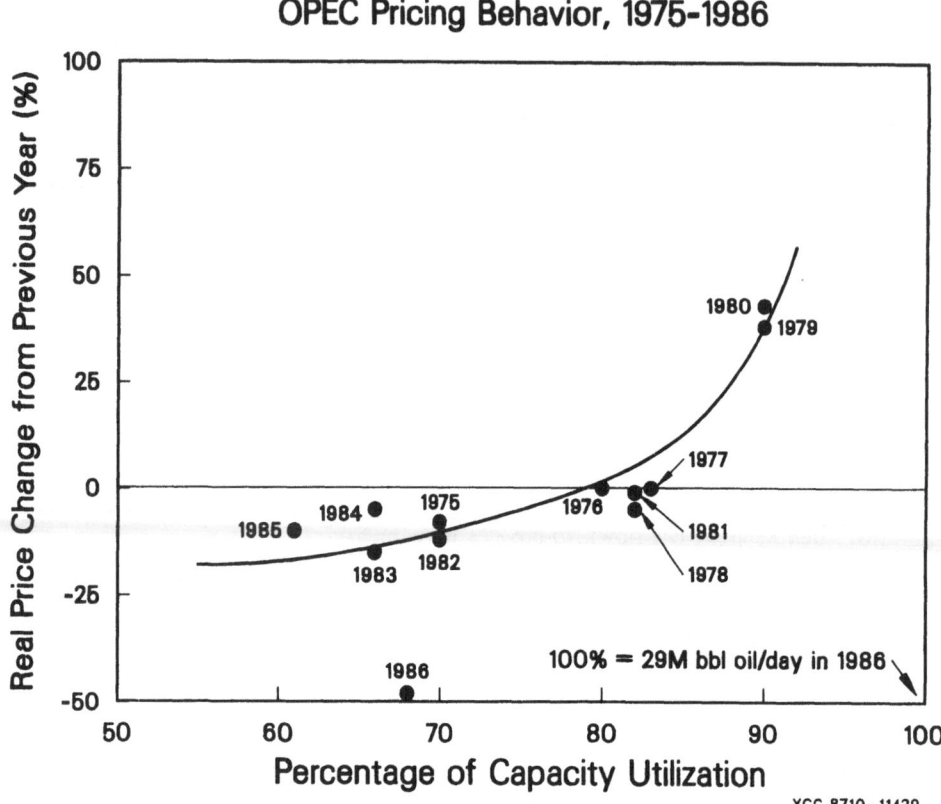

FIGURE 2. OPEC Pricing Behavior, 1975-1986. The 1986 observation, which was not used to derive the curve, reflects Saudi Arabia's decision to switch from providing price support to increasing market share. Figure adapted from: *International Energy Outlook, 1986.* Energy Information Administration, U.S. Department of Energy, page 10.

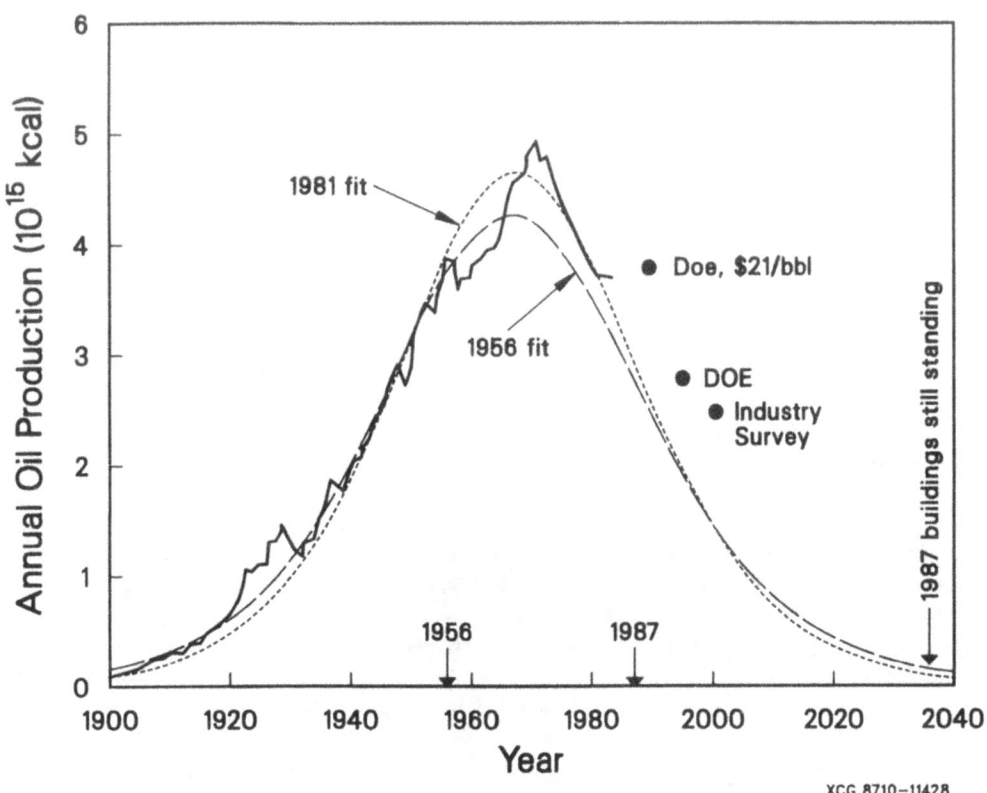

FIGURE 3. Comparison of Hubbert Oil-Depletion Curve to Actual Production, excluding Alaska. Solid line is actual U.S. oil production; dotted line is Hubbert's 1956 curve; dashed-dotted line is Hubbert's curve updated by authors of *Beyond Oil* based on 1981 data. Source: *Beyond Oil*, by the Complex Systems Research Center, University of New Hampshire, Ballinger Publishing, 1986. DOE Estimates for 1990 (6.9 Mbod) and for 1995 (5.2 Mbod), and Oil Industry Survey for 2000 (median = 4.5 Mbod), are from Energy Security (DOE S-57, 1987) and "U.S. Oil Production," U.S. Office of Technology Assessment (Report E-349, 1987).

century, we'll have to build the equivalent of 100 otherwise unnecessary standard 1000-MW power plants at a cost that will probably exceed $1.5 Billion each.

Figure 4 (taken from *Electrical World*) shows the effect on electric utility construction expenditures of the conservation success of Figure 1. During the power plant overbuilding spree around 1980, we invested $50 B/year (12% of our annual capital investment in plants and equipment). Now we have 50 baseload plants (about 1000 MW each) in excess of current need, and utility construction is predicted by *Electrical World* first to fall to $17 B/year (leaving another 10% more of our capital formation for other productive investment), and then to rise to $40 B/year as electrical demand continues to grow at 2%/year. Conservation R&D today, leading to more-efficient use of electricity in 1995, can greatly delay and mitigate the need for this looming $40 B *annual* investment.

Figure 1 showed that conservation is now saving the U.S. $150 B/year, and we have cut our energy bill to "only" 11% of our GNP. But the Japanese only 5%. "Least-cost" calculations show that optimal investment would halve our energy use by the turn of the century (see Figure 5). This suggests the following analogy: if we were stuck at 1973 efficiency, we would be pouring $590 B worth of energy into a pipeline each year and getting out only $220 B in energy services. The rest—$370 B—would have leaked out. But we've already plugged more than a third of the leaks, and we now waste only $220 B/year, so we pour in "only" $440 B worth. To be fair, we are adding something like $15 B/year in retrofit costs—a modest amount yielding something like a one-year payback. We can save the remaining $220 B that is wasted—and cut in half what we currently spend on energy—three to five times more cheaply than continuing to pay for wasted energy. So our first priority should be to finish plugging the leaks, before we invest more in new supply. The longer we let the leaks continue, the quicker we will exhaust cheap, secure sources of oil and gas. Seeking new supplies—"draining America first"—while we continue to waste energy and backslide on auto efficiency, just hastens the depletion of our reserves; heightened efficiency saves the energy until it is really needed.

And how much has it cost to plug the leaks? So far, because we have been skimming the cream, conservation has typically been five times cheaper than purchasing energy. So to save $150 B/year, we have probably invested $30 B/year, leaving a net savings of $120 B/year. In terms of incentive programs by governments or utilities, we can do even better than 5:1. PG&E, the giant Northern California utility, boasts that in 1985 it spent $0.25 B on conservation programs, but avoided committing $1.75 B to new supply, a benefit/cost ratio of 7:1. To save the next $200 B/year, some of the cream will be gone, but least-cost analysts estimate that conservation will still be three times cheaper than supply.

2.2. We are losing the efficiency race with Japan

In 1985, the U.S. used 11.2% of its GNP for energy; Japan used 5%. Figure 6 clarifies this point and puts the efficiency—as measured by energy use per GDP—of other countries in perspective. The details of the figure are explained in the caption, but the summary is that we spend about 6% more of our GNP on energy than do the economical Japanese.

Japan is beating us not only in absolute energy efficiency, but in the rate of improvement. In the period plotted in Figure 6, Japan has improved its energy use

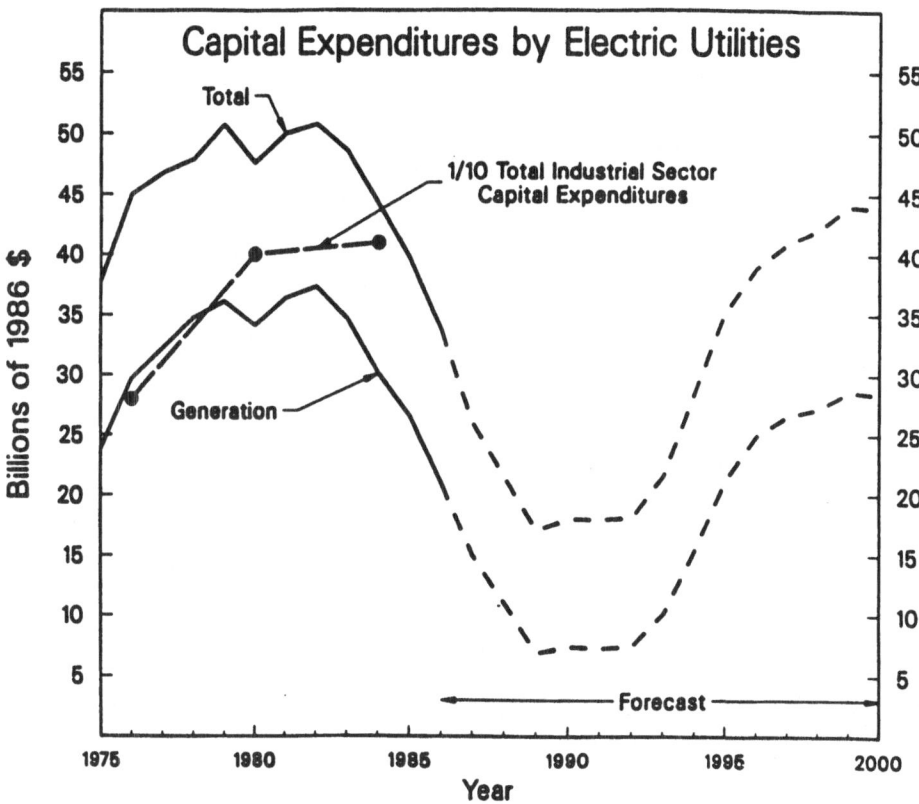

FIGURE 4. Electrical Industry Annual Investment in Plant and Equipment, in 1986$. The equivalent investment by all of industry is about $1B per day, so that the electric fraction has dropped from about 15% ($50B) to a minimum that will be about 5% ($17B). The utility investments do not include cogeneration, which is running at about $2B/year. Source: *Electrical World,* McGraw-Hill, Inc., September 1986. Figures for total industry investment are from *1986 Statistical Abstract of the United States,* 106th Edition, Table 901, p. 529, using GNP implicit price deflators to convert to 1986 dollars.

24

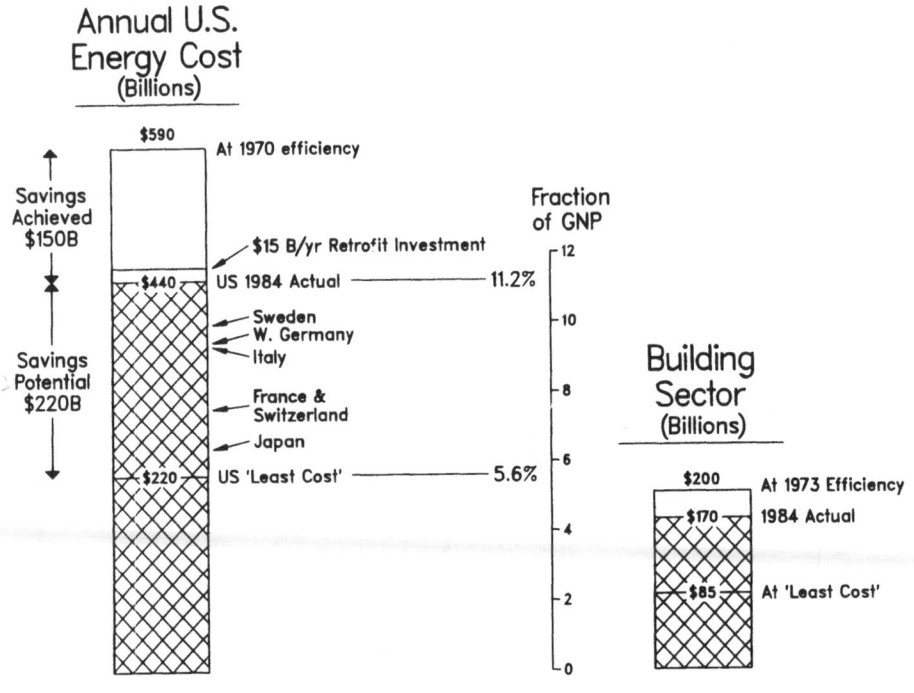

Annual U.S.
Energy Cost
(Billions)

$590 — At 1970 efficiency

Savings
Achieved
$150B

$15 B/yr Retrofit Investment

$440 — US 1984 Actual ——————— 11.2%

Sweden
W. Germany
Italy

Savings
Potential
$220B

France &
Switzerland

Japan

$220 — US 'Least Cost' ——————— 5.6%

Fraction
of GNP

12

10

8

6

4

2

0

Building
Sector
(Billions)

$200 — At 1973 Efficiency

$170 — 1984 Actual

$85 — At 'Least Cost'

XCG 8710–11435

FIGURE 5. Annual U.S. Energy Cost. By 1984, the energy use per dollar of GNP
(in constant dollars) has dropped to 74% of the 1970 level. If efficiencies had stayed
frozen at 1970 values our $440-billion annual cost today would instead be $440
B/0.74 = $590B. On right scale, "Fraction of National GNP," are lines representing
1984 fractions for European countries and Japan. These lines show what the 1984
U.S. economy would pay for energy at various foreign efficiencies.

Energy Consumption and GDP: 1970-1985

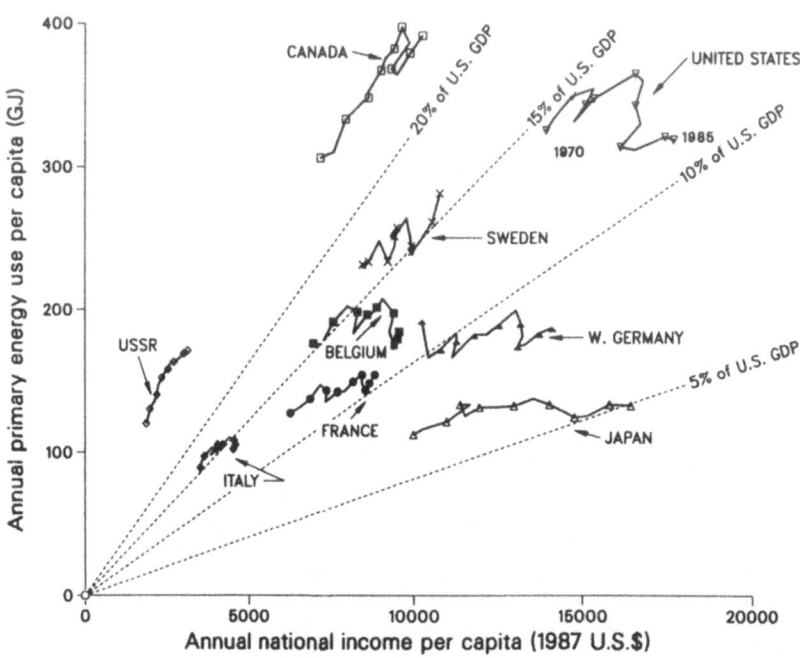

FIGURE 6. Resource Energy use vs. GDP (both per capita) for 9 Industrial Countries. Each country is represented by a sequence of points connected by straight lines, beginning in 1970 and ending in 1985. The conversion from local GDP to dollars depends only on the July 1 1987 exchange rate; earlier points are plotted using individual national deflators. For the lines labeled 5%, 10%, 15%, and 20% of U.S. GDP, we use an average 1987 price of resource energy of $6.14/GJ. Hydroelectric and nuclear electricity are converted to resource (primary) energy using IEA's standard generation efficiency of 38.5% (except in Japan, 35.1%). Data for the USSR is unfortunately site energy. Sources: Price - DOE/EIA 0376-1984 (updated to 1985 by phone to EIA). Income and Population - IMF International Financial Statistics 1986. Energy Consumption - the OECD/IEA volume *Energy Balances 1970-1985* (it should be noted that we use the "Total Energy Requirement" data as opposed to "Total Final Consumption"; the former is resource (or "primary") energy and the latter is site energy, where the losses in electricity generation are ignored. Soviet Data - "UN Demographic Yearbook," 1985. Exchange rates are for July 1, 1987. 1 TOE = 42.6 G.

per dollar of GDP by 31%, while we have improved by only 23%. At the same time, Japan's per-capita income has come up from behind and is now passing that in the U.S.. Nevertheless we can be proud that we have the second best record of the nine countries pictured, thanks to appliance and automobile labels, automobile standards (*and* imported cars), building standards, federal and utility conservation programs, a vigorous R&D program, and of course the market.

How has Japan become as productive as the U.S. on less than half the energy? Between 1973 and 1985, energy use per pound of steel produced had declined by 15%, electricity used to operate new refrigerators had dropped by 73%, and electricity used to run room air conditioners has dropped 42%. [1]

New Japanese cars now average 29 mpg, their national policy is heading for 50-60 and their prototypes have already hit 100 mpg. In contrast, U.S. policy has been to backslide from 27.5 to 26 mpg—even when the incremental cost saves gasoline at 50¢/gallon—and instead to emphasize drilling for oil off the California coast, or in the Arctic National Wildlife Reserve. Ironically, if we can drain the controversial 3.5 ± 1.5 Billion barrels off of California in 30 years we will produce only 0.32 Mbod, just the amount lost in the fuel economy backslide.

Of course to achieve this efficiency they had to make investments, whose repayment eats into about 20% of their savings. So instead of having 6% more of their GNP available than we do, they really have gained only about 5%. We assert (and will explain below) that this differential of 5% of GNP means that, even if all else were equal, our products cost on average about 5% more than comparable Japanese products, thus impairing our balance of payments, the dollar/yen ratio, and our life-style.

Some readers may find this assertion obvious and can skip this paragraph, but those who are surprised at a 5% cost penalty should consider this argument. The total energy cost of any product is the sum of the direct energy cost (significant for iron and steel, insignificant for most high-tech products) *plus* the indirect cost embedded in wages. (For the same life-style, a U.S. worker who commutes in a gas guzzler and lives in a poorly insulated dwelling needs higher wages than his Japanese competitor.) Effectively, U.S. manufacturers pay a total 5% energy tax. But unlike other taxes, which arguably provide government services, this 5% just goes up in smoke and pollution.

The defense version of this tax is already much discussed. Thus, if the Japanese had a GNP equivalent to ours, they would avoid a tax of $300 B (7.5%) for defense, giving them a 7.5% competitive edge. Now we have added a $200 B (5%) energy-efficiency differential tax, for a total handicap of 12.5%—and as energy prices rise, this gap will widen.

Let us examine in more detail what will happen in 10 or 20 years if OPEC regains control and energy prices double. Without a continuing, vigorous conservation program, our energy bill could zoom from 10% to 20% of GNP. We predict that the Japanese will continue to invest in efficiency even during the glut, get down to 2% of GNP at today's cheap prices, and later climb back to only 4%. And they will be experienced at manufacturing and exporting energy-efficient products, which seem likely to be in demand. The competitive outlook begins to look bleak.

3. PUBLIC R&D FOR ENERGY

3.1. Conservation R&D compared with other economic sectors

Table 1 disaggregates our $440 B annual energy bill according to the buildings, industry, and transport sectors and **Table 2** compares our total expenditure in several economic sectors with our publicly supported research and development effort in them. Despite the prominence of our national energy bill (the largest single sector), we invest barely ½ of 1 percent of that amount in research aimed at meeting our energy needs. If we consider R&D effort on construction and conservation (which can meet our needs at one-third to one-fifth the cost of new supply), we invest less than one-tenth of one percent. By comparison, for Defense, Health, and Agriculture R&D we spend anywhere from 1% to 12% of total expenses, or 10 to 100 times more than for conservation. But if we look at what really works, it is conservation that has (literally) fueled our post-OPEC economic growth.

Figure 7 shows the U.S. Department of Energy (DOE) budget over the last few years. It grows from its foundation in 1976 to a peak of $15 Billion a year, then comes down sharply under the Reagan administration. And only a small fraction of this has gone to improving energy efficiency. Most of this is military and in terms of raising our oil or coal production we haven't done anything. This misplacement of priorities isn't a sickness for which DOE is solely susceptible; it's a general trend in our society. If one looks at the Electric Power Research Institute's (EPRI) budget, for example, one sees that it's 6% demand-side and 94% supply-side research. We're very much a society that pays more attention to big aqueducts than to fixing leaky faucets; to large power plants than to many small and efficient lamps; to a rail line than to a flexible fleet of busses in urban areas; to a freeway than to well-timed street lights; to a hospital than to preventive health care.

In 1980, DOE was spending $100 million on Buildings and Community Systems research, or $1.20 per home in the United States. Remember the potential savings are around $2000 a home, aside from commercial buildings where a large savings potential also sits untapped. The Reagan administration thought that DOE was spending too much and requested zero budgets by 1983. Congress helped a bit and things haven't been zeroed out yet. We're now at 50 cents per home per year with Reagan asking for half of that for next year in the face of a 100-to-one return potential.

3.2. Technological triumphs of DOE-supported R&D

Technical successes of DOE-sponsored Buildings R&D were well documented in a 1986 Conservation White Paper [2], so we will summarize only a few points and reproduce its main table (**Table 3**).

In the White Paper, case histories were presented for three important technical developments: high-frequency, solid-state ballasts for fluorescent lamps (Figure 8), "heat-mirror" (low-E) window films, and improved refrigeration. The paybacks on federal R&D funding were typically 5000:1, but the delay times are long, partly because the buildings industry is so fragmented (see Figure 9). Thus, at LBL we started to develop the heat mirror film in 1976, but as shown in Figure 10 it will not reach 50% market penetration until around 2000 (13 years from now), and the majority of existing windows will not be replaced until 2020 (33 years from now). So to save scarce energy for our children, we need to support R&D today.

TABLE 1. U.S. Energy Expenses, 1985*			
Sector	Fuel ($10B)	Electricity ($10B)	Total ($10B)
Buildings	**60**	**110**	**170**
Residential	40	60	100
Commercial	20	50	70
Industry	**70**	**40**	**110**
Buildings	3	7	10
Transport	**160**	**0**	**160**
TOTAL	**290**	**150**	**440**
Percent of GNP			**11.2%**

* Excluding Federal subsidies and rounded to the nearest $10 billion. Source: *State Energy Price and Expenditure Report 1985,* October 1987.

TABLE 2. Comparison of Energy Expenses and R&D With Other Economic Sectors		
	Total Expenses (Billions 1984$)	Publicly Supported R&D + Tech Transfer
		(Billions of 1985$) / Percentage of 1984$
Energy	$440B	
Total Supply R&D		2.50 — 0.5%
Conservation R&D		0.16 — <0.1%
Health	400	
N.I.H.		6.20 — 1.6%
Construction	340	0.01 — <0.01%
Defense	300	37 — 12%
Education	200	<0.1 — <0.1%
Federal Deficit	200	--- — ---
Trade Deficit	200	--- — ---
Agriculture	140	--- — ---
Experiment & Extension		1.70 — 1.2%

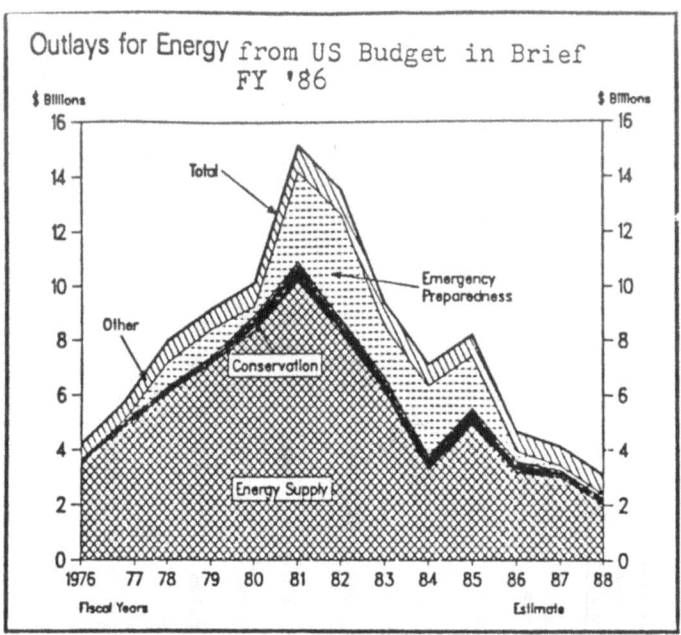

FIGURE 7. Trends in Outlays for Energy. Source: H. Richard Heedy, Rocky Mountain Institute, Testimony to Subcommittee on Energy and Agricultural Taxation, U.S. Committee on Finance, 21.VI.85. The total 1984 outlay was equivalent to 11% of the U.S. energy bill in that year. For detailed 1984 data, see *Energy Conservation Digest,* June 24, 1985.

TABLE 3.

Lead-times and Net Savings for Successful DOE-Sponsored Buildings Energy R&D Projects

	Solid State Ballasts	Low-E Window Films	Residential Absorption Heat Pump	Advanced Electric Heat Pump	High Efficiency Refrigerator Compressor	High Efficiency Refrigerator-Freezer	Heat Pump Water Heater
1. DOE Project Duration	1976-1980	1976-1990D	1978-1988	1977-1986	1977-1981	1978-1983	1977-1982
2. Est. 50% Penetration of Sales	1995	2000	2001	1998	1990	1996	2000
3. Years by which DOE advanced commercialization	5 yrs.	5 yrs.	5 yrs.	2 yrs.	2 yrs.	2 yrs.	2 yrs.
4. Cost of Conserved Energy, (CCE)	2¢/kWh	$2/MBtu	$2.50/MBtu	$2.75/MBtu	1¢/kWh	3¢/kWh	5¢/kWh
5. Cost of DOE Project	$3M	$2M	$6.8M	$2M	$1M	$0.8M	$0.7M
6. Net Annual Savings in 1985	$11M	$14M	$0M	$0M	$0.4M	$0.2M	$0.3M
7. Net Annual Savings at Saturation (i.e. 10-15 after 50% penetration)	$5,000M	$3,000M	$2,400M	$2,500M	$1,100M	$850M	$1,800M
8. Cumulative Net Savings (Line 7 x line 3)	$25,000M	$13,000M	$12,000M	$5,000M	$2,200M	$1,700M	$3,600M
9. DOE Project ROI (Return on Investment, =Line 8 ÷ line 5)	8,000:1	7,000:1	1,500:1	2,500:1	2,000:1	2,000:1	5,300:1

Source: "Federal R&D on Energy Efficiency: A $50B Contribution to the U.S. Economy, a White Paper on the Consequences of Proposed FY'87 Budget Cuts," by the American Council for an Energy-Efficient Economy and the Energy Conservation Coalition. March 4, 1986.

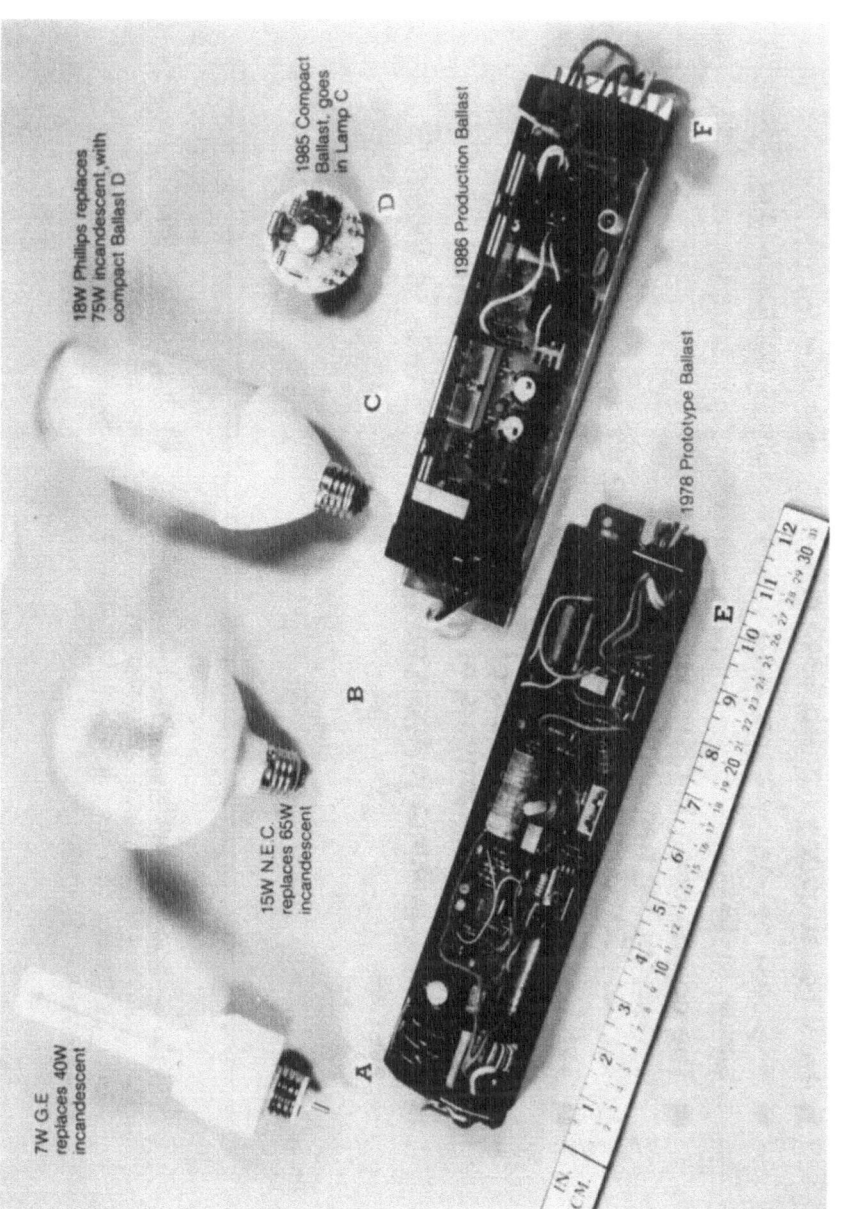

FIGURE 8. Energy Efficient Lamps and Ballasts.

Development, starting in 1978, of high-frequency, solid-state ballasts for fluorescent lamps; first for long tubes, later for compact fluorescents to replace less efficient incandescents.

Size & Diversity of the Buildings Sector

■ New construction totaled $340 billion in 1985 (9% of GNP)

■ The cost of energy consumed in the buildings sector totaled $165 billion in 1985

■ The construction industry is:
  – Over 28,000 homebuilders
  – Over 150,000 special trade contracting establishments

■ The construction material and component manufacturing industries are also fragmented:
  – Over 600 manufacturers of non-electric heating equipment
  – Over 100 manufacturers of mineral wool insulation

■ Each new building requires inputs from more than 50 industrial sectors

FIGURE 9. Size and Diversity of the Buildings Sector. The high degree of fragmentation delays the diffusion of conservation practices. Source: DOE *A*ssistant Secretary's Review of Office of Buildings and Community Systems, Lawrence Berkeley Laboratory. October, 1986.

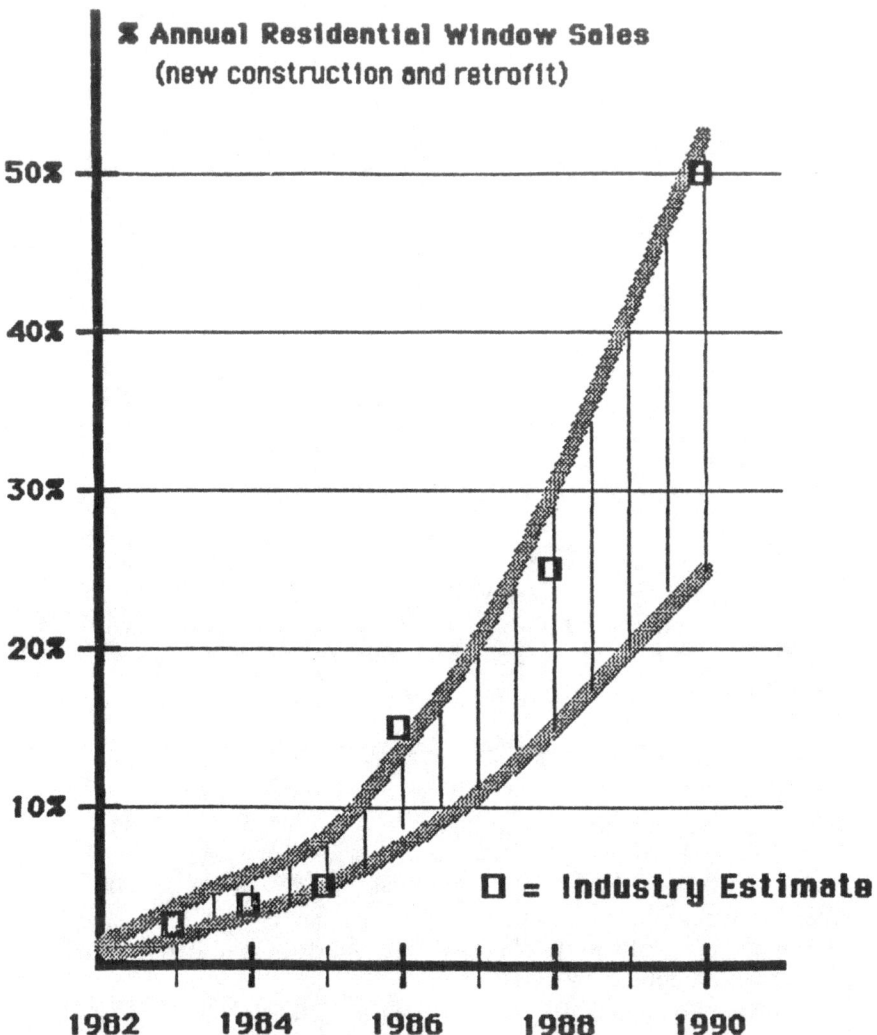

FIGURE 10. Industry Estimate of Low-E Windows Market Share based on Annual Sales. With a projected 20% market share in 1987, sales will be over 100 million square feet. Savings from cumulative installed window area will be approximately $60 million in 1987. At saturation of existing residential stock, savings will be $4-5 billion per year at current fuel prices. These savings will be equivalent to one-quarter of the output of the Alaska Pipeline.

The savings from these three completed projects are astounding, nearly $17 B/year when they finally saturate the market (even at today's prices). This equals the yearly output of about 25 baseload power plants, and an oil & gas saving equivalent to half the yearly output of the Alaska pipeline.

In addition to saving energy, conservation has also saved some U.S. industries and created others. The $1 B/year U.S. ballast market would have been invaded by the Japanese and the Europeans had it not been for U.S. development of the solid-state ballasts. In California, we have two new industries based on the "heat mirror" films: Southwall Technologies sells low-E coated plastic to window manufacturers, and Airco Solar Products sells multi-million dollar plants for sputtering the thin films on glass.

We conclude from the data in Table 3 that DOE-developed technologies have paid off very well, and that they will become commercial in America several years sooner than if we had waited for either domestic or (more likely) overseas industry to develop them. In the case of the examples above, several years' acceleration of the savings of $17 B/year (the sum of Row 8 of Table 3) represents a savings to U.S. ratepayers of $62 B.

With good R&D, we can advance the times when new demand-side options become available by five or ten years. Things are slow now with the low oil prices yet, if we don't keep R&D alive we won't have solutions when we need them. Let us look more closely at one technology—lighting—and why innovation and diffusion has taken so long.

3.3. Lighting: Why it takes 30 years to saturate the market

Early on in our energy-efficient buildings research efforts, it was clear that we should go after incandescent lighting. We knew that if by exciting fluorescent lamps at a very high frequency—30,000 Hertz or so—that we could gain about 15% efficiency. Regular fluorescent lamps cycle at only 60 Hertz. Also, before the oil crisis ballasts were made out of copper wire laminates which ran very warm and thereby dissipated 13 watts of waste heat. So, for every 100-watt fixture a total of 28 watts could potentially be saved.

Along came solid-state electronics which made possible the solid-state ballast, or more accurately the solid-state oscillator. Because they dissipated only a few watts of waste heat, the net efficiency gain was 20%. To this we added a photocell that looked down on the workspace and adjusted the intensity to make up the difference between the available daylight and the preferred lighting level.

Since lighting in the U.S. required 30 power plants, savings potential was phenomenal. We went to GE and were disappointed with their reluctance to adopt such innovations—they much preferred to wait for small companies to take the risks; when the technology and marketing were proven, the big boys would move in and buy the small guys out. Sylvania, Westinghouse, and Advanced Transformer (Phillips) told us the same story.

So, we went to ERDA—DOE's predecessor—and convinced them to support small R&D efforts with small companies. With 60 interested firms, we selected two and within six months had our first prototype; this was 1976. Then we went to our local utility, PG&E, and convinced them to let us showcase the ballasts and daylighting controls throughout three stories of their skyscraper in San Francisco. By the time the installation was debugged, we were saving 40%. Then we waited for two years and nothing happened. To our dismay, the underwriting labs—whose committees were chaired by representatives of the four big lighting companies— hadn't granted any approvals for the new technologies. Then, Beatrice Foods decided to become the fifth big actor and they bought out one of the small companies and within two weeks Westinghouse turned around and bought them back out for two million dollars. Then things started to move. Now we expect these new ballasts to saturate the market by about 1995.

The other lighting success story is the compact fluorescent, screw-in lamps like the Phillips SL-18 shown in Figure 9. The SL-18 replaces a standard 75-watt incandescent and provides the same lighting service for only 18 watts. What's more, it lasts 7500 hours, outliving the incandescent by 10 to 1.

What are the economics? Say we save 33%, or 33 watts for each 100-watt fixture. The cost of conserved energy is 2.5 cents per kilowatt hour and the payback is less than a year. During its 10-year life, it will save five barrels of oil and costs is a lot less. The combination of the new ballasts and compact fluorescents will save about $10 Billion a year when saturated into buildings in the United States.

What have we learned? For one thing, this business takes a long time. A ballast lasts about 10 years, so before they actually saturate all the buildings—the last of the old electricity guzzlers won't burn out until about 2005—it will be 30 years. Federal acceleration of R&D can make a huge difference in the implementation time and increase savings by billions of dollars.

4. COMPETITIVENESS NEEDS MAJOR ATTENTION

Despite the successes of the DOE R&D program that we have just described, the outlook for the U.S. energy-efficiency industries is clouded by our general inattention to new product development. DOE's conservation R&D program is far too small, and we have nothing in the U.S. comparable to Japan's MITI (Ministry of International Trade and Industry) or to the EEC's BRITE (Basic Research Industrial Technology for Europe). For many reasons, including its perception of the market, U.S. industry is not producing energy-efficient products: not cars, manufactured homes, air-conditioners, etc. (Aircraft are a notable exception). As we mentioned earlier, at LBL we had disappointing experiences in trying to interest large U.S. manufacturers in high-frequency ballasts or heat mirror films for windows.

The pattern is quite different in Japan, where R&D budgets are comparable to ours, but MITI can step in to manage and support commercialization of new, beautifully engineered, efficient, exportable products. Sometimes the original R&D was Japanese, but often it was American, acquired by licenses or technology agreements. It is well known that despite U.S. R&D on electronics, Japan has taken the front seat in the world market on VCRs and compact disks.

A similar pattern exists in another high-technology product line: efficient electric motors and controls. U.S. industrial, commercial, and residential consumers pay about $80 B/year for power used to run electric motors. Recent advances in magnetic materials and power electronics are greatly improving the efficiency of these motors and motor-driven systems, reducing costs to consumers. For example, permanent-magnet motors can have 20% lower losses than the best induction motors, run cooler, are smaller and lighter, and can be more precisely controlled.

Current applications include machine tools, robotics, computer peripherals, and home appliances.

A 1986 study [3] points out that:

> U.S. competitiveness in this rapidly growing market for new motor technologies is of concern, however. As pointed out the National Materials Advisory Board, 'The fundamental work leading to the REPMs [Rare-Earth Permanent Magnets] was done largely in the United States ... but after government support ceased, materials R&D in the U.S. magnets industry deteriorated. Practically all recent PM materials have been developed to commercial maturity in Japan.'

> Thus the NMAB concludes that 'despite the critical importance of magnetic materials, the U.S. is rapidly losing its competitive position.' And this in a market that is expected to reach $2 Billion annually next year.

The fast-growing market in power electronics (electronic devices which control power-consuming equipment) is also facing intense foreign competition. For example, electronic adjustable-speed drives (ASDs), which control the speed of electric motors subjected to varying loads and reduce electricity use by 20 to 30%, use basic components that were first developed by American companies. Nonetheless, foreign penetration of the U.S. market for ASDs has grown from 15% in 1980 to over 40% in 1985. Foreign companies have not only taken over the lead in production of ASDs, they have taken over the lead in innovation and product development.

Ralph Ferraro of the Electric Power Research Institute (EPRI) estimates that the U.S. manufacturers' share of the domestic power electronics market will erode from its present level of 50% to about 25% within five years. According to the Federation of Materials Societies, "if the current trend continues, it can be anticipated that the U.S. will be a minor force in the world market for electronic materials and systems by the 1990s".

A final example of competitive problems is in the area of housing technology, an industry that is traditionally seen in the U.S. as fragmented and slow to accept technical innovation. Contrast our situation with that of Sweden, where the government supports an ambitious R&D program in all aspects of basic and applied building technology. [4] Total funding is similar to that in the U.S., even though the Swedish market is only about one-twentieth the size of ours. Swedish researchers have produced a host of technical innovations that are already used in "superinsulated" homes around the world. Applications of R&D results to an industrialized building sector have made high-quality, energy-efficient homes the norm in Sweden, rather than the exception. Several firms are now exporting their factory-built housing to the U.S., and are beginning to compete successfully in upscale markets.

5. TEN YEARS OF CONSERVATION IN CALIFORNIA

On a more positive note, let's consider the experience we have had in California in attempting to institutionalize energy efficiency at the state government and utility level—an experience that has seemed slow at times but that has produced lasting results.

5.1. Innovative rate design

One of the first things that the utilities tried (after they were dragged into it by the Public Utilities Commission) was to invert their declining block rate structure and start charging more for electricity the more the customer used. This spurred a lot of conservation, because people's electric bills were now finally sending them the right signals: the cheapest block represented the utility's old, cheap power; the next block was the average cost; and the highest tier reflected the cost of building new plants or operating the utility's most expensive ones. Since then, because of a temporary surplus of generating capacity, average and marginal cost have veered so close to each other that there are only two tiers, but that situation probably won't last.

5.2. Appliance standards

One of the most successful programs we have tried in California are our appliance-efficiency standards. In 1976, right after the embargo, the state legislature passed a bill under their Title-20 jurisdiction that set maximum energy consumption levels for a variety of household appliances; those standards have since been adjusted to reflect vastly improved technology. We'll consider refrigerators, which, mundane though they may be, nonetheless account for 10 percent of the electricity used in the United States.

The time scale for Figure 11 starts in 1977, when the first standards were adopted, and runs through 1993, when the most stringent standards will take effect. These figures apply to a typical 15- to 18-cubic-foot top freezer with automatic defrost. The number of kilowatt-hours required to run the refrigerator each year has been ratcheted down from 1900, where it started out in 1977, to 1500 then to 1000, and will go down to 700 in 1993. The most important thing is that there is no change in the service the refrigerator provides. This is all for refrigerators which are designed to stay at 40 °F (4.5 °C) in the food compartment and 0 °F (-18 °C) in the freezer compartment. All that we've done is to double the efficiency.

How much does it cost (retail) to go from the old, inefficient refrigerator to the 1993 juice-sipping model? The California Energy Commission estimates $100. That doesn't mean $100 in production costs—the actual increase in factory cost of the refrigerator is only about $35. The typical mark-up in the refrigerator industry is 2.7 times (including marketing and advertising), plus a certain premium because this refrigerator is now marketed as "efficient".

What does society save by that $100 investment? At 8 cents per kilowatt-hour, the less-efficient refrigerator costs $150 per year to operate and the efficient one costs about $55 per year. The savings is about $100 per year on something which costs the manufacturer $35 once every 20 years to make. The payback period for society is either 1 year or 1/3 year, depending on how one looks at it.

With those annual savings of $100, the customer can pay off the retail surcost in the first year. The refrigerator lasts 19 more years, so the savings for the following 19 years are pure profit for the consumer.

Critics have charged that these refrigerator standards are "coercing the American public". Well, yes, they do coerce the public, but not very strongly. They force the public to do things with a one-year payback, not a ten-year or thirty-year payback as is the case with new power plants. Californians, at least, don't seem to mind being coerced to this extent.

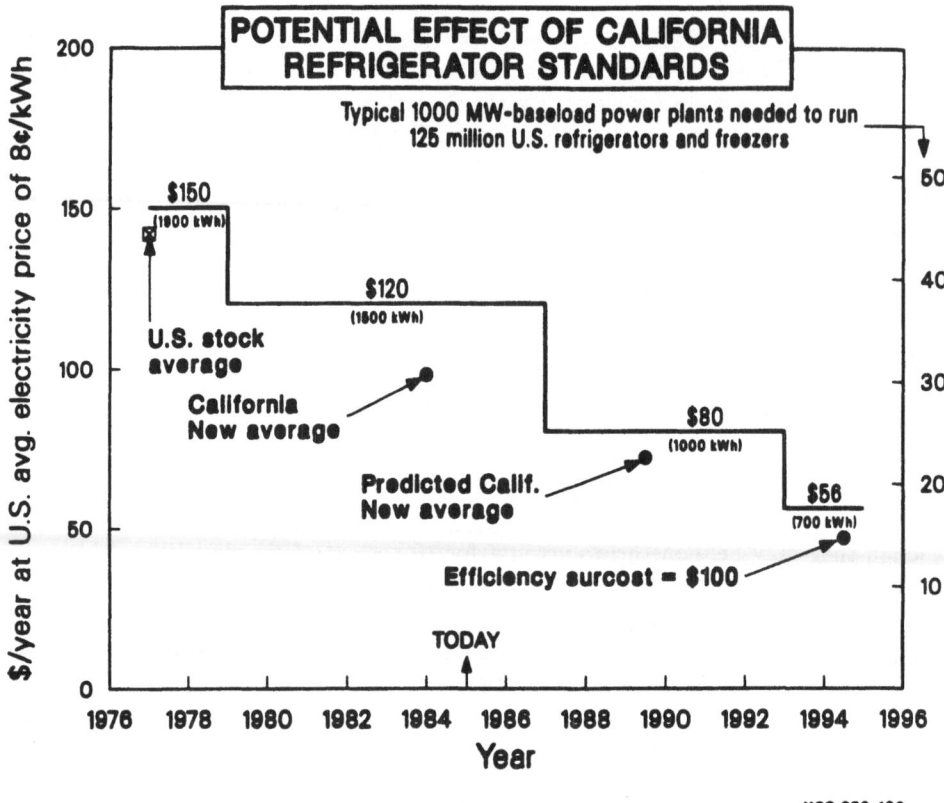

FIGURE 11. Potential Effect of California Refrigerator Standards if Adopted Nationwide. A typical 1000-MW baseload power plant produces 5 BkWh/year. The 125 million U.S. refrigerators consume 240 BkWh or 48 plants, assuming the average refrigerator today uses 1900 kWh/year. If the U.S. stock were to reach the average efficiency of 1994 refrigerators in California, we would be down to 700 kWh/year or 18 plants. Sources: California refrigerator consumption (stock and projected) - California Energy Commission press release, 21 December 1984 (Margaret Fjelsted). Refrigerator Stock - *1984 U.S. Statistical Abstracts*, pp. 755.

These standards have had a significant effect on electricity use throughout the United States. Since manufacturers can make a refrigerator that conforms to these standards for $35, they didn't bother to build new assembly lines to keep making crummier refrigerators than they could sell in California. So every new production line that's been set up since the California standards were enacted conforms to those standards. There is a little "dumping" at first. When California tightens its laws, then the manufacturers dump the bad ones on Texas (a favorite), but eventually all refrigerators tend to conform.

In the United States there are 90 million dwellings, but there are 125 million refrigerators and freezers. That is, the average saturation of refrigerators plus freezers is about one and a half. If refrigerators still used 1900 kWh/yr, it would take 50 power plants of the standard gigawatt size to run all of them* (right-hand scale, Figure 11). By the time we reach the 1993 standard, we'll be down to 18 plants or 18 gigawatts. The savings, then, are 32 GW, or 32 large central-station power plants. Recall that the payback for these savings is only one year.

Air conditioners tell much the same story and are now included in the national standard, discussed in section 7. Air conditioners in small commercial buildings are subject to the same sort of economics we talked about earlier. The California codes are usually based on trying to get a two- or three-year payback. Figure 12 shows the decrease in energy use and therefore dollar use by a three-ton air conditioner coming down from $410 a year for Fresno in 1977 to $290 a year. The COPs are improving from seven to ten. The surcost being about $300 to save 1.5 kilowatts with a cost to conserve power running at about $200 a kilowatt. If you can buy a lot of power at less than that, more power to you, but this seems to be a good way to do it compared to $1,750/kilowatt which is the cost of new capacity for the utility that serves Fresno.

5.3. Effects on statewide energy use

These and other strategies have had a drastic effect on energy consumption and peak demand in California over the last 10 years. Figure 13 shows how far demand has fallen below earlier forecasts. California electric growth had been at 6% per year from 1965 until OPEC. Gas prices went up, and the California utilities shook their heads and said, "Oh dear, that's probably going to slow down growth". They speculated that it would slow growth down from 6% to 5% per year. The 5% line on the figure was the utilities' prediction for what California would need. With that rate of increase, the state's peak demand would have swelled from 30 gigawatts in 1975 to 50 GW by 1985. More than $25 Billion would be needed to build these new plants, nearly half of which were to be nuclear.

We had begun to learn a little bit about buildings, and decided it was really cheaper to turn out the lights, particularly when people are not around. Remember, we lived in a society in which our lights always ran 8760 hours a year even though the buildings were only occupied for 3000. We at LBL came up with the idea of a "conservation potential" which said that instead of growing at 5% per year, an economic optimum was 1.2%. Of course, we knew people wouldn't invest optimally,

---

* The conversion from kilowatt-hours/year to average gigawatts assumes that a typical 1000-MW plant sells 5 Billion kWh/year, the U.S. average.

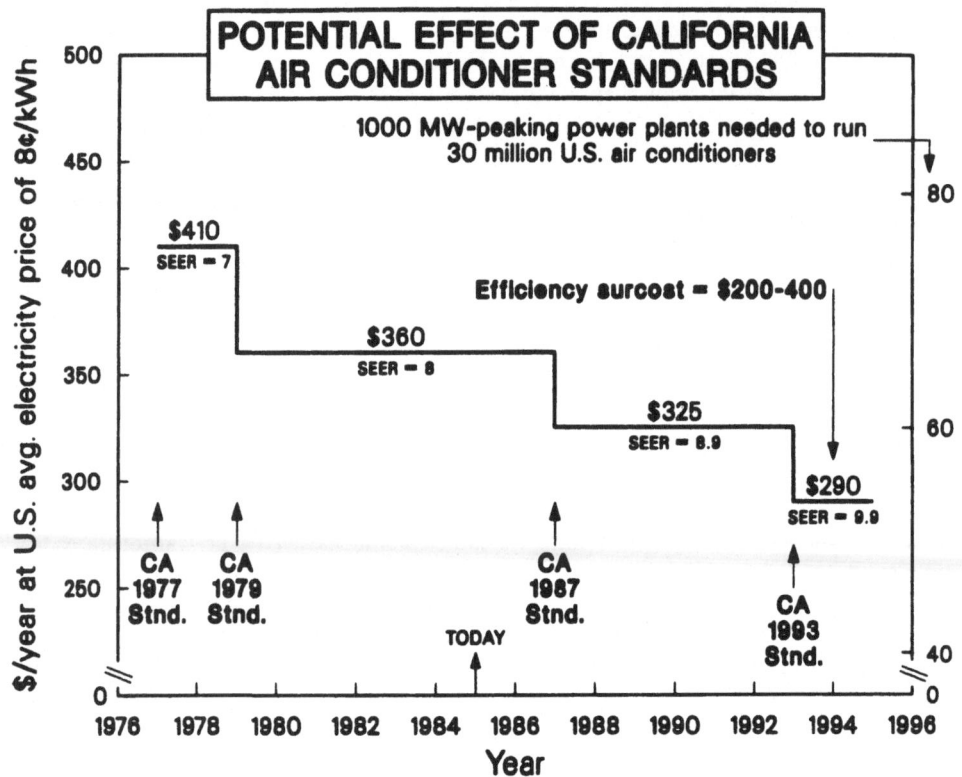

FIGURE 12. Potential Effect of California Air Conditioner Standards if Adopted Nationwide. Annual operating costs are based on SEER (Seasonal Energy Efficiency Ratings) specified by the standards for central air conditioners. The 30 million equivalent U.S. air conditioners represent 22 million actual 3-ton units and an additional 24 million room air conditioners. A typical 1000-MW baseload power plant produces 5 BkWh/year. Assuming 100 hours annual operation, the 30 million U.S. air conditioners consume 750 BkWh or 48 plants given a diversified load of 2.6 kW. Sources: Consumption - The LBL residential forecasting model (Jim McMahon). Air conditioner stock - *1984 U.S. Statistical Abstracts*, pp. 755.

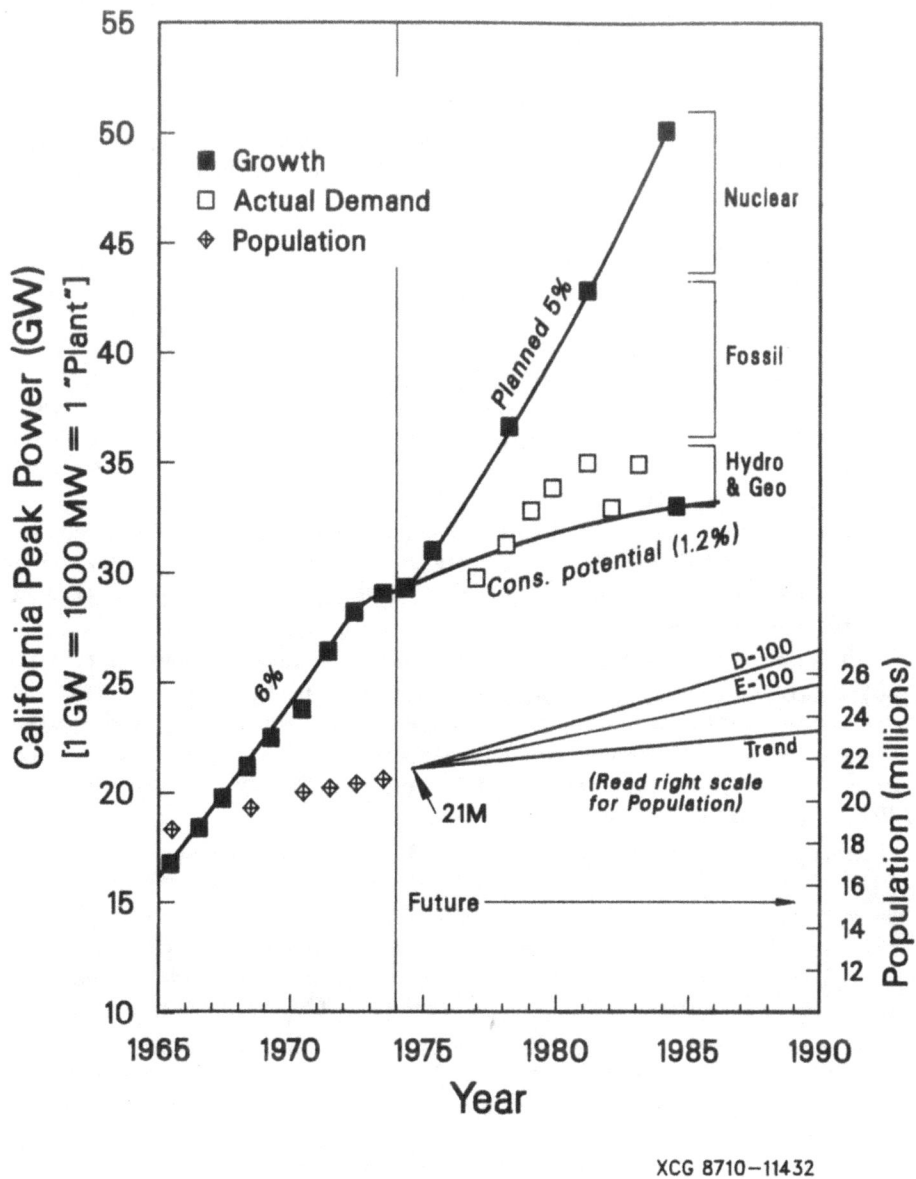

FIGURE 13. Coincident Peak Power in California. Utility projections vs. conservation potential as calculated by A.H. Rosenfeld for Warren Committee Hearings, California House of Representatives, December 4, 1975. Actual subsequent demand is plotted as □'s.

so we predicted consumption would fall somewhere in between, and suggested we could get along on 2 or 3% growth.

The □'s on Figure 13 show what actually happened—about 2% growth. Obviously it really is cheaper to turn off some lights. This figure was controversial in 1976. The president of PG&E tried to get us fired. He has since been fired himself, and we're on good terms with the utilities at the present moment; we do conservation studies for them.

5.4. The utility's perspective

Given the success stories we've shared, one may ask why do California utilities really want to conserve? Doesn't it interfere with their growth? It certainly disrupts one kind of growth. But California utilities have been hassled so much with standard power plants that they don't want to grow in that direction any more. PG&E paid $5.5 Billion for their Diablo Canyon nuke. Now they want to put it in the rate base and the PUC's own Public Staff recommends that the Commission only approve $1 Billion of that. It suggests that PG&E will have to "eat" the rest. No one is really going to be that tough on them, but you know that when somebody asks you to eat $4.4 Billion you're not likely to be terribly enthusiastic about investing in another big power plant. Californians just don't seem to appreciate big power plants the way they used to.

What do we do instead of building new plants? Figure 14 shows the 1981-1983 PG&E conservation program funding as approved by by the California PUC. The budget called for $124 million of programs under the headings of residential conservation service, home appliances, community consumer services, commercial and agricultural programs, and program evaluation. In 1983, $124 million was about 1.5% of PG&E revenues of $8 Billion.

Other conservation programs comprised $148 million, with another $48 million on R&D. These are big numbers. All told, PG&E was plowing 3.5% of its annual revenue back into conservation. Although that has changed somewhat now that PG&E has a temporary surplus of generating capacity, most of the programs remain intact. The most important lesson of this figure, though, is that utilities now know that they can affect demand, and they have the tools when the need next arises.

As a result of these efforts, the planning picture has taken some turns for the better. Figure 15 shows PG&E's 20-year resource plan, looking forward from 1983. The three sets of bars show a transition away from high rates of building new capacity. Hydro is forecast to grow for the first decade, geothermal grows quickly, and cogeneration is taking off. Wind and solar are increasing at a good rate but their overall contribution remains small. As for nuclear power, because of their Diablo Canyon fiasco they're not going to build any more nukes. The most interesting resource is conventional oil and steam. No new plants are planned. It turns out that no utility on the whole West Coast intends to build another thermal plant in the foreseeable future. Ten or twenty years ago, it seemed that 20% of all utility efforts went into buying land, building plants, and so on.

Figure 16 is PG&E's corresponding 20-year plan for conservation. PG&E says that its "business-as-usual" forecast with natural market improvements in conservation would cut growth down to 2.8% per year. Then state-mandated programs of various sorts will save about 1%/year more. The top layer in the graph—the very

Summary of Estimated Conservation Expenses for 1981, 1982, and 1983

(thousands of dollars)

Energy Conservation and Services Programs	1981	1982	1983
1. Residential Conservation Services (including solar)	$16,577	$34,600	$39,400
2. Homes, Appliances and Systems	5,820	9,487	10,527
3. Community and Consumer Services	4,303	6,800	7,500
4. Commercial-Industrial-Agricultural Conservation Service	15,283	45,000	65,000
5. Program Evaluation	1,052	1,100	1,200
6. Sub-total Energy Conservation and Services	$43,035	$96,987	$123,627
Other Conservation Programs			
7. Conservation Research, Development, and Demonstration (R,D,andD)	$2,674	$9,705	$3,051
8. Load Management and Load Management R,D,andD	25,257	60,190	39,791
9. Cogeneration and Solid Wastes (including R,D,andD)	11,297	19,998	36,916
10. Conservation Voltage Regulation	1,122	821	851
11. Energy from Biomass (Gas Production Only)	7,272	17,287	39,833
12. General Office Departments	18,605	25,100	27,842
13. Street Lighting Conversion	6,053	3,208	—
14. Sub-total Other Conservation Programs	$72,280	$136,309	$148,284
15. TOTAL	$115,315	$233,296	$271,911

FIGURE 14. Summary of Estimated Conservation Expenses for 1981, 1982 and 1983 Pacific Gas and Electric Company (thousands of dollars). In 1983, PGandE revenues were $8 billion in 1983 and allocations for conservation amounted to 3½% of this total (excluding the zero-interest weatherization loan program). These percentages spent on conservation are still tiny compared with 10% Federal incentives and Tax credits available through 1986.

44

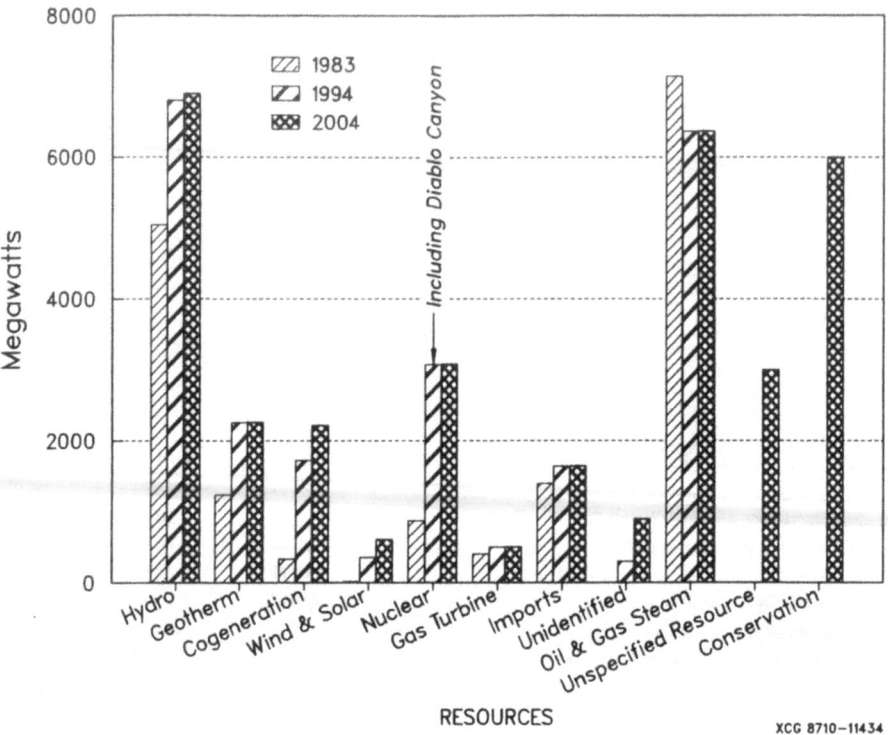

FIGURE 15. Electric Generation Capacity: Pacific Gas and Electric Company (PGandE) 20-year Plan.

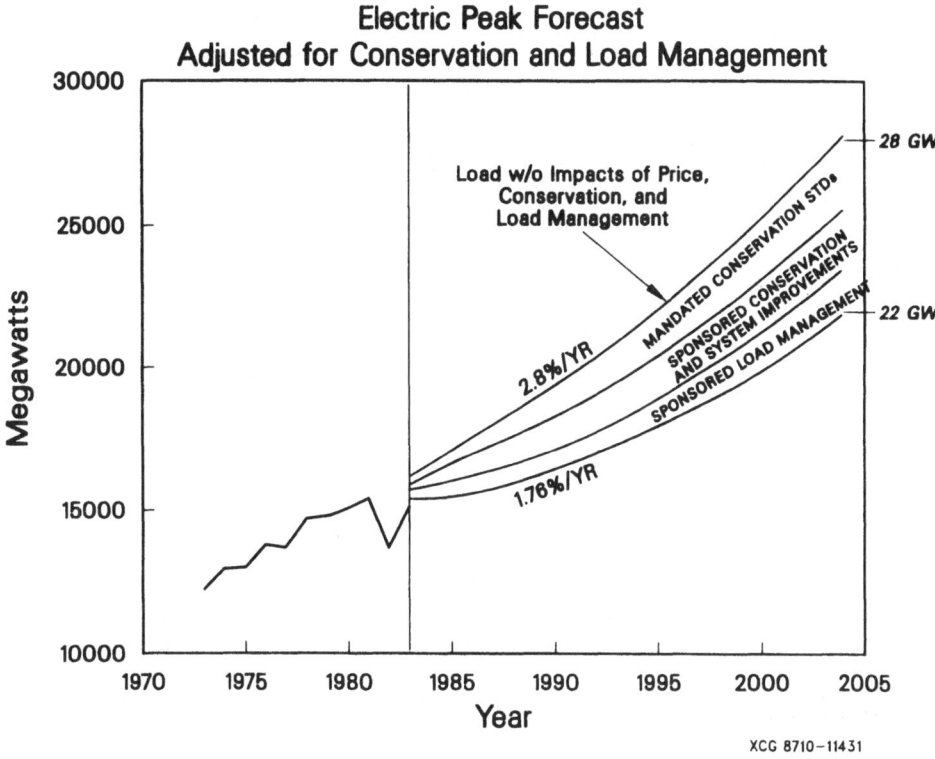

FIGURE 16. Pacific Gas and Electric Company (PGandE) Electric Peak Forecast Adjusted for Conservation and Load Management. Conservation and LM slow growth from 2.8% per year to 1.76% per year and save 6 GW (equivalent to 6 large power plants) over the 20-year planning horizon.

successful mandatory conservation standards for buildings and appliances—seem to be about a third of the story. Sponsored conservation programs are another part of the story, and so is load management (thermal storage, peak shaving). Thermal storage reduces peak load and should be good for about two gigawatts.

6. THE COST OF CONSERVED ENERGY

How do you calculate the cost of conserving energy with a more efficient energy-consuming device so you can compare it with energy from a power plant? Say you've got to pay $100 for the increased efficiency, but only once every 20 years. If you go to the bank and get a consumer loan in real (non-inflating) dollars at 7% real interest, for 20 years, the banker will consult a capital recovery table and charge 9.4% of the principal each year. Thus the annual cost is about $10, but the annual savings are 1200 kilowatt-hours. If you divide $10 by 1200 kWh and cancel the years, you find that the cost of conserved energy is about eight-tenths of a cent per kWh. Yet the average cost of residential electricity is about 8 cents, so in this example efficiency is ten-times cheaper than producing more electricity.

Computing the cost of conserved energy can tell us how much conservation is economically worthwhile. As we carry conservation further, the return from each improvement will tend to diminish. When the cost of conserved energy for the last increment of improvement is equal to the cost of buying new electricity, we will have conserved as much as is economically warranted. The notion of ranking conservation measures by increasing cost of conserved energy gives rise the idea of the "supply curve" of conserved energy.

6.1 The cost of conserved energy in refrigerators

The improvements to-date in refrigerators are far from this point of maximum cost-effective conservation. Recent studies by the American Council for an Energy Efficient Economy show that a 460 kWh/yr refrigerator would be easy to build, at a cost of conserved energy around 3 or 4 cents/kWh, and that more advanced technologies could bring the consumption down to 175 kWh/yr. As far back as 1977, A.D. Little showed that a 600 kWh/year refrigerator could be built for a $120 surcost. (Figure 17). We'll say more about refrigerators in the next section.

6.2. The cost of conserving one gallon in more-efficient cars

Low costs of conserved energy are not unique to home appliances. The U.S. CAFE (Corporate Automobile Fuel Economy) standards have raised the new fleet-average from 14 miles per gallon (0.17 l/km) in 1975 to 26 mpg (0.09 l/km) in 1985, but the '85 cars with their efficiency features (more forward speeds, lighter materials, etc.) retail for about $300 more (in real dollars) than cars cost in '75. Probably only $100 is for fuel efficiency and $200 is for the catalytic converter and other features to reduce emissions. The new car saves 350 gallons per year. At a dollar per gallon, the payback time is one year. If you calculate the cost to conserve gasoline, it is 10 cents per gallon. We assert that most of us are quite happy to conserve gasoline at 10 cents per gallon rather than buy it at a dollar per gallon.

The savings we have achieved at that cost are substantial. Figure 18 depicts the declining fuel consumption of the increasingly efficient U.S. car fleet. Considering that the average car is driven 10,000 miles a year, the graph shows how the

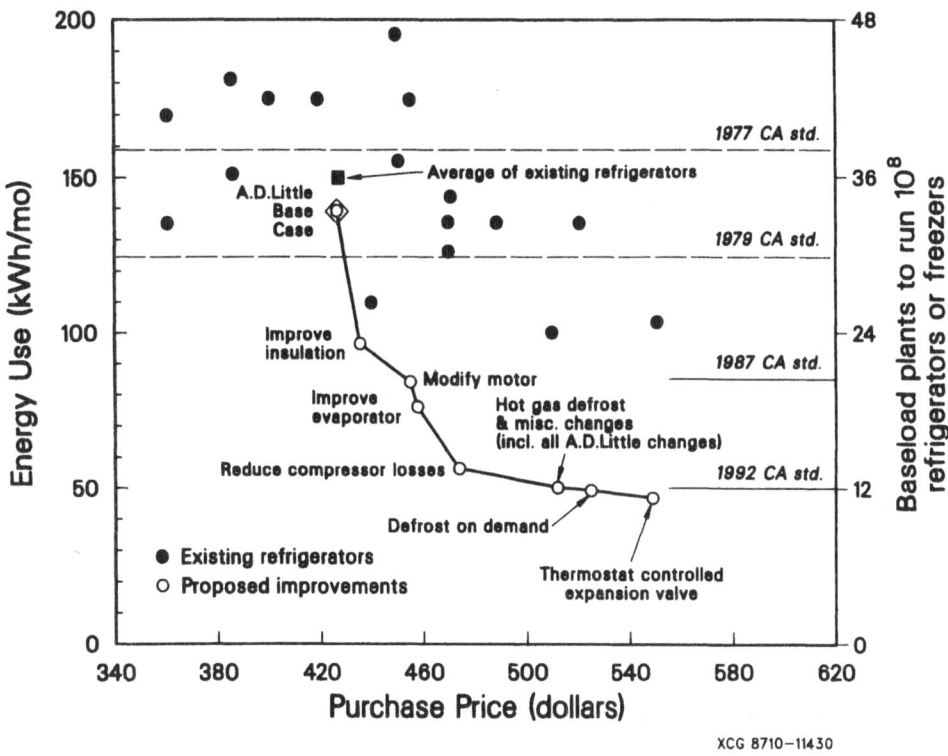

FIGURE 17. Electricity use vs. Purchase Price for Existing and Proposed Refrigerators. The closed circles in the upper half of the figure represent 17-17.5 cu. ft. top-freezer, automatic defrost models sold in California in 1976. The open circles joined by a heavy line are improved design steps proposed by A.D. Little (May 1977). All U.S. refrigerators plus freezers in 1980 used about 140 BkWh, so the vertical scale can also be read in BkWh, for the U.S. The potential savings of 85 BkWh is equivalent to the output of 17 1000-MW baseload power plants.

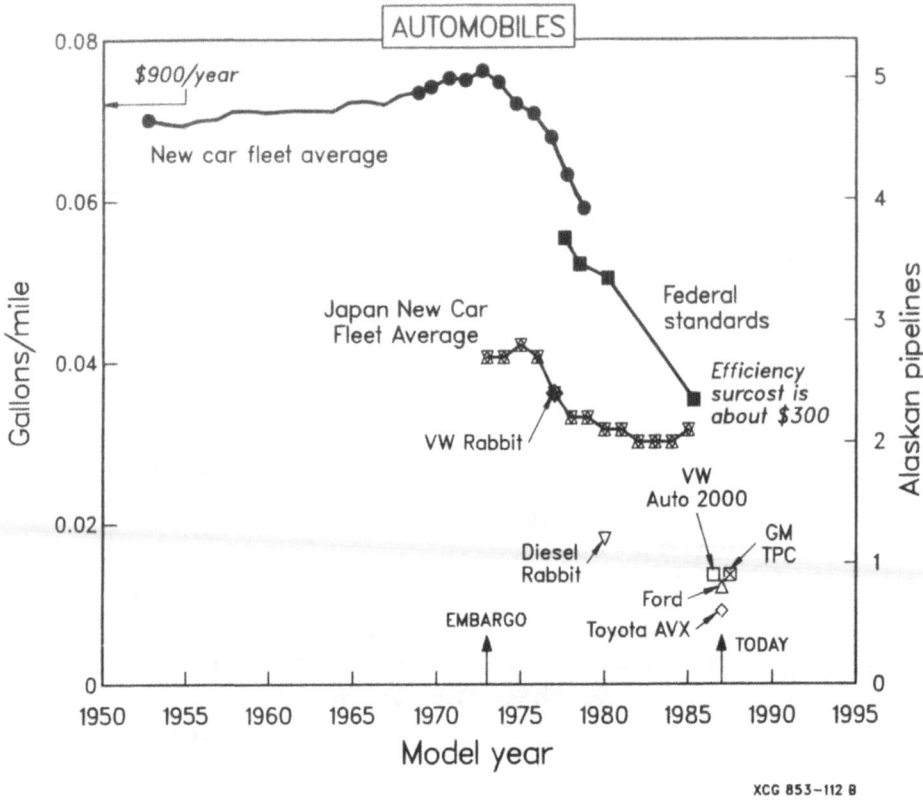

FIGURE 18. Automobile Energy-Use Trends: 1952-1985. The right-hand axis shows the equivalent number of Alaska pipelines required to fuel the U.S. auto fleet at various average efficiencies. The 1983 average fuel efficiency for autos was 16.7 mpg or 0.06 gpm, or 6.7 Mbod for the 1983 auto fleet. The Alaska Pipeline produces 1.7 Mbod; so in 1983 autos were using nearly 4 Alaskas. One Alaska corresponds to an average fleet efficiency of 0.015 gpm (or 60 mpg). Since the oil embargo, Japan's average fleet-efficiency been about twice as good as in the U.S., but the gap has narrowed slightly. The left scale reflects annual operating costs assuming $1/gallon and 10,000 miles driven per year. Federal standards will continue the trend of voluntary efficiency improvements which took place during the late 70's. Some of the more-efficient cars on the market or at the prototype stage are shown for reference. Source: Efficiency - "The Fuel Economy of Light Vehicles," Charles L. Gray, Jr., and Frank von Hipple, *Scientific American*, May 1981 vol. 244 No. 5. Alaska Pipeline production - Annual Energy Review (EIA-0384) 1984. US fleet Average (passenger cars) - Monthly Energy Review (EIA-0035) June, 1987. Japan Data - "Energy Conservation in Japan," The Energy Conservation Center, Tokyo 1986.

annual cost of gasoline to fuel that amount of driving has declined and how much of an improvement that translates into nationwide. We like to use a convenient large unit to think about oil use, which is the amount of oil moved through the Alaska pipeline—about one and three-quarters million barrels per day. By that yardstick, if American cars and light trucks still operated at their 1973 efficiencies, we would need five Alaskas just to run the auto fleet. As it is, in another five years, by the time the pre-CAFE cars are off the road, we'll be down to about two and a half Alaskas. And when we get to where the Volkswagen and other prototypes are taking us, we'll be using no more than about one Alaska. Of course, if it's not Volkswagen, it will be Fiat or Volvo or some other car-maker based in a country with a $1 to $4-per-gallon gasoline tax. Volvo has a car that has passed the California crash tests that gets 65 mpg (0.04 l/km), and Fiat is working on 130 mpg (0.02 l/km). We won't hold our breath for Ford.

7. APPLIANCES: PROGRESS AND POTENTIAL

The difference between the "fleet-average" of existing appliances in the U.S. can be improved by 50-75%. The new national standards will eliminate the real gas guzzlers, but the standards are well exceeded by the best models on the market and by the technical potentials for appliances. Figure 19 was made by Howard Geller of the American Council for an Energy Efficient Economy. The lower half examines the potential for improved central air conditioners, room air conditioners, water heaters, freezers, and refrigerators.

Let's look at central air conditioners. In kilowatt-hours per year, the highest column represents "stock". That is, it's the average unit in use at the latest survey that Geller could find—probably 1985. It's labeled around 3700 kilowatt-hours per year. The next block, at about 2900, shows how much energy was used by the average new unit on the market in 1985. This is partly thanks to the California standards, and partly because electricity prices were going up. Air conditioners are getting better, and at quite a clip: about 22% in 10 years. The next block—1800 kWh/yr—represents the best on the market in 1985. If you had done some comparison shopping, you would have ended up getting that one. The lowest block—1000 kWh/yr—is the best on the Japanese market or the best on the drawing boards. You can see that there's still a lot of progress to be made.

Electric water heaters tell the same story. The tallest bar is the average stock electric-resistance water heater at 4000 kWh/yr. The next smaller, at 3500 kWh/yr, is a better insulated one. Below that (1650 kWh/yr) is a heat pump and the smallest (1200 kWh) is a more-efficient heat pump. The story is the same for refrigerators and freezers, as well as lighting. In clothes drying, enormous improvements are possible even without switching to gas.

Following California's lead, the federal government has promulgated its own standards. The history behind these standards was somewhat messy. The U.S. Congress first required the Administration to set national appliance standards in the waning years of the Carter presidency. The standards were all written, but Mr. Carter lost his nerve and Mr. Reagan decided they were a poor idea.

This produced a backlash in which various states started passing their own appliance standards. That worried the manufacturers because they were faced with a patchwork of 50 sets of contradictory appliance standards instead of one national

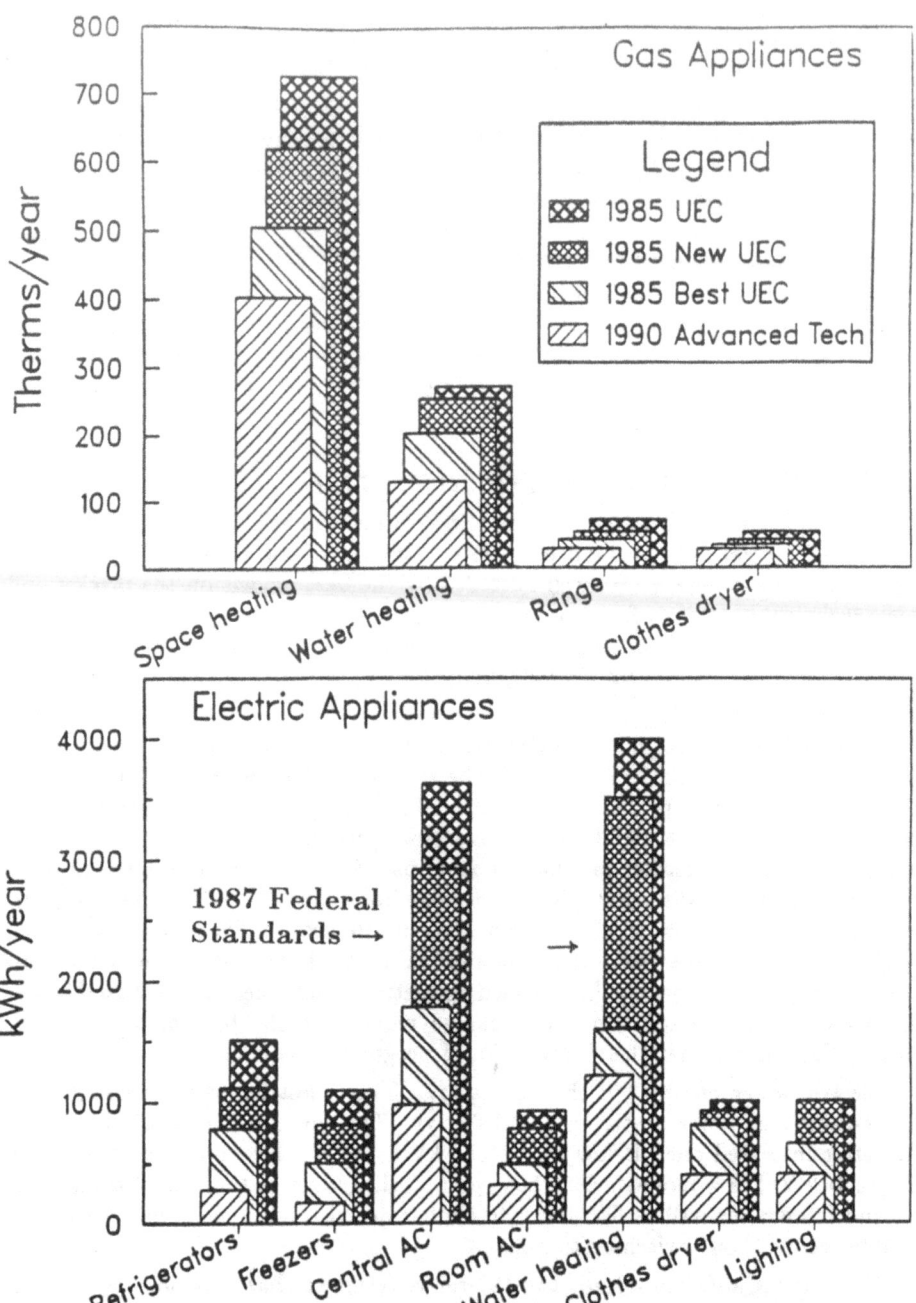

FIGURE 19. Unit Energy Consumption of Gas and Electric Household Appliances in the U.S. Each set of bars compares energy use by the average model in the 1985 stock, the average new unit and best new unit sold in 1985, and the best technology expected to be available in the 1990s. For scale, note that the electric appliances sold in one year use as much electricity as is produced by six large, 1000-MW power plants. This would drop to 2 power plants if all new models were as efficient as the best 1990s technology. Source: American Council for an Energy-Efficient Economy.

standard. They then appealed to DOE to recommend national standards, but DOE refused. Amazingly, the manufacturers then sat down with the Natural Resources Defense Council (a leading environmental and clean-energy group) and agreed on appliance standards very similar to the ones currently in effect in California. Congress passed the package unanimously in June 1986, but Mr. Reagan vetoed it. Then it came up again in spring 1987, and Mr. Reagan decided the handwriting was on the wall and he'd better pass it. We now have *national* appliance standards which will gradually come into force.

Arrows and the notation "1987 federal standards" on Figure 19 mark the levels of the new federal standards for central air conditioners and water heaters. You can see that the standards are not particularly stringent, and don't require manufacturers to develop new, unproven technologies. The payback time on those standards is fairly short—about two years. They don't place any onerous demands on consumers or manufacturers; they simply keep the worst junk off the market and prevent it from loading down the utility grid for the next twenty years.

8. ENERGY CONSERVATION IN THE BUILDING FABRIC

Naturally, the way a building is designed has a lot to do with how much energy it will use over its lifetime. In contrast with the rest of the United States, California is one of the few states that actually have performance standards for new buildings. In the U.S., the American Society of Heating, Refrigeration, and Air Conditioning Engineers (ASHRAE) passes voluntary standards. These are upgraded every few years and are called ASHRAE Standard-90 Series. They tend to be adopted by about half the states, but we're not sure if there is any actual enforcement. In California these standards are enforced, and have been strengthened substantially with a statewide code called Title-24—the most flexible and forward-looking in the country. The code allows builders to either conform to one of a series of packages, or to design a building and simulate its energy use to show that it would not use more energy than the allowed budget. There's even a "point system" that allows builders to come up with innovative designs without having to use computer models to show that they meet the standards.

Earlier, we showed something that most Americans are proud of, which is the progress in automobiles. Now, we want to show that the progress in commercial buildings is more astounding, cheaper, and conserves more fuel. But we're all much more aware of automobiles because we've all waited in line at the pump. We want to point out that the resource energy used by automobiles (motor gasoline) in the United States is about 10 quads or maybe 12, and commercial buildings is 12 quads also. So they're both serious gas guzzlers.

Figure 20 shows how energy use in large office buildings has changed since World War II. In the years of abundance from 1950 to 1973, resource energy use in new American buildings roughly doubled—that's the grey band. After the oil embargo of 1973, a few commonsensical measures led to significant improvement: ideas like not trying to heat and cool the air simultaneously or turning out the lights when no one is in the building. With the aid of voluntary federal standards and mandatory ones in California, we have pared energy use back even further by trying to design buildings that actually make sense. Designers have used daylighting, improved ventilation, intelligent fenestration (instead of simply enclosing the entire building in glass) to reduce the resource energy use of a building to almost a third of

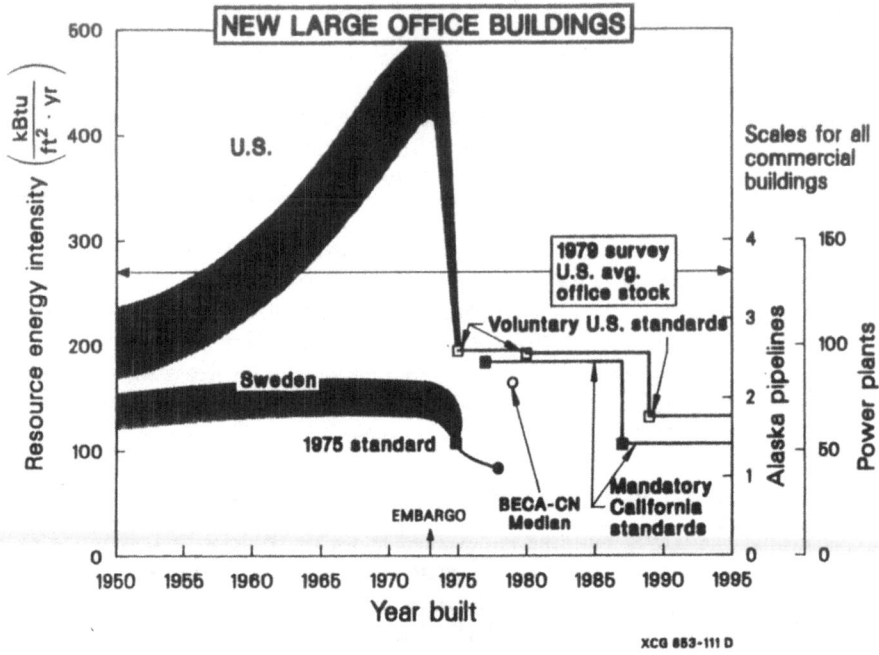

FIGURE 20. Trends in Annual Energy Intensity (use per ft^2) of New Office Buildings. Electricity is counted in resource energy units of 11,600 Btu/kWh. Dots represent data from real buildings. Squares are computer simulations from prototypes. Thus, the U.S. sequence is represented by a broad shaded band and is a crude measure of New York City office buildings by Charles W. Lawrence, Public Utilities Specialist for the city of New York (1973). The 1973 (pre-embargo) square is a simulation by A.D. Little for FEA; the later squares are simulations of buildings conforming to the indicated standards. Sources: A.D. Little, FEA Conservation paper 43 B (1976), and ASHRAE Special Project 41, Vol. III. DOE/NBB 51/6(1983). LBL ID No. XCG 853-111 D.

what the average office tower required in 1979. When these changes capture the entire market, we will be saving another two and a half Alaskas, or about 90 power plants. By way of comparison, the figure shows the experience in Sweden where in fact they tried a few American-style buildings, but they didn't sell. And so they've run with the same amenity for about half of the energy use.

It is also interesting to plot this picture in two dimensions and look at fuel versus electricity use during the 70s. What we have in Figure 21 is electricity in kilowatt-hours per square foot on the horizontal axis and fuel, in thousands of BTUs per square foot, on the vertical axis. Note the United States office stock as taken from a survey in 1979. The typical building used about 17 kilowatt-hours per square foot annually, but a fair amount of fuel as well—nearly 70 thousand BTUs per square foot annually. In dollars, that's about $1.20 worth of electricity every year and another 80 cents' worth of fuel.

Use came down quickly under the ASHRAE standards to less than 15 kilowatt-hours per square foot and virtually no oil. The progressive California standards—to take effect this year—will cut the electricity use again in half. The figure also depicts the residential stock where electricity use has always been low and the savings are only about 20%. However, the use of insulation in homes has saved enormous amounts of fuel.

9. THE UTILITY ROLE IN CONSERVATION

9.1. Marketing and incentives

Let us say a few things about marketing and incentives. First of all, what our utility—PG&E—has discovered is that if a utility wants to get something done they have to do more than just rely on prices or tax incentives. To *really* encourage conservation, PG&E has implemented incentives for refrigerators and new housing.

For new efficient refrigerators, they pay a $25 incentive to the buyer and another $25 to the seller. In addition, there has been an enormous advertising effort. In fact, PG&E took out full-page ads in the San Francisco Chronicle and only ran them for two days because that's how long it took for the efficient refrigerators to clear off of the shelves.

On energy-efficient homes, the utility program beat the Title-24 standard we described earlier by 10%. By beating the standard by up to 10%, the builder got a plaque; by beating the standard by more than 10%—and they had the builders exceeding Title-24 by more than 30% with a payback time of only one year—the builder got a label for the home and points for utility bill savings and was paid 15 cents per kilowatt hour, per year saved (Figure 22). An extra and unexpected benefit was that builders found that the labeled homes sold better, the more points the faster the home sold. PG&E found that this approach worked much better than the flat $175 incentive that they previously had offered to builders.

Why incentives when standards are available? The Title-24 standard took several years to implement. The buildings constructed under the new standards would be much more efficient than the previous generation. Their peak power load was going to be 4 to 4.5 watts per square foot, instead of something like 6 or 7 watts per square foot. PG&E knew that it was much cheaper to invest a few percent more in a building to get that efficiency rather than build the power plants to go with the

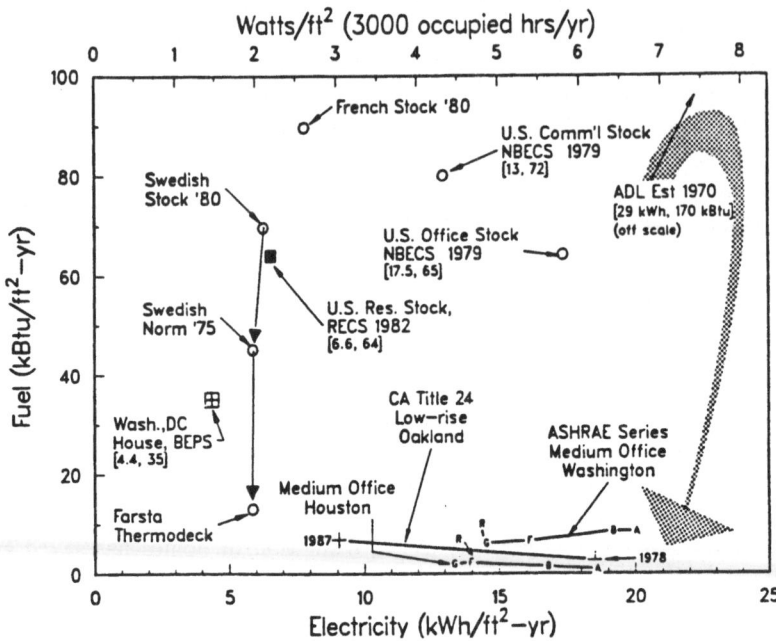

FIGURE 21. Office Building Fuel and Electricity Trends. Office buildings have such a large amount of "free" heat from equipment and people that they now need almost no space heating, even in climates as cold as Sweden's. So, modern office buildings are becoming almost entirely electric. Thus, the sequences labeled A, B, F, G, and R representing *modern office building* prototypes (conforming to the voluntary ASHRAE Standard 90 Series) *are almost lost at the bottom of the figure.* Similarly for the two-point sequence representing the California Title 24 mandatory standard. Real buildings have been found to use 10-20% more energy than that called for by standards. For comparison, residential trends are shown at the left. Sources: NBECS, the "Non-Residential Building Energy Consumption Survey," DOE/EIA-0318(79) and RECS, the "Residential Energy Consumption Survey," DOE/EIA-0321(81). The various standards are described in ASHRAE Special Project 41, DOE/NBB-0051/6.

Key to symbols: Open circles represent measured data, +'s and letters are calculations based on prototypes.

PG&E's Labels for New Homes

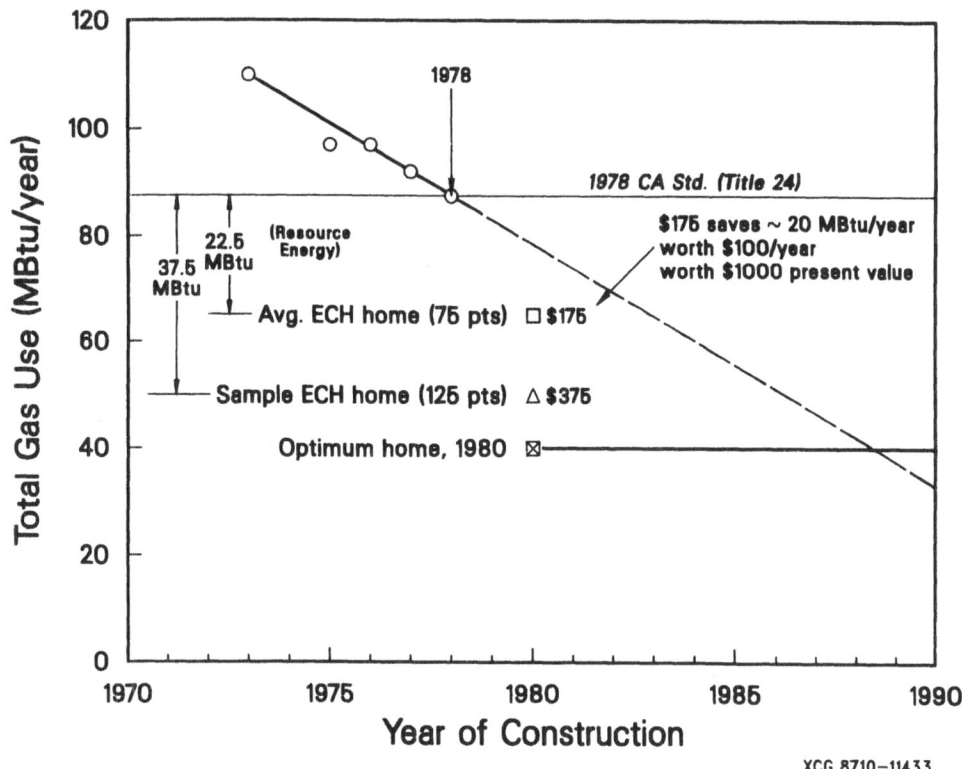

FIGURE 22. PG&E's Labels for New Homes. The y-axis shows total use of gas. The open circles are average billed use of gas and the solid dot is the calculated gas use of the average home qualifying in 1980 as an Energy Conservation Home; 60% of all new homes qualified. The average home in the program saved $100/year at an incremental cost of $175—these savings are worth $1000 in terms of net present value. The "+" is the sample Energy Conservation Home's score according to the labeling system. The "x" is the estimated use for a home built today in Fresno, California's climate that minimizes its lifetime costs. In 1980, Presley Homes currently advertised that its homes were as good as this least-cost optimum. The thick horizontal line is the economic optimum energy use, on the assumption that gas and gas conservation costs remain constant in real dollars. Given the trend suggested by the measured data, new homes would not reach the optimum until the late 1980's. Buyer-information programs, such as labeling, can accelerate the shift to more efficient home design. Economy and the Energy Conservation Coalition. March 4, 1986.

building. PG&E said, "Well, if we can save a kilowatt on a new building, we save ourselves $1500," which was their estimated marginal cost for production, transmission, and distribution. They could easily afford to bribe the building owner $300 per kilowatt to do that, and they would still be way ahead with the rate payers.

Of course, the building lasts a long time. Power plants only last 30 years, or with good luck, 50 years before replacement. Buildings last 50 years and homes last 100 years. It's very important to build right to start with. It's hard and more expensive to fix them later. So PG&E said, "Okay, if you build buildings which beat existing codes and conform with the new Title-24 early, we'll pay you $300 per kilowatt up to $50,000. That should be enough to get your attention". And so was launched a very successful program. They encouraged thermal energy storage, too, because it is very attractive to the utility as a way to sell off-peak power at night and it avoids the construction of new power plants. The rebate was about $300 per kilowatt saved, up to $150,000 per building.

Incidentally, the payback times are very short. It's like shooting fish in a barrel. Some of the payback times on these thermal storage projects are well under a year. In fact, some of them are negative, since builders can save more by downsizing their chillers to run around the clock than it costs to install the thermal storage itself. Despite the essential attractiveness of these investments, the utilities were still willing to offer the $300/kW incentives.

Actually, San Diego Gas and Electric is even cleverer. They said, "Why should we be spending ratepayers' money to encourage people to make an investment that will pay back in three months anyway?" Instead, they guaranteed a three-year return on investment. They figured that people would undertake improvements with two-year paybacks without any utility incentives. By guaranteeing the three-year payback, they bribed customers to go further and pay for measures that are not as dramatically cost-effective, (the two- to three-year paybacks) in order to get the utility to cover the three-year and longer paybacks.

PG&E's incentives for the commercial sector are all very carefully tuned. They don't want to pay more than necessary; they just wanted to get the audience's attention. The program was aimed at encouraging energy-efficient motors in small commercial or residential buildings. If it's a small one, they'd pay up to $60, or about $300 a kilowatt. If it's a larger one, they'd pay $10 per horsepower, which worked out to $100 per kilowatt in terms of savings. This was a very well-tuned program with lots of feedback in which they try to avoid wasting too much of the ratepayers' money and still get the customers' attention.

We've talked a lot about California. But demand-side planning is much more widespread and we will mention one other significant U.S. effort. In the Pacific Northwest, under the auspices of the Bonneville Power Administration, the utilities are also testing and recommending building standards. And they have many incentive programs. A program much like PG&E's was described in a recent July 1986 "Northwest Energy" newsletter, where incentives are held out specially for early-adopters of a new buildings standard. That is, they have some standards ready, they want people to experiment with them ahead of time, and they will pay builders and developers order of magnitude, $10,000 to $100,000 for experimenting with the new codes as they build new buildings. They'll pay up to $100,000 for training county code supervisors to keep up with their new codes. And we think planners will find this investment in training and enforcement to have a payback closer to

TABLE 4.
Project Merlin:
The Potential for PGandE to Defer a Residential Power Plant

ENERGY			
	1985 (BkWh)	2005 (BkWh)	Fraction of PGandE Model
All End Uses			
PGandE End-Use Model	22	28	---
7 Main End Uses			
Same Model	15	21	100%
Potential (current technology)	---	16	77%
Technical Potential	---	12	56%
SUMMER PEAK POWER			
	1985 (GW)	2005 (GW)	Fraction of PGandE Model
7 Main End Uses			
PGandE Model	3.7	5.8	100%
Potential (current technology)	---	4.0	69%
Technical Potential	---	2.5	43%

Notes:

Measures were included if their cost of saved electricity was less than 10¢/kWh. Power-conserving measures were included if their avoided cost of peak power was less than $1165/kW. This avoided cost was annualized over 20 years (the assumed life of a new power plant). A typical 1-GW baseload plant sells about 5 BkWh/year. Source: "Residential Conservation Power Plant Study," American Council for an Energy Efficient Economy. February 1986.

58

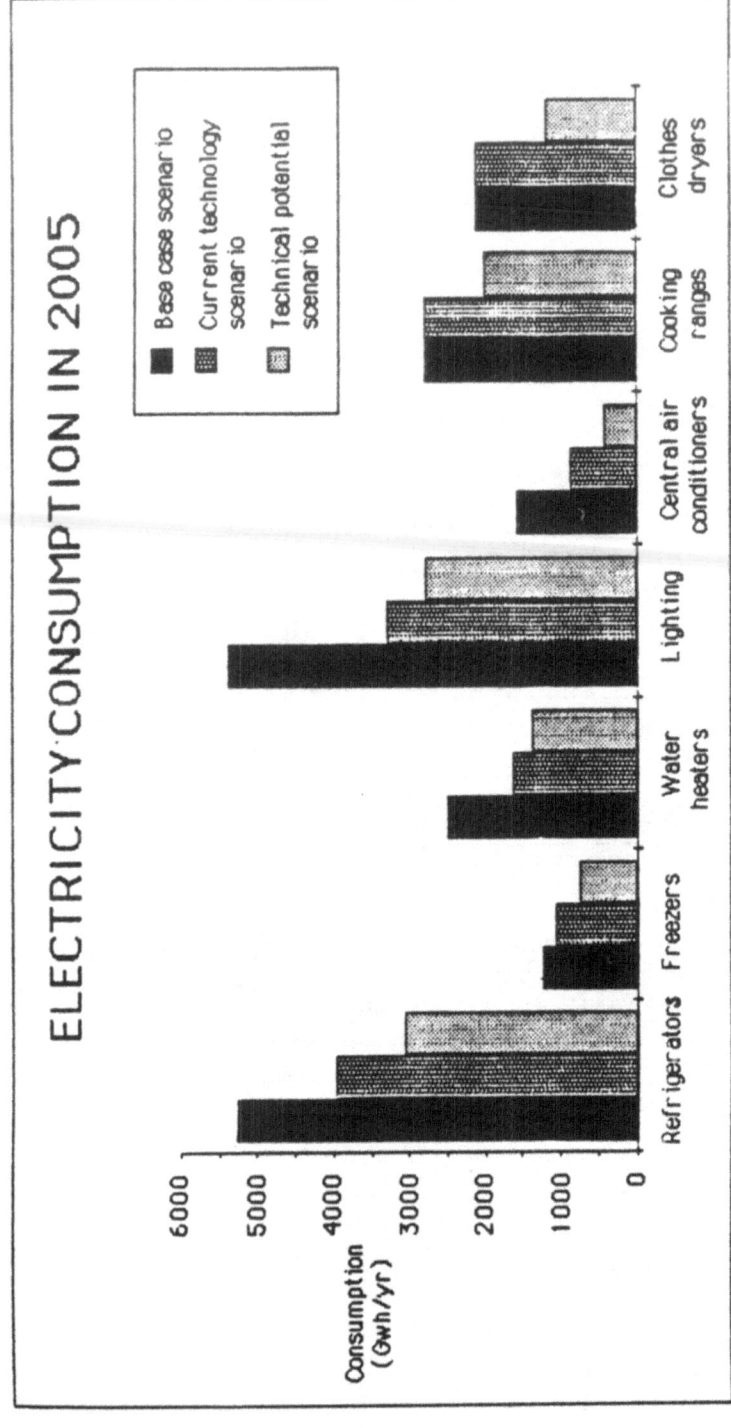

FIGURE 23. Project Merlin Results: Pacific Gas and Electric Company (PGandE) Residential Conservation Power Plants. The y-axis shows electricity consumption in the year 2005.

one-month. Because if they advance a major program by a year, they've saved a lot of energy very cheaply.

9.2. Merlin the Magician

The residential sector offers opportunities for large-scale demand management. PG&E had two planned unspecified power plants—which they called "Merlins"—to meet expected growth in residential demand (the bar labeled "unspecified resource" in Figure 15). Merlin is a nice name, because Merlin was a magician who sometimes appeared and sometimes didn't. And it turned out these Merlins don't have to appear, thanks to a couple of gigawatts' worth of conservation potential that the Merlin study team uncovered using nothing but currently available, off-the-shelf technologies. As shown in **Table 4**, they could save another gigawatt or so in the residential sector once some of the technically feasible efficiency measures are commercialized and implemented. In fact, the chief Merlin planner at PG&E, Lee Calloway, now has big pictures in his office of virgin forest, one's labeled "Merlin Site I" and the other's labeled "Merlin Site II".

The Merlin project plan was to invest in conservation measures that had a cost of conserved energy up to 5 cents a kilowatt-hour or in conserved power up to $1500 a kilowatt, both of them being cheaper than existing supply prices. We started with PG&E's latest resource plan and their end-use model, which is what they thought had all the latest technology from standard sources like EPRI. We couldn't look at everything in the model, we didn't have the resources to do that, but we looked at the seven main uses of residential electricity. Figure 23 shows the savings scenarios for each home appliance.

We found that, in fact, PG&E had missed some 5/16 or 20% of the available conservation options, and that they could cut their energy demand by about another third. And if one took what was known in technology, that is, what's on the drawing board now and probably will be available by the end of the century, in fact the gains were about 50%.

So this is a dynamic field, and it's an argument that shows that if you conduct studies like Merlin that you'll find that a lot of things can be done. If you don't conduct the studies, your plans will surely be wrong and your ultimate cost of providing energy services will be higher than necessary. The point is brought home when, in their 1984 annual report to stockholders, PG&E describes how they spent $250 million per year on conservation. It adds up in the last five years to over $1 Billion. But they say it has avoided the need to commit $7 Billion to new plants. They think the customers and stockholders should be happy. That's basically what we've learned to do: concentrate on conservation where it provides energy services at least cost.

9.3. Michigan's Merlin

All the pieces of this least-cost resource planning approach were recently brought together in LBL's *Michigan Electricity Options Study* (MEOS). We did a thorough evaluation of the residential sector in the Detroit Edison and Consumers Power territories [for more details see the case study by Krause, Colborne, and Rosenfeld in Chapter 7]. We found a great gap between frozen efficiencies and the least-cost potential. We filled this gap with a supply curve of conserved energy

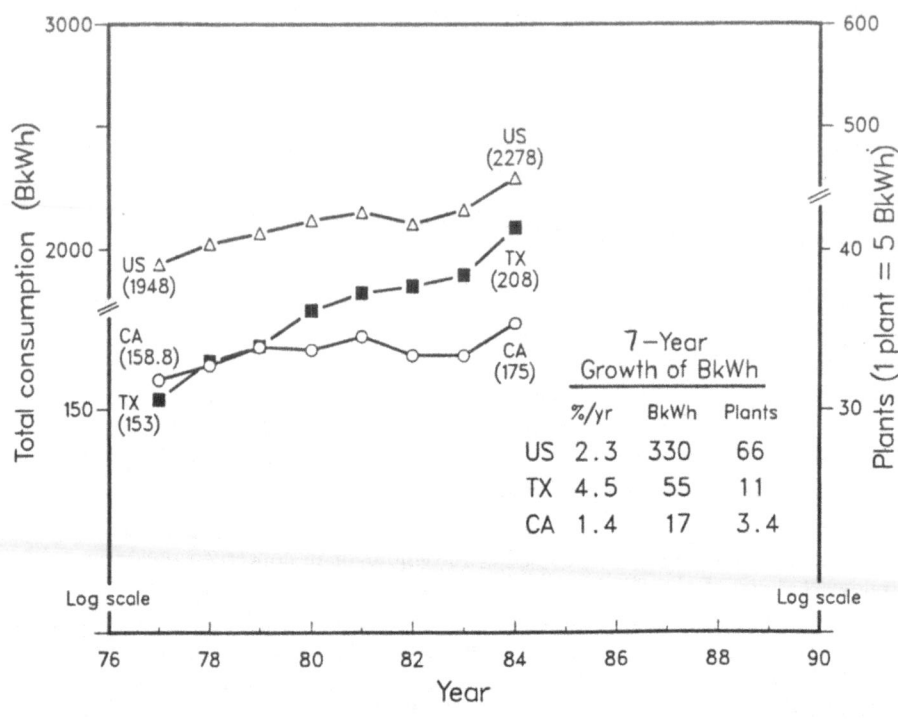

FIGURE 24. Total Electric Consumption by All Customers For CA, TX and the U.S. Comparative populations in 1984 were: CA 25.6 M and TX 16.0 M. In the seven years 1977-1984, annual growth rates were: CA 2.0% and TX 2.8%; TX - CA = 0.8%, whereas for annual BkWh growth, TX - CA = 3.1%. Electricity price increases in TX were less than one percent greater than in CA. The y-axis is logarithmic and the U.S. is shifted down one decade. BkWh are converted to 1000 MW (1 BW or 1 GW) using "1 Plant" = 5 BkWh/year. Sources: Consumption—Electric Power Annual (EIA 0348(84) p. 124 and EIA 0348(82) p. 167). GW—Annual Energy Review (EIA 0348(83) p. 195 and 201.)

amounting to nearly 700 Megawatts. In our program scenario, we deployed the measures until their cost of conserved energy rose to the cost of new power supply. This is the efficient way to do it. We then incorporated the lags in consumer response based on our experience with other conservation programs around the country and accounted for appliance retirements and new additions.

10. CLOSING THOUGHTS

What has it all added up to? Figure 24 sets the stage for a final comment on California conservation. We made this plot when we were trying to sell conservation plans to the Texas Public Utilities Commission. Whether it makes Texans feel bad or Californians feel good, it does seem to show there's a difference. From 1977 to 1984, California's use of electric energy was almost flat, while Texas's consumption grew at 4.5% per year and consumption in the United States as a whole increased at 2.3% per year. (U.S. consumption is divided by 10 so we could get it on the same scale). We observed that in two states which are very similar—both sun-belt states, both growing in economic productivity about 4% per year, both growing in population about 1% per year—the only important differences are that California has a serious energy policy and its electric needs are barely growing. Texas is a very laissez-faire state that hardly enforces policies on anything. It's not just energy. Texas, the day we made this plot, was the only state we know of where you could drive legally with an open bottle of beer in your car. When it comes to energy, one of the prices they seem to have paid is that for the same economic growth, we in California in seven years added the need for 3.4 power plants; the Texans needed 11. [5] That's a difference of seven power plants for a total of over $10 Billion, and that's the price you pay for sticking your head in the sand. Let this be a lesson to us all.

ACKNOWLEDGEMENTS

We thank David Wood and Seth Zuckerman for their help with research, editing, and preparing graphics, and Henry Kelly for his continuing insights and contributions.

REFERENCES

1. "Energy Conservation in Japan," 1986. The Energy Conservation Center, Tokyo, Japan.

2. *Federal R&D on Energy Efficiency: A $50 Billion Contribution to the U.S. Economy*, a White Paper on the Consequences of Proposed FY 1987 Budget Cuts, by the American Council for an Energy-Efficient Economy and the Energy Conservation Coalition. March 1986.

3. Baldwin, S. 1986. "New opportunities in electric motor technology," IEEE *Technology and Society Magazine*, March.

4. Schipper, L., Meyers, S., and Kelly, H. 1985. *Coming in from the Cold: Energy-Wise Housing in Sweden*. Seven Locks Press: Cabin John, Maryland.

5. Mills, E. and Rosenfeld, A.H.. "Managed Versus Unmanaged 7-Year Electric Growth: Californians Needed 3 New Plants, Texans Needed 11." Excerpted in Physics and Society, Vol. 16, No. 2, April 1987. LBL-22932.

Changing Patterns of Electricity Demand in Homes:
An International Overview

Lee Schipper, Andrea Ketoff, Stephen Meyers, and Dianne Hawk
International Energy Studies
Lawrence Berkeley Laboratory

1. INTRODUCTION

Residential electricity use grew rapidly in most OECD countries through 1973. After this, growth was slower. The reasons for the changes in growth rates include saturation of equipment, changes in consumer behavior, electricity conservation through more efficient equipment, substitution of fuels for electricity (or vice versa), and other reasons. This report analyzes the implications of the differences among countries, and the changes each country has undergone.

Changes in residential electricity use patterns can have a significant impact on total electricity use. The share of residential electricity in total electricity, which has risen slowly in most countries, generally is near 30% today, with the highest shares in countries with low industrial use (Denmark, England), and the lowest in Japan, where industrial electricity use is high. In general, residential electricity gained share in the 1970s and early 1980s both because of stagnation in industry and because of the increases in electric appliance and heating ownership. Representing nearly one third of total electricity sales in nearly every country, residential electricity use represents both market opportunities as well as great potential for electricity savings.

1.1. Differences Among Countries

Annual electricity consumption per household in 1983 ranged from less than 3000 kWh in Japan to 16000 kWh in Norway. Growth rates have also been quite different. Of the 11 countries, growth in the 1973-83 period was highest in Sweden and lowest in the U.S. (Table 1.1).

To understand these differences, it is necessary to look closely at how electricity is used, which means looking separately at the main electricity end-use markets. The importance of these markets differs among the countries (Table 1.2). Space heating is very significant in Canada, Sweden, and Norway, but is insignificant in Japan and the Netherlands. Appliances accounted for over half of electricity demand in 1983 in Denmark, Japan, the Netherlands, and the U.K.

1.2. Why International Study?

Although the changes in use patterns are occurring at different rates in different countries, many of the underlying factors or causes are similar. Therefore, study of these patterns across countries could reveal to us both more about the components of change in electricity use (i.e., the technologies, behavioral changes, etc.) as well as the

A. T. De Almeida and A. H. Rosenfeld (eds.), Demand-Side Management and Electricity End-Use Efficiency, 63–84.
© *1988 by Kluwer Academic Publishers.*

causes. In the end, the international comparison can be applied to understand any one country's consumption patterns. And energy and electricity-using technologies are now traded internationally; what is common in Japan or Germany today might be a household word tomorrow in the United States. International studies aid in predicting what changes might occur as technologies move from one country to the next.

The following chapter (2) looks in more detail at changes in each end-use market after 1973. Chapter 3 focuses on the differences among countries and the reasons for those differences. The final chapter discusses some issues of importance for understanding the evolution of residential electricity demand.

2. ELECTRICITY END-USE MARKETS: CHANGES SINCE 1973

Major changes in the use of electricity have taken place since 1973, sparked first by increased oil prices, and later by increased electricity prices, slower income growth, and new electricity-using technologies. Growth rates for electricity use varied among the countries, as did the shares going to each broad end-use category. Part of this change was structural, arising because people did different things with electricity, and part was due to changing intensity -- use of more or less electricity to perform certain tasks. How were these two components related to total change? This chapter outlines the key features of change.

2.1. Changing Importance of the End-Use Markets
The overall residential market can be seen as being composed of many smaller markets. These smaller markets correspond to household demands for particular services. In our work we disaggregate the residential market into five end-use markets: space heating, water heating, cooking, air conditioning[1] and other appliances. The end-use markets' relative importance within a country changes over time. Table 1.2 and Figure 1 contain data which describes the changing shares of end-uses in total electricity consumption between 1973 and 1983. Growth in the share of space heating was substantial in Canada, Denmark, and Sweden.

2.2. Space Heating
Electricity consumption for space heating grew considerably in most of the countries. Great Britain and the Netherlands, countries where gas is readily available, and Italy, where electricity is used mainly in warmer regions, were exceptions to this trend. Growth has been strongest in Sweden, Norway, and Canada, countries with low electricity prices, and France, where electric heating was aggressively promoted. Most of the growth took place after 1978.

The growth in electricity use for space heating was driven by increase in the number of dwellings heated electrically. Increased use of electricity for secondary (i.e. backup) heating also played a role. These changes were moderated by a decrease of some 10-15% in the intensity -- electricity use per household -- of electric space heating. This reflected both conservation in existing homes and the entrance of new,

---

[1] We do not discuss air conditioning in this section, as it is significant only in Japan (though less than 10% of sales in 1983) and the U.S.

tighter homes into the stock.

Market Penetration of Electricity. The share and number of homes with electric heating grew several-fold in Canada, Scandinavia (though remaining relatively unimportant in Denmark), and France (Table 2.1). Growth of electric space heating in the U.S. and Germany has proceeded at a slower rate, while the market penetration has actually contracted in Great Britain.[2] Use of primary electric space heating remains unimportant in Japan. In Italy its share has grown in non-central-heated homes and second residences, mainly in the South where the mild climate does not justify the installation of a central heating system. Electric heating with fixed radiators ("central") remains insignificant, as the national utility strongly discourages its installation.

The main reason for the increased market penetration of electric heating was its popularity in new homes. However, in Canada, Sweden, and Norway, many homes converted to electric space heating from oil or other fuel-based heating systems.

Electric Heating in New Homes. In most of the countries studied, the increased penetration of electric space heating is primarily due to its popularity in new homes. In some countries, this share has surpassed 50%, while in others it has grown slowly, fallen, and/or still remains low (Denmark, Germany, Netherlands, U.K.). The penetration of electric space heat depends upon forces besides just the relative price of electricity. These include different technologies, tariff structures, availability of gas or district heating, and building traditions.

The penetration of electric heating differs in new single family dwellings (SFD) and multi-family dwellings (MFD). In the U.S., Canada, Germany, and France, electric heating has a higher penetration in MFD than in SFD; in Sweden and Denmark, the reverse is true (primarily because of the popularity of district heating).

Conversions to Electricity. The role of conversions was small except in three countries, Sweden, Norway, and Canada. Those have the lowest electricity prices relative to competing fuels,[3] and, as it turns out, the coldest climate. In Sweden and Norway, conversion to electric space heating by existing homes was the dominant structural change. At least 1/3 of the homes using electricity in 1983 used a different fuel in 1975. Conversions dominated the scene so thoroughly in Sweden after 1979 that in years when only 20,000 new SFD with electric heating were built (1980-1983), between 70,000 and 100,000 SFD joined the stock of electrically-heated homes each year through conversion. In Norway many of the homes now using electricity as a

---

[2] Growth in market penetration in the U.S. was slower than in some other countries because the U.S. had one of the highest penetrations already in 1973, despite the low price of gas or oil relative to electricity. This was due in part to the prevalence of electricity in the Northwest and the Tennessee Valley Area. The prices of oil and gas moved up less rapidly than in other countries and electricity was expensive in the important area of oil penetration, the Northeast. Not surprisingly, there were few conversions to electricity.

[3] These countries have the largest share of hydroelectricity in their systems.

principal fuel used it as a secondary fuel before 1978. In Canada, 40% of the additions to the electrically heated stock from 1981 to 1983 were the result of conversions, primarily from oil. In France, the number of conversions in SFD after 1975 amounted to 20-25% of the number of new electric-heated SFD. For Denmark and Germany, conversions to electricity were a minor part of the increase in electric heating penetration.

Changes in Electric Heating Equipment. The stock of electrically-heated homes was heated with different equipment in 1983 than in 1973. In the U.S., the share of electric warm air furnaces and direct resistance heat fell as heat pumps gained ground. In Great Britain, France, and Germany, the role of storage systems, which were the most common mode around 1970 (taking advantage of low off-peak electricity rates), diminished in the 1970s. As incomes permitted more central heating, storage heating yielded its share in new construction to gas systems in Germany and England. In France, low prices and tighter building shells made direct-acting heating attractive and affordable after 1975. Heat pumps entered the market too, reaching 6% of the electrically-heated homes. It seems that storage heating, perhaps popular with lower income groups or in apartments, may yield to more comfortable, unrestricted electric systems if electricity is cheap, or lose ground to gas when living standards permit conversion.

In Sweden, the desirability of a system that could use electricity and/or oil and/or wood boosted the role of hydronic electric systems. These electric boilers are now found in one third of the electrically-heated stock. Many of these systems were originally oil boilers that were simply converted to electricity using plenum heaters. The hydronic systems came to dominate in new construction. This suggests that *source flexibility* might be an important factor in determining the success of electric heating technologies elsewhere.

Electric Backup Heating & Supplementary Fuels. An important aspect of the electric heating scene was the increasingly frequent use of electricity as a secondary fuel. By 1983, roughly 5% of U.S. homes, 10% of Canadian, 11% of Italian, 18% of French, 23% of British, 25% of Danish, 33% of German, nearly 90% of Dutch homes and virtually all in Japan used electricity -- almost always small, portable heaters -- to supplement main heating. The lower values for the U.S., Canada, and France reflect the availability of wood. The use of other fuels (especially wood) as supplement to primary electric systems also increased. In Sweden and Norway, more than half of the homes with electric heat use wood and some kerosene as well. In Denmark some wood, kerosene, or trash is used in about 1/3 of the homes with electric heating. In the U.S., around 25% of the electrically-heated homes also used wood, and 7% kerosene. About the same share of electrically heated homes in Great Britain used a second fuel -- mostly bottle gas or kerosene -- in 1983.

Average electricity consumption for space heating in homes with primary electric heat is 10-15% lower today than in the early 1970s. This is due mainly to the rapid entry of new, well-insulated homes into the stock and decreased consumption in existing homes. The use of supplementary fuels also played a role. Changes in the dwelling mix and in dwelling occupancy probably had only a small role in the change.

Conversions to electric space heating have affected the average heating intensity in Norway, Sweden, and France. The intensity in these countries did not decrease to the extent that it did in countries who's electrically-heated stock was formed through new construction. This was because the shells of these homes were not as tight as those used in electrically-heated buildings constructed after 1975. This effect is apparent in Sweden and France. The relative improvement in new homes varied among the countries. In France, the pre-1975 stock of electrically-heated SFD used 11.2 MWH/home in 1983, while the post-1975 stock averaged only 9 MWH. A similar drop occurred in Sweden, and, taking account the use of wood, in Norway as well. Since electrically-heated homes were initially the best insulated in most countries, these further improvements are noteworthy.

2.3. Water Heating

Market Penetration of Electricity. The penetration of electric water heaters is significantly higher than that of space heating except in Sweden and Denmark (Table 2.2). Electric water heating was common in European homes long before space heating. Small units, either tanks or instant recovery types, were placed in bathrooms and kitchens. Electric water heating was installed in many dwellings independently from the heating fuel choice, especially in non-centrally heated homes. This has been very common in Italy, Holland, Germany, the UK, and Japan. Homes where electric space heating was chosen frequently used the same fuel for their water heating systems. This is common in Scandinavia, the U.S., and France. Finally, increasing numbers of homes with oil-based space-and-hot-water heating systems (in United States, Canada, Germany, Sweden, Denmark, Norway, and more recently England) have chosen electric water heaters because these allow the oil system to be turned off in the non-heating months.

Intensity. Estimates of electric water heating intensity are very rough.[4] Measured and estimated evidence suggests that the change in intensity differed among countries. Intensity appears to have declined in Canada, Denmark, and Great Britain over the past decade. In general, the decline in household size probably pushed consumption down, while greater penetration of boilers has pushed consumption up.

Intensities for water heating are difficult to interpret; they reflect quantities (and temperatures) of water consumed, not simply differences in efficiency. It is clear, however, that unit consumption for electric water heating is much lower in central Europe than in Scandinavia and N. America (Table 2.3). The evidence suggests that the differences reflect the use of instantaneous heaters (as opposed to central hot water tanks) in central Europe as well as different bathing habits.
dent of the main system.[5]

---

[4] The large differences among countries reflect diversity in technologies as well as in habits.

[5] The reasons for this difference are several. U.S. homes had centrally-supplied hot water long before clothes and dishwashers became popular. In Europe, electric water heating was already more important because of the lack of central heating and central hot water. The hot-fill systems in the U.S. take less time than the cold-fill systems because of limitations on the wattage of the heating elements.

Additionally, most washers and dishwashers in Europe heat their own water indepen- Also, use is significantly higher in countries that have the highest incomes, and, with the exception of Denmark and the U.S., the lowest electricity prices. In the "low user" countries tanks are smaller and point-of-use much more common than in the high-user countries.

2.4. Cooking

Market Penetration of Electricity. Electric cooking gained market share at the expense of solids, gas and LPG. The penetration of electric cooking equipment is largely a result of fuel choices in new homes; electric connections are typically less expensive than gas connections. Electric cooking gained everywhere over city gas and even over natural gas in Germany, Canada, and the U.S., but lost in importance to gas in the U.K. In France, electricity's share rose rapidly in the 1970s, but in Japan and Italy, gas cooking has always dominated the market.

Electric cooking is spreading from a single principal stove/cooker to a variety of specialized devices. In Holland, Italy, and France, mixed electric/gas systems are common. Due to the proliferation of small appliances, the overall utility of gas cookers is reduced. Electric juicers, egg cookers, coffee makers, deep fryers, toasters, and a hoard of other devices appear on surveys in most countries. Since many of these produce cooking heat or hot water, they relieve the main stove of much of its work. Microwave ovens have come to be found in over 20% of homes in the U.S., and have grown in popularity recently in the U.K. (11% of households in 1984), as well as in Sweden. They have not caught on universally, however.

Intensity. The few estimates quantifying electric cooking intensity show a slow but constant decrease over time. Manufacturers report improvements in the efficiency of new stoves and ovens. Family size, habits, and house occupancy (i.e., the number of meals cooked at home) have changed since 1970 and these changes could well have caused a greater effect than that of higher efficiency of new equipment, which turns over slowly. In Sweden, unit consumption of city gas for cooking in homes with only cooking stoves decreased by 60% over 25 years. Smaller families are eating at home far less today than in the 1960s, and eating simpler meals too; both spouses work, and children get hot meals in school. Microwave ovens are now popular in most of Europe; these could have reduced use somewhat (or increased electricity use in homes with gas cooking equipment).

2.5. Electric Appliances

By electric appliances we mean devices that run principally on electricity, or "electric specific" devices, for which fossil fuels do not generally compete. Many uses of electricity fit into this category, including major consumers such as food cooling, clothes washing and drying, dishwashing, and lighting, and smaller consumers such as TV and other electronics, small kitchen devices, and small motored devices (tools, vacuum cleaners, etc.). The category includes lighting.

Electricity consumption for these uses played the main role in the growth of electricity use through the mid-1970s (until the rise in electric heating). Since 1978, growth has slowed in all countries both because of saturation of the appliance market and because new appliances are more efficient.

Market penetration. The percentage of homes with the major appliances grew considerably in many countries between 1972 and 1983 (Table 2.4). Growth in Europe and Canada was higher than in the U.S., where there was less room for growth. The relative increases in penetration have been greatest in France, Germany, Japan, and Italy. As a result, appliance ownership pushed up electricity use significantly in these countries. The surge in Sweden was caused principally by growth in the fraction of households occupying SFD. Saturations are always higher in SFD, not simply because incomes tend to be higher, but because living space is greater.

The market penetration of smaller appliances also increased. The most important is television: color TV reached above 90% of the homes in most nations, and in all there is more than one TV (color or BW) per household. Other electronics uses have proliferated at a high rate. The penetration of small kitchen appliances has been substantial in most countries.

Along with growing penetration, there has been increase in appliance size. A survey in Denmark found that in 1981, 70% of the refrigerators built before 1970 were less than 175 liters in volume. But only 50% were that small in 1980/81. Similarly, 54% of combis were smaller than 175l in 1970, but only 31% that small in 1979/80. Based on this comparison of models that survived to 1981, new models of these two appliances grew by 22% in volume between 1970 and 1980. Danish households were already well appointed in 1970; we can presume that such growth was even stronger in G. Britain, Central Europe, and Japan.

Intensity. Electricity intensity for appliances as a group grew because the average household gained more appliances. Moreover, certain important appliances, notably refrigerators, freezers, and clotheswashers, grew in sizer and features, increasing electricity use significantly. However, after 1975, as new, more efficient appliances entered the stock, average consumption began to fall (Table 2.5). The average consumption of new appliances fell markedly, although size and features still increased in some cases. The increase of the early 1970s slowed noticeably by the end of the 1980s. This shows the effects of slower structural growth and reduced unit consumption of new appliances. In Germany, Denmark, Holland, and the UK, estimated use per household stagnated and even fell in some years between 1979 and 1983. In countries with lower consumption in 1973, like France and Japan, increases in appliance size or number of features have led to higher consumption, even as efficiency has increased.

Changes in the average consumption of particular appliances can be seen in estimates of unit consumption. The average consumption is determined by the efficiency, size, and features of the appliances as well as by usage patterns. For the most part, changes in the latter have not been a major factor. Average consumption of refrigerators and freezers has increased in many countries because greater size and use of auto defrost has had more effect than improved efficiency. The U.S. is the exception to this, as these appliances were already large in the early 1970s, and the considerable efficiency improvement in new appliances brought average consumption down. For the other major appliances, the available estimates show either decline in average consumption or no change.

New appliance efficiency. The average efficiency of new appliances has improved in Europe and North America. During the course of this study, we discussed the evidence of more efficient appliances with authorities in Sweden, Denmark, Germany, France, Holland, and the U.K. All European manufacturers (Philips of Holland, Electrolux of Sweden, and those in ZVEI, the Central Group of the Appliance Industry in Germany) report significant improvements in their appliances since 1975. These are paralleled in Japan for TV, air conditioning, and refrigerators.[6] Table 2.6 presents data on the percentage reduction in the average electricity consumption of new appliances.

3. EXPLAINING THE DIFFERENCES AMONG COUNTRIES

The large variation in electricity consumption per household was highlighted in the previous sections. In this section we look to the end-use markets, considering both structure and intensity, to uncover the reasons for this variation.

Some structural elements of the residential sector as a whole affect more than one end-use market. Dwelling size and type affect consumption for space heating as well as for appliances. The average house size is far larger in the U.S. than in the other high income countries, and larger houses admit larger appliances, particularly in kitchens.

Families in multi-family dwellings (MFD) are always smaller than those in SFD and, with smaller incomes and significantly less living area, tend to have fewer appliances. In the U.S., Canada, the UK, and Holland, single-family dwellings (SFD) dominate (roughly 70% of the stock), although these tend to be the semi-detached types in the latter two countries. In the other countries, SFD and MFD tend to have roughly equal shares, with Sweden and Italy having the highest share of apartments (54% and 70% of occupied dwellings, respectively).

Many of the differences in both structure and intensity are related to incomes and electricity prices. For example, in the U.S., higher incomes permitted families to own large homes, and permitted people to live alone, rather than remaining longer in the family nest. Additionally, the larger size of American homes permits larger kitchen appliances, an important factor in explaining the size of American refrigerators and freezers. Within any given end-use market, intensity varies by roughly $\pm$ 30%, with N. America at the high end, Scandinavia in the middle, and Central Europe, Italy, and Japan at the low end. Efficiency appears to depend on price -- rising prices provoke manufacturers to build more efficient appliances. Prices also have an effect on the size and features of appliances. In Germany, Netherlands, and Denmark, appliance energy use is low, and prices are high. In the U.S. Norway, Sweden, and Canada, appliance use is higher and prices are lower . Figure 2 shows the evolution of average residential electricity prices since 1960 (in constant 1982 US $, using 1981 exchange rates).

---

[6] Energy Conservation in Japan, 1984. Tokyo: The Energy Conservation Center.

3.1 <u>Understanding the Differences in Space Heating.</u>

Market penetration. The penetration of electric heating varies significantly. Not surprisingly, the countries with the highest electricity prices, show the lowest penetration (Denmark, Germany, Japan, the Netherlands), and the countries with the lowest prices show the highest penetration (Sweden, Norway, Canada). The U.S. and France are intermediate.[7]

The high penetration in Sweden and Norway is partly due to the massive *conversions* to electricity after 1979. In Canada, conversion also contributed to the rise of electric heating. In Germany, France, Denmark, and the U.S., electric heating rose principally because of fuel choices made in new homes.

The *presence of gas* is very important to that choice. In the UK, nearly 50% of the electric heating systems lay where gas was unavailable in 1983. In the U.S., this percentage was 65% in 1982 (among single-family dwellings). Thus interfuel competition may depend crucially on availability; electricity does not compete directly with other fuels everywhere in a country.

Social and economic factors also influence electricity penetration. In many countries income or social class is associated with particular heating systems. In Great Britain and Germany, for example, storage systems based on electricity are found mostly in rental housing, and more among lower income classes than are oil- or gas-based heating systems. Survey data from France show that electric heating is more diffused in smaller apartments, occupied by lower income households.

Electric heating has penetrated different strata of the dwelling stock in different countries, depending on gas availability, and probably depending on *housing policies* and financing as well. In Sweden before 1973, electric heating, which had lower initial costs, dominated new construction initiated by builders, while oil heating dominated new construction initiated directly by final buyers, and had considerably lower running costs. Similarly, penetration of electric heating in new apartments in many countries was higher than it was in new single-family dwellings, suggesting that builders and homeowners have different criteria.

The penetration of different heating technologies differs considerably among countries. In the 8 countries where electric heating as a principal source is found in more than 6% of homes, storage systems, furnaces/boilers, heat pumps, and direct resistance all share the total picture, with each strong in one or two countries.

Intensity. Differences in average intensity reflect differences in the thermal integrity of homes, the occupant behavior (ie., indoor temperature, heating hours), the type and efficiency of the heating system, the mix of dwellings in the electric-heated stock,[8] and the use of secondary heating fuels. These differences in the structure of

---

[7] The French and U.S. heating markets have many similarities: climate, high share of electric heating among new homes, high price of electricity compared with fuels. In both, the growth in new homes heated with electricity is the major factor adding load. A difference is the relative high share of heat pumps in the U.S.

[8] In the Scandinavian countries, France, and the U.S., SFD dominate the electrically-heated stock, while in Germany, Canada, and the UK the share of MFD lies between 45% and 80%.

electric heating mean that one must be careful in comparing electricity use per electrically heated dwelling (Table 3.1).

Estimated unit consumption for electric heating, adjusted for climate and home living area,[9] is lowest in Sweden. Given that most of the electrically-heated stock in Sweden is composed of detached houses, this low value is testimony to the tightness of Swedish homes. The U.S. is next lowest, followed by Denmark. Tight building shells are responsible for the low values in Sweden and Denmark. In Denmark, many homes with electric heat also have kerosene or wood stoves for secondary heating. England is low because electric heating is primarily storage heating and is mostly found in multi-family housing, and because indoor temperatures are very low (electricity is used mainly among lower income groups). In the U.S., the widespread use of heat pumps may be partially responsible for the low value. The U.S. also has a lower share of single-family houses than the Scandinavian countries.

It is difficult to tell the degree to which behavior (low temperatures, intermittent heating, use of secondary fuels) rather than efficiency is the cause for low intensity in some countries. It appears that electricity prices may affect behavior more than efficiency. Surveys suggest that Swedish households heat to the highest indoor temperatures (20.5 C averaged around the house during the winter), with Norwegian and Canadian a few degrees lower. By far the lowest reported temperatures were in British (15-16 C) and Japanese (13-14 C) dwellings; electricity is expensive in both countries (though there are also other reasons for the low temperatures).

Historically low electricity prices do not always lead to low efficiency in electric heating. Swedish building practices evolved far faster than in any other country towards efficiency over a long period of time, even as electricity prices remained relatively low. Canada and Norway, also low-price countries, and colder than Sweden, might be expected to use more insulation but this is not the case. In 1984, new electrically heated homes in Sweden used 20-30 cm of mineral wool in their walls, vs. 15-20 cm in Norway and 10-15 cm in Canada. The reason is that Swedish authorities have since 1963 included financing of insulation in new-home loans; no such features were included in home-loans in any other country (often homes supported by state grants did require a degree of insulation higher than otherwise used, but these homes never represented the high share they did in Sweden (80-95%). Significantly, Swedish homes were built to *better* than code requirements since the early 1960s. Indeed, we believe that the evolution of tight houses encouraged the development of electric heating when oil was otherwise significantly less costly.[10]

Electricity prices may affect behavior more than efficiency. Surveys suggest that Swedish households heat to the highest indoor temperatures (20.5 C averaged around the house during the winter), with Norwegian and Canadian a few degrees lower. By far the lowest reported temperatures were in English and Japanese dwellings, where

[9] We have estimated home living area for some countries based on data for all homes, but data from Denmark, France, Norway, Sweden, and the U.S. refer to homes with electric heating.

[10] See Schipper, L., Kelly, H., and S. Meyers, 1985. *Coming in From the Cold.* Cabin John, MD: Seven Locks Press.

electricity is expensive.

3.2. Electric Appliances

Electricity consumption for appliances as a group is a function of the penetration of different devices, their efficiency, size and features, and their usage. The price of electricity and income shape these factors. Electricity consumption for appliances and lighting now ranges from a high of over 4000 kWh/dwelling in the U.S. to a low of around 1600 kWh/dwelling in Germany (Table 3.2). Appliance ownership does not differ enough to explain the difference in consumption. Most of the difference is explained by the fact that N. American appliances each use considerably more electricity than those in Europe. After examining catalogues and studies of appliances, we conclude that about half of the difference in use per appliance results from efficiency; the other half results either from habits (wash frequency, etc.) or on the size or features of the appliances (such as the frost-free option in refrigerators).

Market penetration. Differences in saturation contribute to differences in electricity use per household. Saturations of refrigerators, washers, and TV are high in all countries, while those of freezers, dishwashers and dryers vary significantly. In general, the higher the income, the greater the stock of appliances. Saturation of dishwashers in particular appears to depend on income levels. The high income countries -- U.S., Canada, and Sweden -- have the highest saturations (37%, 34%, 31% respectively). Germany, France, and Italy follow with levels ranging from 25 to 15 percent. The saturation of clothes dryers varies greatly around the countries in the study, from a high of 45% in the U.S. to a low of only 10% in Denmark and Germany (and close to zero in Italy, France, and Japan). Habits and climate are important in determining penetration of these devices.

Electricity intensity. Size and features, efficiency, and usage all affect average consumption of appliances. Many appliances in N. America are larger than in Europe, and those in Scandinavia are larger than those elsewhere in Europe. In N. America, 18 cu. ft (510 liters gross) is the average size for refrigerators; in Scandinavia new models generally fall between 300 and 400 liters, in England and on the continent, somewhat less, and in Japan around 200l. (The largest model sold by Philips in 1985 held 450l, including freezer compartments.)

Food cooling appliances have different features in different countries. In Europe, a classification system separates freezers that freeze fresh food solid and keep it for long periods from freezer compartments that can only keep already-frozen food a few days. Roughly half of the refrigerators in Scandinavia, Germany, and England have the former, vs. almost all in N. America. These consume considerably more electricity than do simple models that cool but do not freeze. Automatic defrost is the rule in Scandinavia, but the exception in the rest of Europe and Japan.

Clotheswashers are different in Europe, where nearly all models warm water within the machines. There are differences in hot water requirements/kg clothes, and a variety of wash cycles and detergents that accept lower temperatures. At least half of all European machines are front-loading, which reduces water needs significantly. Thus energy per kilogram of clothes washed is reduced.

In comparing particular N. American and European/Japanese appliances (principally refrigerators, freezers, and clothes washers), we find that there are differences in technologies that probably account for as much of the difference in electricity use as do size and features. It appears that N. American appliances are less efficient than those in Europe, particularly refrigerators, freezers, and washers. This point of view was elicited by representatives of Philips, Electrolux, and ZVEI, and found as well in the Energy Report of the Dutch Appliance Industry, from VLET, in Tilburg, Holland. European manufacturers cited motor efficiency, compressor design, insulation, and overall design as contributing to the lower specific consumption of European appliances.

Refrigerators are an example. The scatter in estimated average annual consumption, from 200-300 kWh/yr for refrigerators in Italy and Japan, to 600-800 kWh/yr in Scandinavia, up to over 1000 kWh in the U.S., represents differences in size and features as well as efficiency. Yet a 400 liter Philips model with automatic defrost and freezer compartment uses about the same amount of electricity per liter as the average for all refrigerators sold in the U.S. in 1982. But the U.S. models are roughly 30% larger, and therefore should use less electricity/liter, because their surface/volume ratios are lower. As this is not the case, we infer that the U.S. models, on the whole, are less efficient than this Philips model. After conversations with experts in Denmark (Philips), Germany (Appliance Association), Holland (Philips), and Sweden (Electrolux), we judge that about 50% of the difference in consumption/liter between American and European refrigerators or freezers represents efficiency differences, with the rest representing features.

While there are great differences between N. America and Europe, there are few apparent differences in efficiency of appliances around Europe. This is because appliances are made and offered by the same firms. For example, most of the same models of refrigerators appear in the 1985 Philips catalogues for France, Sweden, Denmark, and Holland. Electrolux, Sweden's largest appliance maker, offers the same models in high-priced Denmark as in lower priced Sweden and Norway. Our conversations with utility and appliance experts in each country reinforce the indication that it is the offering of the manufacturers, more than the choice of the consumer, that influences the efficiency of appliances the most.

Occupant Behavior. Household occupant behavior primarily affects lighting, washing, and drying. Estimates of average use per home for lighting vary from a few hundred kWh/year (Germany, Japan), to 1500 kWh/yr (Norway), with the U.S., Canada, and Sweden lying towards the high end of the range and Japan, France, Germany, and Denmark lying towards the lower end. Home size is one reason for the difference in levels, but habits, related to the price of electricity, are such that people use lighting more sparingly in Central Europe than in N. America.

4. THE DIRECTION OF RESIDENTIAL ELECTRICITY DEMAND: SOME ISSUES

Future residential electricity demand will be shaped by changes in existing end-use markets and the penetration and intensity of new uses for electricity. In this chapter we discuss some key issues in these areas.

4.1. Old Functions, New Technologies

Electricity can penetrate in the old end-use markets in both new and existing homes. Since growth in the number of households is slowing in OECD countries, the possibility of conversions to electricity is important. Technological change that gives electricity a competitive edge will be important in both instances. Bear in mind that gains in efficiency that help electricity penetrate also tend to reduce unit consumption.

Space Heating. In the U.S., the advance of electric heat pumps has probably allowed electric heating to gain market share that would otherwise have gone to natural gas. What effect might heat pumps have elsewhere? In Europe, electric heat pumps have penetrated the heating markets much more modestly. This is largely due to the colder climates and the lack of demand for air conditioning. Ducted heating systems are also less common; most heat pump systems in Europe are hydronic. (In the U.S. the marginal cost of heat pumps is low because most new homes would have air ducts for a heating system and need an air conditioning device anyway.)

Outside of the U.S., the heat pump has enjoyed its best success in Sweden, where subsidies have helped. Nearly 1% of the total SFD stock converted to heat-pumps in 1984. Swedish units use outdoor air, ground water, ground heat, or even the heat of rocks several hundred meters below ground level, as heat sources. Financially, heat pumps are attractive only in older homes with high heating needs; newer homes are so well insulated that most of the heating requirements come during the few months when the heat pump has its worst performance.

In Germany and France, heat pumps were backed by generous public subsidies, but still gained only a small market share: about 2% of electrically heated homes in Germany (0.2% of all homes) and 5-6% (1% of all homes) in France in 1983. About half of the German systems use fuel during the coldest periods; otherwise owners pay a higher charge for electricity. Indeed, diesel and gas heat pumps are also found in Germany. In France, the dominant system also uses an existing boiler as a complement for cold times. In Japan, heat pumps with variable speed motors now serve as a source of supplementary heat, but the final cost of this heat is still too great to make the devices useful for heating entire houses.

The key point is that in Europe, heat pumps are seen as oil-savers, while in the U.S. they are electricity savers (compared in most cases to a system of electric resistance heating and air conditioning). Systems in Europe will suffer in popularity with the oil-price decline; this is not the case in the U.S.

Water Heating. Two kinds of water-heating technologies stand out as different from those available in the U.S. The first is the "quick-recovery" or "point-of-use" device, which may have a small amount of storage capacity. Such devices provide an energy-saving advantage since there are virtually no standby or circulation/distribution losses. Venting requirements limit their applicability in the U.S. as gas fired systems; no such problem exists for electricity. In homes where hot water for dishwashers and clotheswashers is provided in the machines, these water heaters may be efficient choices.

A newer technology is a heat pump that provides hot water. These have become popular (installed on as many as 25% of all new homes) in Sweden. They operate from the stream of exhaust air from the mechanical ventilation systems found in almost all new SFD and many existing MFD. When the house is outfitted with a hydronic space heating system, then the extra heat from the water heater can be used for space heating. The advantage of this system, which is not expensive, is that it works against a fairly constant load, domestic hot water demand being roughly constant through the year, and it works off the high temperature (20C) of exhaust air.

Appliances and Lighting. Technological improvement has and will continue to increase the efficiency of electric appliances. In the countries with growing saturation (France, Italy, Japan), the relative number of newer, more efficient appliances increases rapidly. In the more saturated countries (U.S., Canada, Sweden, Denmark) growth will be slow, and replacement will dominate change. This means that while unit consumption will fall only slowly, overall growth will be slower.

Electricity use for lighting appears poised to decline as all the major firms market mini-fluorescent bulbs with incandescent-like spectra.

Lessons for the U.S. Electricity-using technologies in use in other countries could reduce electricity consumption in the U.S. Clotheswashers, refrigerators, and freezers, hot-water heat pumps from Sweden, and, most recently, halogen-lamp cooking elements (introduced by Philips and Thorn EMI in Europe) would reduce electricity use in existing applications. If the largest refrigerators, freezers, and clothes washers on sale in Europe today replaced those in the U.S., electricity use for these applications would decrease significantly, probably by 1/3. Very tight Swedish wooden houses with electric heating, now being assembled in many places in the U.S., would reduce electricity intensity for heating, but could also help electric heat to gain market share.

4.2. New Uses for Electricity

There are many new uses for electricity that presently have low saturation, but they are mostly modest consumers of electricity. These include VCRs and home computers. The plethora of small kitchen appliances consume little and appear to save energy and electricity compared to using larger ovens for the same job. In a few colder countries (Sweden, Canada, Norway), electricity is used in growing amounts for saunas, auto engine block or passenger compartment heaters, but these do not appear to be significant elsewhere. Security lighting could see greater penetration, but with the new generation of high efficiency bulbs (11 watts of a Philips or Osram lamp gives the effect of 60 watts) annual consumption would be only a few hundred kWh.

5. CONCLUSION

5.1. Outlook for Residential Electricity

Current trends suggest that we will continue to find more uses for electricity, but each use is growing more efficient. Increased efficiency may encourage greater electricity penetration, perhaps balancing the downward effect of increased efficiency on total sales. New uses are generally low consumers, and tend to be capital intensive.

Despite the attention on new uses of electricity, the main potential growth areas for electricity use in homes are space heating, water heating, and the three major appliances whose saturations are still under 50% in most countries: freezers, dishwashers, and clothes dryers. The major unsaturated market remains space heating. Although electricity enjoys an important role in new construction in many countries, the housing turnover rate is slow. This means that the growth in total kWh sales arising from electric heating will only be rapid where massive conversions are occurring. Even with high world oil prices, this occurred only in three countries. In Germany, Denmark, and even in the U.S., electricity seems to penetrate principally where gas (or district heat) is unavailable, since gas is cheaper. Thus the outlook for electric heating, before the oil price crash, was mixed, and the near-term outlook must be judged as limited.

In the long run, several factors would favor electricity over fuels in competitive markets. In the space heating market, very tight houses reduce the need for a heating system and thereby reduce heating system needs to a few electric resistance heaters. Increasingly smaller households (a trend encouraged by the ageing of the population) may mean more apartments, which are more likely to use electric heat. In the water heating market, point-of-use water heaters eliminate storage needs and pipes, and, compared with gas, have no need for exhaust. Use of specialized cooking equipment to reduce the use of a gas or electric stove or oven probably saves electricity and certainly saves energy. In Germany, Sweden, and the U.S., families in large homes have more appliances than those in apartments. Part of this is due to higher income of those SFD, but younger singles and childless couples (who may or may not have incomes that permit owning dishwashers, clotheswashers and dryers) a result, it should not be assumed that the ultimate level of saturation for the major appliances will approach 100%. From the data we have seen it appears that while refrigerators will reach 100%, washers will only reach 80-90%, freezers, dishwashers, and clothes dryers 70-80% or less.

Demographic and Lifestyle Changes. Families are getting smaller, with single-person households growing rapidly. Populations are ageing. These factors may increase the share of apartments in the housing stock. Apartment dwellers have less need for some appliances. For example, apartments tend to be located in built-up areas where there are more stores selling fresh food, making a free-standing freezer less important.

The work force is changing, as are work hours and workplaces. If these changes keep people at home more, then some uses of electricity in the home could increase, and people might buy more electricity-using goods (with, however, minimal demands on electricity). More do-it-yourself would increase demands for tools and time at home to work with them, but, again, the electricity cost is minimal. If on the other hand, people spend more time away from home -- presumably managing home energy use with small computers, more electricity use would effectively be transferred from the residential to the commercial or transportation sector.

5.2. Implications for Planning and Forecasting

Both changes in the market share of electricity in each market and changes in intensity will determine future electricity demand. An understanding of future electricity use should always take into account the dynamics of each individual end-use market. Top-down econometric methods can give us some insights into the direction of electricity use. But only a detailed market-by-market comparison of trends and possibilities can illustrate the most realistic directions for electricity consumption.

Table 1-1. Residential Electricity Demand in OECD Countries

	Consumption per Household 1983, in kWh	Tot. Ann. Ave. Growth Rate 1973-83, %		Consumption per Household 1983, in kWh	Tot. Ann. Ave. Growth Rate 1973-83, %
Canada	11905	5.4	Denmark	3565	3.7
France	3580	8.2	Germany	3620	4.6
Great Britain	4085	-0.4	Italy	693	3.9
Japan	2850	5.6	Netherlands	3055	3.4
Norway	16000	4.7	Sweden	7940	6.8
United States	8990	2.8	AVERAGE	5867	3.4

Table 1-2. Share of 1983 Residential Electricity Consumption (%)

	Space Heat*	Water Heat	Cooking	Appliances	Air Conditioners
Canada	37	22	8	32	..
Denmark	13+1	8	15	63	..
France	36	12	6	46	..
Germany	21+4	22	9	44	..
Great Britain	18+1	13	15	53	0
Italy	3	23	4	71	..
Japan	..+8	14	5	68	5
Netherlands	1+1	26	3	69	..
Norway	31+25	23	3	18	0
Sweden	36+4	16	8	36	0
United States	17	16	7	47	13

* The first number refers to consumption for primary heating, the second to consumption for secondary heating.

Table 2.1. Percentage of Homes with Electricity as Main Heating Fuel

	1972/73	1983		1972/73	1983
Canada	7	23	Denmark	2	6
France	5*	15	Germany	4	9
Great Britain	13	12	Italy	6	8
Japan	..	2	Netherlands	1	..
Norway	31	49	Sweden	6	24
United States	10	18			

* 1975

Table 2.2. Electric Share of the Water Heating Market (percent of homes*)

	1972/73	1983		1972/73	1983
Canada	48	51	Denmark	4	10
France	18	28	Germany	46	45
Great Britain	40	33	Italy	35	51
Japan	4	7	Netherlands	21	18
Norway	80	90	Sweden	9	26
United States	27	32			

* Based on share with respect to all homes.

Table 2.3. Electricity Intensity for Water Heating, 1983 (MWh/dwelling)

Canada	5.1	Denmark	3.0
France	1.5	Germany	1.1
Great Britain	~1.0	Italy	0.8
Japan	~4.0	Netherlands	1.9
Norway	4.0	Sweden	3.5
United States	4.7	AVERAGE	2.8

Table 2.4. Market Penetration of Major Electric Appliances (% of homes)

	Refrigerator*		Freezer		Clotheswasher		Dryer		Dishwasher	
	1972/73	1983	1972/73	1983	1972/73	1983	1972/73	1983	1972/73	1983
Canada	99	100	37	55	45	67	42	62	11	34
Denmark	97	100	46	61	44	62	1	11	7	21
France	87	100	8	34	65	87	-	-	5	21
Germany	89	100	25	52	71	86	2	10	6	23
Great Britain	71	96	3	32	67	80	26	38	1	2
Italy	87	97	1	14	62	80	0	2	6	11
Japan	100	100	0	0	96	98	~0	9	0	1
Netherlands	91	97	19	46	86	88	6	12	3	11
Norway	89	97	57	75	72	79	22	30	3	17
Sweden	94	99	60	83	49	60	9	21	11	30
United States	99	100	34	37	70	70	38	45	25	36

* By 1983 some households had more than one refrigerator in most countries.

Table 2.5. Unit Consumption of Major Electric Appliances (kWh)

| | Refrigerator | | Freezer | | Clotheswasher | | Dryer | | Dishwasher | |
	1972/73	1983	1972/73	1983	1972/73	1983	1972/73	1983	1972/73	1983
Canada	-	-	-	-	-	-	-	-	-	-
Denmark	730[a]	700[a]	700	900	545	480	450	400	435	475
France[b]	425	540	740	720	300	300	-	-	470	440
Germany	375	480	680	580	350	280	450	290	-	400
Great Britain	300	300	975	780	200	200	300	300	500	500
Italy	180	220	490[c]	470	450	410	540[c]	500	1000	1050
Japan	700	600	-	-	-	37[d]	-	-	-	-
Netherlands	440	400	520	550	455	275	-	400	875	475
Norway	600	600	750	750	500	500	600	600	-	300
Sweden	600	510[a]	1040	900	500	350	400	225	-	295
United States[e]	1400	1290	1390	1220	90	85	1110	1080	250	250

(a) Includes Combis. (b) First year is 1975. (c) 1980. (d) Japanese washers do not heat water. (e) First year is 1977.

Table 2.6. Reduction in Electricity Consumption of New Electric Appliances

Country	Refrigerator	Freezer	Washer	Dryer	Dishwasher	Oven	Cooling	Years
Denmark	-25-35%	-30%	-25%	-10%	-24%	-15%	-	70-84
Germany	-24%	-	-15%	-	-27%	-14%	-	78-84
Japan	-60%	-	-	-	-	-	-32%	73-84
Netherlands	-30%	-27%	-70%	-	-26%	-	-	71-83
Sweden	-	-56%	-	-	-33%	-	-	75-83
United States	-42%	-37%	-	-	-	-	-23/20%*	72-84

In Germany, the Netherlands, and the U.S., figures represent sales-weighted averages (expected annual consumption for different models of each kind of appliance, weighted by the actual appliances sold). For Sweden, the improvements reflect those demonstrated by Electrolux (the largest seller). For Denmark, a rough estimate of sales-weighted average improvement was made by Statens Husholdningsraad.

(*) The figures refer to central and room air-conditioners respectively.

Table 3.1. Electricity Intensity for Space Heating, 1983 (per degree-day)

	kWh/dw	kJ/m^2	SFD Share		kWh/dw	kJ/m^2	SFD Share
Canada	4.2	150	55	Denmark	2.5	95	73
France	2.9	122	56	Germany	3.0	144	44
Great Britain	2.0	102	35	Norway	3.1	124	74
Sweden	2.3	69	91	United States	2.6	78	56
AVERAGE	2.8	110	60				

Long-term heating degree-day values, with base 18 C, are as follows: Canada - 4580, Denmark - 3122, France - 2450, Germany - 3113, Great Britain - 2823, Italy - 2140, Japan - 1975, Norway - 4069, Sweden - 4011, U.S. - 2172.

Table 3.2. Electricity Intensity for Appliances*, 1983 (MWh/dwelling)

Canada	3.9	Denmark	2.2
France	1.7	Germany	1.6
Great Britain	2.2	Italy	1.6
Japan	1.9	Netherlands	2.1
Norway	2.8	Sweden	2.9
United States	4.3	AVERAGE	2.5

*Includes lighting

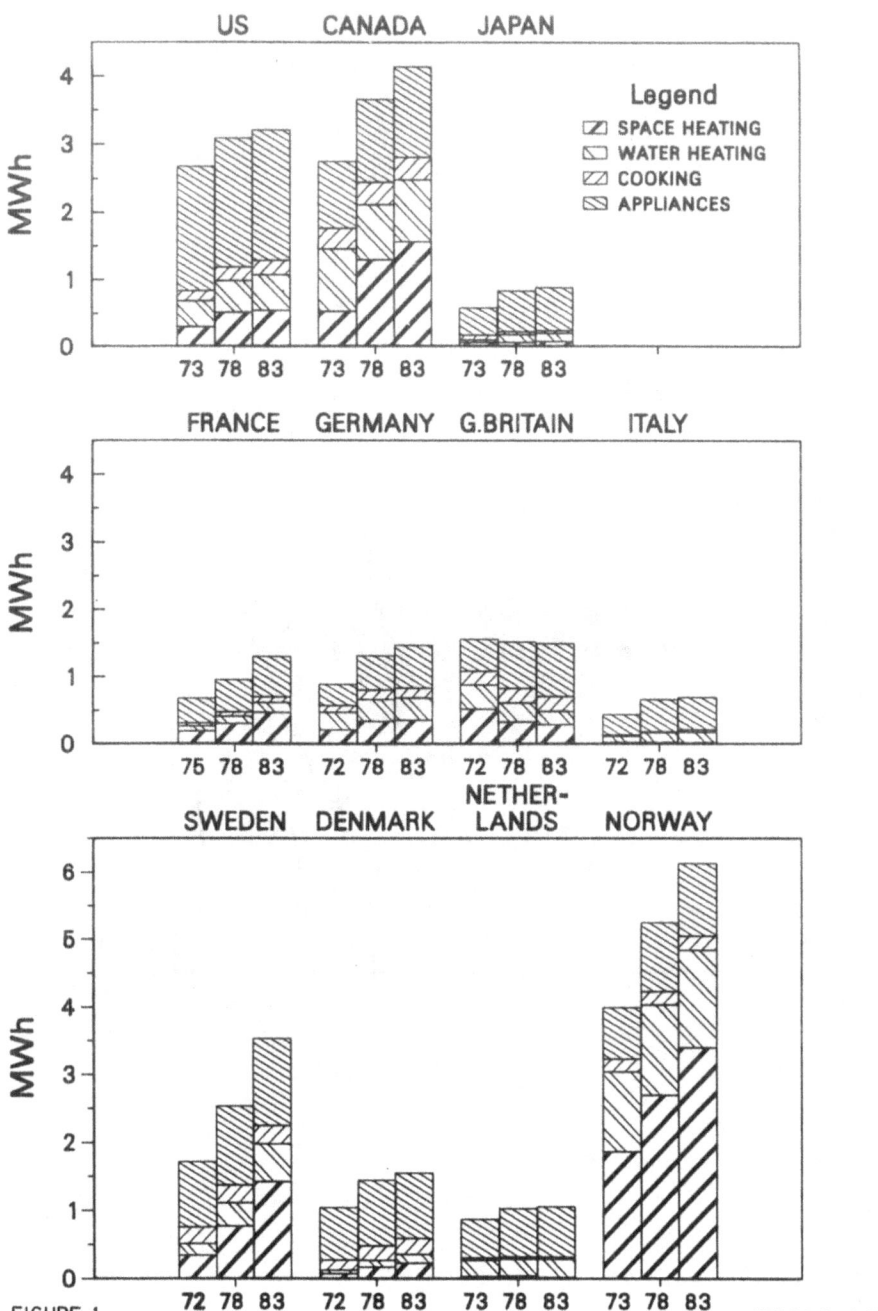

FIGURE 1

XCG 8610−12187

84

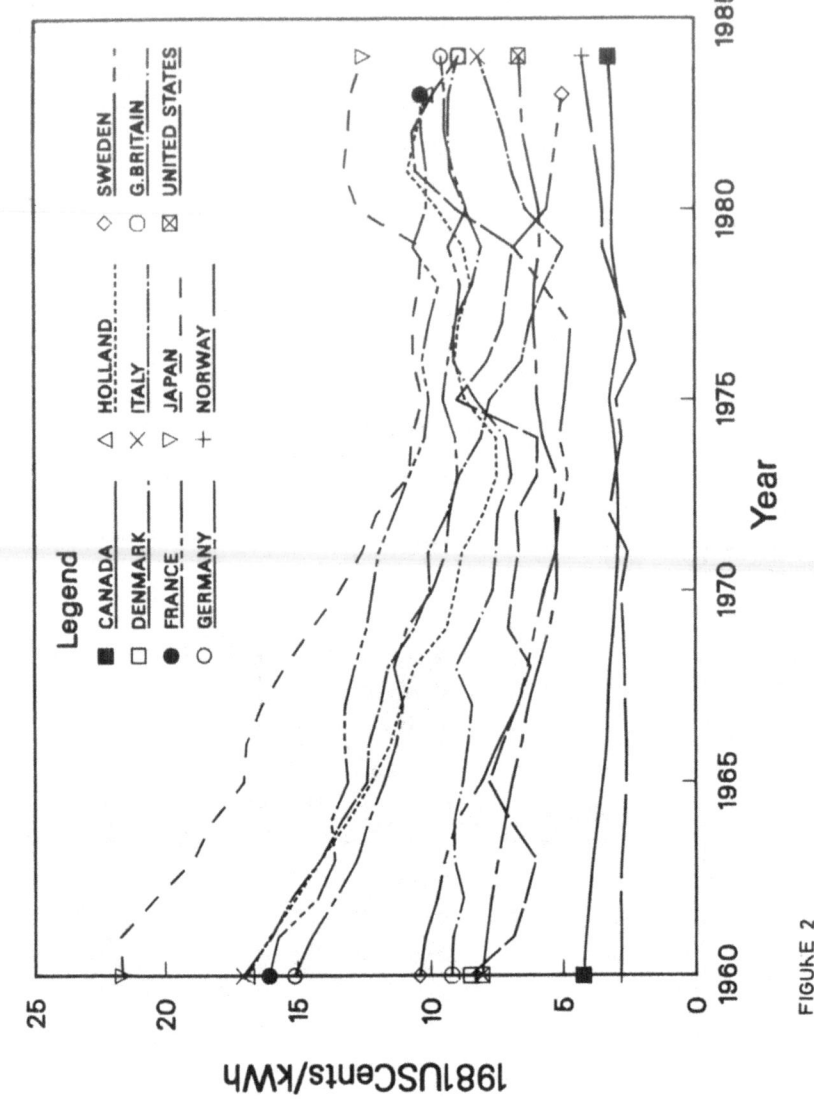

Residential Electricity Prices

FIGURE 2

XCG 8610-12183

POWER SYSTEM PLANNING FOR THE NEXT 20 YEARS

Fred C. Schweppe
Richard Tabors
Laboratory for Electromagnetic and Electronic Systems

David C. White
Energy Lab
Massachusetts Institute of Technology

Most papers which discuss the future of electric power systems contain many plots and tables summarizing forecasts of future load growth, fuel mix, generation technologies and mix, environment impacts, etc. However, past experiences has repeatedly shown that

"The Forecast is Always Wrong".

This paper provides no forecasts. Instead it addresses planning methodologies and issues.

The particular methodologies to be discussed are those we have come to believe are the "best" way to proceed. Thus we are not presenting a forecast of what type of planning will be done. Instead we are presenting what we feel should be done.

We are presently developing an unsolicited proposal for a planning study for a particular, multiple state region of the USA. The proposal is still in its early formulation stage and will evolve further as discussions with the concerned parties progress. However, an initial draft has been prepared. The rest of this paper is a version of that draft that has been modified to be non region specific and more generic in character.

This use of a specific proposal is an unconventional way to write a paper. However, it seems to be as good as any way to present ideas on how power system planning should be done in the future.

A. T. De Almeida and A. H. Rosenfeld (eds.), Demand-Side Management and Electricity End-Use Efficiency, 85–96.
© 1988 by Kluwer Academic Publishers.

A PROPOSAL FOR A LEAST COST PLANNING STUDY

1. Background

Electric utilities are currently in a state of transition. Power system planning no longer involves just deciding which plants to build to meet the demand at minimum cost. Today the "in phrase" is Least Cost Planning; which is supposed to consider both the supply and demand sides.

An ideal least cost planning study would explore:

.   Utility supply side options
.   Customer owned supply side options
.   Demand side management and conservation

while similarly dealing with the existence of

.   Multiple attributes; i.e. many concerned parties, each
with   different objective functions

.   Massive uncertainty; i.e. forecasts of future load growth, costs, technology, etc. are always wrong.

Unfortunately, many present day studies fail to meet all these conditions for reasons such as:

.   Utility supply side options and models are understood the best and therefore may receive most of the attention.

.   The studies may be done by one of the many concerned parties which limits the attributes to be evaluated; or

.   The study team may insist on finding the "optimum" solution and hence circumvent the inherent multiple attribute character of the problem.

.   The effect of the uncertainties may be severely underestimated by thinking that a few "one parameter at a time" sensitivity studies are sufficient.

The least cost planning study being proposed here has three key features:

.   Treatment of demand side-conservation on a equal footing with supply side

.   Use of an analytic framework which explicitly addresses the multiple attributes and massive uncertainties

.   An organizational structure which combines an independent Study Team with an Advisory Panel formed by representatives of all concerned parties; utilities, regulatory commissions, and end users (industrial, commercial, residential).

The proposed study does not include customer owned generation. In practice customer generation could be very important. It is being ignored here simply because we believe it should be addressed in a separate study. If we were to write a proposal for a customer generation study, it would have many features in common with the present proposal. It would use the same basic concepts and framework; only the details would change.

The objective of the proposed study is to provide qualitative information which can be considered by all concerned parties and used by existing decision making entities.

The objective is not to suggest the "best solution" or to establish, either directly or indirectly, still another decision making entity.

2. The Analytical Framework

The analytical framework to be used to explicitly handle the multiple attributes and massive uncertainty has evolved through applications over the years. Only its general characteristics are presented here. Detailed discussions and applications can be found in Refs. [1], [2], [3].

The analytical framework is structured around the definition of three interrelated data sets.

o **Strategic Alternatives**: What are the alternatives open to the region? How could/can these alternatives be combined? Conservation strategies could form one set of alternatives as could natural gas fired combined cycles and imported energy from other regions.

o **Attributes**: What are the critical tradeoffs of interest to the region? Cost of electricity is one attribute but there are many others that are important. Security of supply may be another. Environmental quality is clearly a third. Reliability for high tech industry is probably a fourth. Quality of life is a fifth. Equity cannot be ignored.

o **Uncertainties**: Within any analysis, there are a set of inputs concerning the future which will determine in large part the outcome of the analysis but over which the decision makers have no control. International fuel prices are one example. Under some circumstances, demand for electricity is a second, environmental regulations a third, availability of nuclear power a fourth, etc. The existence of these uncertainties in all forecasts cause the assumptions concerning these inputs to drive the output.

It is assumed that a simulation is available which can evaluate the attributes for a given strategic alternative combined with a given future (a future is some set of values for the uncertain variables). For most applications, the simulation is actually a set of computer models rather than a simple computer program.

The analytical framework can be thought of as the development of sets of tradeoff curves between the attributes when the strategic alternatives are evaluated for a set of uncertainty values which span the range of all "reasonable" futures. Figure 1 shows the outcome of one set of tradeoffs for a case or application in which the strategic alternatives are evaluated against a specific future scenario. Two points become clear.

.   There were a large number of strategic alternatives evaluated and their position on this tradeoff curve identified.

.   Only those alternatives on this tradeoff curve which fall on or near the bottom line are serious contenders. Why? Because for these two criteria, the line reflects the most efficient alternatives. Any alternative above and to the right of the line is dominated by one which is closer to the line.

The tradeoff frontier shown in Figure 1 is two dimensional given the limit of the paper, not the mathematics or the available computer programs. The actual analytical framework allows for the tradeoffs to be expressed simultaneously along a set of axes.

One strength of this method is its ability to inform the decision makers, i.e. it does not define an optimal strategy but rather may be thought of as eliminating bad strategies.

The second strength of this framework is its ability to identify attributes and uncertainties which are important to the outcome of the decision. If all strategic alternatives are ranked the same regardless of the SO_2 emissions of the region, then further detailed evaluation of this attribute is not necessary. Similarly, if all strategic alternatives are ranked the same regardless of the load growth of the region, then further consideration of this uncertainty is not necessary.

The third strength of this framework is its ability to identify strategies which are robust across the set of scenarios about the future. As was stated above, the objective is not to find an optimum given one possible future, but to define the next step or set of steps given a range of possible futures.

A fourth strength of this framework is its heavy reliance on the use of existing models and computer programs such as production cost models, financial models, and end use models. The approach does not involve the recoding of approximations of such models into some new "super simulation". The existing models (and equally important the existing data bases) for the particular region are to be used, as is. The analytical framework provides a way of integrating together the outputs of many separate programs. The analytical framework itself requires special computer codes to generate the tradeoff curves; throw out unimportant attributes and uncertainty; and look for robust strategies. However, such computer codes have already been developed and used. (Power Technologies, Inc. of Schenectady, N.Y. presently has the most sophisticated version.)

3. The Advisory Group

The objective of the Study Team is to provide information (tradeoff curves) for use by the appropriate decision making entities. This objective can only be accomplished if representatives of such decision making organizations are closely involved with the Study Team; from the beginning. Therefore an Advisory Group is to be formed.

The Advisory Group consists of representatives of concerned parties such as

- . Utilities
- . Regulatory Commissions and other state agencies
- . End Users (industrial, commercial, residential)
- . Privately owned generators
- . Conservation and environmental groups

Hopefully there are at least two representatives from each group; one executive level and one staff level. An executive level meeting of the Advisory Group is to be held every 6 months to provide high level coupling and interaction. Sub meetings between staff level representatives and the Study Team members are held as needed.

One initial role of the Advisory Group is to help the Study Team define the attributes, uncertainties, and strategic plans to be considered and to understand the strength and limitations of the simulation model to be used. Later, the Advisory Group's role is to make sure the Study Team's outputs (tradeoff curves, etc.) are expressed in a manner and form that is understandable to the parties the Advisory Group are representing.

The Advisory Group will not attempt to make final decisions on what should be done. Such decisions belong to existing decision making entities operating within existing frameworks.

4. Selection of Variables

Application of the general analytical framework discussed in Section 2 to the present case of interest yields Figure 2.

The Study Team and Advisory Group will work together to explicitly specify and define the utility strategic plans, demand side management strategic plans, uncertainties, and attributes to be used. The following constitutes a starting point.

Utility Strategic Plans: The initial set will start by reproducing the present plans of the utilities. Other options will be added if deemed necessary.

Demand Side Management-Conservation Strategic Plans: These strategic plans are decomposed into those that effect investment and operating decisions. Strategies that effect end use investment decisions include:

. Customer Incentives

       . Cash
       . Information
       . Planning, Design Support

    . Legislation
       . Appliance Standards
       . Etc.

    . Design of rates-contracts

Strategies that effect end use operating decisions are primarily

    . Design of rates-contracts

Uncertainties: A starting list of the uncertainties to be considered
are:

      .   Cost of capital

      .   Availability/cost of fuel

      .   New technologies
           Supply
           End Use

      .   Regulatory and Legislative actions

      .   Environmental standards

      .   Load behavior

      .   Societal constraints

Attributes: The basic attributes to be considered are summarized in
the following table:

	Utility	Customers . By Class . By End Use
Capital Costs	. Present Worth (\$) . Annualized (\$) . Cash Flows (\$)	
Operating Costs/ Benefits	. Fuel Costs (\$) . System Security	. Benefits . Electric Bills (\$) . Quality of Life
Environmental Impacts	. Air . Water . Safety	

This table implies the use of demand side attributes such as customer
financial details, customer benefits and "quality of life" which are not

included in many present day least cost planning studies. Their inclusion is essential if the demand side is to be treated as a equal with the supply sale. However, their inclusion introduces difficult modeling problems which will be addressed in the next section.

5. The Simulation

The Simulation of Figure 2 is obviously a key to the success of the proposed study. The Study Team will specify the simulation with the help of the Advisory Group. The following consists of the starting point.

Figure 3 provides an overview of the overall simulation. The Utility Capital Stock Model and Production Cost Model of Figure 3 are standard utility supply side models.

The three major demand side models

    1.   Customer Investment Decision
    2.   End Use Capital Stock Model
    3.   Demand Response Model

one disaggregated by class; residential, commercial, and industrial. Each class is further disaggregated by end use devise or process: space conditioning, refrigeration, metal melting, etc.

Three major problems which must be addressed in developing the overall simulation are:

I.   Available end use models have major limitations, i.e. have much more uncertainty than the corresponding utility supply side models.

II.   There are many demand side management-conservation plans for investment that must be investigated.

III. There are many possible rate structures.

An approach to overcome the limitations of the end use models (Problem I above) is outlined in Figure 4. By hypothesizing reasonable structures (form of the equations) for the components of the end use models and treating the parameter values as key uncertainties, a robust set of strategies relating to these uncertainties can be developed.

The multiple demand side-conservation strategies for investment (Problem II) is handled by a two stage simulation as outlined in Figure 5. The end use investments, such as rate of penetration for more efficient refrigeration are used as exogenous inputs to the overall model simulations. The strategy to get this end use result is determined as a sub study of the customer investment model.

The problem of multiple rate structures (Problem III) is handled by using a two phase approach as shown in Figure 6. The production cost, rates, demand response parts of the simulated model of Figure 3 are replaced by a single Production Use Cost Benefit Model that is a production cost model coupled to a customer price response model by the assumptions that the customer see "spot prices" (as determined by the production cost model). The methodology to do this is well developed in prior work. The second phase is then a sub study to generate spot price equivalent rates that yield the same or similar demand response.

A summary of the proposed approach to deal with the modeling and simulation difficulties posed as Problems I, II and III is shown in Figure 7. The decomposition of the overall simulation of Figure 3 into the 2 phases and 2 stages of Figure 7 is felt to be essential if meaningful results are to be obtained with a reasonable amount of effort.

6. Discussions of Proposed Study

The study being proposed here is a very ambitious, multiple year and many person-year effort. It is based on the extensive use of existing models along with some limited new modeling on the demand side. The overall tradeoff analytical methodology has been used before; but never in a problem of this complexity. Success will require a lot of work.

We believe the effort is justified by need to take a "quantum jump" in the way electric power planning is done. The demand conservation side has to become an equal to the supply side. The multiple attributes, multiple decision makers, and massive uncertainty aspects of the real world can not be ignored.

References

1. MIT Energy Laboratory, "Strategic Planning for Electric Energy in the 1980's for New York City and Westchester County", prepared for the Consolidated Edison Company of New York, Inc., MIT-EL-81-008.

2. Burke, W.J., F.C. Schweppe and B.E. Lovell, "Trade Off Methods In System Planning", paper SP87-128, presented at the IEEE Power Engineering Society 1987 Summer Meeting, San Francisco, CA., July 12-17, 1987.

3. Merrill, H.M. and F.C. Schweppe, "Strategic Planning for Electric Utilities: Problems and Analytic Methods", Interfaces 14:1, January-February, 1984 (pp. 72-83)

Figure 1

TOTAL COST (1980-1995) VERSUS PEAK 1995 SO$_2$ CONCENTRATIONS
FOR ALL COST-BENEFIT CASES STUDIED

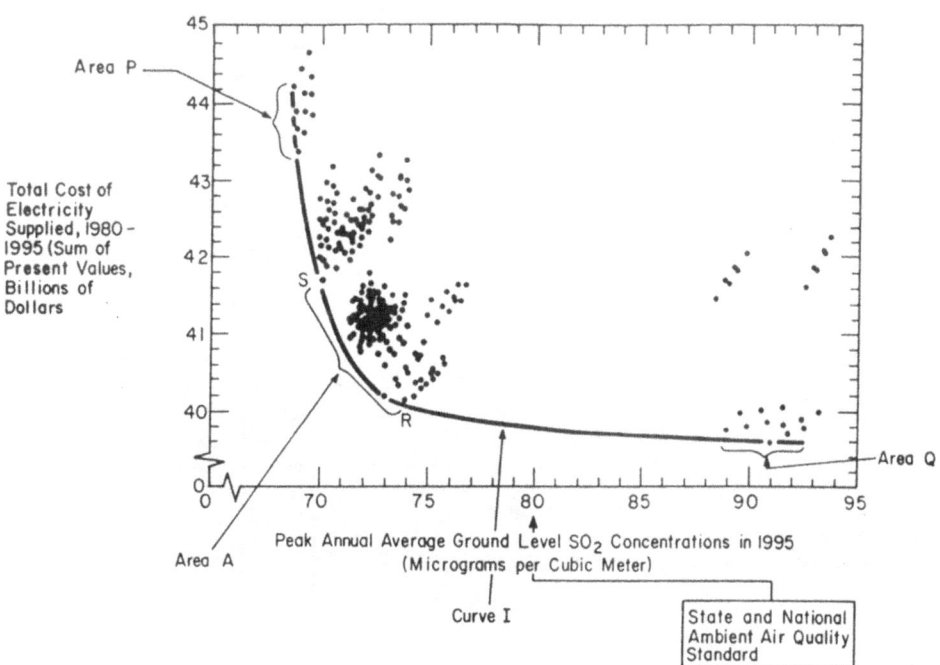

Figure 2
Simulations for Tradeoff Analysis

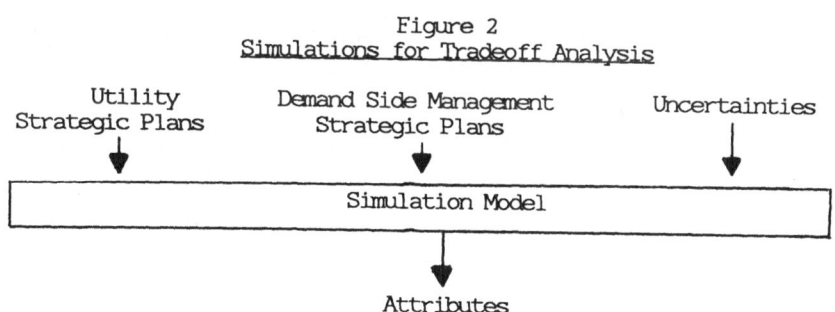

94

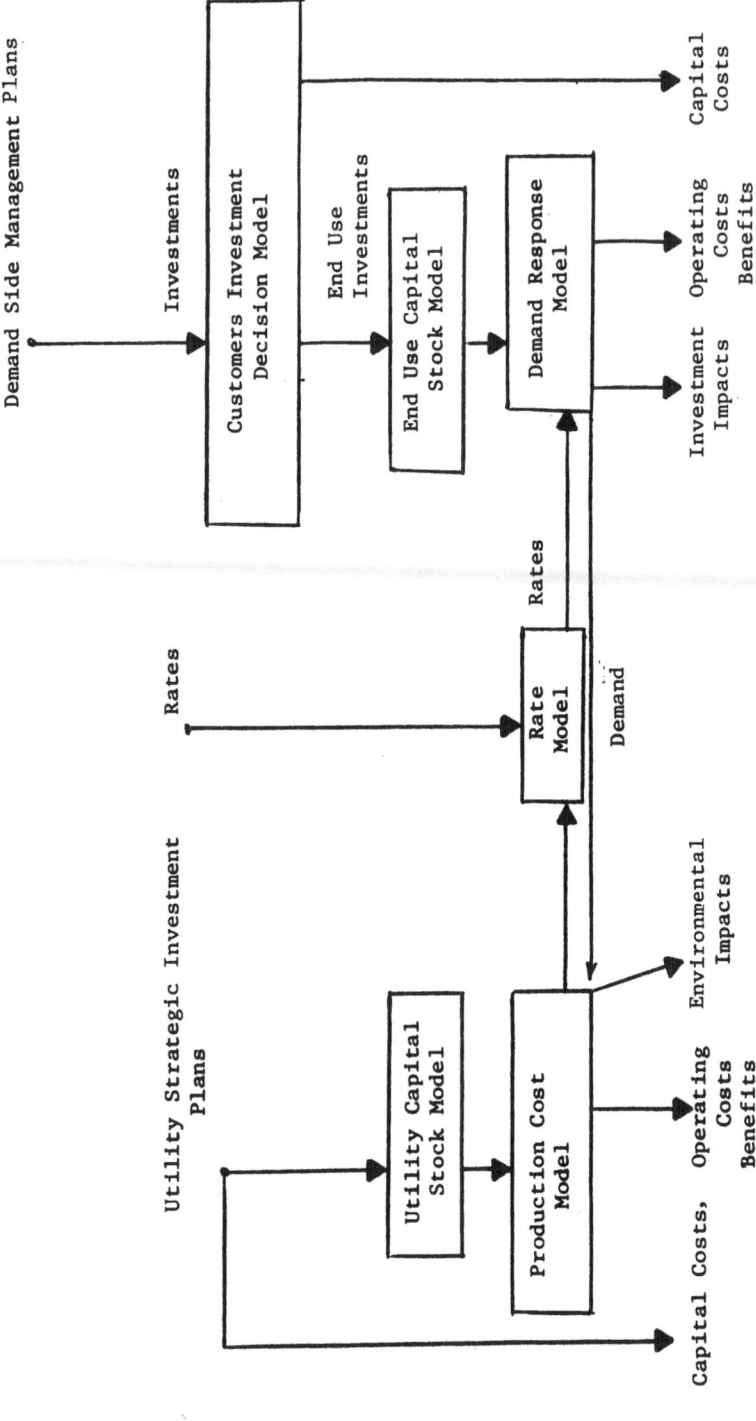

Figure 3
OVERALL SIMULATION MODEL

Figure 4
PROBLEM I: LIMITATIONS IN AVAILABLE END USE MODELS

o   Problem: Inadequate Models

        .   Customer Investment Decision Models
        .   Demand Response Models

o   Proposed Approach

        .   A Model Has Two Parts
                .   Structure (Form of Equations)
                .   Parameter Values (Numerical)

        .   Hyothesize Structure for End Use Models
                .   Reasonable Structures Can Be Developed

        .   Treat Parameter Values as Key Uncertainties
                .   Look for Conservation Plans That Are Robust
                      Relative to These Uncertainties

Figure 5
PROBLEM II: MANY POSSIBLE CONSERVATION PLANS FOR INVESTMENT

Problem: Difficult to Run Full Simulation for all Possible Plans

Proposed Approach: Use Two Stage Simulation

        o   Stage I:

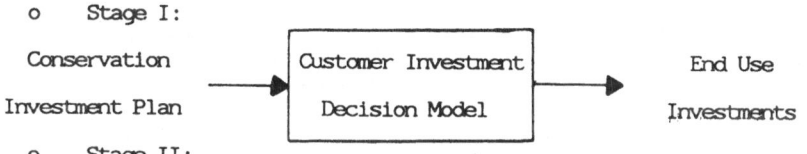

        o   Stage II:
                .   End Use Investments Are Exogenous Inputs to Rest of
                      Simulation

Example: More Efficient Refrigerators
        o   Stage II Input:
                .   Rate of Penetration

        o   Stage I Study
                .   Cash Incentives
                .   Legislation   Rate of Penetration

Figure 6
PROBLEM III: MANY POSSIBLE CONSERVATION PLANS FOR RATES

Problem: Difficult to Run Full Simulation for All Possible Plans

Proposed Approach: Two Phase Study
    Phase I: Hypothesize Spot Pricing

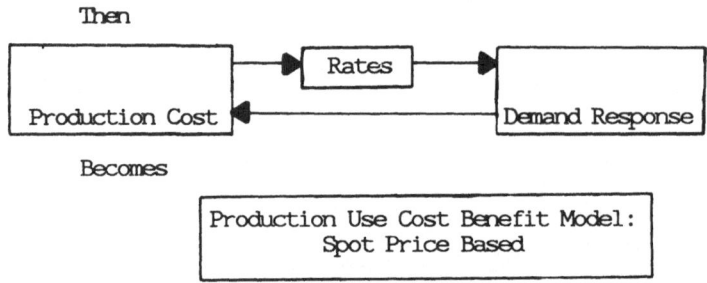

Phase II: Determine "Spot Price Equivalent Rates"
           Yield "Same" Demand Response as Spot Prices

Figure 7
SUMMARY OF PROPOSED APPROACH TO SIMULATION

o   Uncertainty In Parameters of End Use Models

o   Phase I: Two Stage Simulation

    o   Stage I:

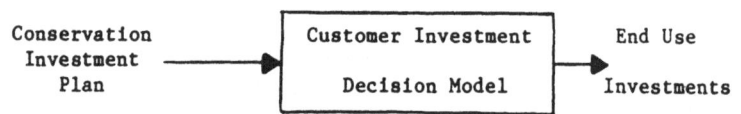

    o   Stage II:

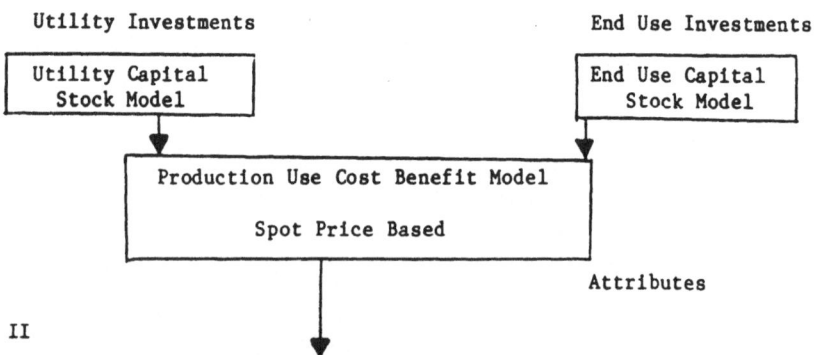

o   Phase II

    o   Find Spot Price Equivalent Rates

II. Load Management Technologies and Programs

THE CONCEPT OF DEMAND-SIDE MANAGEMENT

VERONIKA A. RABL and CLARK W. GELLINGS

Electric Power Research Institute
Palo Alto, California
USA

1. INTRODUCTION

The pattern of electricity consumption varies in the course of a day, typically reflecting the patterns of human activity -- high during the day and low at night. This simple fact has fundamental implications on the electric utilities business. Adequate generating capacity has to be available to serve the demand during the peak periods, even though much of this capacity is idle during the periods of low demand (off-peak). In order to cost-effectively serve this varying demand, with both diurnal and seasonal variations, the utilities use three types of generating facilities: baseload, intermediate load, and peak load plants. Baseload plants serve the portion of the demand which is present most of the time and, as such, they are designed to operate at a constant level for most of the year and use low-cost fuels, such as coal and nuclear, resulting in low operating cost. The capital cost of these plants is, however, high. On the other hand, peaking plants that serve the peak portion of the load and operate for only about 10% of the time, have low capital but high operating costs. They rely primarily on oil- or natural gas-powered units to provide quick start capabilities. Intermediate plants meet the portion of the load that varies daily and, as implied by the term "intermediate," their sizes and costs fall between those of baseload and peaking units.

While this apportionment of the load among different types of generators represents the most cost-effective means of meeting the expected load pattern, it also results in electricity costs that vary depending on the magnitude of the demand. As the load increases, additional plants are brought on-line according to economic dispatch procedures, which aim to keep the cost of delivered electricity as low as possible. This is accomplished by "stacking" the plants in the order of their operating costs -- starting with those that are least expensive to operate and adding successively more expensive generators as needed. For these reasons it is generally more expensive to provide customers' electricity needs during the day (peak period) than at night (off-peak period).

Although some "valleys" in the annual load may be needed to provide time for plant maintenance, the power can be supplied most efficiently if the variability in the load is relatively small, allowing the more efficient plants to carry a greater portion of the load. One of the measures used to describe the variability of the electrical load is the load factor, a ratio

99

A. T. De Almeida and A. H. Rosenfeld (eds.), Demand-Side Management and Electricity End-Use Efficiency, 99–112.
© *1988 by Kluwer Academic Publishers.*

of the average demand to the maximum demand during a given time period. A load factor close to one (100%) describes a load that varies only minimally with time. Conversely, a low load factor implies a demand pattern characterized by sharp peaks and valleys.

The close relationship between costs and load patterns was recognized essentially at the very beginning of the utility industry's existence. It was then only a small step from this recognition to initiation of activities designed to influence customers' use of electricity so as to create a better match between the demand and supply. (For more information, see, for example, References 1 and 2.)

2. LOAD SHAPE CHANGES

The various changes in the load patterns were recently classified (3) into six generic categories shown in Figure 1. The first three of these categories, peak clipping, valley filling, and load shifting, are the techniques used in appropriate combinations to achieve load management. When added to the second group of load shape changes, which emerged more recently, they comprise a set of options for demand-side management.

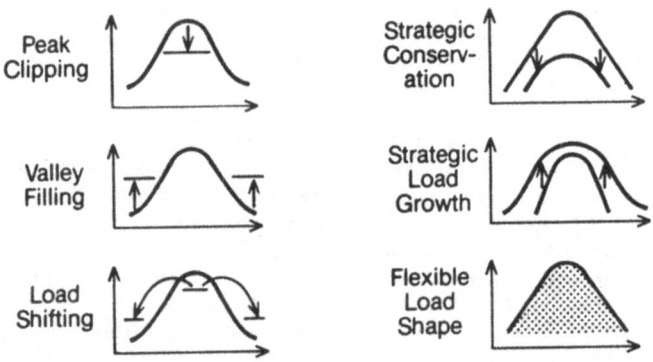

FIGURE 1. Load shape modification objectives.

Peak clipping aims to reduce the peak demand on a utility system by decreasing the on-peak electricity consumption. The principal motivation for this action is to reduce current and future capacity requirements. The technique most frequently used to accomplish this change is disabling or restricting the operation of selected electricity-using appliances during the time of the peak.

The objective of valley filling is to increase load during off-peak periods. Such actions are appropriate to undertake when the incremental cost of serving this load is lower than the average cost of electricity. Adding off-peak load under those circumstances decreases average costs. As a shorter-term strategy, it can also improve utilization of available plants and/or their dispatch.

Load shifting transfers loads that would otherwise occur on-peak to off-peak periods, thus combining peak clipping and valley filling. Thermal storage technologies, such as cool storage, heat storage, and storage water heating, are most commonly used to accomplish this goal.

Strategic conservation involves an overall reduction in the use of electricity. This can be achieved by more efficient appliances, building envelope improvements, or other measures that decrease customers' electricity requirements.

Strategic load growth is aimed at an overall increase in electricity sales. It requires achieving a greater share of competitive markets and/or development of new uses for electricity. Heating and water heating systems that offer competitive cost/performance or substitution of electricity to improve productivity are examples of options involved in this strategy.

Flexible load shape is a concept related to reliability, one of the planning constraints. Once the anticipated load shape, including demand-side activities, is forecast over the corporate planning horizon, the power supply planners study the final optimum supply-side options. Among the many criteria used is reliability. Load shape can be flexible -- if the options presented to the customers include variations in quality of service in exchange for appropriate incentives. The programs involved may take the form of interruptible or curtailable loads or individual customer devices capable of incorporating service constraints into load control actions.

3. DEMAND-SIDE MANAGEMENT IN PERSPECTIVE

While the term "demand-side management" and the classification of the associated load shape changes are fairly recent, the practice of influencing the demand has always been integral to the utilities' business. However, the programs that were emphasized changed in response to the changing technological and economic environment in which the utilities operated.

The first part of the century was characterized by rapid improvements in the technology. Each additional unit of capacity cost less than the previous one and, therefore, along with load factor improvement, load building was the most effective means to reduce the average cost of providing electricity. Initially, the principal load served was lighting, a night-time need, and marketing efforts were focused on building the load during the day. Time-of-use rates, time clocks, promotional rates, and electric appliance sales provided the means for shaping the demand. By the early 30's utilities' peaks had shifted to the daytime period and, with it, new loads to fill night-time valleys were sought, in particular storage water heaters.

In the early seventies, however, major changes in the marketplace brought the rapid growth in electricity demand to a halt. Major increases in energy costs, environmental constraints, and high cost of capital reversed the historical trends and the electricity prices started rising. Concerns about capacity shortages, caused by delays in plant construction, were reflected in a growing number of load management programs (4), this time aimed primarily at controlling peak loads. Meanwhile the rate of growth in the demand, which was about 7% in the 60's, declined to about 3% (5). By the time they were completed, many of the new plants were no longer needed to meet the demand.

At the same time, conservation measures, growth of the service sector, and loss of industrial loads have combined to cause a steady deterioration of U.S. utilities' load factors (see Figure 2). The average capacity factor of coal plants has declined to less than 45%, substantially below the average plant availability of more than 70% (6). This underutilization of capacity is expected to continue into the next decade, during which base load plants represent over 80% of planned capacity additions (see Figure 3). In view of this, it is not surprising that valley filling, or increasing off-peak loads, is now one of the most frequently cited utility load shape objectives.

Load Factor (%)

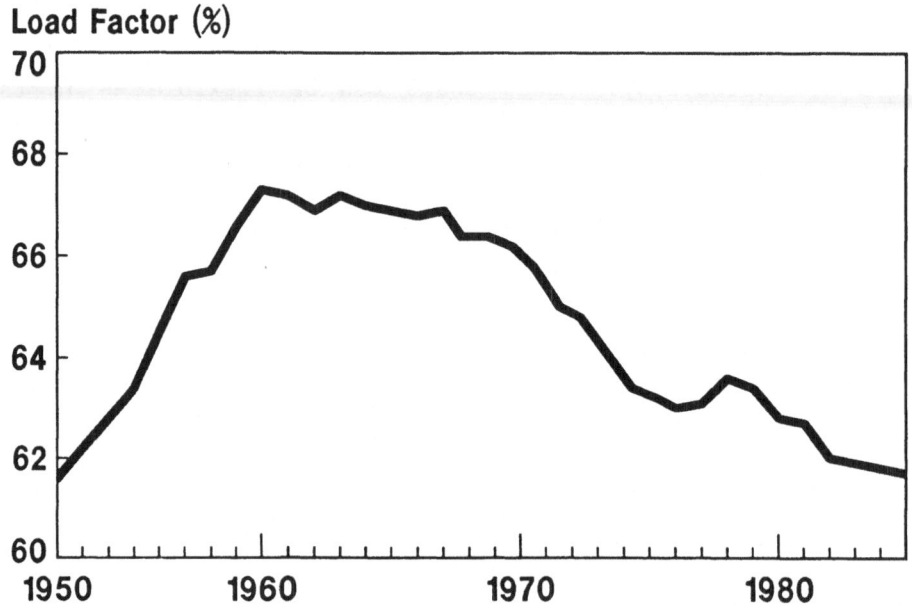

FIGURE 2. U.S. annual load factors; 5-year moving average (6).

The future is characterized by large uncertainties on both sides of the meter. As seen in Figure 4, the annual rate of growth in the demand could range from a low of -0.2% to a high of 4.4%. Some of this uncertainty is associated with the aggressive load management and conservation programs, which are estimated to save about 30,000 MW and are incorporated in utilities' plans to yield an expected annual growth rate of 2.2%. While the planned generating resources could adequately support this 2.2% load growth, timely completion of the planned units cannot be assured. About 30% of over 86,000 MW of planned capacity additions, needed by 1995, represents plants that are

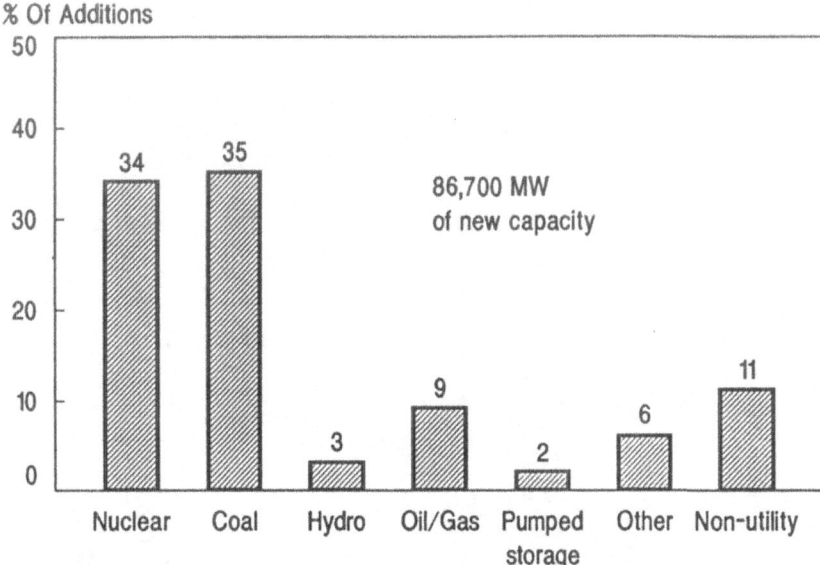

FIGURE 3. Planned U.S. capacity additions, 1986-1995 (7).

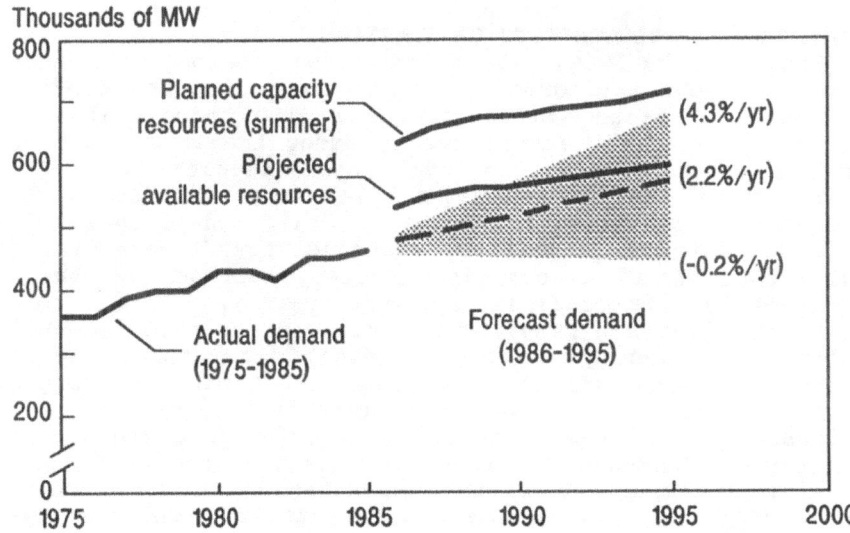

FIGURE 4. Projected U.S. power demand and supply (7).

larger than 100 MW but are not yet under construction (7). In combination, these factors create a serious concern about the ability of the industry to meet customers' needs in the 90's and beyond. The potential problems can be illustrated by the probability of shortages, based on the estimated margins of available supply at the time of system peak, shown in Figure 5.

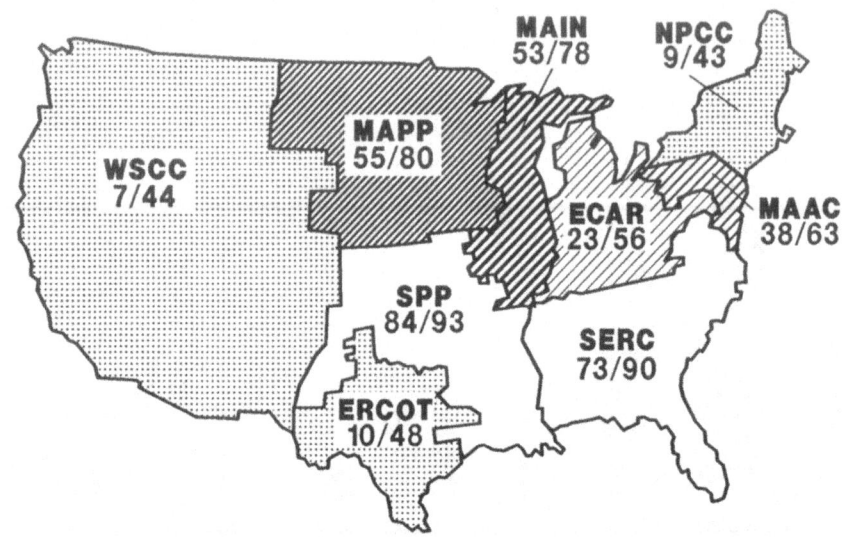

FIGURE 5. Probabilities that operating margins will be below zero; % in 1995/2000 (8).

4. PATTERNS OF ELECTRICITY CONSUMPTION

Allocation of U.S. electricity sales among the end-use sectors is shown in Figure 6. With electricity use growing in the commercial sector and declining in the industrial sector, the use is now split almost evenly among the three major customer types, residential, commercial, and industrial.

Patterns of use in the residential sector by end-use category and fuel are shown in Figure 7. While the largest single energy use category is heating, refrigerators/freezers are the leading consumer of electricity, closely followed by space and water heating, air conditioning, and lighting. In the commercial sector (see Figure 8), the largest electricity end-use is lighting, followed by air conditioning. As seen in Figure 9, primary metals and chemical industries lead both in energy and in electricity consumption in the industrial sector.

Consumption of kWhs, however, is only part of the information needed to understand electricity usage patterns. Equally, if not more, important is the understanding of the contributions of the various end-uses to the utility peaks and system load shapes. Figure 10 shows the typical composition of a load shape on a peak summer day and Figure 11 on a peak winter day. These figures and the regional summary of peak load compositions in Figure 12 show that commercial space cooling is the largest single contributor to the utility summer peaks; this end-use is largely responsible for the fact that the U.S., as a whole, is summer peaking.

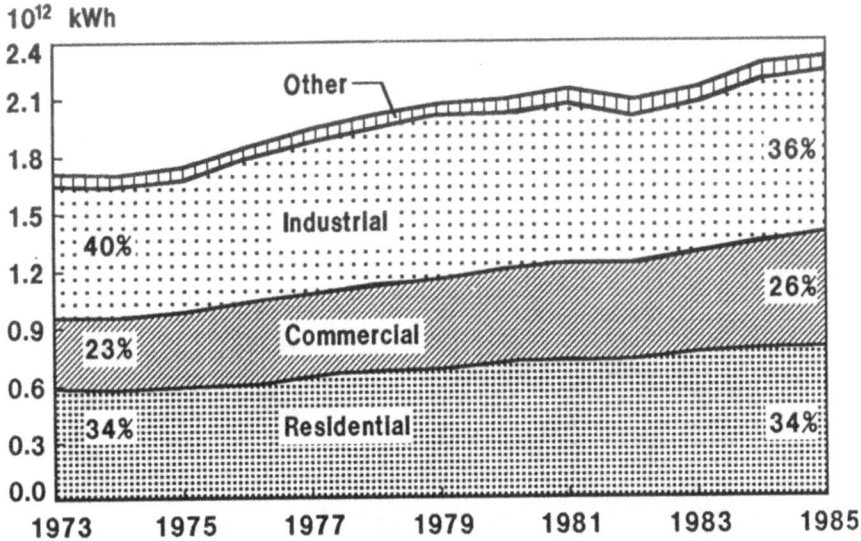

FIGURE 6. Electricity sales by type of customer (9).

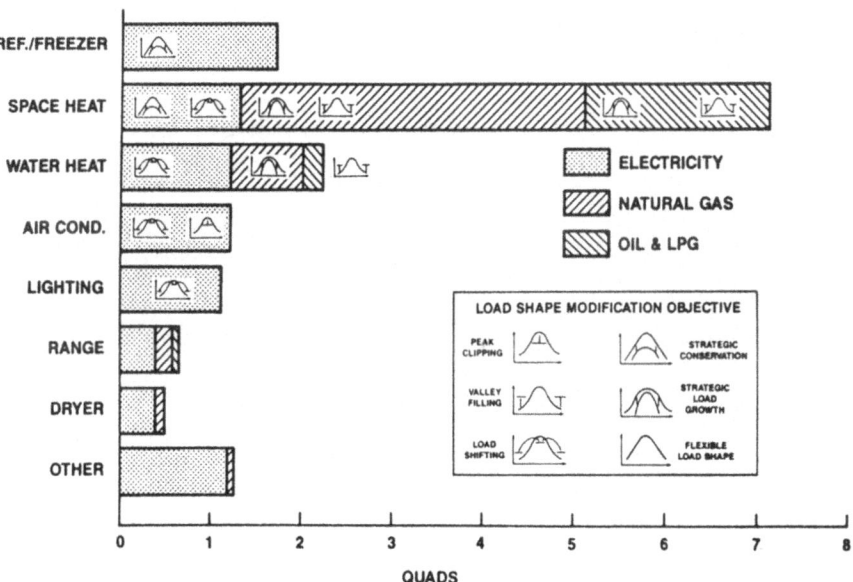

FIGURE 7. Residential energy consumption in 1980 (10).

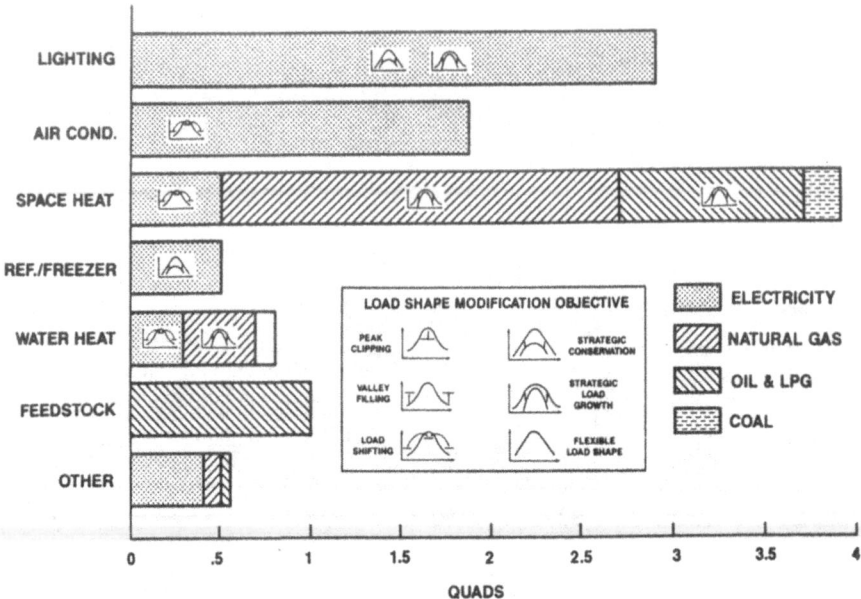

FIGURE 8. Commercial Energy Consumption in 1980 (10).

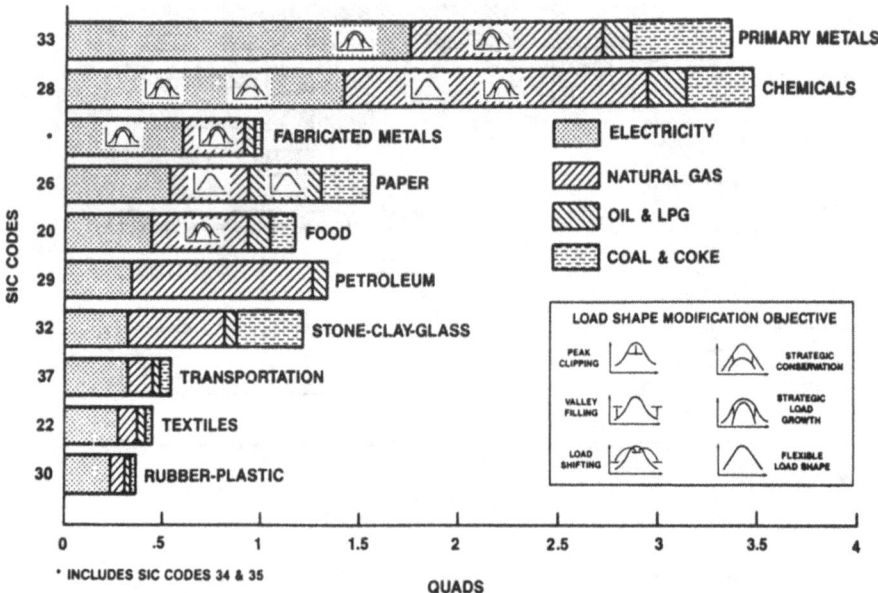

FIGURE 9. Industrial Energy Consumption in 1980.

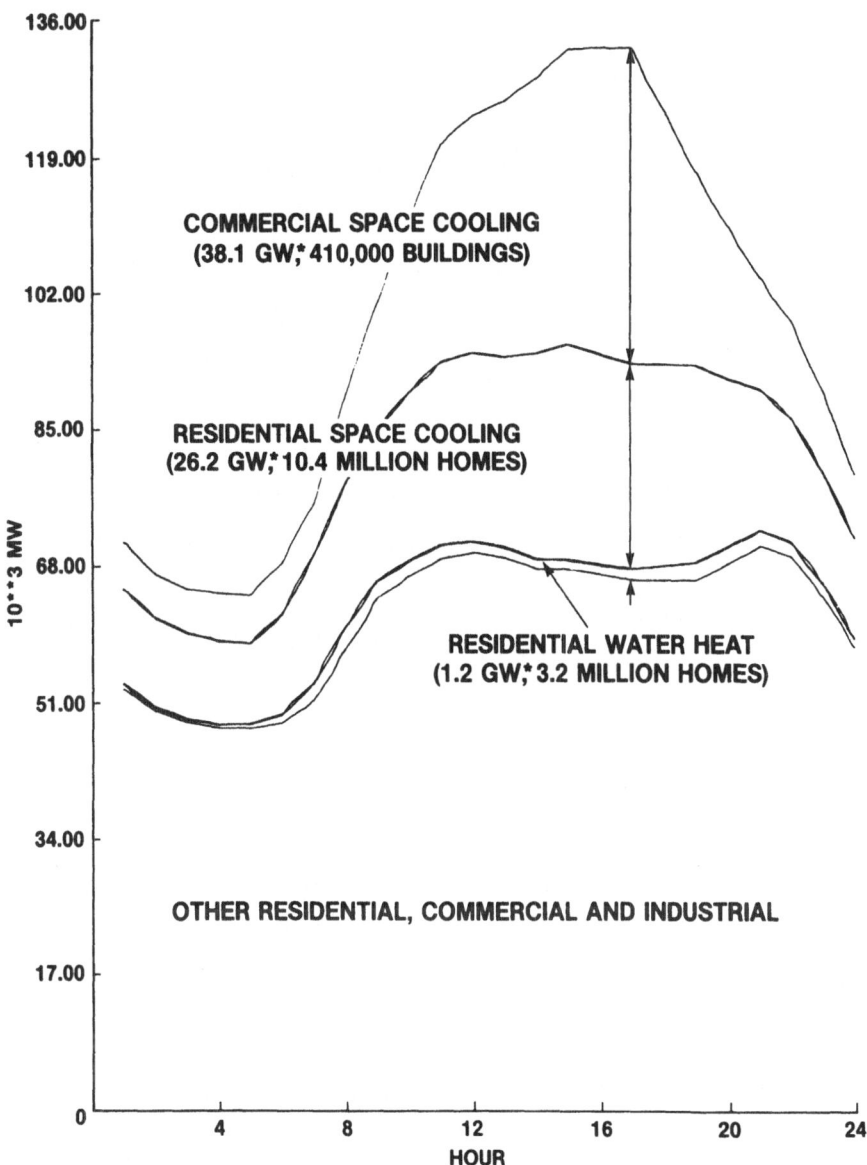

COMMERCIAL SPACE COOLING
(38.1 GW; 410,000 BUILDINGS)

RESIDENTIAL SPACE COOLING
(26.2 GW; 10.4 MILLION HOMES)

RESIDENTIAL WATER HEAT
(1.2 GW; 3.2 MILLION HOMES)

OTHER RESIDENTIAL, COMMERCIAL AND INDUSTRIAL

*CONTRIBUTION TO SUMMER PEAK DEMAND FOR THIS END USE

FIGURE 10. Peak summer day load profile for one of the U.S. regions.

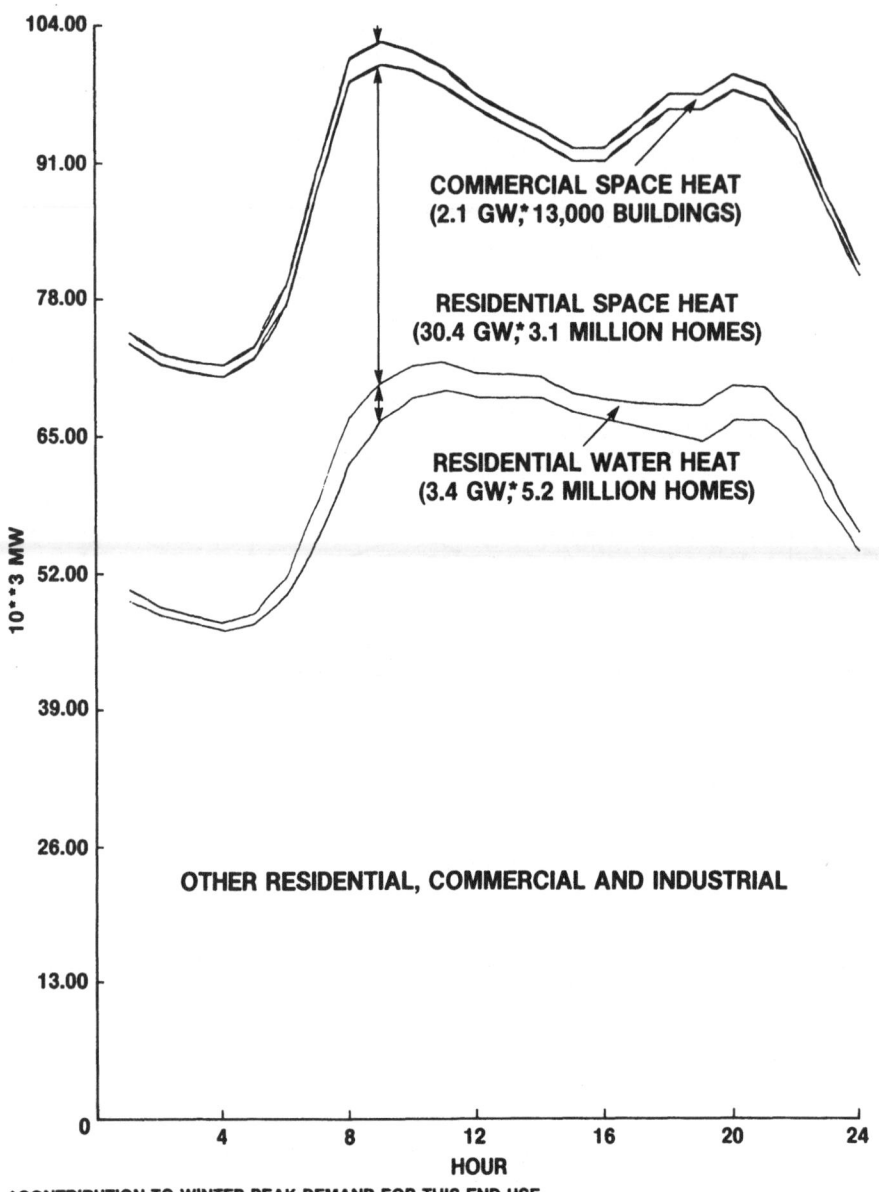

*CONTRIBUTION TO WINTER PEAK DEMAND FOR THIS END USE

FIGURE 11. Peak winter day load profile for one of the U.S. regions.

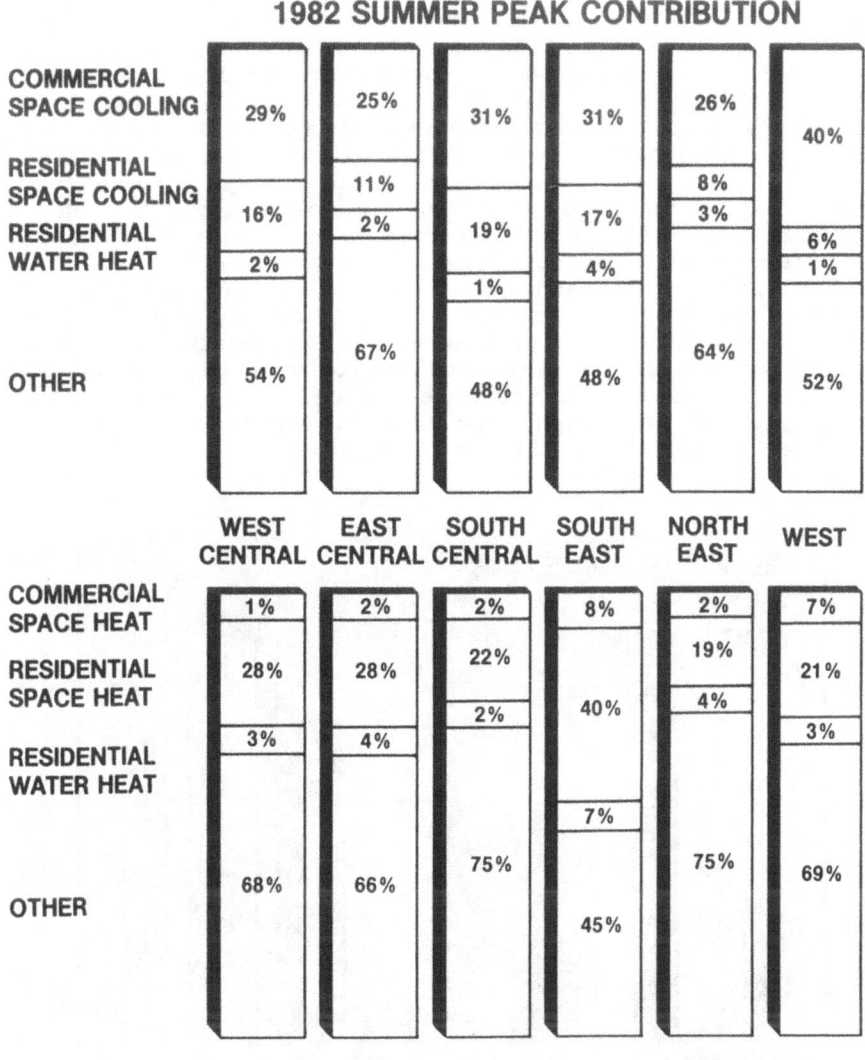

FIGURE 12. Contribution of end-use loads to seasonal peak demands (11).

The commercial sector also exhibits the worst load factor, principally due to the space cooling load. This situation is expected to become even worse in the future. New information processing technologies and automation are estimated to add about 28,000 MW to the commercial sector's demand (12) by the early 1990s, effectively increasing its annual growth rate by over 2% and further aggravating the load factor problem.

5. OPPORTUNITIES, MARKETS, AND PROGRAMS

In principle, the end-use patterns present virtually unlimited opportunities for managing the demand on the customer-side of the meter. In practice, however, the products designed to realize these opportunities have to meet the same criteria as other successful consumer products -- appeal to the consumer through a combination of functional and financial values. While various alternatives appear to offer a good potential for success, it is unlikely that the ambitious goals for demand-side management will be achieved with the currently available set of technologies and programs. Furthermore, because of the recent decline in gas and oil prices, electric options are now facing a formidable competition from fossil fuels in many areas. Cogeneration and self-generation have expanded the competitive markets beyond the more traditional heating and water heating. The fact that these technologies aim to capture the most cost-effective, high load factor, electrical loads is of particular concern to utilities.

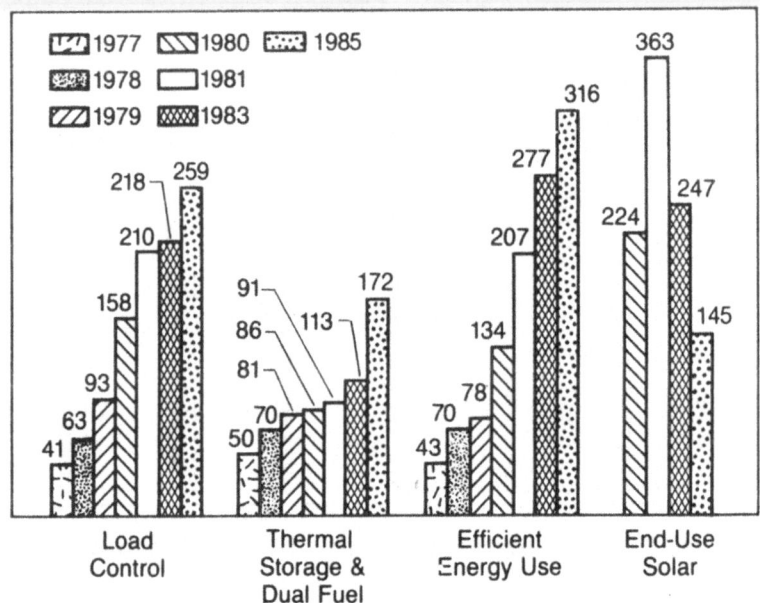

FIGURE 13. Residential end-use projects reported by utilities from 1977 to 1985 (4).

Clearly, new or improved technologies and pricing mechanisms will have to be developed to provide utilities with the driving factors needed to productively utilize available resources and manage the patterns of future loads. The new thrusts will be able to build on a foundation of experience gained in a wide range of programs conducted by utilities over the past decade. For example, as shown in Figure 13, a range of technologies has already been tested and marketed in the residential sector. In the load management category, these include load control and demand management devices, thermal storage (heating, cooling, and water heating), and dual fuel heating systems. Heat pumps, heat pump water heaters, and weatherization/insulation measures were the principal targets of activities in the efficient electricity use category. While the end-use solar category shows a decline and is now essentially limited to solar water heating, it did cover passive solar design and solar space heating during the earlier years. A wealth of data obtained in the course of these programs has been compiled from utilities across the country and integrated into broadly applicable findings (13, 14).

A high level of activity has also been taking place in rate (tariff) design. Table 1 shows the results of an EPRI survey of innovative rates (15) mailed to 157 major investor-owned and 63 publicly-owned and cooperative utilities. As seen in the table, time-of-use rates are the most widely used of all rate types; nearly 70% of all respondents offer this rate form. Participation is generally voluntary, but there are a substantial number of mandatory programs, particularly for customers who use large amounts of electricity. Interruptible and curtailable rates are the second most common rate design. About half the respondents reported offering such a rate, most often to industrial customers. Other common rates are special-purpose incentive, partial requirement, and inverted block. About 15% of responding utilities offered residential demand charges. The remaining rates surveyed -- industrial incentive, vintage, demand subscription, coincident use, and low-income residential -- were offered only rarely.

TABLE 1. Innovative rate structures offered by U.S. utilities (% of respondents).

Rate Form	Investor-owned utilities	Publicly-owned utilities
Time-of-Use	69%	57%
Interruptible/Curtailable	56	46
Industrial Incentive	9	9
Inverted Block	17	20
Special Purpose Incentive	24	23
Residential Demand	9	26
Vintage	2	0
Demand Subscription	2	3
Partial Requirement	21	6
Coincident Use	1	9
Low-Income Residential	7	14
Other	10	11

6. CONCLUSIONS

Most U.S. utilites today are working on a broad set of programs designed to realize the opportunities for demand-side management. A number of technologies, to go hand-in-hand with these programs, are under development by the Electric Power Research Institute (EPRI), the research arm of the electric utility industry. In all cases the goal is to provide new or improved options that meet the needs of all parties -- enhancing the value of using electricity while reducing utilities' costs of supplying the power to their customers.

REFERENCES

1. Electric Power Research Institute, "Electricity: Today's Technologies, Tomorrow's Alternatives," 1987.
2. S. Talukdar and C. Gellings, "Load Management," IEEE Press, 1987.
3. Battelle-Columbus Division and Synergic Resources Corporation, "Demand-Side Management," EPRI report EA/EM-3597, 1984.
4. Plexus Research, "1985 Survey of Utility Residential End-Use Projects," EPRI report EM-4578, 1986.
5. North American Electric Reliability Council, "Electricity Power Supply and Demand, 1983-1992," Princeton, NJ, 1983.
6. F. Kalhammer and T. Yau, "Energy Storage and Management on Electric Power Systems," presented at the VGB-Congress, Power Plants 1985, Essen, Germany, 1985.
7. North American Electric Reliability Council, "1986 Reliability Review," Princeton, NJ, 1986.
8. Adapted from "Energy Report on Electricity Pushes Policy Changes," Electrical World, April 1987.
9. Energy Information Administration, "Monthly Energy Review," 1986.
10. F. Kalhammer, "Utilization and Management of Electric Energy: Current Developments in the USA," presented at the International Electric Research Exchange, Tokyo, 1986.
11. Electric Power Research Institute, "Electricity Outlook: The Foundation for EPRI R&D Planning," 1985.
12. R. Squitieri, O. Yu, and C. Roach, "The Coming Boom in Computer Loads," Public Utilities Fortnightly, December 1986.
13. Analysis and Control of Energy Systems, "Residential Load Management Technology Review," EPRI report EM-3861, 1985.
14. Synergic Resources Corporation, "Review of Energy-Efficient Technologies in the Residential Sector," EPRI report EM-4436, 1986.
15. Ebasco Business Consulting Company, "Innovative Rate Design Survey," EPRI report EA-3830, 1985.

LOAD MANAGEMENT TECHNOLOGIES AND PROGRAMS IN THE U.S.

VERONIKA A. RABL

Electric Power Research Institute
Palo Alto, California
USA

1.   INTRODUCTION

Load management activities are undertaken by utilities to alter the load shape so as to achieve a better match between the customers' cyclic demands and utilities' current and planned generating and T&D (transmission and distribution) resources. A variety of customer-side technologies to control peak loads, shift loads from peak to off-peak periods, or to fill off-peak valleys are beginning to emerge on the market. These technologies include remote control systems, which can affect real-time changes in the load, some of them with smart controls on the customer-side, as well as thermal storage systems for cooling, heating, and water heating, which are designed to shift the corresponding electricity consumption to off-peak periods.

Most U.S. utilities today are testing or implementing one or more load management programs. (See, for example, References 1 and 2. In addition to listing specific utility projects, these reports also provide descriptions of the techniques and technologies available for load management.) Generally, these programs are motivated by the need to improve the load factor, improve the utilization of existing base load plants, and postpone planned capacity additions. It is estimated that, by the year 2000, load management can defer almost 30 GW of generating capacity and add over 40 billion kWh/year to off-peak loads. Several estimates of individual utilities' benefits were presented during the International Load Management Conference, held in December 1985 (3).

Savings on future capacity additions and operating costs are, however, not the only reasons for considering load management. For many utilities, another major consideration is the improvement in customer relations. As part of their load management programs, utilities offer a choice of rates and incentives, which more closely reflect the diurnal and seasonal variations in the actual cost of supplying electricity. The customer, in turn, can take advantage of these rates by installing one of the load management options to save on the electric bill. Compared to flat rates, these options create new avenues for reducing the cost of electric end-uses and could make electricity more competitive in the marketplace. The only cost reduction alternatives available to a customer on a flat rate are conservation or switching to fossil fuels.

A. T. De Almeida and A. H. Rosenfeld (eds.), Demand-Side Management and Electricity End-Use Efficiency, 113–125.
© 1988 by Kluwer Academic Publishers.

2. OVERVIEW OF TECHNOLOGY OPTIONS

In the near-term, cool storage offers one of the most attractive opportunities for load management. Space cooling represents about one-third of the commercial sector's energy consumption and is the largest single contributor to utilities' summer peaks. In the future, new information/automation loads are expected to accelerate the already high growth rate of the cooling demand, currently almost 3%, and further aggravate the load factor problems. Cool storage systems could provide unimpaired service to customers while building off-peak load and offering an effective protection of the electric air conditioning market against growing pressure from gas technologies. From the customers' viewpoint, the systems are installed to reduce demand and save on demand charges.

Because of increased penetration of electric heating and water heating, the growth of utility winter peaks is now expected to outpace that of the summer peaks (4). As a result, a number of individual utilities and two entire reliability regions may become winter peaking around the turn of the century. Here again, heat storage can build off-peak loads and provide a controllable block of loads for the future, while fully meeting customers' comfort requirements. At off-peak rates, heat storage has a good potential for competing with fossil fuels for the heating market.

Thermal storage is particularly well suited for loads with low load factors where addition of storage can reduce the capacity of the equipment that generates the cooling or heating. Conversely, the diurnal HVAC load profiles in the residential sector make the economics of thermal storage less favorable and, in absence of subsidies or advanced technology, the penetration is likely to remain limited.

Energy management/control technologies can be used to shave relatively short-duration peaks and defer operation of some of the loads to off-peak hours. With a real-time link to a utility, they provide a customer-side capability to respond to system conditions. To date, most of the utility programs in this area have focused on direct control of individual appliances on customer's premises. As of 1985, utilities have installed about 2.5 million control points, with another 3.5 million planned for installation by 1990. While direct control is a fairly low-cost solution to real-time control of the loads, its use is largely limited to air conditioners and water heaters. In addition, it is exercised in absence of individual customer's comfort or equipment operating data and, at large saturations, may present distribution and generation control problems due to loss of natural diversity.

Customer-tailored load control options, which provide cost/comfort control and convenience to the customer and real-time dispatch to the utility, are believed to present a more promising load control opportunity for the future. These options add some intelligence on the customer side, creating a system that can offer smart energy management to the building occupants while incorporating utilities' load management objectives. A real-time communication link can be used to transmit control commands, electricity prices, or other information to

influence the actions of the on-site controller. In addition
to dedicated controllers, automation system are viewed as one
of the vehicles for incorporation of these smart energy management function.

3.   COOL STORAGE
3.1. Overview
Cool storage is a technology designed to shift all or
part of the air conditioning load from peak to off-peak
hours. Instead of operating air conditioning compressors to
meet the load when it occurs, during on-peak hours, the compressors operate at night to produce ice or chilled water,
which is then stored and used during the following day to provide comfort cooling (5, 6, 7). The technique has been prac-
ticed since the early 1940's to meet large, short duration
cooling loads. More recently, demand charges and utility incentive programs spurred a major growth in this market. It is
estimated that at present there are several hundred systems in
operation in U.S. commercial buildings (8); for the past several years, the annual installation rate has been doubling each
year. The storage medium is primarily chilled water or ice,
other phase-change materials are used to a limited extent.
Although use of cool storage for residences has been
explored in the past (9), the current market is largely limited
to commercial buildings. Nevertheless, it is estimated that
the present value of a residential off-peak cooling system is
over $3,000, ranging from $1,800 to $6,400, depending on spec-
ific utility's circumstances(10).

3.2. Chilled-Water Storage
Approximately one half of the installed commercial cool
storage systems utilize chilled water as the storage medium,
however, they account for over 75% of the installed storage
cooling capacity and continue to predominate in large buildings. The advantages of chilled-water storage, as compared to
ice, are higher efficiency, standard HVAC practice, familiar,
low-cost refrigeration equipment, and availability of experienced designers and operators. The principal disadvantage is
the requirement for a large expensive storage tank.
The existing installations utilize a variety of methods
to separate the chilled water from the warm water returning
from the building loop. These include labyrinth, baffle, tank
series, one-empty tank, membrane, and stratification
techniques. EPRI laboratory research and full-scale testing
(11) has demonstrated that thermally-stratified systems are
potentially lowest in cost and most efficient. The thermally-
stratified design takes advantage of water's tendency to
separate into layers according to density, which depends on
temperature. Under proper conditions, these density differences create a temperature gradient region -- called a thermo-
cline -- that forms a natural barrier between warm and cold
water. To minimize turbulance and mixing, specially-designed
diffusers are used at the inlets and outlets of these tanks.

3.3. Ice Storage

In the past few years, there has been an increasing trend towards ice as the storage medium. The advantages of ice over chilled water include smaller storage volume, about 1/5 that of chilled water, and the availability of factory-packaged equipment. On the other hand, the operating efficiency is somewhat lower because of the lower compressor suction temperatures.

There are several vendors supplying three different generic types of ice generators. The static ice-on-coil system is the oldest, having been applied in dairies and churches since the 1940's. Ice is formed and stored on metallic tubes which are submerged in a tank of water. Either refrigerant or brine (glycol-water solution) is circulated through the tubes. The efficiency and rate of freezing decreases as the ice thickness increases because ice acts as an insulator.

Another type of ice storage module is referred to as the secondary-loop brine system. A 25% glycol-75% water solution is chilled to below 32°F and circulated through small diameter plastic tubing submerged in a cylindrical water tank. The water in the tank is frozen solid at full charge. Cooling is recovered by circulating the same brine solution through the tubing and to the building. Melting occurs next to the tubing such that charging always begins with bare evaporator surface. The resulting improvement in efficiency is partially mitigated by the two heat transfer steps.

A recent trend has been the application of dynamic ice makers, or ice harvesters, in cool storage. Water is circulated over vertical refrigerated plates suspended above a tank. When the ice thickness reaches about 3/16 inch a hot-gas defrost cycle releases the ice which then falls into the tank below. This technology avoids the decreasing suction temperatures resulting from greater ice thickness, however, this is somewhat offset by the defrost penalty.

3.4. Utility Programs

Cool storage can substantially reduce building operating costs but it generally does command a first cost premium.

The payback depends on the utility rate structure. Generally, it is most sensitive to the on-peak demand charge, and to a lesser extent, the differential between on-peak and off-peak energy charges (6). A utility rate structure having about $12/kW demand charge for a 10 hour on-peak period and a 3¢/kWh on-off-peak differential will provide a 2-5 year payback for a typical commercial building. This is not a sufficient incentive in many cases, particularly for speculative builders and where it involves some perceived risk. Hence, to overcome this hurdle, many utilities are offering up-front cash incentives in the range of $100-400 per kW shifted to off-peak with storage. Some utilities also offer design fee assistance and sometimes special storage rates.

The utility activity has been steadily increasing over the past five years -- as of mid-1987 eighteen utilities offer incentive programs, with four of these initiated since the beginning of the year (12). This count does not reflect a large number of other utilities that do not favor cash incentives and encourage cool storage solely by way of rate structures, some of them with demand charges as high as \$20/kW.

4. OFF-PEAK HEATING AND WATER HEATING
4.1. Heat Storage
Storage space heating systems use off-peak electricity to heat a storage reservoir, discharging the stored heat as needed to heat the space. The most commonly used systems in the United States, derived from designs proven in the European market, contain ceramic bricks heated to 1200 to 1500°F by electric resistance elements. A number of individual room and central systems have been installed in the United States, primarily in areas with severe heating requirements, and have been proven to perform well -- shifting essentially the entire heating load to off-peak hours (9). Other equipment in use includes pressurized or unpressurized water storage and under-slab resistance heater mats. Market penetration of these systems is, however, still limited. This is primarily due to their high costs, compared to conventional heating systems, and the paucity of attractive time-of-use rates.

To overcome the cost impediment, EPRI initiated an effort to assess the feasibility of a lower cost system. It was established that for central forced-air systems (20-30 kW), the major factor affecting the cost are the ceramic bricks used as the storage medium (13). This provided an opportunity for a significant cost reduction, if another, low-cost material could be found to store the heat. Upon an extensive review of various options, common crushed rocks were selected. A system design, utilizing the rocks, was estimated to reduce the installed cost by 30-40%. Prototype units were built and a total of fifteen tested over 2-3 heating seasons in the Northeast, Central, and Southeast regions of the U.S. The experience verified that the furnaces can maintain comfort, even under severe winter conditions, while shifting all of the electricity consumption for heating to off-peak hours.

A new company, CaliDyne Corporation has recently been established in Elk River, Minnesota, to manufacture and market the furnaces under EPRI license. Laboratory testing and UL certification of the production prototype are in progress.

4.2. Storage Water Heating
A storage water heater is a domestic water heater with sufficient storage capacity to provide daily peak period hot water needs with only off-peak operation of the resistance heaters.

The added capacity is generally obtained either by adding a second water heater to an existing installation, or by installing a single, larger-than-normal tank. In residential applications, total storage capacities of 100 gallons or more are typically used. High efficiency models, with relatively heavy jacket insulation, are generally used to minimize storage standby losses; water heater wraps can also be used on standard tanks. Commercial and industrial applications are sized appropriately to store the full peak period load off-peak.

Charging is most often regulated by a timer or time-of-day meter contactor, although remote control techniques are also used. For residential installations, the standard 4.5 kW resistance heater is allowed to operate through the night (generally for an 8 or 12 hour period) and is switched off for the rest of the day. If two tanks are used, the resistance elements in both are interlocked so that the maximum demand is 4.5 kW.

Storage water heating is often used in conjunction with storage space heating, because a customer willing to purchase the larger system will often see the cost-effectiveness of the incremental expenditure.

4.3. Dual-fuel Heating

A dual fuel heating system combines a fossil-fueled with an electric space heating system, both sized to fully meet the building's design heat loss and both capable of functioning independently of each other. The electric heating system provides space heating energy at all times except during utility peak demand periods. During these peak periods the electric heater is turned off and the fossil-fueled system takes over. The electric system resumes operation after the peak period has ended.

The system can consist of the retrofit of an existing fossil-fueled heating system with electric heat; the retrofit of an existing electric heating system with a fossil-fueled burner; or may be installed as a new integral heating system capable of using both fuels. Retrofitting an existing fossil system typically involves the addition of a bank of resistance heating elements in the ductwork of a forced air system or in the tank of a hydronic heater, or the addition of a zoned electric baseboard system. Retrofitting an existing electric system requires the installation of a stand-alone fossil system including any necessary flue and fuel storage tank.

Some mechanism must be used to switch from electricity to fossil fuel and back again. Control options available include timers and outdoor-temperature-activated switches, but the most common method is the installation of a remote control link utilizing one of several available communication media (radio, ripple, power line carrier, and so forth).

4.4. Utility Programs

As of 1985, 172 utilities were testing or implementing programs in off-peak heating and water heating. These programs involved over 68,000 installations. The majority of these were dual-fuel heating systems which accounted for nearly 38,000 installations or 56% of the total. Storage water heating accounts for over 18,000 installations or 27% of the total, and storage space heating accounts for nearly 12,000 installations or 17% of the total.

Off-peak heating systems are significantly more expensive than the conventional heating systems that they replace. Therefore, utilities usually offer the customer an incentive to encourage customer investments in the equipment. Incentives are generally based on a special rate, although the rate is sometimes accompanied by financial assistance such as cash rebates, equipment discounts, or free or reduced-price equipment (such as a free storage water heater for customers purchasing a storage space heating system). Billing credits are also sometimes used, as are other forms of direct incentives such as subsidies and financing. Sometimes the programs rely solely on the ability of the equipment to reduce the customer's bill as the incentive.

Several types of special rates are commonly used: thermal storage rates, time-of-day or off-peak rates, demand rates, and reduced energy rates. While different utilities often have different names for given types of rates and may administer them somewhat differently, most will fall into one of these rate categories.

The most frequently specified direct incentives involve cash rebates; Table 1 summarizes utility-reported rebates made in connection with off-peak heating programs.

Storage water heating projects will often involve a monthly bill credit. Reported monthly credits for storage water heating ranged from $3.75 to $10 with the average being $6.25.

TABLE 1. Ranges of Reported Rebates

Technology Category	Range of Rebate
Storage Space Heating	$150-$2,000/installation
Storage Water Heating	$50-$200/installation
Dual-Fuel Heating	$100-$200/resistance system
	$250-$400/heat pump system

5.   LOAD CONTROL AND ENERGY MANAGEMENT
5.1.   Overview

Load control technology uses a control device to permit or inhibit normal operation of end-use equipment. Load control devices range from simple switches, which operate under direct utility control, to intelligent on-site controllers which are set according to parameters determined by the utility or in response to site conditions and customer preferences. Once installed, the control technology and control strategy determine how load shape is affected. Commercial energy management systems, implemented by larger customers, exemplify one type of load control.

The types of loads targeted for control are typically those that meet the dual requirements for controllability and cost-effectiveness. Basically, loads targeted for control are those that exhibit relatively high diversified demands and can be controlled without significantly impacting customer lifestyles. The loads most commonly controlled include electric water heaters, central air conditioners, water pumps, and, to some extent, space heating systems.

120

5.2. Control Techniques

There are three generic techniques that can be used to control loads: direct, local, and distributed control. These are illustrated in Figure 1.

Direct load control is defined as utility-exercised, real-time control over the operation of selected customer loads. This control is effected utilizing a remote communication system. The parameters of control (when to control, what to control, and how to control) are determined by the utility. After-the-meter hardware consists simply of a communication receiver, signal decoder, and switch for opening or closing the controlled load circuit.

Local control strategies make use of information on local, customer site conditions, such as temperature, time of day, current demand level, or historical energy use, to determine whether and how to exercise control of appliances. The customer installs the device to take advantage of a rate and determines which end-uses will be controlled and how.

Distributed control strategies combine features of direct and local control. There is a real-time communication link between the customer's controller and the utility to transmit commands or information (e.g., price, "activate" control, or "reduce demand" signals). However, the actual control actions are taken by the intelligent customer-side device, which utilizes local conditions, as well as utility signals, as inputs.

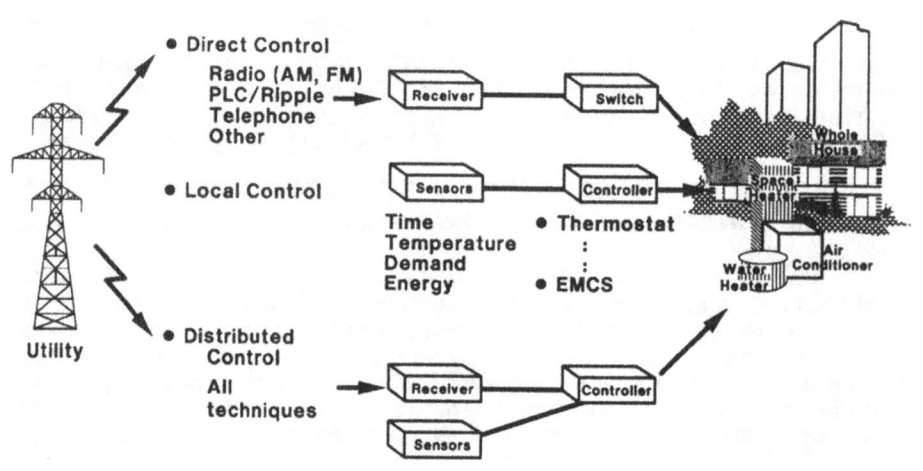

FIGURE 1: Classification of load control techniques

With direct control, the utility is capable of making real-time changes in the use patterns of controlled end-uses. Local control, on the other hand, permits control actions to be set according to the needs and comfort conditions required by each customer. Distributed control incorporates both of these attributes, clearly offering the best opportunity to meet load management objectives in a manner acceptable to both utilities and their customers.

5.3. Control Strategies

Load control strategies characterize the algorithms by which end-use operating patterns are modified during a load control action. Strategies representative of those now being tested and implemented include:

● Restraint requesting - requests that the customer reduce energy use during specific time periods. Response is voluntary.

● Scheduling - disables a controlled appliance based on a preset time schedule.

● Duty cycle limiting - limits the percentage of time an appliance is permitted to operate during a specified time interval (minutes per hour, hours per day, etc.).

● Setpoint alteration - redefines the start, stop, or operating range of a controlled appliance by varying a setpoint parameter, such as temperature.

● Interlocking - establishes an on/off relationship between two or more appliances, allowing selected appliance(s) to operate only when other interlocked appliances are not.

● Demand clipping - limits the short-term (one- to five-minute) demand by controlling one or more appliances to avoid exceeding a predetermined demand threshold.

● Demand managing - limits the medium-term (15- to 60-minute) demand by controlling one or more appliances to avoid exceeding a predetermined demand threshold.

● Load factor managing - regulates the operation of controlled appliances such that medium- or long-term load profile meets or exceeds a self-adjusting load factor set point.

● Price responding - controls appliance use in response to the price of electricity, with the objective of minimizing the customer's electric bill; the utility can set electricity prices in advance or in real time.

Most of these strategies require data gathered at the point of energy use, and have to be implemented using either local or distributed control techniques. Scheduling or duty cycle limiting in the direct control mode are, however, the most common forms of load control in use today. Typically, residential water heaters and residential and commercial air conditioning systems are controlled in this manner, using receivers and remote switch/relay devices. However, in response to demand charges, other devices have recently been introduced into the market or are being developed to implement alternative strategies. These include smart thermostats, interlocks, demand subscription service devices, demand controllers, and load factor managers (14).

5.4. Load Control Emulator System

Tests of load control strategies to date have provided only data pertinent to manufacturer-specific controllers. To better understand the impact of different load control options available to utilities and their customers, EPRI has developed a general purpose energy management/control device, the load control emulator system (LCES), suitable for testing a variety of residential load control strategies before designing and implementing programs for a utility service areas.

The LCES is designed to control end-use electrical loads in several hundred residences, monitor the resulting changes in energy use patterns, collect the data from each residence, and analyze and evaluate the affects of the control strategy. The LCES allows flexible, multi-purpose control of HVAC equipment and eight separate appliances. It will record electricity use, for total household and for individual appliances, at one-minute intervals. Internal and external temperatures, humidity, and appliance status are also recorded.

Data is collected automatically, using the telephone company switched network, and stored in the utility's central computer. Data is examined daily, to estimate load changes, to suggest changes or improvements in the control strategy and to identify problems when they occur. The complete data set is provided in a format compatible with standard utility load research techniques.

Because the load control emulator can use the house wiring for communications, these capabilities are provided with a minimum of rewiring, customer inconvenience, and permanent mounting of equipment.

5.5. Smart House

One of the opportunities to implement intelligent control systems is expected to be provided by the Smart House -- a project launched by the National Association of Home Builders in cooperation with manufacturers, EPRI, GRI, and individual utilities. The project aims to develop a new wiring system for residences, which will combine power distribution with control, data, audio, and video signal cabling and integrate house functions via a common communications protocol. By effectively interconnecting all sensors, switches, control actuators, and appliances in the house, the system will make it feasible to readily implement an array of desirable features, such as home automation, security, and energy management. The ease of communication and control will open up new opportunities for improving current appliances and developing new, more efficient and convenient alternatives (15, 16).

Electric utilities view the Smart House as one of the promising opportunities to create a better environment for productive use of electricity. EPRI is participating in the design to assure that several specific features of interest to the industry are incorporated in the system. These features include intelligent energy management/control functions, such as those described in Sec. 5.3, end-use load data acquisition for remote load research and meter reading, and compatibility with current and planned communication systems to be used for transmitting signals to the house controller and/or retrieving appropriate data from the residence.

5.6. Utility Programs

As of 1985, 259 utilities were involved in the control of more than 2.5 million loads. The electric water heater is the most commonly controlled load. The 1,078,337 controlled water heaters can be further broken down by customer class into 1,063,795 residential and 14,542 commercial loads. Central air conditioners are the second most commonly controlled load. The 953,651 controlled air conditioners include 913,413 residential and 40,238 commercial units.

The 148,756 space heating systems under control include conventional electric heating (108,617 residential, 935 commercial), storage heating (7,742 residential, 511 commercial), and dual-fuel heating systems (29,956 residential, 995 commercial). Essentially all of the reported 311,763 swimming pool pumps under control are residential, and the 17,064 irrigation pumps are agricultural.

A miscellaneous load category includes 20,484 residential, 78 commercial, and 41 industrial points of load control. Included in this category are residential installations where multiple appliances are controlled by local controllers, commercial local load controllers, and miscellaneous industrial loads such as air compressors and water pumps.

The reported points of control are associated with 17 different technologies or technological applications within the three load control categories (see Table 2).

TABLE 2. Reported Points of Control by Technology

Technique Technology	Points of Control
Direct Control	
Radio	1,698,847
Power Line Carrier	302,656
Ripple	146,185
Cable TV	4,926
TWACS	1,215
Broadcast Radio	400
Telephone	140
Distributed Control	
Demand Subscription Device	3,000
Multi-Function Programmable Controller	715
Commandable Thermostat	665
Smart Duty Cycler	60
Chiller Controller	5
Local Control	
Mechanical Timers	295,380
Temperature-Activated Cyclers	58,833
Multi-Function Programmable Controller	9,422
Appliance Interlocks	325
Multiple Technologies	7,400

In the majority of reported cases some type of financial incentive is offered to the participating customers, although 29% of the reported projects carry no direct financial customer incentives. In most cases, customers are offered monthly, seasonal, or annual bill credits. This form of incentive, associated with 45% of all reported load control projects, is particularly common in direct load control programs. The second form of incentive, associated with 24% of all projects, is a special rate, including a demand charge and/or a reduced kWh cost. This type of incentive is particularly common in distributed and local control projects and in direct control projects involving the control of irrigation pumps and off-peak heating systems. Table 3 summarizes the incentive characteristics of the reported load control projects.

TABLE 3. Incentives for Load Control

Type of Incentive	Range of Incentive
Bill Credits	$1 to $5/month for water heaters
	$1.25 to $12/month for air conditioners
	$0.33 to $5.50/kW/month for air conditioners
	$1.50 to $29/hp/year for irrigation pumps
Special Rates	7% to 50% rate reductions per kW
	8% to 20% rate reductions per kWh
	$0.003 to $0.061 discounts per kWh

6. SUMMARY

As discussed above, a variety of load management alternatives offer a promise for improving plant utilization and controlling peak load growth. A growing understanding of the technical issues relating to improved performance, lowered costs, and expanding applications potential is giving rise to a whole new spectrum of technologies capable of bringing benefits to both utilities and their customers. Nevertheless, it has to be recognized that many of these technologies are still in their infancy and the availability of viable load management products is very limited. Commercial development of such products is hampered by lack of rate structures to provide strong market signals and by market fragmentation into individual utility service areas. If load management is to become an effective end-use strategy in this century, a full spectrum of cost-competitive products is needed to offer customers choices compatible with utilities' load shaping and real-time control objectives.

REFERENCES
1. Plexus Research, "1985 Survey of Utility Residential End-Use Projects," EPRI report EM-4578, 1986.
2. Synergic Resources Corporation, "Survey of Commercial Sector Demand-Side Management Activities," EPRI report EM-4142, 1985.
3. Synergic Resources Corporation, "Proceedings: International Load Management Conference," EPRI report EM-4643, 1986.
4. North American Electric Reliability Council, "1986 Reliability Review," Princeton, NJ, 1986.
5. R. Wendland, "Commercial Cool Storage," presented at the 1986 Thermal Storage Conference for Architects and Engineers, Oklahoma City, OK, October 1986.
6. RCF, "Commercial Cool Storage Primer," EPRI report EM-3371, 1984.
7. Electric Power Research Institute, "Commercial Cool Storage Design Guide," Hemisphere Publishing Corporation and Springer-Verlag, 1987.
8. Argonne National Laboratory, "Current Trends in Commercial Cool Storage," EPRI report EM-4125, 1985.
9. Oak Ridge National Laboratory, "Field Performance of Residential Thermal Storage Systems," EPRI report EM-4041, 1985.
10. QLA, "Market Constraints for Residential Cool Storage Systems," EPRI report EM-4722, 1986.
11. University of New Mexico, "Evaluation of Stratified Chilled-Water Storage Techniques," EPRI report EM-4352, 1985.
12. "Utility Inducement Programs for Cool Storage," ITSAC Thermal Storage Technical Bulletin, San Diego, CA, July 1987.
13. Dynatech R/D Company, "Review of Heat Storage Materials," EPRI report EM-3353, 1983.
14. Electrotek Concepts, "Residential Load Control and Metering Equipment: A Summary of Costs and Capabilities," EPRI report EM-5392, 1987.
15. "The Smart House: Wired for Electronic Age," EPRI Journal, November 1986.
16. "Smart House," Professional Builder, December 1986.

ENERGY MANAGEMENT SYSTEMS

A. G. MARTINS

Departamento de Engenharia Electrotecnica
Universidade de Coimbra   Portugal

1. INTRODUCTION

Scarcity of energy resources, higher costs associated to energy production, tariff changes and price incentives, consumer education towards rational use of energy, high number of different types of energy consuming processes and different levels of complexity of the associated control strategies, need for global optimization, microprocessor availability at lower and lower prices, more and more efficient data communication techniques, equipment standardization efforts - all these items, and more that could be mentioned without effort, may be easily related between them. Many cause-effect relations are easily identified, all leading to the conclusion that computer-based systems had to be the natural choice for most energy management applications.

Buildings and industrial plants are excelent candidates for the application of Energy Management Systems (EMS), either small or large, depending on the size and complexity of the facility. Though EMS is an acronym nowadays used also in the world of electric utilities, we shall here confine ourselves and this designation to the customer side.

In the early days of computer applications in control, economy of scale determined that expensive computing hardware had to be used to its full capabilities in order to obtain a good return on capital investment. There were, however, several drawbacks associated to the use of a single machine to perform all the control tasks, namely low overall reliability and difficult expansion of systems capabilities.

The availability of low-cost hardware changed the scenary. It became economic to use more than one machine and to distribute computing power among several processes to be controlled. The coordination problem could be solved with appropriate communications between the several partners in charge of local control, or between them and a supervisor. Reliability and expandability reached much more sactisfactory levels.

It may be stated that EMS have followed the general trends of industrial automation systems. The operational and architectural options are very similar and it is not uncommon to find a single system performing in an integrated fashion,

A. T. De Almeida and A. H. Rosenfeld (eds.), Demand-Side Management and Electricity End-Use Efficiency, 127–144.
© *1988 by Kluwer Academic Publishers.*

controlling energy use as well as other processes.

2. ENERGY MANAGEMENT SYSTEMS CONFIGURATIONS

The early days centralized control architectures referred above soon have been abandoned due to the effective alternatives offered by distributed processing capabilities of more recent systems.

Nevertheless, the premises where energy management measures are to be taken vary widely in dimension and complexity. Hence, in many cases a single microprocessor-based system is enough, as long as it provides the necessary input/output capabilities and the flexibility to adapt easily to the actual needs of control. Usually, these small systems are a standalone version of the terminal outstations of bigger distributed systems. In this manner it is possible to gradually upgrade an EMS, installing on a phased basis several local control stations working in a stand-alone mode and, finally, a central station to increase processing power and integrate local station functions.

2.1. Microprocessor-based pre-programmed EMS

2.1.1. Hardware. These provide a limited capacity of input/output, typically some number equal to a small power of 2 (eg.32 or 64) of load control points, normally requiring suitable power interfaces (relays) to be able to curtail and restore loads. They also provide a simple user interface through a keyboard and a display of LED or LCD type, both used to allow parameter programming to customize the system to the site. The parameters are kept from erasure due to power failures by some means of battery back-up of parameter memory.

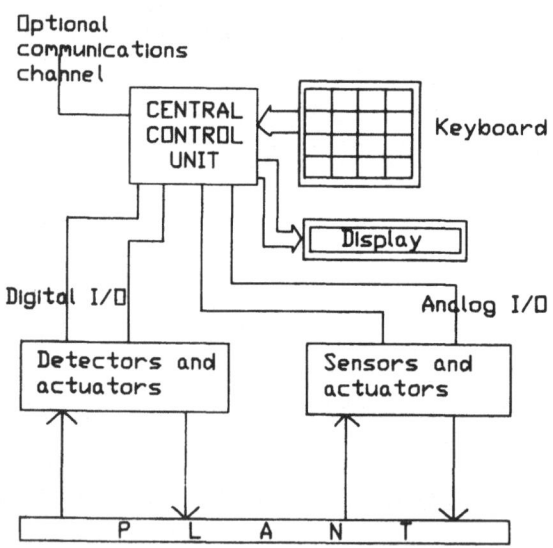

FIGURE 1. Block diagram of a stand-alone pre-programmed EMS.

2.1.2. <u>Software</u>. Pre-programmed EMS have a set of control capabilities as programs kept in ROM and readily available to the user. This one only has to follow an automatically driven procedure to enter the adequate values of control parameters. Among the most commonly available control programs in pre--programmed EMS, that will be referred in more detail later, are: time of day control, duty cycling, demand limiting, optimal start, monitoring. Also available are telecommunications facilities, either to perform remote alarm signaling or to accommodate integration in a hierarchical network under the supervision of a central computer. Figure 1 represents diagrammaticaly a pre-programmed EMS.

2.2. <u>Energy Management Systems with distributed processing</u>
2.2.1. <u>Hardware</u>. In Fig. 2 a possible configuration of such a system is depicted. As can be seen, a considerable set of additional resources is available for interface to the user at

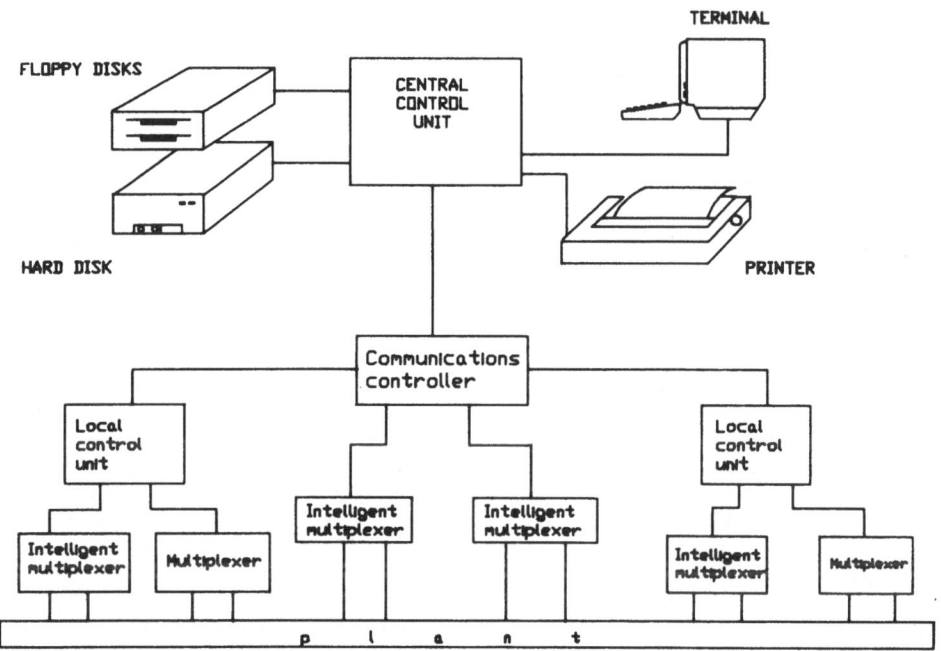

FIGURE 2.Block diagram of an EMS with distributed processing.

the top system level, contrasting with the pre-programmed units above.

The central control unit (CCU) consists of a mini- or a microcomputer and peripherals for data entry and programming, for alarm reporting, for regular data printing and for data and program storage.

The communications controller (COMC) includes modems and data concentrators, managing the communications between the CCU and the rest of the EMS.

The local control units (LCU) are microcomputer based stations that provide interface to the local environment, respond to the CCU orders and are able to perform their job in a stand-alone mode, should the CCU, the CCOM or the communications channel fail. Pre-programmed EMS are typical candidates to LCU in distributed environments, provided that they have the necessary hardware and software support for communications.

Multiplexers (MUX) and intelligent multiplexers (IMUX) are interface devices that stand between the LCUs or CCU and the plant. They perform the combination of data from several sensors in the plant, transferring the information upwards on a single channel. Demultiplexing actions of orders from the upper level are also performed. IMUXes differ from MUXes because the former only report data that has changed since the last report, whereas the latter always send upwards information about all the measuring points they deal with.

The configuration depicted in Fig. 2 is by no means unique, only representative of the philosophy of organization of medium to large EMS.

2.2.2. Software. EMS are usually supplied with application software implementing some fundamental control programs from a list that will be seen later on. While in pre-programmed EMS these applications are in firmware support, in this case they reside in magnetic media. Besides, the user may also develop his own control programs through the use of development tools that come with the system (language interpreters or compilers with capability to interact with the associated hardware environment).

3. EMS MAIN FUNCTIONAL CHARACTERISTICS AND PERFORMANCES
3.1. Supervisory control and direct digital control

Energy management systems may use both categories of control, though the present trend is towards direct digital control alone.

The difference between the two is better understood through the typical example of a local space heating system. In supervisory control mode the EMS, after measuring the external temperature, provides the correct set point to an analog feedback controller located in the zone to be heated. Figure 3 a) depicts this arrangement. The advantage over a conventional thermostat derives from the fact that the set point is a function of external temperature, allowing for savings in heating energy.

Under direct digital control, the same space would be heated without the intervention of any kind of analog controller. The EMS directly measures both interior and external temperatures and drives the actuator in order to maintain inside temperature within limits. In this manner inaccuracy of analog controllers is avoided, reducing discomfort due to long response times and avoiding energy waste due to lack of precision of the interior temperature measurement. However, supervisory control has the advantage of continuous operation even if the computer fails, using a fixed

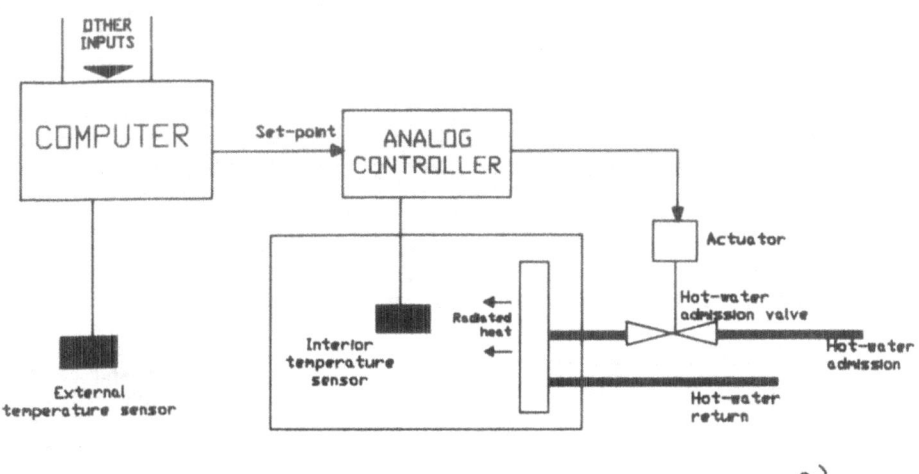

a)

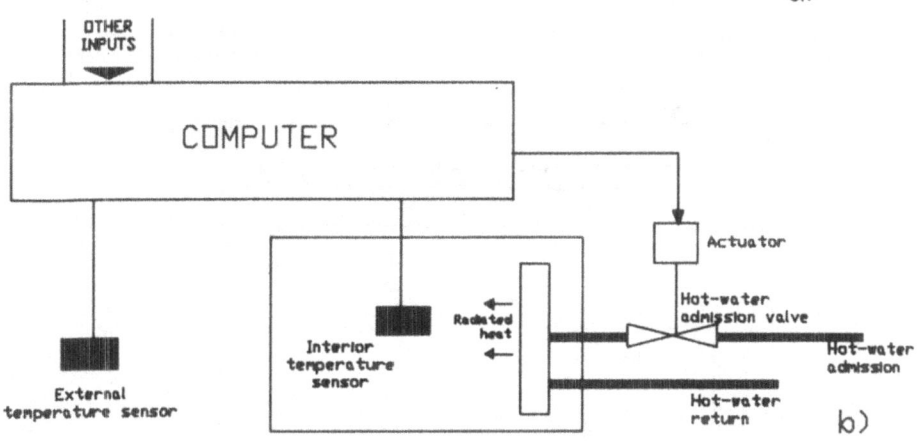

b)

FIGURE 3. Example of a space heating system with a) supervisory control and b) direct digital control by an EMS.

set point (the last issued). In Fig. 3 b) an alternative arrangement to the one in Fig. 3 a) is presented, with direct digital control.

Supervisory and direct digital control (DDC) may, however, exist simultaneously at different levels in an EMS. This is illustrated in Fig. 4, where supervision can be seen to be identified with a station in charge of coordinating the lowest level processing in the hierarchy, where DDC is implemented. It is as if DDC stations performed in a manner similar to the analog control loop in Fig. 3 a), respecting the reference parameters the supervisor imposes, to control the several dependent processes.

3.2 Plant supervision

A certain number of EMS functions primarily devoted to

interface with the operator may be grouped together under this designation. Printed reports of status conditions of previously selected points or of alarm conditions are some of those functions. Also, alarm reporting on the console and redefinition of set points for local control functions. One of the most important features under this category, for it is most of the times the one the user gives more importance, is the capability to remotely activate or deactivate loads throughout the plant. Actually, it may help to save considerable amounts of energy without direct local human intervention. As an example, the remote switching of lights at the end of a building's occupancy period may be pointed out.

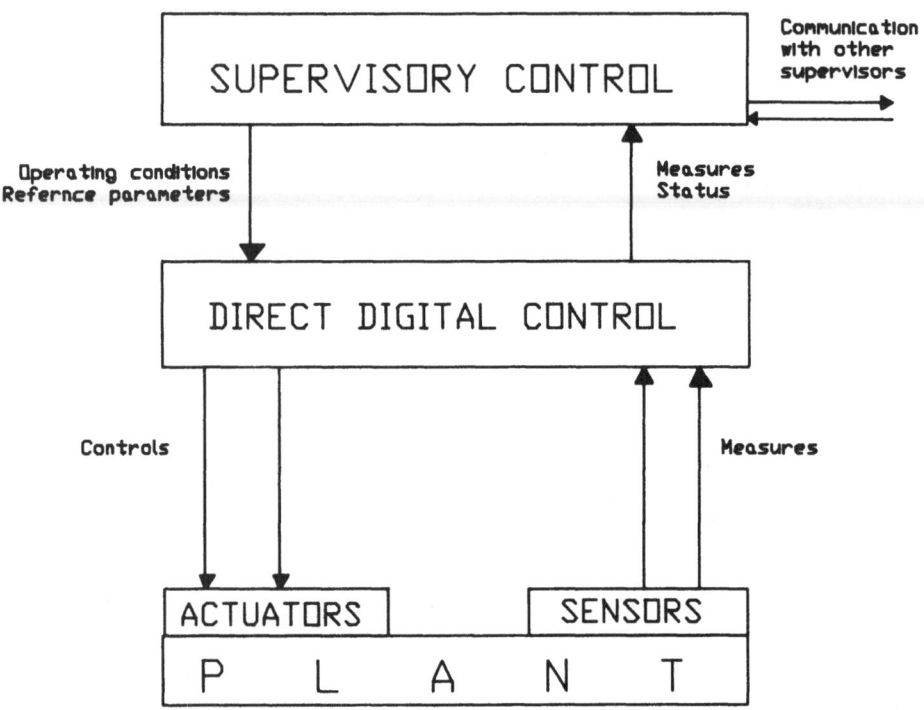

FIGURE 4. Example of a space heating system with direct digital control by an EMS.

3.3 Time-of-day control

The capability of starting and stopping equipment according to the time of day and day of week is probably one of the most cost-effective features of EMS, responsible for most of the savings. It is also one of the most simple to implement. Based on the real-time information resident in EMS, it has the primary advantage of switching off equipment at the beginning of unoccupied periods. The use of EMS is in this case clearly in advantage as regards to the use of traditional time-clocks. Besides being innacurate, these latter must be reprogrammed each time a power failure occurs and there is no possible coordination of actions between different devices.

3.4 Duty-cycle control
Duty cycling consists of shutting off certain equipments for short programmed periods of time during the normal operating hours. It is based on the assumption that it is seldom true that equipments should run allways at peak load conditions, for which they were designed. As it is by nature a discontinuous control process, it must be used carefully in order not to create adverse conditions, usually of comfort. As a matter of fact, duty cycling is usually applied to space conditioning loads, such as heating, ventilating and air-conditioning (HVAC) systems.

Some secondary effects of duty cycling may be beneficial also. If a curtailed fan, for example, does not consume power and, therefore, leads to overall reduction of energy consumption, it is not less true that heating or cooling loads will also consume less because there is less open air to be conditioned. Sometimes, the effect on AC compressors or on boilers is more important than the savings in fan power consumption itself.

Care must be taken in several aspects, when using duty-cycling control. On one hand, when there are other strategies active, such as demand limiting (see 3.5), coordination between them must be very well established in order that their

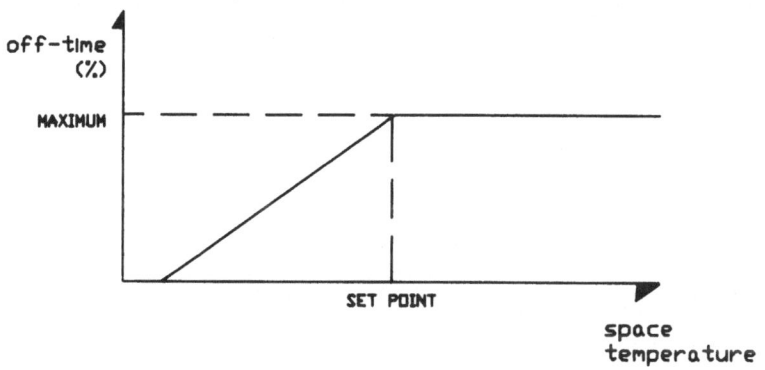

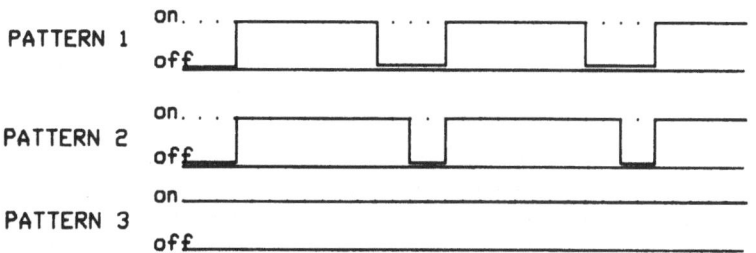

FIGURE 5. Effect of automatic adjustment of on-time/off-time ratio as a function of space temperature.

effects do not cancel each other. Besides, duty cycling on its own should not be applied to loads that will have to compensate for the off periods, such as for instance battery chargers. Finally, duty cycling may impose an excessive stress on certain equipment by allowing too big switching rates. Because of this latter aspect, some parameters should be allowed for programming by the operator, relative to each load, such as maximum and minimum off times, minimum on time and maximum frequency of switching, besides the period of the day during which duty cycling is to be effective.

Some EMS offer an improved version of this feature, that may be called temperature-compensated duty-cycling. It consists of varying the ratio on-time/off-time according to space temperature. In Fig.5 an illustration of this effect is represented.

3.5 DEMAND LIMITING

Big customers are usually charged for the demand they impose on the distribution network. There is a demand control period typically of fixed lenghth, of 15 or 30 minutes, during which the power demand is monitored and the average computed. The monthly maximum average value is recorded, being the base upon which the utility computes the demand charge. Alternative procedures use a sliding time-window approach. In this case, the demand period has not fixed references. Consumption data are recorded during a certain period, being then scanned searching for the n consecutive readings with the highest medium value. n is fixed during the search procedure and constitutes the window itself.

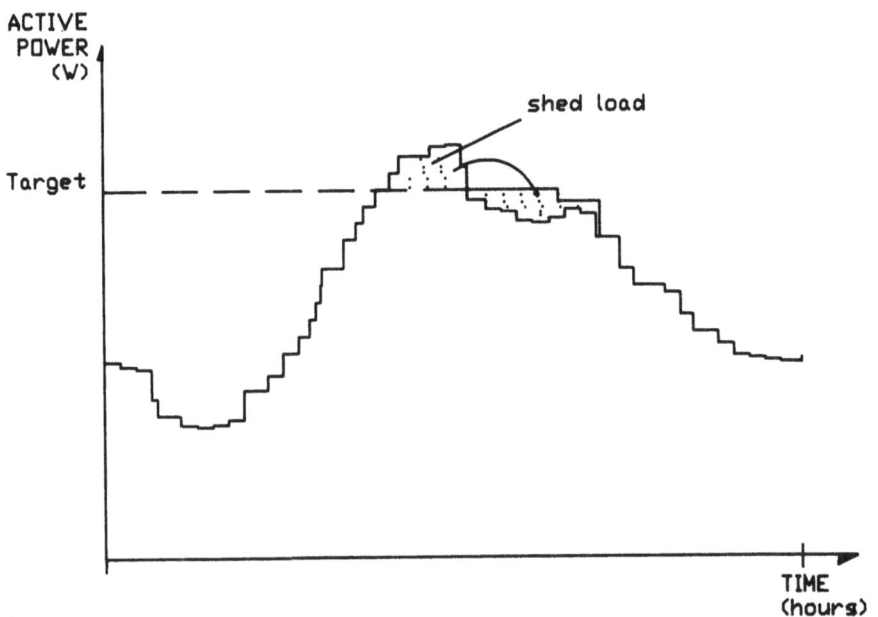

FIGURE 6. Effect of demand control on the daily load diagram.

Demand is controlled by means of shedding loads when predefined demand limits are to be exceeded and restoring them in the opposite case. In Fig. 6 a sample load diagram is used to illustrate the effect of a demand control routine. The upper portion of the uncontrolled diagram, which would occur without any corrective action, is shifted to a later period in the day, thus limiting the average power value during each demand control period.

There are industries and buildings where controlling the demand is a must and reveals to be very cost-effective. The most suitable loads to be controlled by shedding actions are usually characterized by long time constants and examples can be pointed out as battery chargers, cold room refrigeration plants, induction heaters, compressors, arc furnaces, air conditioning plants, fans, etc..

Three main types of demand control can be found implemented in EMS: ideal demand curve, rate of energy use and predictive control. They have in common the fact that they consider loads according to some pre-established priority order, beginning the shedding actions by the lowest priority ones.

3.5.1. <u>Ideal demand curve controllers</u>. These are programmed with an ideal load curve (KWh vs. time). The demand control period is divided into a certain number of sampling periods of

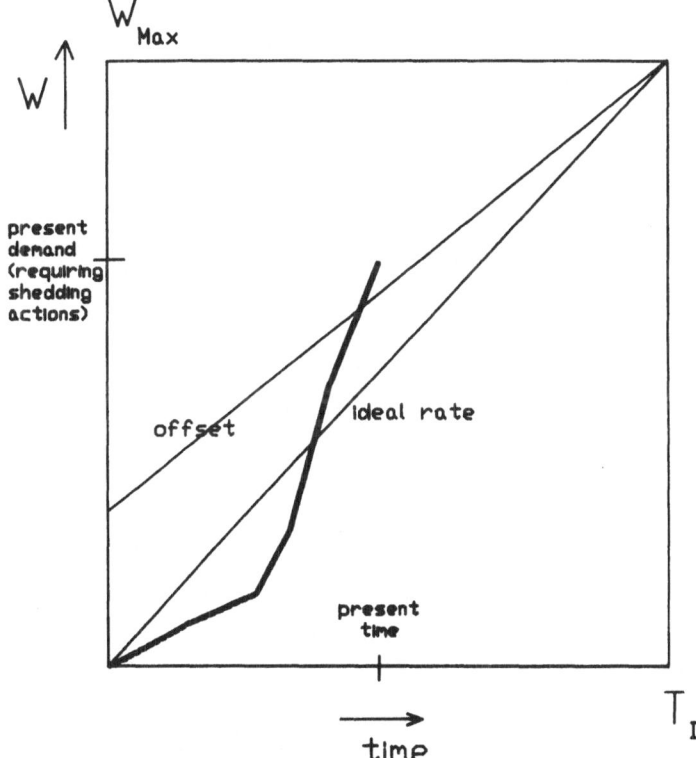

FIGURE 7. Ideal load curve control algorithm.

equal duration. At the end of each, energy consumption is
measured. Shedding actions begin whenever the demand point is
above the curve. Loads are all restored at the beginning of
the demand control period. Figure 7 depicts the operation of
this algorithm.

3.5.2. Rate of energy use controllers. This algorithm is
used when the utility charges demand on the basis of a sliding
window approach. At the end of each sampling period demand is
compared to a pre-defined limit and shedding or restoring
actions are determined if the former is above or below the
latter, respectively. It usually leads to undesirably high
switching frequencies and premature shedding actions. Figure 8
depicts the operation under this strategy, again in a KWh vs
time diagram.

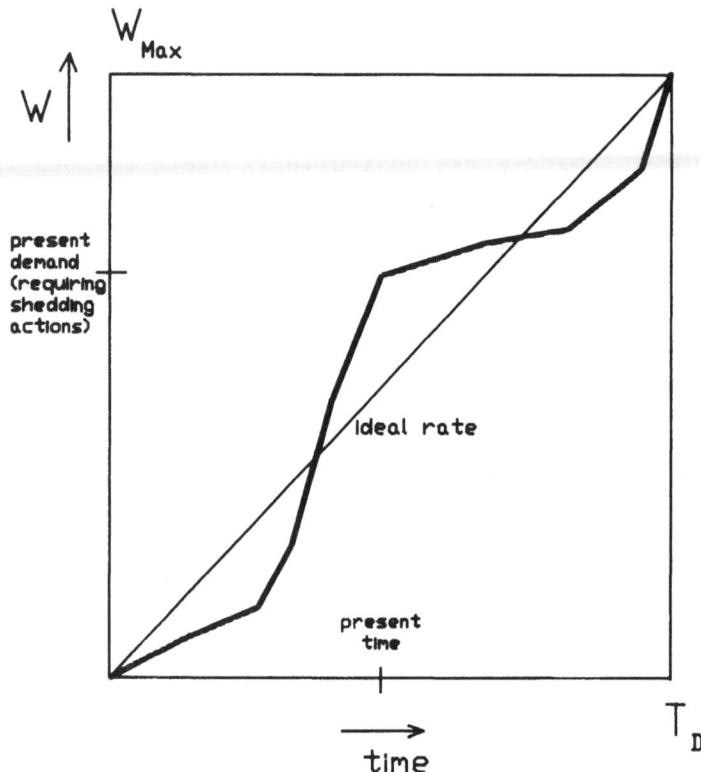

FIGURE 8. Rate of energy use control algorithm.

3.5.3. Predictive controllers. In this case a time
projection of the demand until the end of the demand control
period is computed at each sampling period. Shedding actions
begin as early in the demand period as necessary in order to
avoid demand limit to be exceeded. Also, unwanted shedding of
large amounts of power near the end of the demand control
period, as is current with ideal demand curve controllers, is
avoided. Figure 9 represents the operation of this algorithm.

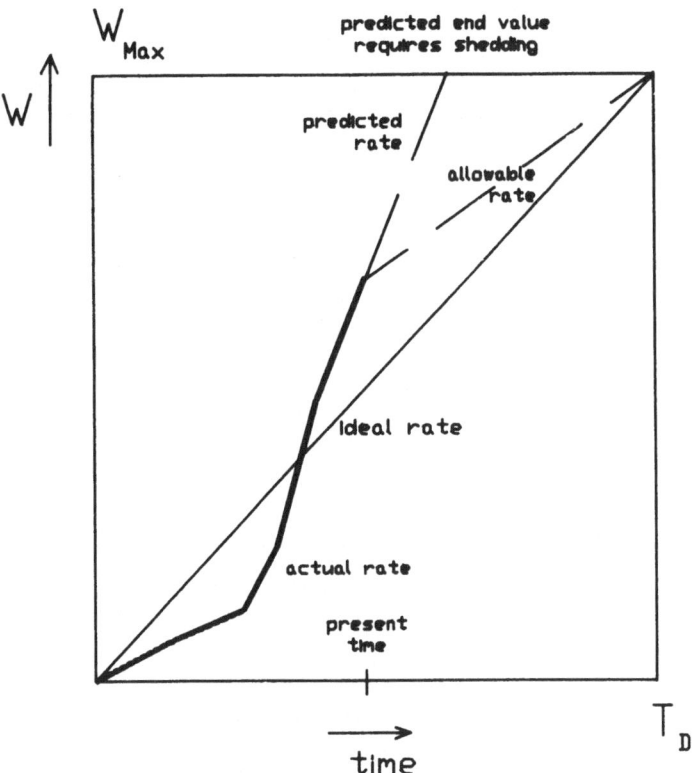

FIGURE 9. Predictive control algorithm.

The considerations stated earlier about duty-cycling routines and their effect on loads switching frequency apply here as well. The parameters then referred as necessary to limit excessive switching are usually available to demand limiting routines also. They may, in certain conditions, modify the normal rule of load election for shedding or restoring, based exclusively on priority order.

When there is provision to program also the rated power of each controllable load, it is possible to implement at the EMS level another criterion for shedding and restoring that searches the more adequate amount of load to switch in order to minimize variations of global load power. This is known as load levelling. Priority order is also affected by this procedure.

3.6. OPTIMUM START-STOP CONTROL

Space conditioning equipment cannot be started at the same time as occupancy of the conditioned zones begin. HVAC loads are usually started prior to occupancy by people, soon enough to allow space temperature and humidity to settle at the desired level. If a simple time-clock is used, as environmental conditions vary, either a waste of energy may

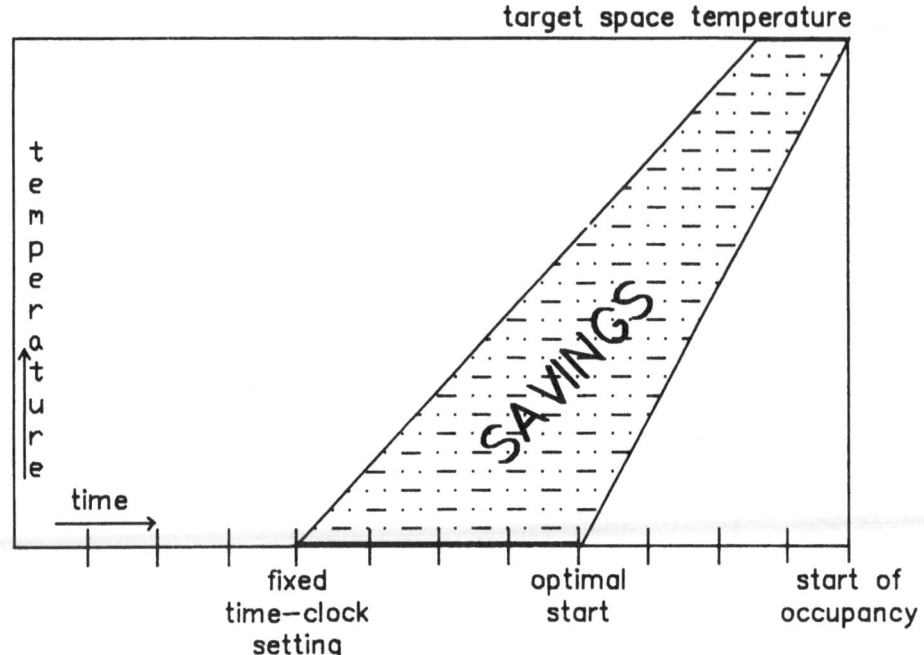

FIGURE 10. Savings due to optimal start.

occur due to a premature start or uncomfortable conditions are found at the beginning of occupancy.

EMS solve this problem by self-adjusting the time at which the HVAC loads shall be started each day to attain the adequate levels of comfort exactly at the desired time. By monitoring ambient temperature and the time required to achieve the target value, EMS constantly update the profile of the facility's heating/cooling characteristics kept in memory, in what may be called a constant learning process.

A similar procedure may be implemented for the shut- n of HVAC equipment prior to the end of building's occupancy. This procedure must ensure that minimum environmental conditions exist at the end of occupied hours. The comparative potential savings as regards to the optimum start routine are small. In Fig. 10 the advantage of optimum start is showed as compared to the use of a time clock with fixed setting.

3.7. DAY/NIGHT SETBACK

When the climatic conditions do not recommend total shutdown of HVAC equipment during unoccupied hours, a usual procedure consists of maintaining the building at a space temperature slightly different than uncontrolled. This is known as the temperature setpoint, and it may be under the EMS control, in coordination with the optimum start routine. The basic function is to bring into service the conditioning equipment whenever the temperature falls below the setpoint in the heating season or rises above it in the cooling season, and to disconnect it otherwise.

3.8. LIGHTING CONTROL
This usually resumes to a time-of-day (TOD) control applied to illumination loads. However, energy savings in lighting are not mostly obtained with automatic control. This one can only help in the situations where it is frequent to find lights on in unoccupied buildings by using TOD control. If other conservative measures are taken such as retrofitting the illumination circuits in order to provide separate control of more than one lighting level, the EMS may also be used to control the switching actions in order to adapt the artificial lighting level to the exterior light, thus contributing to further energy savings.

3.9. ECONOMIZER CONTROL
Outdoor air may be used in certain conditions to reduce the cooling requirements of a building. By appropriate control of dampers and measurement of external and internal temperatures, decisions may be taken by the EMS as to admit (and in which percentage of damper opening) outdoor air for cooling purposes.
The simplest application consists of admiting cool outside air during the early morning hours of Summer days to pre-cool the building and reduce the prospective energy consumption for cooling.
If measures are taken not only of temperature but also of relative humidity of both internal and external air, it is possible to compute the heat content of both and decide to use the one with the lowest to cool the facility. This allows to take into account ambient humidity requirements as well. Decisions may be taken about the mixing percentages of exterior and cooling system return air in order to attain the targets.

3.10. POWER FACTOR CONTROL
Many utilities place a charge on customers' reactive energy consumption. The method used by customers to avoid this charge consists of connecting capacitors in parallel to the main inductive loads or at the premise's entry point in order to implement power factor correction. As active load varies, so varies the amount of capacity needed for compensation. Local control units exist that perform this function sactisfactorily. However, provided that appropriate sensors are used, the EMS can handle also this task.

3.11 ENERGY STORAGE
Energy storage is a technique that uses the lower electricity costs during the night (under a time-of-day rate) to store heat that will be used during the day for space heating. Storage reservoirs may be of several different types, using water, water under pressure, ceramic bricks, etc.. Water is forced to circulate through radiators during the day. Air is forced to circulate through the solid storage medium used,

or it circulates by natural convection. The EMS plays the role of switching on and off the heating elements of the storage system at the appropriate times.

4. DATA COMMUNICATIONS
4.1 Data transmission media
Communications between MUXes and IMUXes and LCUs and between the latter and CCU or CCOM may be performed through different media. Characteristics such as bandwidth, transmission speed or noise immunity are used to compare them. In the following, a brief reference is made to the more commonly used on the EMS field.

4.1.1. Twisted pairs. As the name indicates, a twisted pair consists of two insulated conductors twisted together to improve immunity to external interferences. Unconditioned twisted pairs have a bandwidth 300-3000 Hz, allowing up to 6000 bits/second of transmission speed. The most common speeds lye between 1200 and 4800 bits/sec. If adequate conditioning is provided, high data rates become possible. A particular implementation, for instance, operates at 375 kbits/sec.. Twisted pairs are permantly hardwired between sending and receiving ends.

4.1.2. Telephone lines. They are much like twisted pairs as regards to bandwidth and transmission speeds. Two main categories exist: privately owned and public. In both cases modems are required to implement the interface between the computer-based equipment and the line. Modems perform basically modulation of digital signals to transmit information across the lines and demodulation of the received alternate signal to convert it to digital information. When using the public switched telephone network (PSTN) modems must be of an approved model by the telecommunications authority.

Facilities must exist to allow a sending end to dial the receiving end code number, controlling the dialling sequence and repeating it if access is not obtained at the first attempt. It is also necessary to a receiving end to recognize its own code when it appears on the line. This is implemented through auto-dialler devices. When PSTN are used, rules exist that impose certain restrictions of access to the line that must be respected by the auto-dialling equipment.

4.1.3 Coaxial cable. It is a medium with very good immunity to electromagnetic interference due to its shielded configuration. It allows high data rates, in the range of megabits/sec., though signal attenuation increases with transmission speed. This imposes that beyond a certain interval repeaters must be installed to regenerate the signal to be transmitted.

4.1.4. Power lines. Though power lines are a noisy medium, it is possible to use it for data transmission and many suppliers actually offer this option for EMS. The principle of operation consists of superimposing through capacitive coupling an alternate signal of high frequency (typically 100-200 kHz) to the mains voltage, which is the carrier for a modulated information transmission. This modulation may be FSK or PSK and data rates are usually low, 2400 bits/sec. being a very good performance.

Some problems exist that require special handling when nower-line-carrier (PLC) techniques are used. One of them consists of the strong attenuation imposed by transformers to the carrier signal, due to the high impedance of the coils to high frequency waves. Receiver/transmitter pairs are needed to pick up the signal on one side and re-inject it on the other side of the transformer. Also, capacitive coupling between phases is needed to assure the transmission across the whole facility.

The medium is very noisy due to high frequency spikes generated by commutations across the power network. As a consequence, error detection/correction procedures must be implemented which take account of the medium severity as regards to noise. This adds to slowing down speed performances.

4.1.5. Other media. Radio communications are more commonly found in remote control applications by the utilities. To the customer, they represent normally too high an investment and present difficulties due to legal restrictions on radio frequencies use.

Microwave links are better suited to long distance transmission in line of sight. They allow very high data rates. However, similar considerations as were produced for radio may be made in this case.

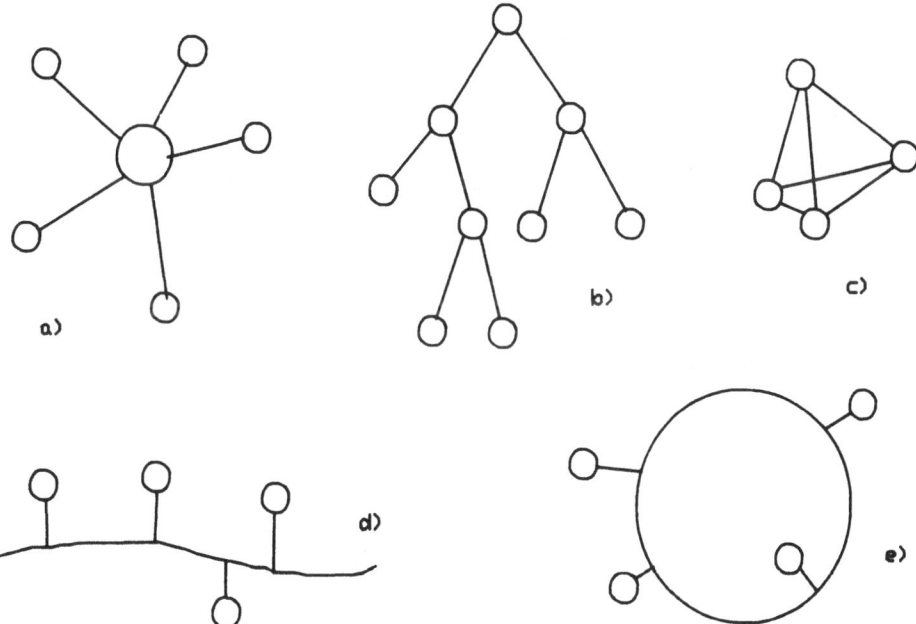

FIGURE 11. Network topologies.

Fibre optic links have virtually unlimmited bandwidth, allowing very high transmission speeds, and are immune to electromagnetic interference because the principle of operation is based on infrared light travelling through transparent fibres. It is, though, still too expensive to be cost-effective in situations where high data rates are not a must, as in energy management applications.

4.2 Network topologies

Two broad categories may be considered as to the use of the communication channels: the case where a dedicated channel exists for each pair of stations; when a single chanel is used by all the stations in the system.

In the first case, the structures may be : star, tree, complete. Figure 11 (a, b,c) illustrates these three types.

In the second, either the bus or the loop structure may be implemented (see Fig. 11 d, e).

4.3 Protocols

The International Standards Organization (ISO) issued a standard consisting of a set of rules to guide the implementation of communication networks for computers or computer-based stations - Open Systems Interconnect (OSI) . It defines seven logical layers for the hardware and software support of communications. In Fig. 12 a block diagram of the OSI model is presented, which is aimed at standardizing communications in order that integration of equipment from different manufacturers becomes possible. There are still many system suppliers using communication protocols of their own

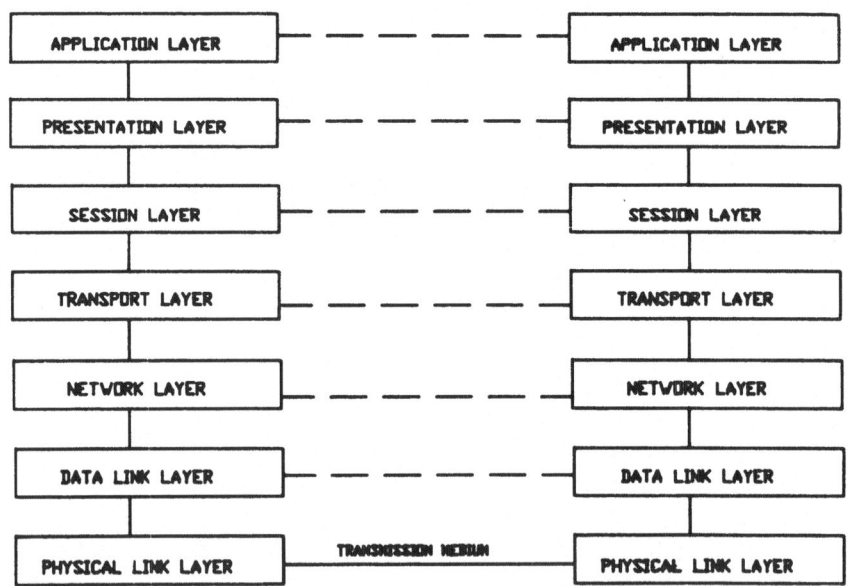

FIGURE 12. OSI standard.

design. The OSI standard will contribute to gradualy modify
this situation.

The physical link is the hardware connection between
stations (dedicated lines, power-line-carrier, or any other)
with the necessary interfaces and conditioning elements. The
physical link protocol is the definition of the medium
mechanical and electrical characteristics as well as the
control signal characteristics and methods requirements. The
physical layer performs all the functions associated with
signalling, modulation and bit synchronization.

The data link layer is the set of procedures that act
-ccording to a protocol, the data link protocol, defining the
transfer of bit frames across the network. This includes
delimiting and selecting frames, detection of noise, error
correction, transmission retries and resolution of contention
for use of the transmission medium. Conceptually, this layer
should provide an error-free environment for the network layer
above it.Examples of data link protocols are the High Level
Data Link Protocol (HDLC) -nd the Synchronous Data Link
Protocol (SDLC).

The network layer solves the problems of routing and
relaying packets of information across a single network or a
set of subnetworks, assuring that a certain message reaches
its destination, in an manner independent of the network
topology. It should free the superior layers of the routing
tasks.

Each layer within a station logically "sees" only the
corresponding layer in the addressed station, across the
network. Actually, data travel through all the layers in a
descending order within the former, then across the physical
medium and, finally, in an ascending order through the layers
of the latter.

In EMS applications, the superior layers are usually not
implemented.

5. PROBLEMS ASSOCIATED TO EMS USE AND SPECIFICATION

The very first step towards automatic energy management
implementation is perform an audit in the facility. Without a
rigorous analysis of the facility's behaviour it is not
possible to establish adequate specifications for what an EMS
should do. Hence, the problem of bad or over specifications
usually comes from an ill-conducted audit or of the total
absence of one. Too big systems or systems with the wrong
functions implemented may well be installed under these
conditions, later revealing a discouraging lack of ability to
save energy and/or money.

Cost-benefit analysis is also a key procedure, both before
and after the purchase of an EMS. Failure to do this usually
leads to surprisingly low rates of return on investments.

Acceptance tests should also be made to the EMS and
handover should never occur before solving all the
deficiencies detected.

Every facility justifying the installation of an EMS should
have an energy manager to solve the operational problems,

144

namely taking care of maintenance aspects.

Monitoring and trend-logging features are very important as EMS characteristics and should always be part of the supply. Without them it is difficult to adapt the EMS to the facility's needs.

Manual override functions are also an indispensable characteristic, that an EMS should never fail to offer. It may reveal very important when unexpected changes in the normal use of the facility arise. Care should however be taken in order to prevent a misuse of this characteristic, that could cancel effectiveness of control actions.

REFERENCES

1. Andrews, W.: How to avoid the ten most frequent EMS pitfalls. Energy User News, April 19, 1982.

2. Koch, E.:Building Services Automation Evolves. Power Engineering and Automation, vol. 8, July/august 1986.

3. Guggisberg, S.: Economies d'Energie avec les Systemes de Supervision des Batiments. Revue Landis & Gyr 27 (1980)2.

4. Akbari, H.: Use of Energy Management Systems for Performance Monitoring of Industrial Load-Shaping Measures. Lawrence Berkeley Laboratory publication, October 1986.

5. Thumann, A.: The Energy Management Systems Sourcebook. The Fairmont Press Inc., 1985.

6. No, J.: Sistemas Distribuidos de Control de Procesos. Automatica e Instrumentacion no. 163, nov. 1986.

7. Fielden, C.: Computer Based Energy Management in Buildings. Pitman Publishing Pty Ltd., 1982.

8. Canfield, K. Energy Monitoring and Control Systems - problems and potential solutions. Naval Civil Engineering Laboratory, 1982.

9. Martins, A. and Almeida A.: Load Management, an Efficient Tool for Energy Conservation. Proc. of the European Conference on Energy Economics & Management in Industry, Albufeira, Portugal, April 1984.

COOL STORAGE IN COMMERCIAL BUILDINGS; EFFICIENT AND COST-EFFECTIVE TECHNOLOGY

Olivier de la MORINIERE
COMPAGNIE GENERALE DE CHAUFFE, FRANCE
118, 120 rue de Rivoli - 75001 PARIS

INTRODUCTION

During the late 1970's and early 1980's, the demand for new electrical capacity leveled off, and in some utility areas, actually decreased. This anomaly was a result of increased emphasis on conversation, higher electric rates, and the general business recession. However, this condition is in the process of reversal. There is a practical limit to the amount by which electrical consumption will be further reduced by conservation efforts, ant it is likely that the impact of conservation may have reached its peak. The nation has experienced a substantial business recovery. The combination of these factors has led to a resumption of more normal patterns in electrical demands throughout the nation.

In the past, utilities have met new demand by building addtionnal generating facilities, with strong reliance, during the past quarter century, on nuclear plants. Nuclear power was considered by the utility industry as the long term solution to the dwindling availability and rising costs of fossil fuels. However, utilities are now faced with the fact that new nuclear plants have been regulated and priced out of the competitive marketplace. Additionally, environmental concerns have led to strong resistance, in many areas of the nation, to the construction of new generating facilities using conventional fuels. Conventional plants will inevitably cost more to construct, due to the measures which must be taken to diminish adverse environmental impacts.

Due to these changing circumstances, the utility industry is now giving more emphasis to alternative methods of generation, and perhaps more importantly, to new technologies which provide viable alternative to the need for additionnal generating capacity.

The technology of thermal energy cool storage is the most advanced, and appears to have the greatest potential of all presently conceived alternatives to the construction of new generating facilities. Optimum application of cool storage concepts will result in significantly more efficient operation of existing generating facilities. Cool storage can provide utilities with the equivalent of new generating capacity, at a substantial reduction in the cost per megawatt. An additional benefit derived from the use of cool storage is substantial off-peak load building. The technology has a most significant potential for improving utility load factors.

A. T. De Almeida and A. H. Rosenfeld (eds.), Demand-Side Management and Electricity End-Use Efficiency, 145–168.
© 1988 by Kluwer Academic Publishers.

UTILITIES TAKE THE INITIATIVE

No knowledgeable consumer is likely to invest in the additional capital costs associated with cool storage, unless he is assured of economic benefit. Since customer operating cost benefits from the use of cool storage are derived solely from the avoidance of demand charges and reduction of energy charges, the amount of savings will be largely dependent upon the electric rate schedules of the local utility. The payback which the customer will receive on his investment will not likely be adequate unless the rate structure has been specifically tailored to accommodate the optimum usage of cool storage. The rate elements which primarily influence customer-side economics are (1) magnitude of the demand charge, (2) length of the peak period during which demand charges are levied, (3) the specific hours encompassed by the peak period, and (4) the energy (KWH) charge differential between peak and off-peak. So customer-side economics of cool storage are largely beyond the customer's control.

In recent years, more and more utilities have come to recognize the benefits of encouraging   installation of cool storage systems in the commercial and industrial sectors. Utilities are taking the initiative by introducing time-of-day rates, demonstration installations, and, in several cases, financial inducement programs.

Utilities can evaluate and quantify the value of peak demand reduction, and off-peak load building, in terms of generating, transmission, and distribution costs —both short term and long term. They can then determine the degree to which it may be economically justified to   provide prospective users of cool storage a financial inducement to partially defray the additional costs of the storage systems. By modifying rate structures to accommodate the characteristics of cool storage, the savings realized by customers are increased, and the magnitude of such financial inducements required to assure customers of a cost beneficial installation is substantially reduced.

While many electric utilities are evaluating cool storage as an alternative to new generating capacity, and/or to build off-peak power usage, several have already initiated aggressive programs by offering financial inducements to customers installing cool storage. Inducement payments are calculated on the amount of demand (KW) shifted off-peak, and are paid to customers during construction or at project completion to help defray added construction costs. Utilities having such programs are:

Utility	Inducement per KW of Load Shift	Maximum per Project
Southern California Edison	$200*	$100,000
San Diego Gas & Electric Co.	Variable up to $350	$300,000
Texas Electric Utilities Co. (3 operating companies)	$350 - first 200KW plus $250 - next 300 KW plus $200 - next 500 KW plus $125 - all KW over 1000	No Maximum
Pacific Gas & Electric Co.	$300	$150,000
Arizona Public Service Co.	1st 500 KW - $250 plus $115 over 500 KW	No Maximum
City of Austin Power & Light	Variable - 3 year payback up to $300 per KW	No Maximum
City of Palo Alto	$350 partial storage*, $400 full storage*	$250,000

*Plus matching funds up to $5000 for feasibility studies.

The cool storage technology is rapidly gaining momentum, largely through the initiative of the electric utility industry, and particularly through the research and development efforts of the Electric Power Research Institute.

Thermal cool storage is a tested and proven technology which is beneficial to all - using customers, electric utilities, non-participating electricity users, and society as a whole. It provides improved utilization of existing resources, poses no environmental threats, and is not subject to regulatory constraints. Under this most favorable set of circumstances, cool storage seems destined to be the wave of the future for HVAC and industrial refrigeration.

COOL STORAGE FUNDAMENTALS

Cool storage is accomplished by cooling a substance (e. g., water) by mechanical refrigeration, using electricity during periods of low electrical demand. By various means, the storage medium is used to provide cooling of buildings or process loads during the daytime work period. Up to 80 to 85% of the electrical demand and consumption for air conditioning can be shifted to off-peak electrical demand periods by use of cool storage technology.

Cool storage can utilize either sensible heat or latent heat characteristics of a substance. Sensible heat is the heat added to or removed from this substance, resulting in a temperature rise or reduction. This heat is expressed in British Thermal Units (BTU). A BTU is defined as the quantity of heat required to raise the temperature of one pound of water through one degree Fahrenheit. Latent heat is the heat necessary to change phase of this substance without a change of temperature. For cool storage, this phase change is the transformation from a liquid to a solid state, or the reverse. The heat required to change from a solid to a liquid phase is called the latent heat of fusion. The latent heat of fusion equates to 144 BTU's required to melt a pound of ice at 32°F.

Materials that are subject to phase change for use in cool storage are called phase change materials (PCM). The most commonly used PCM is water - ice. There are other substances which can be used for both cool and heat storage applications. However, most of these other PCM's do not approach the heat capacity of water-ice.

There is ample literature in the cool storage field, since it has been the subject of extensive research at national laboratories, universities, and private research facilities. The Electric Power Research Institute has provided notable leadership in furthering the development and application of the technology. Numerous technical reports have been published by EPRI on cool storage, and ASHRAE has made notable contributions through its Technical Committee on Thermal Storage (TC 6.9).

The most commonly used methods for achieving cool storage are as follows:

Chilled Water Storage

Water is chilled to 40-42°F at night and stored in a large tank for circulation to fan coil units in the building to provide daytime cooling. The primary disadvantages of this medium are the requirement for a large volume of storage capacity, and the design difficulties associated with maintaining proper separation of the cool water and warmer return water.

Ice Storage

Water is chilled to 32°F and frozen in order to obtain the enhanced storage capacity associated with a change of phase. The volume of the container required for ice storage is only 25 to 30% of that required for chilled water storage. A new segment of the technology has recently received attention, in recognition of the construction cost savings which can be achieved in new buildings from the availability of the colder water output of ice storage. This involves reduced sized pipes, pumps, and air ducts. EPRI has a project underway to evaluate and quantify the designs which would provide optimum benefits.

Eutectic Salt Storage

This is a relatively new storage medium, not yet in common use. It has the advantage of a freeze point of 47°F, which reduces energy requirements. Several eutectic systems have been installed, or are under construction in commercial and institutional buildings in Southern California, and operated successfully. Eutectic systems are also being used to cool commercial aircraft on the ground at major airport terminals in the East and Midwest.

Extensive and continuing research is being performed at laboratories throughout the nation seeking even more efficient storage mediums. The presently available methodologies perform very effectively and provide reliable cooling if properly designed and operated. However, none have all of the ideal characteristics for which researchers have been seeking, and first costs of the systems are normally greater than conventional air conditioning, particularly for retrofits. The cost effectiveness of cool storage (payback/life cycle costs) is largely dependent on electric rate structures - the length of the peak period when demand charges per KW are imposed, the magnitude of demand charges, and time-of-day energy rates. More and more electric utilities are adopting time-of-day rate structures to encourage customers to use cool storage and other demand side management technologies. In many areas, only partial storage systems are cost effective because the rate structures are not yet designed to best accommodate optimum sizing.

COOL STORAGE IS NOT A NEW CONCEPT!

The cool storage technology has been used successfully for over 60 years. But despite the obvious benefits, many building owners and engineers have been reluctant to utilize what they perceive to be a new, relatively untried concept. They do not appreciate that ice storage systems have been in common use in dairies throughout the United States for many years. A small refrigeration system is used to build ice - which is then used to chill milk during a short cooling period. The advantage was the reduction in size and cost of refrigeration equipment - energy cost savings were not an initial economic objective. This same concept was soon applied in theaters and churches in the South and Midwest. Again, small, inexpensive refrigeration systems provided stored cooling to handle proportionately large cooling loads over short periods.

With the advent of substantial increases in electricity rates over the past decade, cool storage has become an especially attractive choice for commercial buildings and industrial processes. Particularly since electric utilities, in recognition of the benefits to the utility industry, are adopting rate schedules and programs which make cool storage even more economically beneficial to using customers.

A DYNAMIC TECHNOLOGY

As a consequence of the mounting interest in cool storage, new and improved products are becoming available. Ice systems are now commercially available in several different forms - static ice-on-coil, water glycol charged solid ice, dynamic ice shuckers, and slush systems. Eutectic ice is now available as a storage medium, both for air conditioning and cold temperature applications (such as frozen food warehouses). A clathrate ice system is expected to be commercially available in the near term. All have differing operating and performance characteristics. A knowledge of the types and characteristics of available storage mediums and equipment will enable users to select the system best suited to individual applications.

ELECTRIC LOAD CURVE IMPACTS

The benefits of cool storage are derived from the impact of this technology on a building's daily electric load curve, primarily by a reduction in total electric demand. The choice of storage design and operating mode is a key decision which influences both first costs and operating savings.

Conventional System Design-Day Load Profile

The cooling load on the design-day, or warmest weather day, is of critical importance. This establishes capacity requirements of major components for both conventional and storage systems. Non-cooling loads are fairly uniform throughout the year. Variable cooling loads are superimposed on the non-cooling loads to create excessive electrical demand (KW).

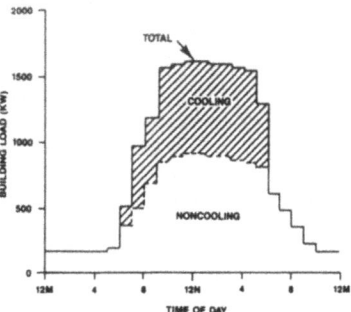

Full Storage Mode

The objective of full storage is to shift all of the refrigeration load out of the period of building occupancy. While full storage will normally achieve maximum demand reduction, it is seldom economic, since both refrigeration and storage components must be significantly larger than required for the other options. Full storage may be cost effective for facilities with cooling loads of relatively short duration (i.e., schools) particularly if utility demand charges are high.

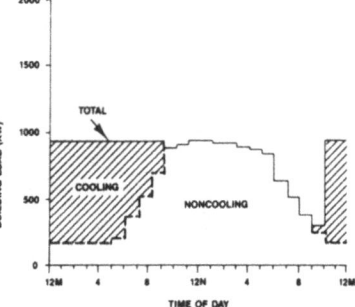

Partial Storage Mode

With partial storage, the refrigeration equipment is sized to meet design day cooling with continuous (24 hour) operation. On hot days, the storage component is charged at night, with daytime cooling provided both from storage and directly by the refrigeration equipment. Cooling electrical demand will be reduced by about 50% on hot days. Under moderate weather conditions, this mode may satisfy the entire cooling load from storage. First costs are the least with partial storage, due to the relatively small size of both the refrigeration package and storage containers.

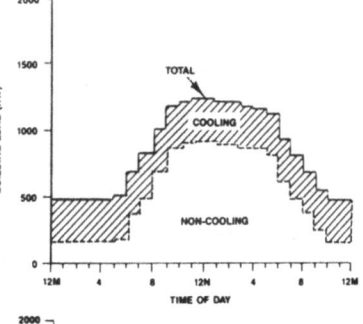

Demand Limited Storage

This mode uses a sophisticated method of controlling output from the refrigeration equipment, to operate during "shoulder periods" at a level adequate to meet part or all of the building load, without causing the total building demand to exceed the peak demand of the non-cooling loads. This strategy can substantially reduce refrigeration and storage capacity requirements, while achieving essentially the same cost savings as are derived from full storage. However, it does require more careful design and complicated load control equipment.

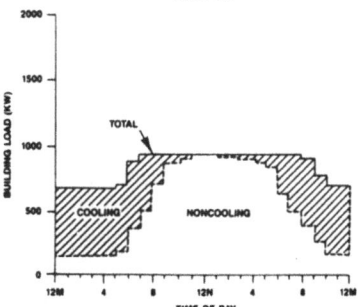

Modified Demand Limited Storage Mode

The modified DLS mode is used to best accommodate to a time-of-day rate schedule, when the utility's "peak period" (with high demand charge) is relatively short (i.e., 6 to 8 hours). This strategy provides full storage during the peak period. During hot weather, the refrigeration equipment charges the storage overnight, and meets the cooling requirements during "shoulder periods", direct to load. During the peak period, refrigeration equipment is "locked-out", and all cooling is provided from storage. Demand and energy cost savings are maximized, both refrigeration equipment and storage capacity are of moderate size, and controls are relatively simple.

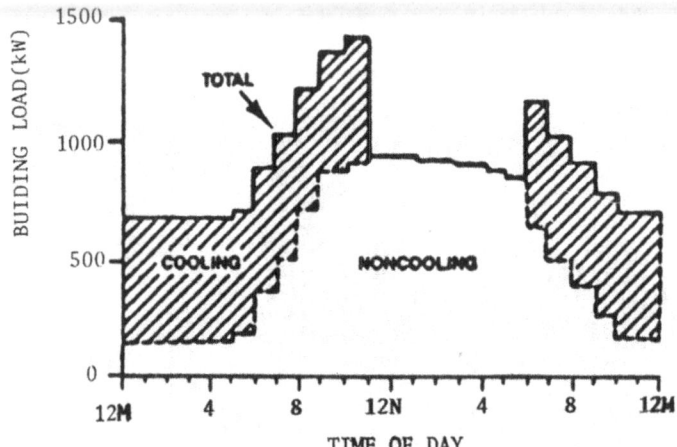

System Comparisons

The following figure shows the relative sizes of refrigeration equipment and storage component, and the refrigeration electric load profile, of each of the storage modes as compared to a conventional system.

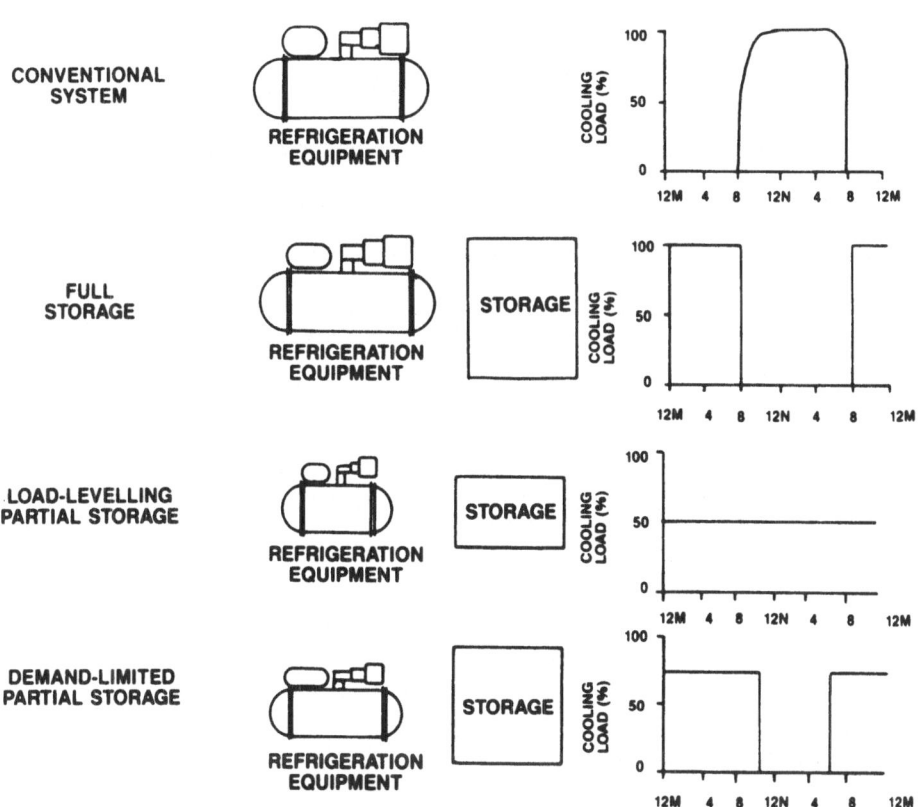

Comparison of cool storage systems

The determination as to which mode is best for a particular application is largely a matter of economics. The foremost consideration is normally the rate structure of the local utility. Other significant factors are length of the cooling season and whether the application is a retrofit or a new construction. Partial storage systems are usually the most cost effective in utility service areas without time-of-day rates, or where the peak period (when high demand charges apply) is long (i.e., 11-13 hours). The modified demand limited mode is favored by utilities with financial inducements, since significantly more demand is shifted off-peak. These utilities typically have time-of-day rates with relatively short peak periods. The short peak period, combined with the financial contributions, will usually make this mode the best design option.

THERMAL STORAGE IN NEW COMMERCIAL BUILDINGS

In order to cool a building, we can choose from among three strategies :

1. A conventional chiller which runs about eight hours daily and mainly when electricity is scarce and expensive.

2. "Partial storage", in which a chiller that is three times smaller runs continually and stores excess coolth for use during hot afternoons. Of these two storage strategies, this has the least first cost.

3. "Demand limited storage", in which a somewhat larger chiller runs only off-peak. This also requires somewhat larger storage and (from the builder's perspective) may not be worth the extra expense.

Choosing among these strategies means comparing three options : (1) installing a partial storage instead of a conventional system, (2) installing a demand-limited storage instead of a conventional system, (3) climbing the step from partial storage up to demand-limited storage.

This paper compares these three options.

Parameters of This Study

Because costs of water or ice storage are roughly similar, we do not differentiate between these storage media.

For cold water storage, piping and ducts whithin the building are of course independent of the presence of thermal storage. For ice storage, which supplies colder water, there is a small saving from down-sizing pipes and ducts, but we have included these by slighty reducing the quoted costs of ice storage.

Design Parameters. These design parameters should be calculated from the precise load profile of the individual building.
Supposing the cooling profile is almost flat, we can easily calculate, as shown in figure 1, significant values of these parameters. We adopt these parameters, which are summarized in Table 2.

Economic Parameters. As in any economic study, the two main parameters are the incremental first cost and the savings in the cost of operation. More precisely we consider the average first cost for a conventional chiller, a partial storage, or a demand-limited storage, and for each local electric utility, the annual dollar savings from shifting a kW of cooling from peak to off-peak.

For conventional chillers, Figure 2 shows that chillers cost \$336/ton (+ a constant) installed. For storage, Figure 3 shows cost quotes based on phone calls to manufacturers an engineers who have installed coolth storage. We see that the cost per ton installed, c, falls in the range of \$40-100/ton hour. In the analysis below, we allow the cost parameter, c, to cover this range. This does not include cost of the chiller.

Annual Dollar Savings. The savings, due to cool storage, should be esti-
mited by a thorough calculation using the cooling load profile of the
building and the electric utility rate and structure. We plan to soon
use a DOE2 simulation to estimate, on a day by day basis, these savings.
But what we will present here, once we know the utility rate and structure, is an easy way to get a precise idea of these annuel dollar savings.

Although energy consumption with storage can strongly influence the cost
of operation (systems using cool storage can have a lower efficiency and
control problems, thus increasing energy consumption, or, on the other
hand, these could take advantage of the lower night outdoor temperatures
and of a chiller used mostly at full load, thereby decreasing energy
consumption) we will consider that the system with a storage has the
same efficiency as a conventional chiller, thus resulting in the same
energy consumption.

The local utility rate influences savings in two ways : by the power
charge in $/kW per month, which we will call "p", and by the differential
cost of energy, between on-peak periods and off-peak periods, in
cents/kWh, which we call "e". We will use a parameter that is a
special combination of both, and is representative of how much the
utility rate favors the installation of cool storage.

Let us suppose that one kW of coolings, used 6.5 hours on-peak, is
shifted to 6.5 hours off-peak for six months, 20 days per month. The
annual dollar savings, expected, are :

1. annual power charge savings (in $) : 6 months x p = 6 p

2. annual energy charge savings (in $) : 6 months x 20 days x 6.5 hours
                                          x e / 100 = 6 x 1.3

3. annual total savings (in $) : 6 p + (6 x 1.3 x e) = 6 (p + 1.3 e)

We will now use this new parameter, "Su" = 6 (p + 1.3 e) named "annual
dollar savings due to one kW of cooling shifted from on-peak to off-peak".
"u" (for utility) indicates the variation of Su from one utility to
another. Figure 4 shows that, among neighbouring electric utilities,
wide differences apperar in this parameter.

For example, an investment leading to the shifting of one peak kW for a
utility (LADWF) rate schedule will induce annual savings of $32, or $57
for another utility (PG&E) and $67 for a third one (SDG&E).

This new parameter "Su" could be made more precise by taking into account
the real peak period, in months, days, and hours. We think the above
defined "Su" is more accurate and easier to use.

Economic Evaluation

We calculate three economic criteria : (1) Simple Payback Time (SPT), (2)
Net Present Value (NPV), both of which are popular with customers, and (3)
cost of a conserved kW, which is popular with utilities and policy-makers.

Simple Payback Time (SPT). Using the design parameters of Table 1, a
range of "c" from Figure 3 and "Su" from Figure 4, the payback times for
options 1 and 2 are set forth in Table 3.

For option 3, the hardest hurdle (going beyond Partial all the way to Demand Limited), SPT = Δ n/Δ d where n = numerator of row 10 and d = denominator of row 10, so SPT = (2,95 c + 52) / 0.396 Su. Numerical values for three utilities are given in Table 4. For other utility rates ("Su") and average cost of storage ("c"), the payback time can be found in Figure 5.

It appears that the payback time of demand-Limited storage is always bigger than that of partial storage and the difference, in years, is close to $170 / Su (so 5 years fors LADWP, 3 yeras for PG&E, and 2.5 years for SDG&E), irrespective of the average cost of storage.

Net Present Value (NPV). Although convenient, SPT can not compare projects with long equipment lifetimes. Since cool storage can last for 20 years, the net present value criterion, in this case is, more appropriate.

The net present (NPV) values are calculated as follows (ignoring rate increases in the future years) :

	option 1	option 2	option 3
Incr. First Cost	5.25 c - 120	8.2 c - 158	2.95 c + 52
Annual savings	0.659 Su	1.055 Su	0.396 Su
Net Present Value	0.659 s (fi)	1.055 s f(i)	0.396 s f(i)
	- 5.25 c + 210	-8.2 c + 158	-2.95 c - 52

Where $\quad f(i) \quad = \quad (1 = \dfrac{1}{(1 + i)^{n}}) \dfrac{1}{i}$

In Figure 6, we take representative values of Su = 55 $ /kW an c = 55$/ton hour, then we can drax the net present value of partial storage and demand-limited storage according to the real discount rate. For these values, we see that partial storage almost always has a positive net present value (even if i = 25 %), that demand-limited storage has a positive net present valeur for a real discount rate lower than 19 %, and that for a comnonly used real discount rate (8 %) the two strategies compete.

But what will happen, if for example, we choose a lower storage cost (&40/ton hour) and have a more attractive electric rate ?

Figure 7 shows, according to varying "c" and "Su", the real discount rate which equates the net present values of partial storage and demand-limited storage. It is clear that in most cases the net present values of a partial storage project will be higher than the net present value of the demand-limited storage.

Investment Per Peak kW Saved. Electric utilities are interested in saving peak power for two main reasons. It leads to improvements in overall electric generation efficiency, by increasing the demand on baseload, and by delaying the need for new generation, which is estimated to cost in the future at least 1200-1500 $/kW.

Figure 8 compares, as far as first cost per kW saved in concerned, all the three options. Even with a high cost of storage ($100/ton-hour) and the most expensive strategy of demand-limited storage, cool storage is far more attractive for utilities than building new generation.

Our analysis so far has used an engineering parameter, c, for the first cost of storage on which our conclusions are based. But what about real buildings? We are compiling a data base (BECA LM), which we are confident of the incremental cost of about of the buildings. This is a biased sample since proud owners are more willing to talk than unhappy ones. Unfortunately, we do not as yet have energy bills for all 12, so we cannot vouch for their efficiency. Despite these uncertainities, their first cost is plotted in Figure 9 and we see that about half of these projects are clustered around an incremental first cost of zero.

COMPARISON OF WATER, ICE, CONCRETE, AND PHASE CHANGE MATERIALS

In the analysis above, we have focused on water and ice, because they are the only widespread successes in the U.S., but we should not ignore storage materials that will perhaps turn out to be cheaper in the near future, e.g., concrete floor/ceiling slabs (where room air is circulated through hollow cores), and phase change materials (PCMs). In Stockholm today, most new buildings have thermal storage : one-third of them use the hollow cores of concrete, one-third use water, and only one third still hook up to the Stockholm district heating system (Anderson et al. 1979). Swedish proponents of thermal storage assert that there is no increase in first cost for either the concrete or the water strategy.

PCMs are in their infancy but will of course be in use when their reliability is established and their price is competitive. We point out further that the value of heat capacity is nonlinear, the most accessible mass (e. g;, the surface of a wall or a duct) being far the most useful. Hence, we have no doubt that the cheapest systems will use all of the materials shown in the columns of Table 5. In particular, we are enthusiastic about loading containers of PCM into the hollow cores of contrete floor-ceiling slabs. This material should of course change phase at 70-74 F (21-23 °C), so as to lock the slab at this temperature. Then during occupied hours, both winter and summer, the air will be locked just a few degrees above the slab temperature.

We have ignored rock-bed storage. We approve of using the heat capacity of rock or concrete but feel one should try to exploit the mass that surrounds each space and not a distant rock bed.

Table 5 summarizes the results of our study of water and ice and gives cruder values for PCMs and hollow core slabs. The following comments need to be made on this table.

Comments on the Columns

Column C : the optimum available PCM is a mixture of 50 % NPG (Neo Penthyl Glycol) and 50 % TMP (Tri Methylol Propane) with a heat capacity of about 35 Btu/lb and is available currently at about $0.55/lb Dr. Dave Benson in SERI hopes to develop a "Warm Ice" mixture of water and TME (Tri Methylol Ethane) as a cheap PCM with a heat capacity of about 80 Btu/lb.

Column D : see Anderson et al.(1979). Each 10-inch-thick slab has a heat capacity quoted in the paper : 100 Wh/(K*sq.m), but again we add 20 % for facade, partitions, and furniture. We assume a daily heat rise of 8 F over eight hours.

Comments on the Rows

Row 3 : this is the thickness of the thermal mass if it were spread uniformly over the floor space and responded with 100 % efficiency. For hollow-core slabs this is, in fact, the thickness of plank needed.

Row 3b, Column D : for partial storage 1.1 slab is of course ridiculous, so we took one slab. For demand-limited storage, 1.4 slabs of thermodeck is also unreasonable ; instead the cores of one slab would be partially filled with phase change material to increase the thermal capacity of the single slab.

Row 5 : the zeros in column D for hollow-core slabs have the following meaning: The additional first cost of using the hollow cores instead of conventional ducts seems to be insignificant (Shelley, Dean and Fuller for EEB Program at LBL, Feb. 1980.). Even in the case where we have to add some phase change materials, the savings from down-sizing the chiller should leave a negative first cost, which we call "zero".

Row 6 : (no kWh savings) assumes that the summer is so hot that a chiller has to be run   at night, and any
increase of COP at night is offset by thermal losses, and controls are 100 % efficient. So the savings come only from cheaper nighttime electricity.

Row 7 : (cheap night cooling) assumes that nights are cool during spring and fall (or just because the climate is mild). Cooling may still be needed, but it can come from running the cooling tower without the compressor, installing an evaporative cooler for the hollow-cores, etc. For hollow-cores this strategy can be used when the wet bulb is below 65 F, for water storage, the wet bulb must be below about 50 F, and for ice, this strategy will not work.

Row 6 and 7 : for intermediate climates, the theoretical savings should lie between lines 6 and 7. Allowance for realistic control problems will reduce the savings to 80 %.

CONCLUSIONS AND POLICY QUESTIONS

Partial thermal storage, using water or ice, is already cost-effective and should be used wherever there is the incentive of time-of-use meters and demand charges. More complete "demand-limited storage" is only attractive to the builder where peak charges are high. Water and ice will perhaps be supplemented or superceded by the use of hollow cores in concrete and by PCMs.

TABLE 1

U.S. Peak (and % of peak) for Cooling in buildings. Col. A (cooling) comes from Fig.3 of V. Rabl's paper, Col. B (Total Peak) comes from assuming that So. Cal. Edison's peak distribution is typical, and then scaling up to the U.S. total peak which we have rounded to 500 GW. Col C. (annual % of new construction) is % of square feet for comercial. % of dwelling units for residential. New construction is defined as net rate of growth of our stock, plus retirements). Col D. (New Annual Targetable GW) is the product of Col. A (GW for cooling) x Col. C (annual rate of construction), but the format 9.3. --➤ 6 takes into account that new buildings will use only 2/3 rds as much power for cooling. as our current stock.

	(GW)	A. Cooling Only (fraction of 500 GW)	(GW)	B. Sector Peak (fraction of 500 GW)	C. Annual Construction (% /year)	D. New Annual Cooling : targetable GW A. x C. in GW
Residential	(64.4)	(13 %)	(175)	(35 %)	2 1/2 %	1.6 ➤1
Commercial	(153.0)	(31 %)	(185)	(37 %)	5 %	7.7 ➤5
Buildings	(217)	44 %	360	72 %	—	9.3 ➤6
Industrial	?	?	140	28 %	—	?
Total	217	44 %	500	100 %	—	9.3 ➤6

TABLE 2
Sizing Cool Storage and Estimating Peak Reduction

	Chiller Size (tons)	storage Size (ton-hours)	Reduction in chiller Size	Reduction in Peak demand
Conventional chiller	L	0	0	0
Partial storage	0.375 x L	5.25 x L	62.5 %	62.5 %
Demand limited Storage	0.53 x L	8.2 x L	47 %	100 %

TABLE 3

Simple Payback Time (SPT) of Cool Storage

		Units	Conventional Chiller	Partial Storage	Demand limited Storage
(1)	Chiller Size	tons	1	0.375	0.53
(2)	Cost of Chiller (=(1)x336)	$	336	126	178
(3)	Size of Storage	ton-hs	0	5.25	8.2
(4)	Cost of Storage	$	0	5.25 c	8.2 c
(5)	Total First Cost (=(2)+(4))	$	336	126+(5.25 c)	178+(8.2 c)
(6)	Increm. First Cost (15 - 336)	$	0	(5.25 c)-210	(8.2 c)-158
(7)	Peak tons saved	ton	0	0.625	1
(8)	Peak kWe saved assuming COP=3.33 (=(7)x1.055)	kWe	0	0.659	1.055
(9)	Annual Dollar savings (=(8)xs)	$	0	0.659 Su	1.055 Su
(10)	Payback Time (=(6)/(9))	yrs	/	$\frac{(5.25c)-210}{0.659Su}$	$\frac{(8.2c)-158}{1.055Su}$

Notation: c= cost of 1 ton-hour of storage;
Su= annual dollar saved by shifting 1 kW from peak to off-peak.

TABLE 4				
Utility	Option	c=40	c=60	c=80
LADWP (Su=32$/yr)	1	0	5.03	10.06
	2	5.03	9.89	14.75
	3	13.41	18.07	22.72
PG&E (Su=57$/yr)	1	0	2.82	5.64
	2	2.82	5.55	8.28
	3	7.52	10.14	12.75
SDG&E (Su=67$/yr)	1	0	2.40	4.80
	2	2.40	4.72	7.04
	3	6.40	8.63	10.85

Simple Payback Time, SPT in years, for three utilities, and three values of c=cost of one ton-hour of storage. Notation: Su=annual $ saved by shifting 1 kW from peak to off-peak.

TABLE 5
Thickness, Costs and Savings for Several Thermal Storage Media
(*Means see notes below tables)

Medium	Units	A Chilled Water $\Delta T=18°F$ 40°-->58°	B Ice H=144 Btu/lb $+\Delta F=20°F$	C Phase Change Materials H=35 Btu/lb	D Concrete Hollow-Core Floor Slabs (1 ft. thick)
1. Peak cooling load (thermal)	Wt/ft²	10	10	10	10
2. Electric cooling load (COP = 3)	We/ft²	3.33	3.33	3.33	3.33
3. Thickness needed of thermal mass a) partial	inch	2.2	0.22	1.2	1.1 slab*
b) demand-optimized	inch	3.15	0.31	1.73	1.4 slabs*
4. Remaining peak demand a)	%	37.5	37.5	37.5	37.5
b)	%	0	0	0	0
5. First Cost 1. $\frac{\Delta s}{kWe}$ displaced a)	$/kWe		0-340	250	0
b)	$/kWe		170-500	420	0
2. $\frac{\Delta s}{ft^2}$ a)	$/ft²	0.56-1.65	0-0.70	0.52	0
b)	$/ft²			1.40	0
6. Annual electric $ savings assuming SDGE tariff and NO Kwh savings a)	$/kWe	55	55	55	55
b)	$/kWe	55	55	55	55
7. Annual electric $ savings assuming cheap night cooling a)	$/kWe	100	Not Applicable	100	100
b)	$/kWe	100		100	100
8. Annual return on investment (SDGE) a)	%		17-∞	22	∞
b)	%		11-32	13	∞

$$\left[8 = \frac{6}{5.1} \right]$$

160

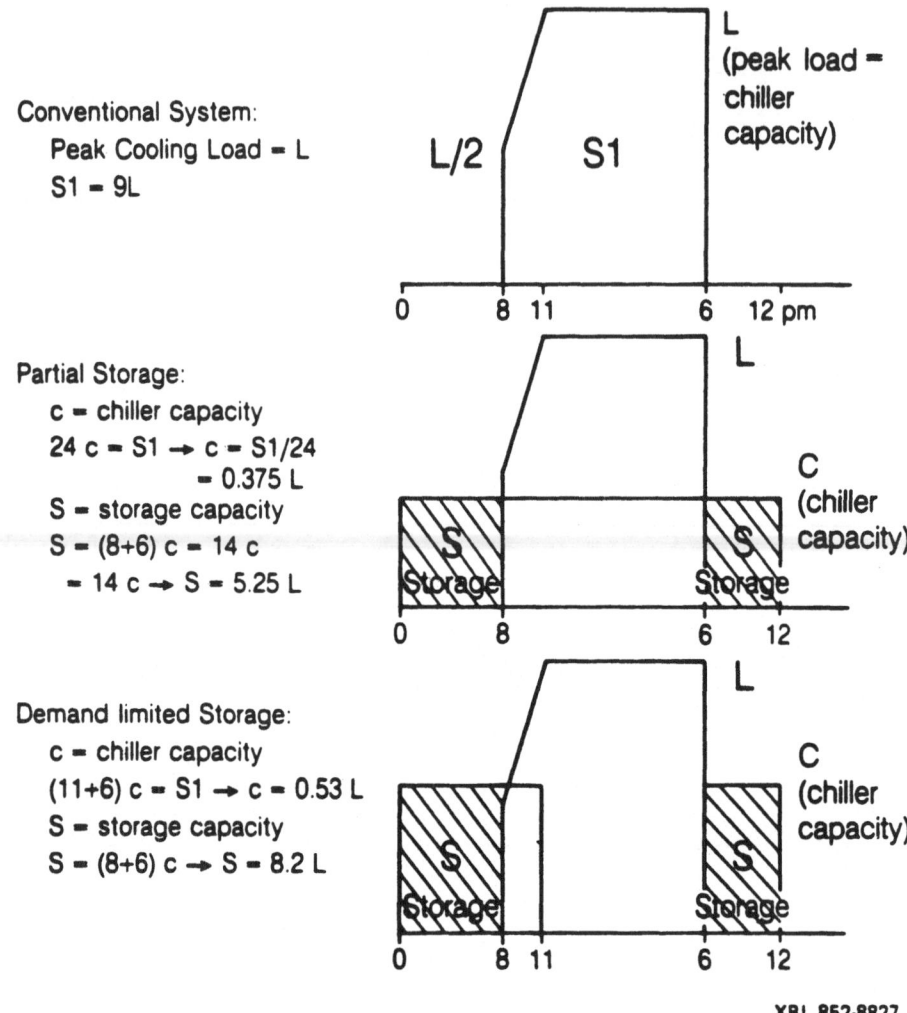

Conventional System:
  Peak Cooling Load = L
  S1 = 9L

Partial Storage:
  c = chiller capacity
  $24 c = S1 \rightarrow c = S1/24$
  $\qquad = 0.375 L$
  S = storage capacity
  $S = (8+6) c = 14 c$
  $\quad = 14 c \rightarrow S = 5.25 L$

Demand limited Storage:
  c = chiller capacity
  $(11+6) c = S1 \rightarrow c = 0.53 L$
  S = storage capacity
  $S = (8+6) c \rightarrow S = 8.2 L$

XBL 852-8827

Figure 1. Calculation of the capacities of chiller and storage according to the load profile.

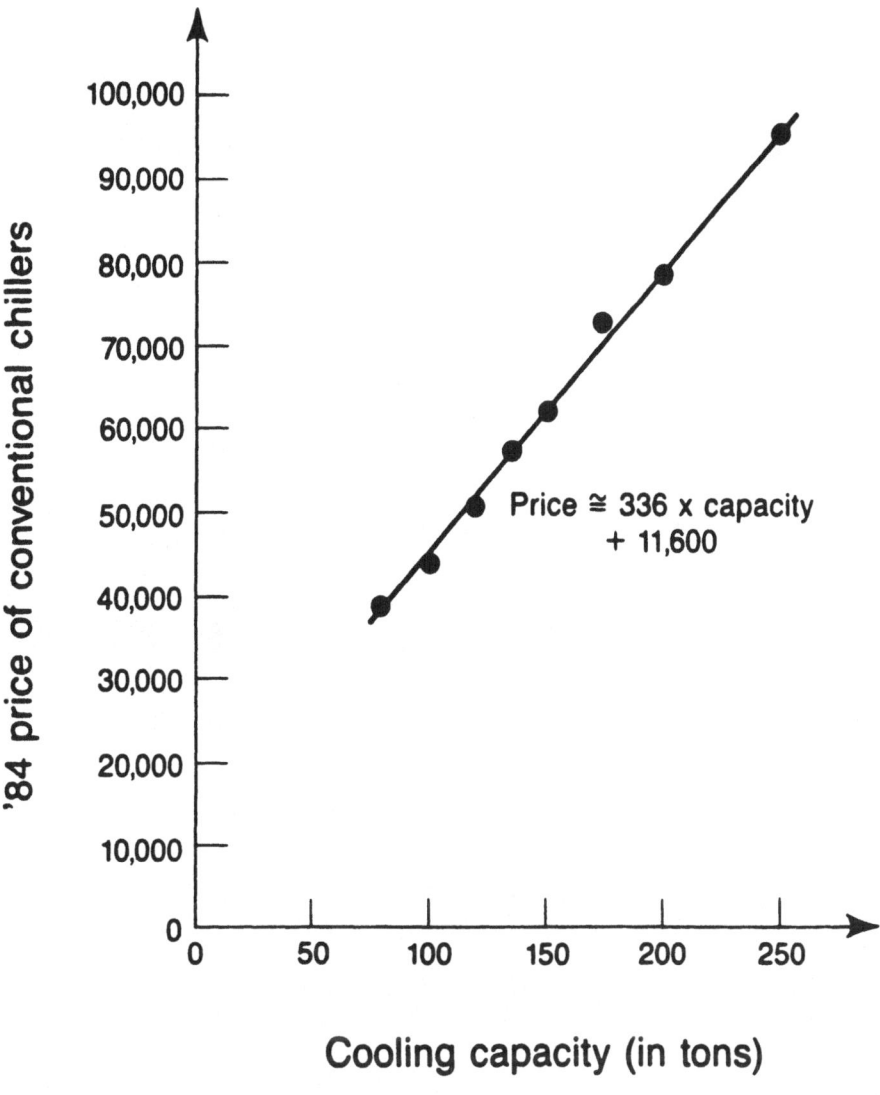

Price ≅ 336 x capacity + 11,600

Cooling capacity (in tons)

XBL 852-7066

Figure 2. Installed average cost of conventional chillers (in ´84$), including profit, according to the capacity of these chillers.

Source: "Means Construction cost data 1980" and x 1.25 for inflation from 1980 to 1984.

162

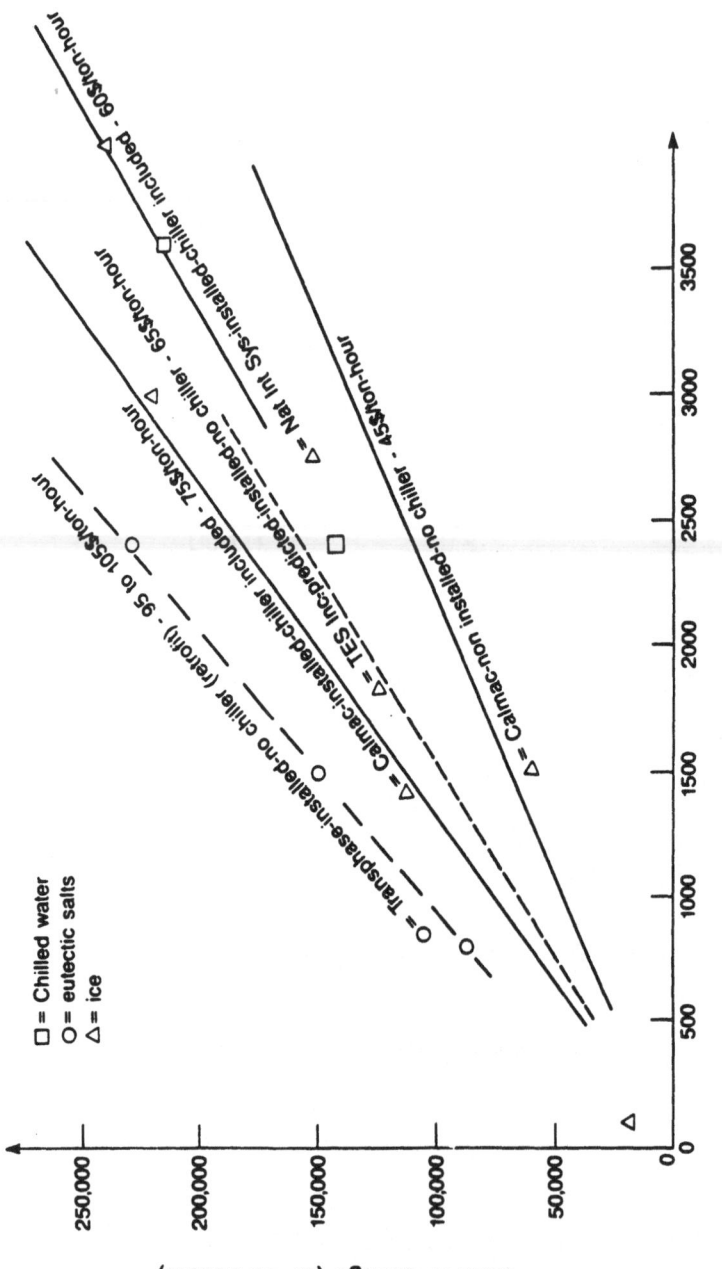

Capacity of the storage (in ton-hours)

XBL 852-7067

Figure 3. Installed Cost of Storage, quoted by manufacturers (lines) and engineers who have installed coolth storage (dots). Note that some lines include the chiller, some do not, some include installation, some not.

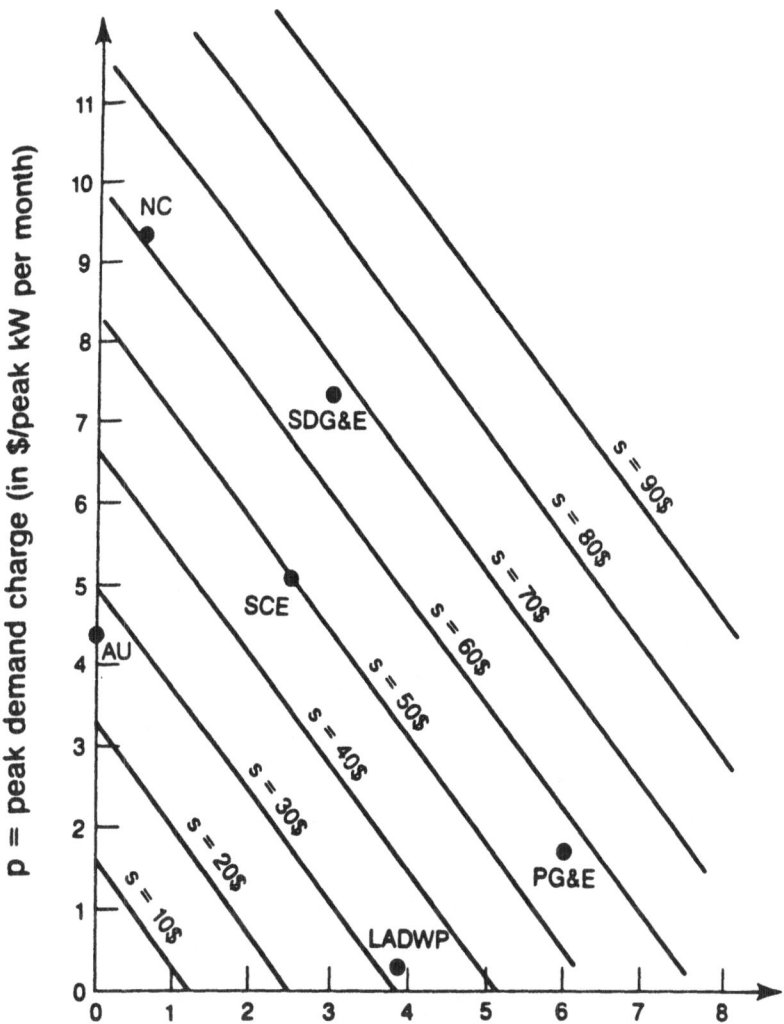

e = differential energy charge (on peak - off peak) (in ¢/kWh)

XBL 852-7060

Figure 4. Annual dollar savings by shifting 1 kW for 6.5 hours on-peak
to off-peak for many utilities, characterized by their monthly power
charge "p" and their energy on-peak off-peak differential cost "e".
Equation used was Su=6(p + 1.3 e).

164

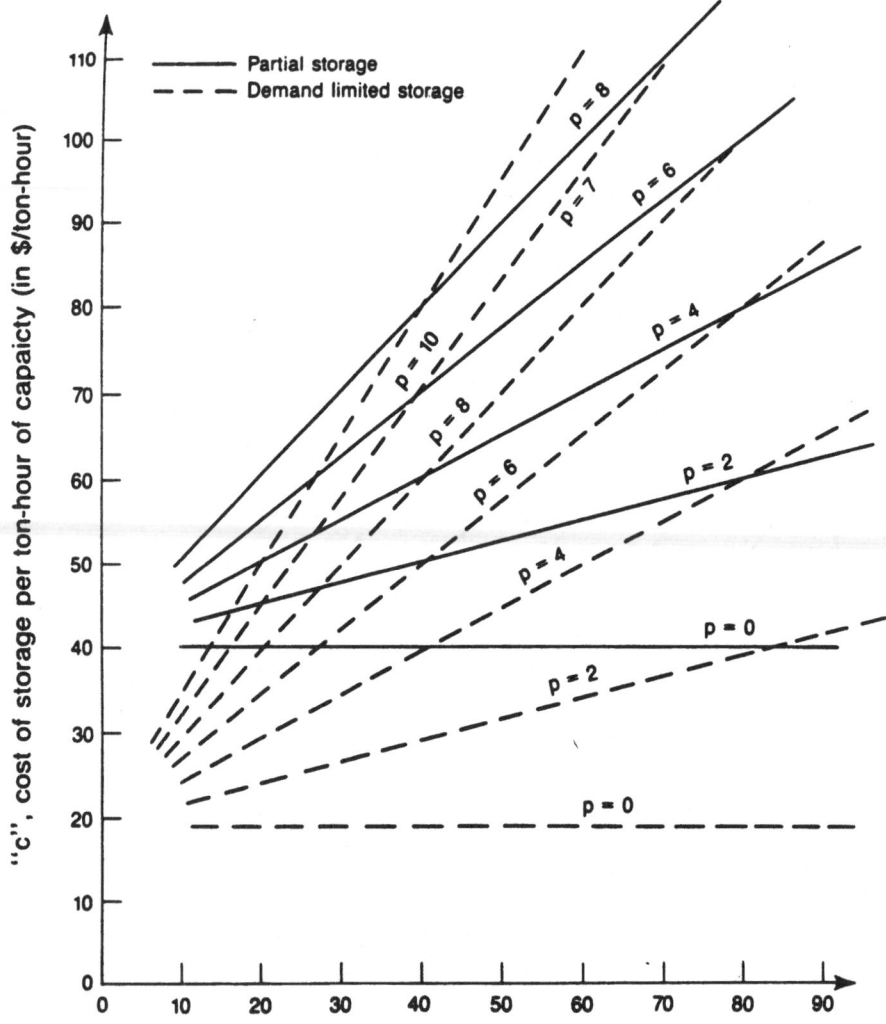

"s", annual dollar savings due to 1 kW shifted from peak (in \$/kW)

XBL 852-7063

Figure 5. Simple Payback Time for both strategies of table 2, ie plots of line 10, solid lines for partial, dashed for demand-limited.

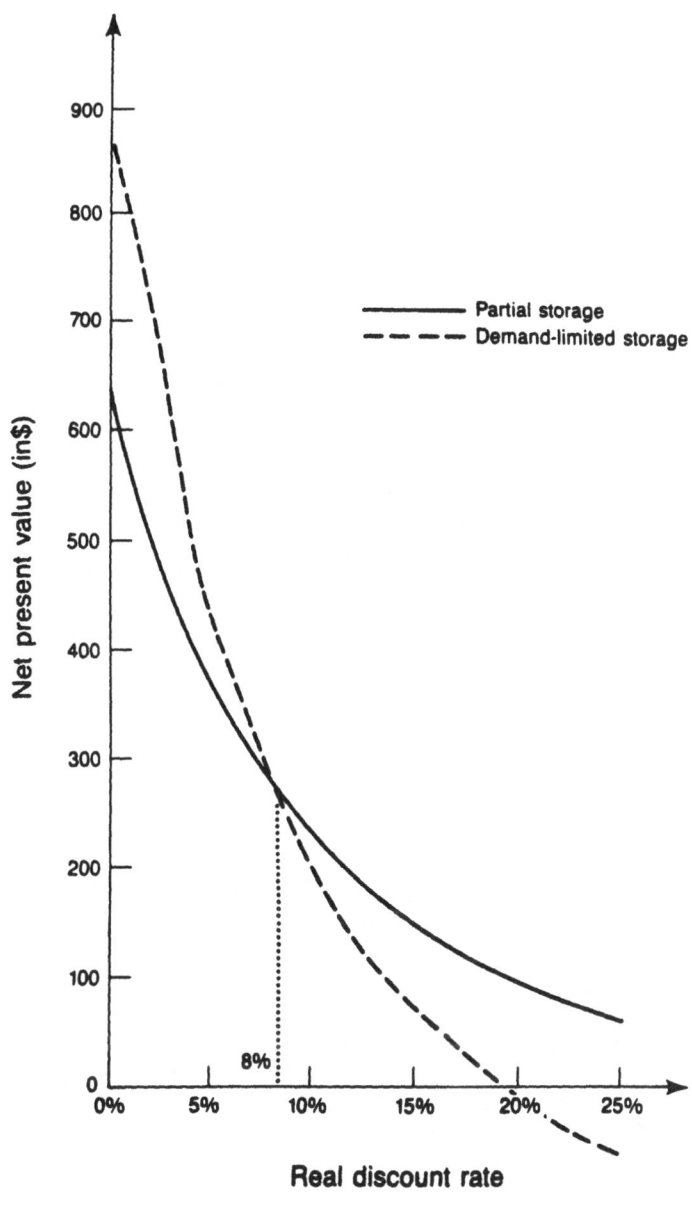

Figure 6. Net Present Values of a partial storage and of a demand-limited storage, assuming the first cost of the storage was 55$/ton-hour, not including the savings due to downsizing the chiller, and the annual dollar savings due to one kW of cooling shifted from peak was 55 $/kW. Storage was supposed to last 20 years.

166

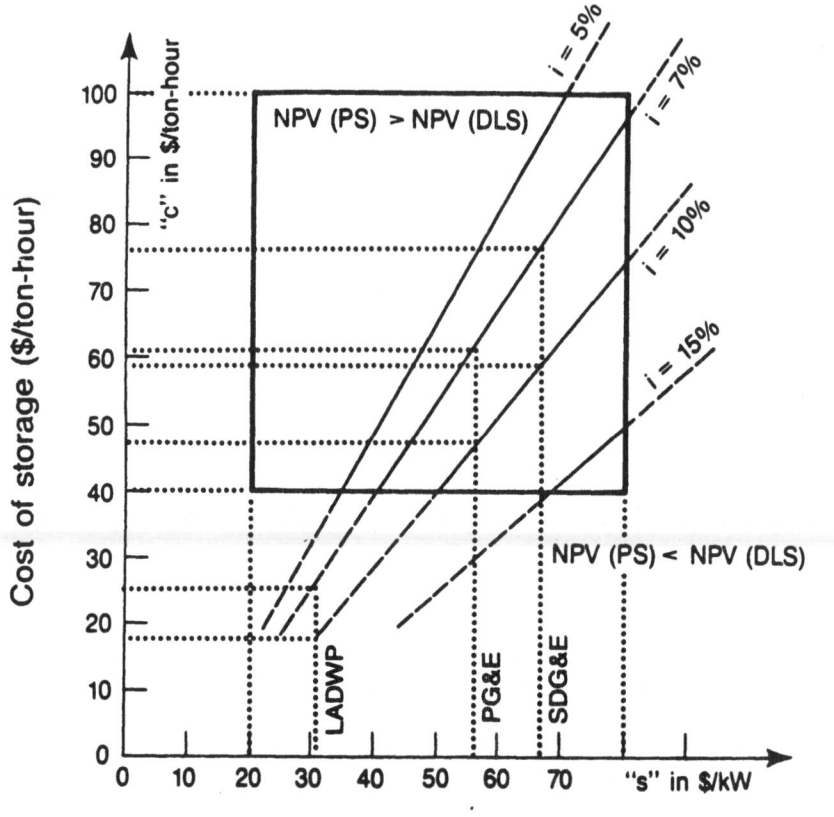

XBL 852-7061

Annual dollars savings from shifting 1 kW to off-peak ($/kW-year)

Figure 7. Net Present Value of partial storage (NPV(PS)) and demand-limited storage (NPV(DLS)) according to the real discount rate ("i"), to the average cost of storage per ton-hour of storage ("c") and to the annual dollar savings due to the shifting of one kW of cooling shifted from on-peak periods to off-peak periods ("Su"). 20 to 80 $/kW is today's range for "Su", and 40 to 100 $/ton-hour for "c".

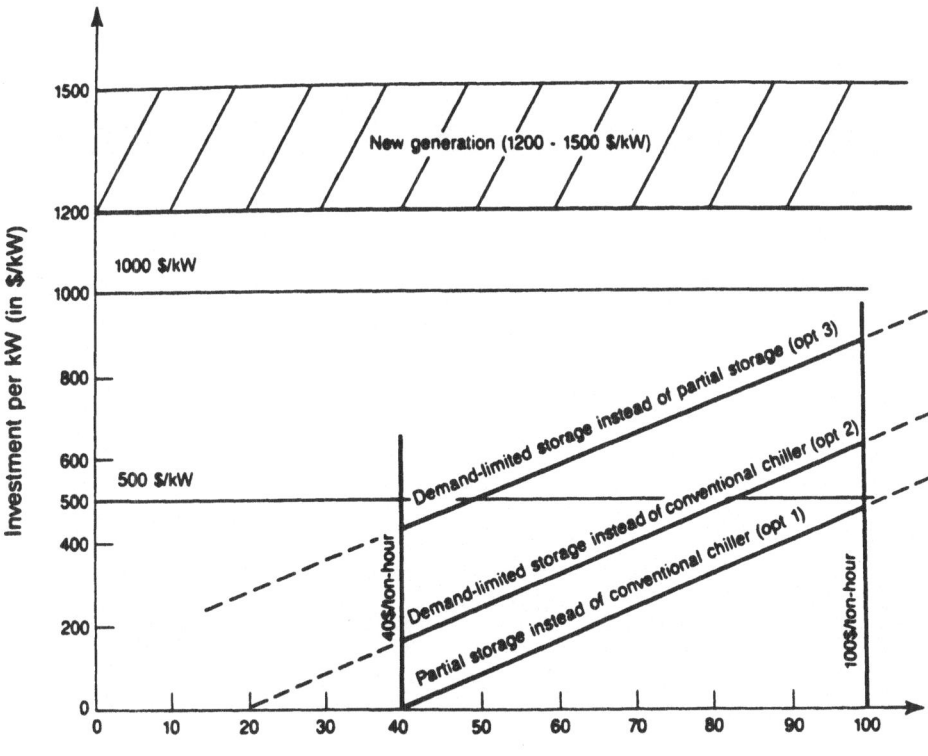

"c" = average cost of storage per ton-hour of capacity (in $/ton-hour)

Figure 8. Investment necessary to save one peak kW by using cool storage technologies, according to the average cost of storage not including the savings from downsizing the chiller. 40 to 100 $/ton-hour is a representative range for today's available technologies. These costs are compared to the estimated cost of new generation (sources PG&E).

168

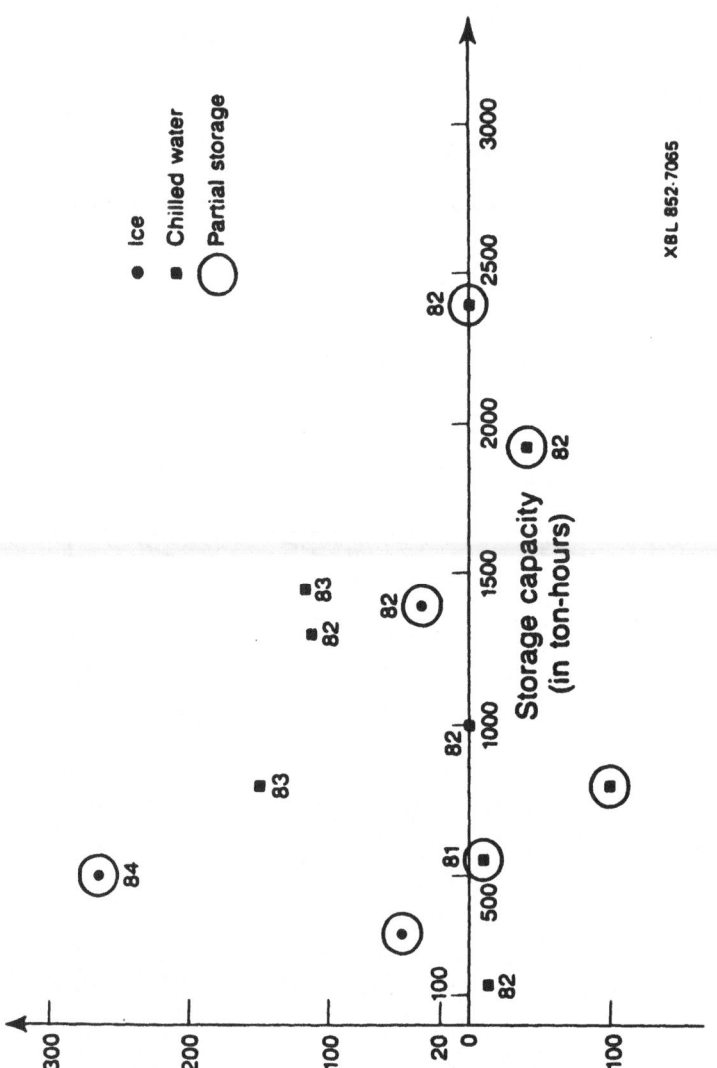

Figure 9. Incremental first cost (installing storage instead of a conventional chiller) from case-histories in real buildings. This costs
are not representative of the average first costs faced by owners when
installing a cool storage systems, but show that compensating the first
cost of the storage only by the savings from downsizing the chiller, is
already possible on the field. The data comes from a data base being
built at LBL, in the BED Group.

III.   Prices and Metering

SMART METERS

R. A. PEDDIE
Consultant. (Previously Chairman S.E. ELec Board, England.)

1. INTRODUCTION.

With electrical energy such a integral part of modern
living and so necessary for economic progress, higher energy
costs, both actual and projected, have adversely impacted on
the public's perception of utility management. In response
utilities have taken increasing interest in demand side
management and end use efficiency with the dual objectives of
improving the economics of supply and reducing the customers
energy costs. This in turn has created a need for flexible
metering systems which can communicate between the customer /
utility.

The expression "smart meters" have been coined to cover
this new requirement, whilst such expression may be of use in
the marketing or advertising arenas it is of little use in
formulating management policy to cover the years ahead. In an
era of rapid social, economic and technical change it is
essential to carry out a thorough examination of (1) all the
present and projected influences on the supply and utilisation
of electrical energy, (2) possible changes in customer /
utility relationships, (3) identify elements which will not
change, and (4) postulate time scales, before attempting to
formulate future policy on which the specification of a "smart
meter" would be based.

2. INFLUENCES ON THE ECONOMIC SUPPLY OF ELECTRICAL ENERGY.
2.1 Criteria for the supply and sale of electricity

The fundamental criteria which have evolved over the years
for the supply and sale of electrical energy are unlikely to
change these are:-
* Prices should be related to costs
* No price discrimination between customers
* A safe an reliable supply
2.2 Evolution not revolution.

Because it involves the whole of society the
characteristic of any change at the customer / utility
interface is that it will be evolutionary in nature with an
implementation phase of at least a decade, this characteristic
has a bearing on the economics of many policies deploying
transient technological "fixes" to current problems.
2.3 Types of change

The future changes which may affect the long term
economics of electricity supply can be grouped under three
headings (1) Social (2) Economic (3) Technical briefly these
are

A. T. De Almeida and A. H. Rosenfeld (eds.), Demand-Side Management and Electricity End-Use Efficiency, 171–180.
© 1988 by Kluwer Academic Publishers.

2.3.1   Social. The rise and political effectiveness of numerous pressure groups which impinge on the economic operation of energy utilities and have the power to influence the political and regulatory framework within which the utility has to operate, e.g. consumer protection, environmental, anti nuclear, etc..

2.3.2.   Economic. Increased use of diffuse sources of production, either utility owned, joint or private ventures under the influence of legislation such as PURPA in the U.S.A.. Increased exploitation of renewable resources for economic or political reasons etc..

Political and institutional changes e.g. denationalization of publicly owned utilities, rejection of nuclear power etc.. Changes in the method of selling energy, the unit of sale being kVAh instead of kWh, with only energy in the fundamental frequency being charged for, generators of harmonic pollution being penalised, introduction of dynamic pricing.
Continued uncertainty in the supply and costs of fossil fuels

2.3.3.   Technical. Limited technological innovation in the fields of generation, transmission and distribution foreseen for the next three decades which would reduce the unit cost of energy. Increases in plant complexity and production costs arising out of social & environmental constraints. Demise of the electro mechanical meter and associated equipment. Increased penetration of Information Technology into the home and industry driven by the falling cost and increase sophistication of micro electronics and communication technology.

3.   DESIRABLE CHARACTERISTICS OF FUTURE SYSTEMS

In view of all these and many more uncertainties it is prudent to try and identify some of the broad characteristics any future system should have.

One such characteristic, because of the slow and unpredictable way change will take place and the time scale of implementation, must be that any change does not vitiate investments made, or constrain management's ability to rapidly adapt to changing circumstances.

Another relates to the unit of sale, the kWh, which was reasonable in the early days when customers had largely resistance loads and which has been carried forward by the restrictions imposed by economical electro mechanical metering. With the increase of inductive loads the kVAh more equitably reflects the capital equipment required and the supply losses incurred by each customer.

In recent years there has been a slow but steady deterioration in the quality of supply due to the rise in harmonic energy in the electrical supply, this pollution cannot be fully utilised by the customer in many appliances and is harmful to others. As the implied contract of supply relates to a stated frequency (50 or 60 Hz.) it is only a matter of time before customers demand to pay for only the energy at the stated frequency, this will result in a requirement to measure energy both at the fundamental frequency and in the harmonics, in both the import and export mode, so that polluters can be penalised.

Customers increasingly call for more information about all aspects of their supply, financial, commercial, technical. This can easily be made continuously available to the customer through their terminal with the added advantage that much of this information would also be utilised by the utility to provide an improved service. As this last aspect is of importance to the design of any future system installed on the customers premises it bears more detailed examination, so in reply to the question "What additional information or facilities would allow the energy utility to manage electricity supply more efficiently ?" some (not all!) of the answers were as follows.

1.   Recording and recall of phase and line voltages at each customers terminals at any time.
2.   Recording and recall of the current at each customer terminal at any time.
3.   The phase relationship of 1 & 2 and between the voltage vectors themselves across the network
4.   The price elasticity of demand of each customer under all conditions with the ability to automatically update as circumstances change.
5.   For the system operator to have continuous control over the total system demand in addition to control of production.
6.   Electronic banking to permit more flexible methods of payments and improve the cash flow.
7.   Ability to implement any tariff or pricing system without additional equipment at the customer or utility interfaces.
8.   Ability to use any communication system which is economic at the time with minimum system change.

In addition the question was posed "What changes would the customer consider beneficial ?".

9.   An improved range of choice in the methods of purchasing energy
10.   Easier and more flexible methods of payment.
11.   Facilities and information to improve their utilisation efficiency
12.   Less interference with their supply by the utility
13.   To be able check on usage and costs simply by interrogating the terminal.

In view of the wide range of uncertainty which permeates all future utility activities coupled with the spectrum of desirable features associated with a customer terminal it is small wonder that the expression "smart meter" has been coined. It is clear that modifications or "add ons" to the existing meters based 100 year old technology can only constitute expensive palliative, it is also clear from the above that the expression "smart meter" is itself narrow and dated. The requirements postulated above go well beyond the simple functions of measuring and recording energy used, for this reason the acronym ACCET (A communicating customer electronic terminal) has been coined. This I will use from now on.

4. THE DESIGN CRITERIA FOR AN ACCET

Reformulating the above characteristics into a form useful for the writing of a users specification the following emerge.

1. The basic function of measuring any electrical variable to be performed reliably over the terminals "on circuit" lifetime to an equal, but preferably higher, accuracy than any current equipment.

2. To be capable of implementing any tariff structure or pricing system, present or future, without any change to the installed hardware.

3. To be able to retain all data, programmes, and other information indefinitely irrespective of the status of the electricity supply.

4. Have the capability of implementing all present and future management systems which support the customer / utility interface.

5. Have the capability of supporting any customers energy management system

6. To have both a stand alone and communications capability.

7. To be able to operate through any communication system with minimum site modification and cost.

8. To be capable of being configured to any present or future management system without any modifications to the installed hardware.

9. Not to constrain in any way managements ability to adapt, as circumstances change, across the whole range of utility activities.

5. PIPE DREAM OR REALITY ?.

To the majority of utility professionals the picture outlined in part 1 will have some of the unreality of De Quincy's novel "The Opium Eaters". In reality, after the expenditure of several million pounds and field trials of Mk 1 & 2 devices, it can confidently be asserted that there are no technical problems to the manufacture of a communicating customer terminal (ACCET) capable of meeting all the requirements previously outlined.

The impediments to the adoption of such a system are the inherent conservatism of public utilities and regulatory bodies, coupled the symbiosis between traditional suppliers and the electricity industry, each waiting for the other to innovate.

6. EXPERIENCE FROM FIELD TRIALS

The concept outlined in Part 1 evolved out of the experience gained in the design and field trials of the Credit and Load Management System (CALMS) carried out by the U.K. South Eastern Electricity Board. For those not familiar with the system CALMS consisted of a Credit and Load Management Unit (CALMU) installed in the customers premises designed to provide a customer programmed energy management system and be in communication with the utility management system through both one way radio using the B.B.C. commercial radio broadcasts and two way using the public telephone system. The CALMU was however designed to operate over any nominated communication system through plugging in a simple daughter board.

Field trials of the Mk 2 units were conducted with 300 domestic customers over approximately two years, during which time the units performed reliably and more importantly were greatly valued by the customers almost all of whom saved on their energy bills. The unit used was deliberately designed as a hybrid, it had all the capability of a three phase design but the sensing and associated circuits were configured to provide three single phase outlets. In this way experience with the production problems of both single and three phase designs could be studied, in addition the unit was used to study customers reaction to three different tariff structures, a four level time of day - varying at weekends, a utility interruptible tariff and a night only supply. No comprehension problems were encountered by the customers in the use of the tariffs or the touch panel through which information of the supply was given. Both the one way BBC and the telephone communication performed well.

Out of the experience with the Mk 2 the following modifications were made to the design philosophy to reduce cost, increase reliability and make the design more insensitive to external changes.

1. Flexibility to be obtained through Minimum hardware / maximum software this design policy has two beneficial consequences 1) a small component count which both reduces cost and increases reliability. 2) The use of a resident software compiler in the terminal to improve the operational flexibility.

2. To measure only the fundamental values I & V per phase, their phase angle and time, from which all other electrical values can be computed.

3. Separation of the unchanging internationally standardised items (I, V, phase angle) from those man is always changing (prices, tariffs, tariff structures, accounting etc.). The former to be "built into" the unit the latter to be remotely programmable.

4. To increase customer information and reduce billing disputes, all units used to be recorded with time, date and price, thus creating and "audit trail" for both the customer and utility.

5. Customers to have access to all relevant data in the terminals memory via an international standardised communications interface. (RS 232) to allow the interactive use of home computers, energy management systems etc. as well as simple touch panel display of basic information.

6. Encryption of all data transmissions to improve the security of communications.

7. FREQUENT QUERIES

Turning now to the subjects which provoke the most queries by power engineers

7.1 Reliability

This generates more emotion than rational thinking amongst large numbers of utility engineers. They demand "That an electronic meter shall be as reliable as existing meters !" without reference to ownership, usage or anticipated "on circuit'" lifetime of the new designs - all of which could be dramatically different from those applying to existing installations. With ACCET the criterion of acceptability changes, meter reliability is related to customer confidence that the device is "good value" & not to the historical performance of electro mechanical meters. The factors determining the on circuit life of electronic meters can be grouped into two categories, technical and social, with the latter being of overwhelming importance.

7.1.1   Technical change. With all microprocessors it is reasonable to assume that those currently in use will be technically primitive within 5 years, obsolete in ten years and unobtainable shortly afterwards. This rapid change is not necessarily a disadvantage as increased value added services will be offered to the customer in later designs.

Because reliability is a pivotal criteria in the electronic component industries, enormous advances have been made in the reliability of both components and systems, such that they are now considered sufficiently reliable to be used in the most hostile environments of noise, high temperatures, humidity and vibration, e.g. the braking and control systems of commercial aeroplanes.

The elements of failure in electronic systems can be grouped into three categories, component failure, mechanical failure and system complexity. Component failure has been extensively studied and documented so that component reliability can be readily determined. Operating reliability can further be improved by production techniques which identify and remove the infantile mortality amongst components. By careful selection of components and production techniques lifetimes of 10 - 15 years are readily obtainable.

The dominant cause of failure in electronic systems is mechanical:- wires breaking, faulty plugs, poor soldering, failure of auxiliary components e.g. current transformers. It is the elimination of this type of failure which underlies much of the interest in very large scale integration (VLSI) of electronic systems onto one piece of silicon. Good design and production control can also avoid many of these weaknesses enabling an economical customer terminal to be produced with a nominal life of 15 years.

Clearly reliability is related to the complexity of system, the less components used the higher the reliability. Nevertheless careful component selection and design can produce good reliability in complex systems. A transistor radio, which withstands much abuse from young people, gives remarkable service from its average of 60 components. The component count for a flexible communicating customer terminal is currently about ten and can be expect to decrease to half that number over the next few years. Existing in a static, vibration free environment a six fold improvement over a transistor radio could reasonably be expected for ACCET.

On the basis of the above it can be shown that the mean time between failures (MTBF) of an electronic meter over a period of 15 years could equal and later exceed that achieved by existing electro mechanical meters.

7.1.2   Social change. The humble electricity meter is no more immune to social change than the horse, so that as customer requirements change so will the way a meter is used and its location in respect to other customer services. The most significant change arises because only a fraction of the power of the microprocessor is used for unit measurement and recording, so many other value added services can be given through the same unit e.g. storage, display and communication of all the data pertaining to other utilities, customer controlled energy management systems, ability to pay any of the accounts electronically, etc. In addition home alarm services, electronic mail, telephone directory search and anything else which calls for the storage, display and communication of data becomes feasible at negligible incremental cost. In the future the humble meter will be hidden in the core of a very flexible electronic consumer terminal and it is the lifetime the customer places on the terminal which will determine the period over which "the meter" shall be required to reliably operate. From the above it follows that the choice of the criterion of reliability and the economic "value" of the services made available ceases to be solely determined by the utility.

Of equal importance when considering "on circuit" lifetimes is the fact that customers also have a lively appreciation of the improvements in their lifestyle which technical innovation can bring and will readily change reliable domestic appliances as new products and facilities become available. It is this characteristic which will be the main determining factor of the time over which "the terminal" will be required to be reliable and hence "the meter" accurate. Against this background a judgement has been made that the average period of use of such a terminal, before it is replaced by the customer, would be approximately ten years.

7.2 Accuracy

Due to the non linear characteristics of commercial electro mechanical meters the existing accuracy standards have had to be based on arbitrary criteria (e.g. basic currents etc.) (Ref 1) Electronic meters with their essentially linear characteristics permit a more logical and tighter specification of accuracy to be applicable under all operating conditions. As an interim measure until the formulation of national and international standards it is proposed that the definition of Class Index should take the form used by high quality test instruments (Ref 2) i.e.

Class Index = +/-(X% Full scale reading +/- Y% Reading) Where X = 0.001/0.002/0.003 and Y = 0.1/0.2/0.3 for classes 1/2/3. This accuracy to be maintained over a temperature range of -20^0 C to $+ 75^0$ C with no corrections.

It should be noted that the existing IEC standards for metering accuracy could be met by the primitive electronic designs of 5/7 years ago with wave form sampling rates of 8 per cycle. Current ACCET designs sample the wave form 120 times per cycle, this makes them more accurate than best standards currently in use in test stations.

7.3 Data storage

Currently the information derived from integrating unit registers is of little use in itself and has to be supplemented by a recording / administrative system in order to derive useful information (store last reading, subtract from present reading, multiply by unit cost etc.). Electronic systems remove these constraints so an appraisal of information which could be of use has to be undertaken. Such consideration leads to a bifurcation of register types, the first being one into which all measured energy is stored on a continuously integrating basis, as at present. Such registers being "read only" & designed to be incapable of being changed or reconfigured by anyone, including the utility, over the life of the meter.

The second type are the working or derived registers which will vary in number depending on the pricing system, these are capable of being reconfigured as the customers change their contracts. At the end of a defined cycle of prices or billing, the information in these working registers would be transferred to another part of memory and stored for as long as required, the working registers would then be cleared for the next cycle.

Useful information to the customer can be considered in two parts, firstly information to enable them to manage their use of energy more effectively and secondly the status and trend of their indebtedness to the utility. Both of the above should incorporate sufficient detail that an "Audit trail" can be followed. The audit requirement results in the units recorded in the working registers being tagged with both date and time, so that the incidence of cost & usage, at any time, can be ascertained.

With communications, price changes within any one customer's billing period creates no problems in administration or equity, as the new prices could be applied to every one simultaneously.

7.4 Security

Electronic meters with or without communications must be proof against intelligent manipulation. For this reason the total units register must be inaccessible to everyone. Whilst all the existing administrative checks on the accounting main frame would continue, with communications a new dimension is introduced into the security of the system, as both programmes (e.g. a new price structure) and data (e.g. changing prices) can be remotely transmitted to and entered into ACCET. In order to protect such information from being altered encryption is essential. Fortunately due to the extensive work carried out on secure communications for both the military and commercial establishments sophisticated data encryption systems (DES) are available (Ref 3) these have been implemented, at low cost, in software and so a base line for any secure ACCET communications specification has been established.

7.5 Communications

Having established a secure system for communications, the type and economics of systems to be used have to studied. With the extremely rapid advance being made in communication technology a simple technical study will demonstrate that it is impossible to specify a single system which would be not become obsolete over as short a time scale as ten years. A study of the economics of communicating with all a utility's customers shows that a dedicated system to each customer would not be viable. This immediately creates the requirement that ACCET should be capable of being adapted to any communication system as both the economics and technology change.

The way forward is for the communications with any ACCET to be parasitic on an existing system which is economically viable through its primary use, e.g. commercial radio broadcasts, power lines for one way communications, telephones, cable t.v. etc for two way. The exception to this generalization is the use of mobile two way radio.

The estimated incremental production cost per ACCET of the various parasitic communication systems, which are currently available and proven, assuming large scale implementation, would be small. For one way systems the production costs would be approximately:- Mains borne (Emerson Electric TWACS) <£1, Phase modulation of commercial broadcasts £2, telephone system (analogue) £5 + the cost of the exchange modifications possibly £2.. Local systems would consist of a hand held reader costing around £10 which would constitute an overhead on the meter reading costs. (the latter being approx £5-7 per annum in the U.K.). A mobile radio van having a capability of reading between 5,000 and 15,000 meters per day depending on the customer concentration, would add about £4 per ACCET. To the above must be added the operating costs but in the main these are small on an annual basis.

The implementation of a communication system between the customer and the service providers, would start with the use of a variety of systems e.g. initially with one way radio broadcasts handling the outward bound common messages, complemented by the meter reader and his hand held unit for both return and individual messages, the latter would be gradually phased out in favour of the using a two way radio van. In turn the work of the radio van would gradually be taken over by two way systems involving the telephone or cable, resulting in 95% coverage by outward bound one way systems, 90% coverage by fixed two way systems with the shortfall being covered by mobile two way radio systems. At no point would the utility management be committed to a fixed rate of change over or forced to continue to use any one communication system.

The important point to note is that through out all this evolution the customer's terminal would be easily and cheaply adapted to the various combinations of communication systems through simple plug in daughter boards and software changes.

8. ECONOMICS

It will be noted that, as with the existing meters, nowhere in the forgoing has any attempt been made to produce a spurious economic appraisal. Any "comparative" economic appraisal would have to quantify the "value" the customer placed on the increased information and improved services the ACCET provides - no sensible utility executive would be so arrogant. The way forward is for the utility to accurately describe ACCET and the services it provides and leave the customer to request its installation, They would then apply the same economic judgments they use when purchasing other durables. In the industrial case where the cost are the similar the question does not arise. Currently the cost of a three phase ACCET would be comparable with the simplest single rate three phase electro mechanical meter, whilst the cost of a single phase terminal would be similar to a two rate single phase meter.

9.   CONCLUSIONS

The impermeable economic membrane at the customer / utility interface, the 100 year old electro - mechanical meter, which prevented utilities offering non discriminating, cost related prices can now be replaced by a communicating customer electronic terminal (ACCET) which removes all restrictions on customer / utility relationships. Such a terminal not only places the customer in full control over their energy use and removes the need for utility controlled demand management schemes which, because they impinge on customer sovereignty, run counter to the social trends of the times, but also render limited electronic meters uneconomic.

The only barrier to the widespread introduction of ACCET's and the consequential improved utilisation of the nation's energy resources is the lack of management appreciation of their potential, leading to the absence of policy.
Like all changes when it is introduced it will present interesting challenges to utility managers, not only in the technical (system control) administrative (electronic banking) and other spheres of existing operations, but in the way they approach the wide range of non traditional services which a single home terminal can provide.

10. REFERENCES

1   Peddie R. A., I.E.C. Standard 521, a critique, Jan 1986 *
2   Peddie R. A., Proposed standard for electronic meters, Jan 1986 *
3   Electronics,(U.S.A), LSI based data encryption discourages the data thief, June 21st 1979
*   1 & 2 Have been circulated to interested parties for comment.

APPLICATION & POTENTIAL OF ACCET's

R. A. PEDDIE
Consultant. (Previously Chairman S.E. ELec Board, England.)

1. INTRODUCTION
    If communicating customer electronic terminal's are bought
and installed against a user specification, drafted in such a
manner that the majority causes of obsolescence (Changes in
technology, management systems etc.) are virtually eliminated,
then the manner of their utilisation is bounded only by the
imagination of the utility's management and the historical and
cultural environment in which they operate. With the
elimination of the majority of the restraints on the provision
of a safe and reliable supply, through the application of non
discriminating cost related pricing, the immediate problem is
managing the transition. Each utility will adopt a different
management strategy from the wide spectrum of alternatives in
the light of the local circumstances. This paper outlines one
strategy which has been pragmatically evolved with the ACCET
design,

2. ACCET DESIGN REQUIREMENTS
2.1. The Terminal
    A key aspect of the electronic design has been to meet the
general requirements that technological change does not vitiate
the investment, or constrain managements ability to adapt
rapidly to changing circumstances, which effectively rules out
traditional meter hybrids.
    From this the following management requirements arise:- 1)
Those characteristics which are internationally standardised
(measurement of amps, volts, time etc.) to be built into ACCET.
2) Those system characteristics of ACCET which are subject to
change (units of sale, pricing systems, accounting and payment
facilities and other services - security , fire etc. as well as
services to other utilities) to be capable of being changed by
installing revised software. 3) Changes in the communication
path to be accomplished by the simple exchange of a small "plug
in" daughter board. 4) As none of the system changes can be
defined in advance, a requirement arises for them to be
programmed in at any time, to achieve this the following have
to permanently reside in ACCET:- a resident language compiler
and an encryption system for the secure communication of
programmes and data.
    From the above a design evolves consisting of a simple
electronics card with no more than ten components, interfacing
with (1) a communication system through plug in daughter
boards, (2) the electrical sensors (e.g. V.T's & C.T's)
(3) A standardised communication interface (RS 232) to enable
either the customer or the utility to communicate with the

A. T. De Almeida and A. H. Rosenfeld (eds.), Demand-Side Management and Electricity End-Use Efficiency, 181–190.
© 1988 by Kluwer Academic Publishers.

terminal (Local customer display or energy management system)
(4) A controllable mains switch (for those customers without
one.). In addition the microprocessor itself has a number of
interrupts which could directly control customers equipment to
their requirements if so required.

Although the electronic card is common to all
installations, the supporting peripherals now offer a range of
nine basic choices. The communication path can be (1) Local -
hand held reader (2) One way broadcast communication (Utility
to Customer) (3) Two way communication. The switching of the
customers supply can either be (1) none (2) electro -
mechanical switch (3) Electronic switch. Each will have a
different cost and a commensurate capability so the actual
design adopted by a utility would be a matter of judgement.

The management strategy described in this paper assumes
that the industrial customer would have an ACCET with both one
& two way communications, which would provide a price stream to
their energy management system which in turn would regulate
their demand. The smaller customer's ACCET with their
capability of implementing any tariff or pricing structure,
would receive their data via one way broadcasts supplemented by
either a hand held communicator or a mobile radio system for
two way communications. In addition their supply would be
controlled via an electronic main switch to allow for the
widely different customers choice, and for the better control
of safety, debt and the electrical system.

2.2. Communications.

Communications are an essential component of the design of
any customer electronic terminal, so the characteristics of
this element needs to be described.

A study of the economics of communicating with all a
utility's customers shows that a dedicated system to each
customer would not be viable. The way forward is for the
communications with any ACCET to be parasitic on an existing
system which is economically viable through its primary use,
e.g. commercial radio broadcasts, power lines for one way
communications, telephones, cable t.v. etc for two way. The
exception to this generalization is the use of mobile two way
radio.

As the relative economics of the different communication
systems are also in a continuous state of flux it is necessary
for the communication interface in ACCET to be capable of being
simply and economically changed. Such system independence also
helps with another management requirement, namely that the
utilities management systems can be changed without
invalidating the investment in either the customers terminal or
the communication system.

To commercially exploit the communication systems
available and to achieve minimum cost commensurate with the
desired service it is essential to specify the management's
communication requirements, these can be studied under the
following headings (1) Broadcasts (2) Programme and data
traffic to the terminal (3) Data acquisition by the utility (4)
Status monitoring by the utility (5) Customer initiated signals
to the utility (6) terminal to customer communications.

Each of these have a different volume of traffic and time
dependance, for example the transmission of prices or actuation
of emergency limits, to all customers simultaneously, calls for
short response times (seconds) and involves small amounts of
data. Status monitoring of each customers system is
intermediate in data volume and timing, whilst data acquisition
by the utility from each customer could be large in volume but
is not time dependant due to the memory in the ACCET.
Having evaluated the communication traffic by volume and timing
the economics of each communication system has to be studied
under the headings of (1) remote systems, (radio, power line
carrier, telephone,) (2) local systems, (hand held
communications, mobile two way radio.) (3) Systems internal to
the customer's premises. The latter traffic is further sub
divide into two types, communications to other utilities
equipment (gas, water, telephones) and service companies (fire
& security alarm services) generally these connections will
require to be hard wired; the second type relates to the
customers energy management system and can be hard wired, or
use the mains wiring for signalling. Reliability and accuracy
of transmission is important and mitigates against the use of
two way, wide area use of power line carrier systems, however
such techniques have been shown to be more reliable when used
one way from the utility to the customer and can be easily be
used within the customers premises.
      There is no unique solution to the most economic way to
conduct the above traffic, in the U.K. the economics of this
spectrum of demands has given rise to the provision of more
than one communication system to each terminal, use of the BBC
phase modulated 200 kHz commercial radio for all broadcast
messages, whilst the two way traffic can use one of the other
two way systems i.e. mobile radio, telephone, power line
carrier.

3.   MANAGEMENT STRATEGY.
      Having regard to the title of this programme it is
appropriate to show how Demand Side Management (DSM) is
influenced by the use of ACCET's.
3.1. Evolution of DSM
      Literature surveys show that DSM can take two forms. In
the first the utility takes the initiative and exercised
control over the customers energy usage through a variety of
techniques, (a) tariffs based on differential prices and
physical interruption, (b) low interest loans to purchase
selected hardware or effect modifications to existing
installations, (c) customer education, advertising etc. (Ref
1). As this route can be followed with negligible change in
management systems and installed equipment, it is the one most
utilities are pursuing. Regrettably it is economically flawed
as it manipulates the customers demand to only optimise the
utilities INTERNAL economic efficiency and does not meet the
criteria of prices being cost related and non discriminatory.
      The second form starts from the premise that if the total
energy cycle is to be economically managed, the customer must
be solely responsible for the demand side management and end

use efficiency. The utility in this scenario plays a true
service role by responding, in the most economic manner, to the
demand generated by the customer, after the latter has been
informed of the true cost / unit of energy at that time.
Although the literature advocating this approach dates back the
early 1970's (Ref 2,3,4) its application is only now being
studied by a small number of utilities, mainly in the U.S.A..

3.2. Critique of DSM techniques

Before a management strategy is formulated it is useful to
appreciate the advantages and disadvantages of the two main
forms referred to.

3.2.1. DSM - Utility managed. The impact of the changes
in the early 1970's focussed the utility management attention
on improving their own economic performance, which they hoped
would simultaneously improve customer relations. The actions
undertaken by most utilities aimed to increase system
utilisation, defer new plant, as well as reducing dependency on
oil, through the modification of the load shape by a variety of
means:- chopping the peak, filling the valleys, and shifting
the load. A more flexible load shape was also sought through
conservation and the promotion of specific loads. This has
given rise to a plethora of new tariffs / rates, variously
referred to as:- demand, time of use, off peak, seasonal,
inverted, interruptible, promotional, conservation etc., which
the utilities currently offer in order to implement a DSM. As
the tariff prices quoted are an amalgam of extrapolated
historical data, management guesstimates, and political
expediency, they exhibit all the well known and documented
defects of electrical tariff systems, so they do not and cannot
reflect the true costs of production at the time of
consumption. Hence their role in improving the overall economic
efficiency of the energy cycle in a non discriminatory manner
is very limited.

Additionally the use of existing tariff systems and
equipment to implement a utility controlled DSM poses a number
of problems. Consider the cases of just two of the tariffs, the
first to illustrate a technical disadvantage, the second a
potential social one. Time of day tariffs are currently
utilised by only a minority of customers, so the effect of
synchronising their demand changes at the price / time
boundaries is not significant, if this type of tariff where
universally applied the rapid change in system demand at these
boundaries could impose unacceptable repercussions on both
electrical system stability and frequency control. Staggering
of these boundaries across customer groups could be challenged
on the grounds of discrimination and equity. Secondly operation
of utility interruptible rates take no account of the effect of
the loss of supply on the customers. In view of the social
trends in recent years such infringement of the customers
sovereignty must eventually be unacceptable, especially when
the customers become aware that non intrusive systems are
available.

The effectiveness of utility controlled DSM is also
circumscribed by the electro mechanical meter which, because of
its limited register capability and lack of real time

communication between the customer and the utility, is unable
to economically implement flexible cost reflective pricing.
This defect gives rise to a web of tariffs which some customers
have difficulty in understanding & exploiting. Such limitations
are the root cause of the administrative and technical
complexity of the tariffs and their support systems and have
spawned a market for "add ons" and other methods of increasing
the flexibility of the 100 year old metering technology,
unfortunately due to the rapid advances in flexible electronic
metering such modifications and / or additions are effectively
obsolete before being installed !.

So it must be concluded that whilst utility controlled DSM
has improved the economic efficiency of <u>the utility's</u>
<u>operations</u>, it is only a economic sub optimization of the
energy cycle because it does not dynamically include the
customers response to the true production costs at the time of
consumption.

3.2.2   <u>DSM - Customer controlled.</u> As customer demand
varies continuously and electricity cannot be economically
stored in large quantities, it is quite clear that a utility's
costs of production also vary continuously. In this environment
the only way the criterion of non discriminatory, cost
reflective pricing can be met is by adopting a policy of
offering the customer a continuous price stream related to the
costs at that time. Such a system of dynamic PRICES eliminates
those elements which are the subject of guesstimates in the
compilation of TARIFFS :- weather, ambient temperature, plant
breakdowns, fuel cost, future political and economic factors,
interest rates etc.. Such a policy calls for a more flexible
metering system having the capability of receiving dynamic
prices and recording the energy used in separate price (unit)
registers. One essential proviso is that the customer has to
install an automatic energy control system responsive to the
price stream. Such equipment is currently available and
installed in many commercial and industrial premises, whilst a
simple, cheap price / energy controller has been designed for
use with domestic equipment.

With the customer responding in this manner to a stream of
cost reflective prices, the "value" they place on energy <u>at</u>
<u>that time</u> is revealed and so for the first time electrical
utilisation is incorporated into the economic use of energy.
Further being aware of the range and incidence of prices, the
customer takes a positive interest - not only in the incidence
of consumption - but also in the energy characteristics of the
equipment they buy. A year long study on the application of
dynamic pricing conducted by the CEGB and two manufacturers of
industrial gases reported that whilst monetary savings had been
achieved in using such prices, even larger savings would occur
through economically viable investment in more energy efficient
processes. (Ref 5). For the above reasons a customer controlled
demand management system using ACCET's is now examined in more
detail.

3.3 <u>Planning and System Operation.</u>
As stated in a previous paper the transition from the
present management systems to dynamic pricing is evolutionary

in view of the time scales involved in economically changing
the installed equipment. Fig 1 illustrates the manner in which
this change can be effected in the U.K. without abortive
expenditure or an irrevocable commitment to one strategy.

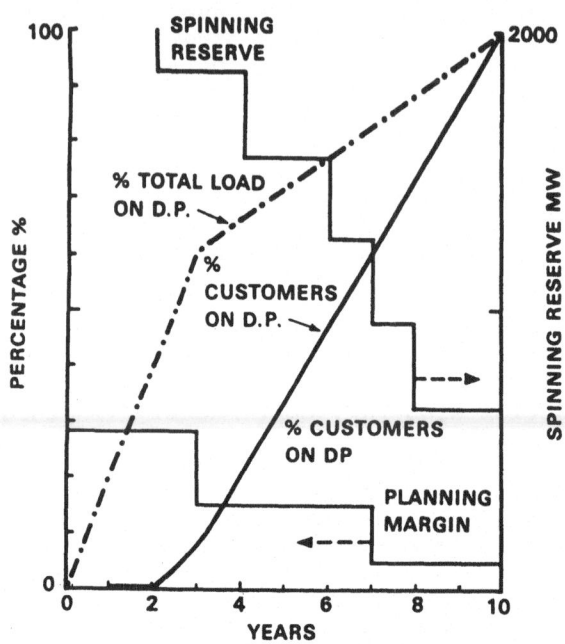

<u>FIG 1 Possible introductory time scale for</u>
<u>Dynamic Pricing and the ensuing operational changes</u>
      The figure reflects the fact that:- (1) the average
turnover of customers meters, from all causes, is about 10%, so
changing the whole stock over a 10 year period would be well
within the capability of existing staff. (2) Industrial
customers are 0.86% of the total and consume over 40% of the
annual GWhs. (3) Non domestic customers are 8.6% of the total
and consume over 61% of the annual GWhs. (Ref 6). After two
years 40% of the total load could be controlled by dynamic
pricing, whilst still retaining all the existing supply control
facilities.
      With communications the price elasticity curves,
reflecting all the changing influences, for each day of the
year could be obtained to any detail required, by geography,
industrial classification etc. After the fourth year and the
inclusion of all the non domestic users the continuously up
dated price elasticity data would be highly reliable.
Consequently as confidence was built up so the level of
spinning reserve could be reduced and likewise the planning
margin. The above curves are not greatly altered if it is
assumed that a domestic customers do not initially move
directly to dynamic pricing but evolve to it through say a four
level, time of day tariff, with a subscribed maximum demand
(which would only be invoked at the time of the <u>actual system</u>

<u>peak</u> or abnormalities). Appendix 1 sets out the System state and Terminal capability in such a customer controlled demand regime.

The reference to catastrophic failure in Appendix 1 requires some explanation. Each ACCET can be given two specific frequencies, the lower is the one at which it automatically disconnects the customer supply, the higher one being the value at which supplies are restored. System stability studies determine the shape of the disconnection/restoration curves and the customers location on them. (Fig 2).

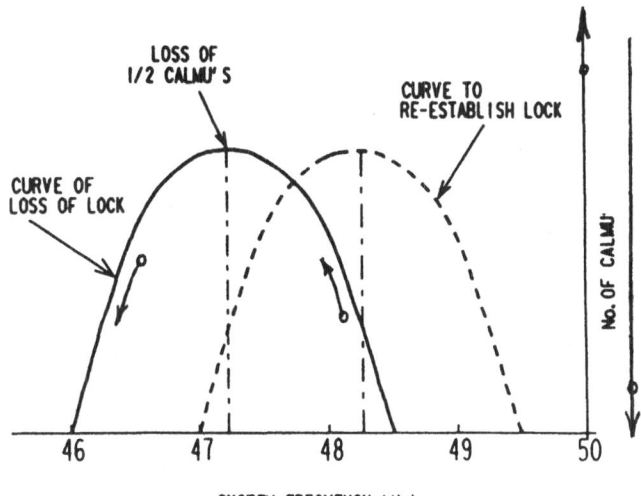

SYSTEM FREQUENCY (Hz)

<u>FIG 2 Automatic ACCET disconnection & reconnection curves</u>

This automatic load shedding under conditions of catastrophic system collapse frees the control engineer from trying to initiate and control load shedding to diagnosing the cause and eliminating it, in this way the nightmare of total system collapse is eliminated. There are two exemptions to auto - load control (1) Essential services (water, gas, sewage, hospitals, home kidney machines etc) are exempt from auto disconnection (2) For safety reasons auto restoration of individual supplies is not given to commercial / industrial supply unless requested, the ACCET can simply actuate an audio / visual indication of supply restoration.

3.4 Financial

In the U.K. the daily energy sales are over £ 27.5 million / day (1985) so cash flow is of considerable interest, four years ago the Chief accountants showed that with more imaginative methods of payment made possible with ACCET's, savings of £ 150 million / year could be made.

Debt management is a traditionally difficult social problem which incurs a disproportionate administrative effort, ranging over people in temporary financial difficulties (illness, redundancy etc.) to people who appear to be incapable of managing their income. With the ability to control supplies against units used or money, coupled with electronic payment facilities in both the credit and debit mode, more help can be extended to these people with a significant reduction both outstanding debt and its administration.

With the existing systems theft is virtually unquantifiable, over a decade ago estimates for the U.K. put losses at over £ 40 million / annum. To deal with this aspect ACCET only utilises valid encrypted data and programmes, continuously monitors its own functioning and that of the supply, disconnects the supply if the enclosure is tampered with and reports all discrepancies on being interrogated by an authorised message. With these precautions a considerable reduction in this loss could be anticipated.

Billing queries is another source of customer irritation and administrative cost, to reduce both of these ACCET can be configured to record not only the units used at each price but to tag them with the date and time. In this manner both the customer and the utility has an "audit trail" in contrast to the uninformative total units currently displayed on th customers bill.

4. BENEFITS

Summarising - a selection of some of the benefits which a communicating customer terminal brings are as follows :-

(1) Prices which truly reflect the rapidly changing marginal costs of production can be simultaneously transmitted to all customers.

(2) All changes in costs are immediately incorporated into prices, i.e. fuel, interest rates, taxation, borrowing, wage rates, Government policies etc..

(3) All changes in operating conditions are likewise immediately incorporated into prices, i.e. weather conditions, plant availability etc.

(4) Non-discrimination is clear, the differing costs of supply can be allowed for in the individual dynamic price. (This is important for industrial users taking supplies at H.V..)

(5) Tariff harmonisation is automatic if all are on dynamic pricing.

(6) With each dynamic price reflecting the utility's total costs at that instant the customer is for the first time able to make economic decisions - not only on usage - but on the energy characteristics of the equipment they purchase, resulting in conservation and a rational use of energy.

(7) Improved information and help can be given to customers in genuine difficulties, whilst theft is minimised

(8) Under dynamic pricing as the system is primarily electrically balanced through customer demand control, planning, operational plant margins & spinning reserve could be reduced as experience of system response dynamics and customer price elasticity is acquired.

(9) Price elasticities of all customers under all

conditions can be determined to any degree of detail.

(10) Stability and control of the electrical system under all conditions, steady state, emergency and catastrophic is considerably improved.

(11) Data on all aspects of the distribution system is easily obtained leading to improved economic operation and planning

(12) The financial books of account can be continuously balanced with any chosen margin, as the majority of the present unknowns in the revenue balance sheet are eliminated using dynamic pricing.

(13) As dynamic prices would be based on marginal costs this would simplify the contractual relationship with autoproducers and alternative energy sources as the utility would buy and sell at the margin.

5. THE POTENTIAL OF ACCET's

The benefits which both the customer and the utility derive from the general installation of ACCET's are substantial, of equal importance are the new management and organisational challenges it presents to utility management. Ignoring the electrical sensors consider what has been devised:- A customer terminal capable of meeting a wide ranging set of utility objectives through the use of a resident compiler backed by an expandible non volatile memory and the capability of receiving intelligence or engaging in an electronic dialogue with third parties through any known communication system.

Thus in meeting the electrical utilities objectives ACCET has been transformed into a generalised terminal capable of providing a wide range of services e.g. (1) to other utilities :- Gas, Water, Telephones.(for example the cost of the last telephone call, who it was made to, the time, telephone number enquiry service, total itemised account etc. etc.), (2) security & fire alarm services, (3) provision of electronic banking facilities etc. One very interesting aspect is that the basic impediment to the nationwide introduction of electronic mail, lack of person to person communication, would also be removed.

As noted in the previous paper even the most sophisticated measurement requirements and pricing systems utilise only a fraction of the power of the microprocessor, so in the future the meter will constitute only a small part of a very flexible electronic consumer terminal. In many utilities there will be a natural antipathy to diversification into other unfamiliar services, but the customer will become intolerant of the physical presence and cost of a multitude of dedicated terminals, spawned by the advances in information technology when it is obvious a significant proportion of the services could be provided much cheaper through one terminal. This raises the spectre of the ownership of "the meter" passing out of the hands of the utility. There will be plenty of entrepreneurs (but it could equally well be a subsidiary company) willing to manage a terminal service of which metering is one. Even the possibility of the customer owning a terminal "sealed and certified" by a regulated supplier cannot be ruled out. So a fascinating range of challenges unfold for the managements of electrical utilities.

6.   REFERENCES

1   Demand side management - overview of key issues
     by Battelle-Columbus & Synergic Resources for E.P.R.I. &
     E.E.I. November 1983
2.   Responsive Pricing & Public Utility Services, Vickery W.,
     Bell Journal of Economics & Management Sciences Vol 2. No.
     1 Spring 1971
3.   Optimal Spot Pricing : Practice & Theory, Schweppe F.C.,
     I.E.E.E., Power Eng. Soc. Winter meeting 1982.
4.   The application of economic theory utilising new
     technology for the benefit of the customer, Peddie
     R.A.,Frewer G., Goulcher A., South Eastern Electricity
     Board, May 1983
5.   A Short Notice Pricing Experiment. Neal R.J., Friend J.F.,
     Hall R.G., I.E.E. Metering and Tariffs Conference,
     Edinburgh, April 1987
6.   U.K. Electricity Council, Handbook of electricity supply
     statistics 1985.

APPENDIX I.   Hierarchy of System State and Terminal Capability in a Demand Control Regime

System State	System Control	Customer Reaction	
		Domestic	Industrial
Steady state (no fore-seeable difficulty)	Invoke subscribed MD at times of peak	Load limiting against load or price criteria. Customer-chosen priority interrupt system used in both cases to eliminate disconnection	10-minute price-stream coupled to customer-automated load control
Non-standard (predicted fuel or plant shortage — sufficient time to react)	Publicise load limit via media and display on customer's terminal. Immunity capability for hospitals, life-support systems, etc.	Reduce load to displayed level. Individual dis-connection if limit not observed. No area dis-connections	Reaction by reducing against a higher price-stream
Emergency (serious trans-mission system fault or major plant failure)	Pre-set load limit stored in terminal memory activated by radio broad-cast or telephone link in seconds	Prompt reaction to reduce load to desired level so as to preserve supply assisted by the priority interrupt switches where so installed. Area dis-connections very unlikely	Load reduction via immediate high price message. If not adequate, MD limitation as per domestic.
Catastrophic (grid severence, major blackout potential)	Automatic disconnection and reconnection of customer terminals against frequency. Hysteresis on the disconnection/re-connection curves avoids system oscillation	NIL but benefit of improved stability under fault conditions; less likelihood of system judgement error and much more rapid resoration	

MARGINAL COST PRICING : AN EFFICIENT TOOL TO ENSURE ELECTRICITY DEMAND SIDE MANAGEMENT

B. LESCOEUR - JB. GALLAND - E. HUSSON

ELECTRICITE DE FRANCE
2, rue Louis Murat
PARIS - FRANCE

I. INTRODUCTION

The constant adaptation between electricity supply and demand can be achieved in two ways : On the supply side, through the construction of additional facilities, and on the demand side, by implementing tariffs, load management schemes and a commercial policy. Changes in demand clearly imply changes in the supply system in terms of both the installed capacity and the operating conditions of the system which need to be taken into account.

Like many other utilities, EDF has always held that the problem was to control the total system to reach an overall optimum for the community as a whole, and to define the most appropriate tariffs and load management schemes by comparing costs (including implementation costs) and benefits for both the supplier (reflected by marginal generation and distribution costs), and the consumer.

There are two main principles underlying the French electricity pricing policy : equality of treatment and economic efficiency.

Equality of treatment means that all customers with the same characteristics of use pay the same price, or, in the case of systems with options that they have the same tariff options. This principle allows wide tariff differentiation. Indeed, as regards the importance or the nature of the services they imply, there is a great difference between a kWh supplied at a very high voltage to an industrial customer and a kWh supplied at a low voltage to a domestic customer.

For the community, the principle of Economic Efficiency leads to selling of electricity at its marginal cost, which is the cost to the community of a customer who decides to consume an additional unit of a good.

A. T. De Almeida and A. H. Rosenfeld (eds.), Demand-Side Management and Electricity End-Use Efficiency, 191–205.
© 1988 by Kluwer Academic Publishers.

II. NATURAL MONOPOLY AND MARGINAL COST PRICING

Any activity using networks is more efficient the more the networks are used. In particular, it is difficult to imagine that there will be an advantage in building two electric networks, always very costly, which would provide the same service in the same area. This explains why, in all countries, the authorities recognize only one distributor for a geographically homogeneous zone and give it the right to perform its activity within a legal framework defining its duties. The point is then to define "public service" activity of an electricity distributor. But the customer may fear that the monopoly distributor may use forms of "dumping" on some markets. Consequently, rules must specify the objectives of the monopoly and the way prices will be fixed for the different services offered. Economic theory guides the choice of these rules.

As regards production, theory clearly shows that to achieve a collective optimum, the producer must meet demand and minimize production cost in the broadest sense of the word, including a standard of service.

The last, but a real problem, lies in the necessary coordination between the utility's and the customers' decisions. Customers make their choices in terms of their own interest, i.e. according to the tariffs they are offered. Therefore, the utility has to inform them on the economic consequences of each of their decisions.

Economic theory suggests a solution for all these points : which is selling at marginal cost. Moreover, this principle is consistant with the principle of equality of treatment since the cost that a customer entails for an elctrical system will be reflected in the tariff applied to him.

III. THE EDF'S APPROACH OF MARGINAL COST PRICING

EDF's pioneering efforts in implementing tariffs based on marginal cost in the electricity sector are well known (see [1] and [2]). The favourable conditions in France have contributed to the success of this implementation. Institutionnaly, EDF has had the possibility of defining its pricing policy with the approval and support of the French Regulatory Authorities. In addition, through tariffs based on long run marginal cost, EDF has been able to meet its financial requirements, and only minor financial adjustments have been necessary. Indeed EDF has always been characterised by a rather high rate of growth of electricity consumption (see [3] and [4] on this point of compatibility between marginal cost pricing and financial constraints).

Tariffs cannot reflect all the differences in costs, or the cost of all the various kinds of supply. Equalizations are therefore necessary to avoid excessive complexity of the tariffs and to limit metering and installation costs. The optimal complexity of a tariff will result in balancing higher costs of metering and implementation by an advantage for the community as a whole, obtained by a change in the consumption pattern which a more precise and efficient tariff signal brings about. The electricity producer must consequently study energy requirements for the most important customers for whom electrical solutions may be possible. On the basis of the overall cost for the community, when an electrical solution is competitive and has a large development potential, it is then appropriate to draw up a tariff which reflects its cost most accurately.

EDF has adopted this method of analysing electricity uses for defining its marketing and pricing policy. This is an attempt for taking into account the long term price elasticity of electricity demand. For instance, when it appeared in France, that hot water production by electricity using storage water heater was very competitive and less costly than other alternatives, EDF created an optional two-period time-of-day tariff for residential customers : normal hours and low-load hours ; 7 million customers have taken this option since 1965.

Another example is given by the many studies relating to direct space heating which have led to defining the optimum degree of insulation, and have made it possible to check that a two-part tariff system with one price for subscribed demand and one or two prices for energy could give an accurate indication of the cost of this application. EDF has been applying this kind of two-part tariff to every residential customer over the last two decades.

To limit metering and implementation costs, EDF has been led extensively to propose optional tariffs, instead of a single but more complex tariff. All the options are incitative, i.e., they are designed for each customer to select the option which best reflects the cost of his supply. EDF considers the use of this method of optional tariff as the best approach for decentralizing the cost-benefit analysis of a more complex metering system. For instance, the demand charge in the low voltage low-load hours tariff is higher than for the basic tariff, in order to cover higher metering costs, so that the customer chooses this option if and only if the advantage of this system is greater than the additional metering cost.

The recent revision of EDF's tariff aims at improving the efficiency of control techniques of the French electrical system on the demand side and illustrates the application and the possible development of the principles laying the foundation of EDF's pricing and load management policies.

IV. RECENT DEVELOPMENT IN THE FRENCH ELECTRICITY SUPPLY-DEMAND SYSTEM

The characteristics of modulation of electricity demand in France have changed considerably over the last decades.

The daily load curve at system level has flattened out considerably, particularly under the effect of the tariff policy and the development of electricity uses (direct space heating and water storage heating) deriving from it. The daily load factor of the busiest day is now 90 % in comparison to 85 % ten years ago. Figure 1 indicates the past changes in the profile of the daily load curve.

By contrast, changes in the working pace and the development of electrical heating have enhanced the seasonal aspect of the demand, which on a winter day is almost twice that of a summer day.

In addition, the sensitivity of the load to random conditions, particularly temperature, which has less effect on the shape of the daily load curve than on its level, has increased to a considerable degree. Demand at low load hours (night-time) on a cold winter day is now greater than at peak hours on a mild winter day (see figure 2) :

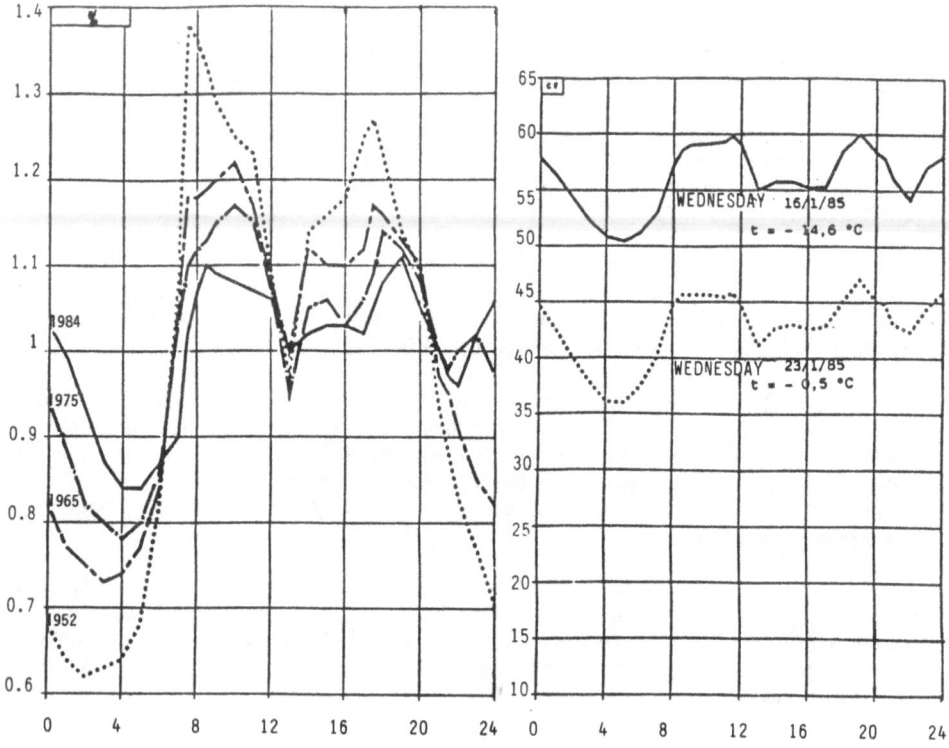

Figure 1   Figure 2

Whereas in the past, the periods of high demand were concentrated on a few hours per day, over a large number of days, they are now concentrated on a large number of hours on the same day on just a few days during the winter at unpredictable dates.

As regards supply side, the economic advantages of nuclear derived electricity as a base or semi-base load has become quite clear over the last decade in France. The break even plant factor in terms of hours of plant use when substituting nuclear for coal is equivalent to less than 3000 hours. Another modification is the relative reduction in hydro electric storage facilities. The hydro-electric plants available to EDF, including pump storage facilities, are appropriate for the regulation of the daily load curve so that the use of thermal facilities is almost constant for 15 to 18 hours of the busiest days, but are inadequate to transfer energy from a mild to a cold day. Specific peak thermal (oil-fired) units are required which operate for a very short period of time in a year. Table 1 lists the mix of plants of the French system :

	1986		1995	
	GW	TWh	GW	TWh
fuel oil	12	10	10	7
coal	12	30	11	15
hydro-electric	23	64	25	73
nuclear	45	241	63	365
Total	92	345	109	460

- Table 1 -
Mix of plants

The main consequence of these modifications is a very substantial difference in marginal costs between low-load periods, in which the marginal cost is equal to the fuel cost of nuclear power stations (these plants alone are then sufficient to meet demand), and periods in which peak units with a very high running cost must enter into service, as well as when meeting an additional demand requires the development of new equipment. The marginal cost may therefore vary by a ratio of 20 to 1 between these two extreme situations.

Because of change in the structure of the world energy market since 1973, and the economic advantage of nuclear electricity in France, and despite the marked widening of the range of marginal costs, electricity has had to play an increasingly substantial role on the french energy market and is rapidly replacing other, more expensive energy sources in all sectors, by using technologies which have already proven appropriate, as well as new technologies (bi-energy systems, thermal plasma, etc...)

Greater possibilities of load management emerge from the increasing recourse to electricity on the energy market.

V. THE EDF TARIFF SYSTEM

V.1 High voltage customers

The former green tariff, created in 1957 was applied to the 150 000 EDF customers connected at medium, high and very high voltages. Five periods were defined with different prices : 3 periods per day in winter (October to March) and 2 in summer. The impact of this time-of-day tariff has often been described ([5], [6], [7]).

For these customers, the aim of the revision of the tariff (see [8]) was to adapt the prices to the change in marginal costs, and especially to reflect, the increasing seasonality of the costs. In summer, prices are much lower, and the winter period now only covers 5 months. In addition, for the 500 largest customers (with subscribed demand higher than roughly 10 MW), the tariff signal is more detailled and now offers 8 different price periods, distributed over the 4 seasons and according to the time-of-day.

In 1985, this revision has been completed and the customers' response regarding their seasonal consumption is already substantial : the 300 largest customers' subscribed demand during normal hours in summer is roughly 1 500 MW higher than the subscribed demand during normal winter hours which corresponds to a total of approximately 8 000 MW.

Customers respond by scheduling maintenance in winter, or by using the production capacity more in summer than in winter (iron alloys, chlorine or zinc electrolysis). But the main impact of the new seasonal price differentiation is to promote the substitution of electricity for fossil fuel in steam production in the summer, using the bi-energy system.

V.2 Low Voltage customers

Since 1965, the "Universal tariff", applied to Low Voltage (LV) customers, is a two-part tariff with a demand charge relating to the subscribed demand, which is scaled in 3 or 6 kVA steps. After selecting a level of subscribed demand, the customer is held to respect the contract which is controlled by a circuit-breaker.

The universal tariff also offers a tariff by time-of-use with two-period as an option : normal hours and low-load hours. This option is implemented with the help of a two-dial meter.

This optional time-of-day tariff, chosen by 7 000 000 customers today, has encouraged the use of water storage heaters which now correspond to approximately 12 000 MW of diversified demand under EDF control, and promote the development of around-the-clock uses of electricity such as direct electric heating which, in the case of France, is an economical solution for substituting electricity for more expensive (and imported) fuels.

The customer average daily load curves, with and without the low-load hours options (figure 3), has been drawn up by taking a sample of approximately 1 000 customers in an extensive permanent load research program [9].

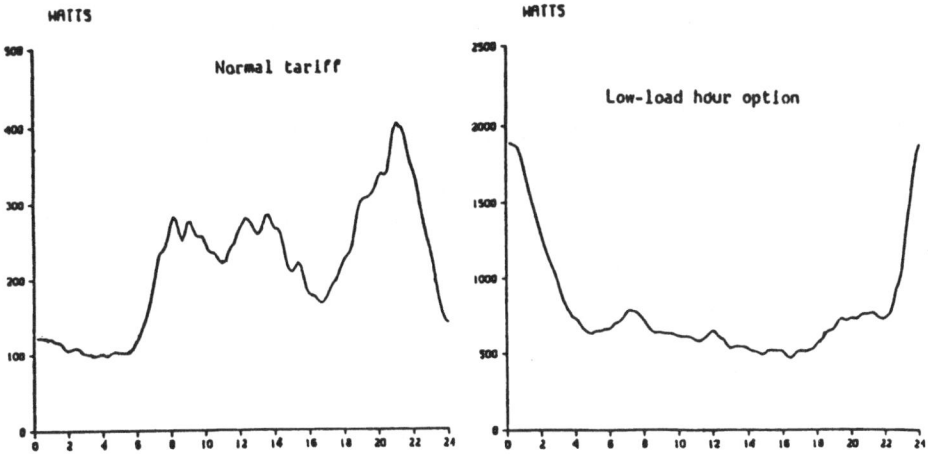

- figure 3 -
Average load curve for the average LV customers

The success of this optional tariff explains the remarkable increase in the daily load factor of the total load curve for France (approximately 90 % currently).

Two different techniques have been used for implementing this time-of-day tariff : time-switches, or ripple control relays. These devices change the dial of the meter, and may also be used by the customer for controlling part of his load (often a water heater). Table 2 indicates the proportional use of the two techniques :

LV customers	25 000 000
Low-load hours option	7 000 000
with time-switches	3 700 000
with ripple control	3 300 000

- Table 2 -

No new time-switches are installed today, since more than 85 % of all LV customers supplied by EDF may now be under ripple control, the reliability and cost-effectiveness ratio of the 175 Hz ripple control system is better. This is the effect of the economies of scale : 400 000 new relays are installed per year in France.

The current tariff revision does not change the main lines of the LV tariff structure. A new tariff has been created for customers whose subscribed demand exceeds 36 kVA. This new "Yellow Tariff" offers 4 price periods : winter and summer, normal and low-load hours. For the smaller customers, prices have been adjusted to the new conditions of marginal costs, and the definition of low-load hours has been made more flexible. Because of the success of this optional tariff, differences in marginal costs within the day are now smaller. Furthermore, it is essential to prevent the development of electricity uses during low-load hours from

198

causing a local peak on the distribution network, so that the benefit at generation level is offset by a higher cost of strengthening the network. The differentiation of tariff periods by category of customer solves the problem. There are still 8 low-load hours per day for low voltage consumers, but the timing may vary from one customer to the next, and may not necessarily be continuous. For instance, there may be two hours in the middle of the day and 6 hours in the night. This diversity in timing among the customers is now substantial with the increasing adequacy of the ripple control system : by super-imposing a second frequency of 188 Hz, the system will offer 900 different additional commands in comparison to the 40 commands of the 175 Hz system.

The low-load hours option is a good example of the use of a time-of-day tariff as an effective load management scheme. And the creation of the "peak day withdrawal option" illustrates the possible extension of this approach.

VI. FROM TIME-OF-USE TO REAL TIME TARIFFS

VI.1 Theoretical principles

During the last decade, EDF's tariff system was only based on time of use tariffs. This type of tariff is well suited to reflect the marginal cost of all the various kinds of supply when the load duration curve during each tariff period is not too much modulated. Other way equalizations which are necessary to avoid an excessive complexity of the tariffs, and to limit metering and installation costs, may lead to a non efficient signal for some type of uses.

TIME OF USE TARIFF

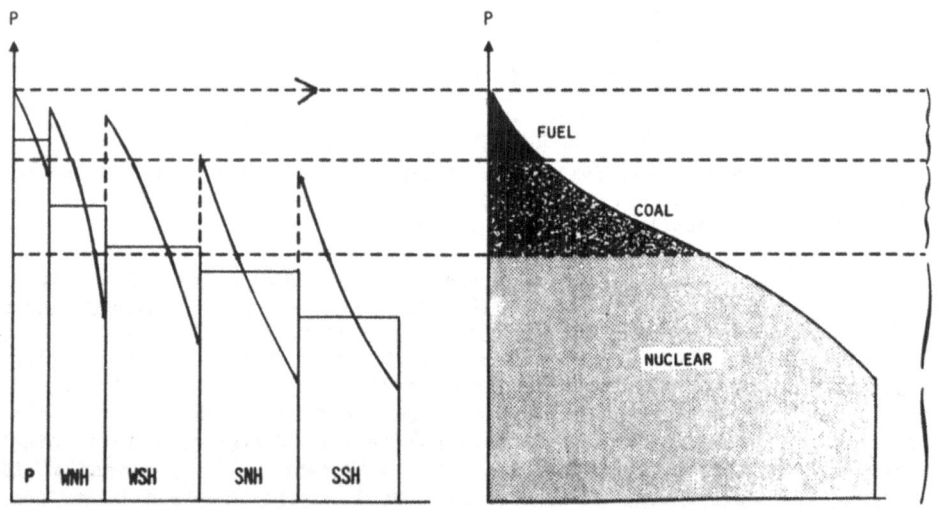

LOAD DURATION CURVE DURING EACH TARIFF PERIOD   LOAD DURATION CURVE

- figure 4 -

In France, the evolving nature of the time of the peak, and the changes in the mix of plants, has led to such a situation. These facts, combined with the potential development of dual energy system, made it necessary to draw-up a tariff structure which reflects more accurately the differences in costs and leads to operating a dual-energy equipment in the most economical way.

The solution was found by implementing optional real time tariffs which allows to distinguish low load periods in which the marginal cost is equal to the fuel cost of nuclear.

Real time tariffs can be considered as a major step towards an effective implementation of "spot pricing", but the advantage is that the customer is aware of the specific duration of the price period, even if dates are random. As a consequence, he can take investment decisions about electric equipment without any uncertainty on his electric bill.

REAL TIME TARIFF

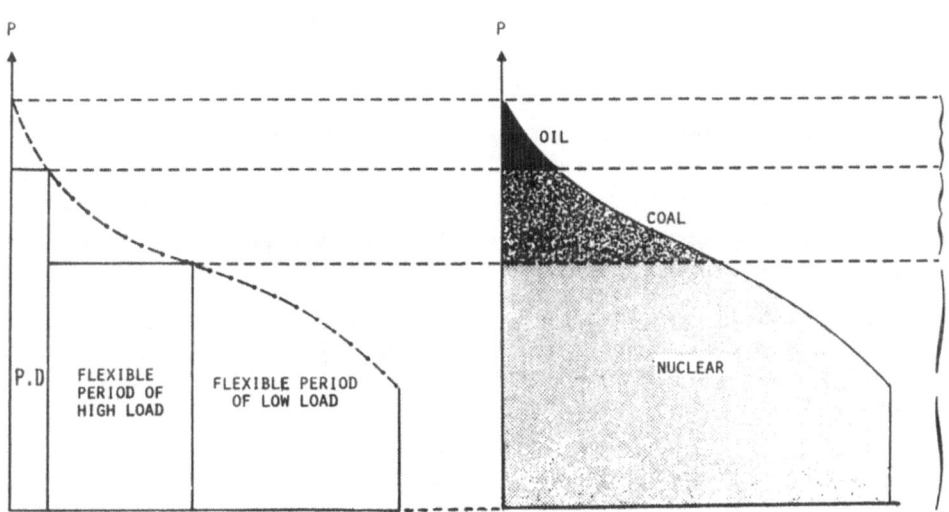

LOAD DURATION CURVE DURING FLEXIBLE TARIFF-PERIOD   LOAD DURATION CURVE

- figure 5 -

VI.2 The peak day withdrawal option

This new optional tariff has been introduced to reflect the evolving nature of the time of the peak. As indicated above, the peak period now covers a large number of hours per day distributed over a small number of days of the year. However, the date of these peak days is unpredictable. This new phenomenon calls for a specific response.

The peak day withdrawal option, offered to small as well as large customers- includes a flexible peak period consisting of twenty two, 18-hour days, which EDF chooses in real time. Their choice allows them to select periods in which, with a high degree of probability, the load is such that specific peak production units have to be installed and commissioned. The energy prices vary much more widely than in the standard tariffs. For Low Voltage tariffs, table 3 shows that the price of energy may vary from the ordinary period, to the peak day period, by a ratio of 10 to 1.

	Demand charge F/year	Energy cF/kWh	
Standard tariff	1118.16		49.59
Low-load hours option	1555.92	Normal hours : Low-load hours :	49.59 28.28
Peak day with- drawal option	681.00	Peak days : Off peak :	289.13 31.58

- Table 3 -

The different options offered to a customer subscribing 12 kVA (effective July 1986)

The signal is given with a very short notice (of half an hour) to the LV customer using the ripple control system .

For the green tariff applied to large supplies, the differences in prices are even more acute. For HV and VHV customers, the signal is given through the switched network (they are also informed by phone the day before).

The peak day withdrawal tariffs are particularly suitable for reflecting the cost of electricity supplies for bi-energy systems, and give the customer enough information to operate his equipment in the most economical way. These tariffs are a strong incentive for the installation of a heat pump or an electric boiler combined with an oil boiler as a back-up system. In this way, EDF expects a large substitution of electricity for fossil fuels on the market for space heating appliances.

The figure 6 presents the load curve of a residential customer equipped with an electric/fuel oil boiler, for one week during which 5 peak days (from 7am to 1am) have been decided by EDF - and illustrates the possible large response of some customers to this type of tariff signal.

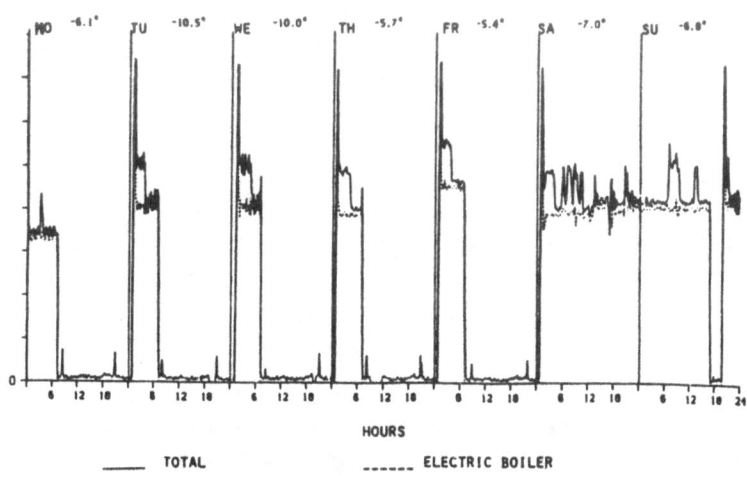

WEEKLY LOAD CURVE
WEEK : 7.1.85 TO 13.1.85

HOURS

_____ TOTAL   ------ ELECTRIC BOILER

- figure 6 -

This tariff is also interesting for a large number of industrial sectors : for example arc furnaces and zinc or chlorine electrolysis can withstand a partial or total shutdown during the peak days. The use of self-generation is also another means of response to this tariff signal.

In 1985, after four years of different experiments, these options are offered to all categories of customers. The impact of these options at the beginning of 1986 is indicated in table 4 ; results are encouraging, considering the short period of application : one year for LV customers, and three years for MV and HV customers.

	LV customers	MV and HV customers
Number of customers	100 000	1100
Reduction in diversified peak demand	350 MW	1200 MW
% of total system peak (1987 ; 60 GW)	2.3 %	

- Table 4 -

The reduction in peak demand is expected to increase to 5 000 MW in 1995, which will represent approximately 6 % of peak-demand of the system.

VI.3 The "modulatable" option

In low-load periods, electricity is generated at particularly by low cost, especially when nuclear capacities are only partly used. However, a strict definition of tariff periods does not represent accurately this type of phenomenon, since prices are necessarily the mathematical expectation of costs which are largely of random character, particularly with respect to the availability of equipment or the level of demand.

The modulatable tariff is presently experimented by HV and VHV customers. This tariff is proposed as an option and is based on the same principle as the peak day withdrawal option. This option offers 4 tariff periods of fixed duration but with a flexible timing defined in real time by EDF. The different periods are :

- The peak day period : 22 18-hours days, with the same definition and prices as for the peak day withdrawal option,

- the flexible winter : 9 weeks (except the possible peak days) during which the marginal cost of generation corresponds to the fuel cost of the most expensive units but with almost no capacity cost,

- the flexible intermediate season : 19 weeks (except the possible peak days) ; marginal costs are mainly fuel costs of coal-fired stations,

- the flexible low-load season : the remaining 24 weeks in which it is highly probable that the marginal cost is limited to the nuclear fuel cost.

Table 5 presents the standard tariff and the modulatable option for customers whose power exceeds 10 MW, and indicates that with the latter option, prices are more differentiated and closer to the variations of marginal costs according to the state of nature.

1 - Standard tariff

Demand charge (F/kW/year)   242.46

Energy charge (cF/kWh)
Winter :
peak hours :   89.71
normal hours :   63.18
low-load hours :   42.30

Intermediate season
normal hours :   31.52
low-load hours :   23.44

Summer
normal hours :   14.28
low-load hours :   9.29
July-August :   6.87

2 - "Modulatable" tariff

Demand charge (F/kW/year)   242.46

Energy charge (cF/kWh)
Peak day :   187.13
Flexible winter :   45.18
Flexible
intermediate season :   17.12
Low-load season :   8.00

- Table 5 -

**Standard tariff and modulatable option for HV customer
whose power exceeds 10 MW, and with an average load factor
(general variant)**

This tariff is very effective for bi-energy systems. Table 6 indicates the additional sales of electricity during 1985 and 1986 associated with new bi-energy systems purchased by HV or VHV customers, on the grounds of the tariffs.

	Consumption per year	
	1985	1986
$ / barrel	28.0	15.0
Summer bi-energy (Standard tariff)	2.3 TWh	2.0 TWh
Modulatable bi-energy (experimental option)	2.7 TWh	1.6 TWh
Total	5.0 TWh	3.6 TWh

- Table 6 -

**Additional sales of electricity
for 1985 with new bi-energy systems**

After one year of testing, the results point clearly to the fact that this modulatable option really improves the tariff system.

For residential customers, this type of modulatable tariff could be a better signal than a seasonnal time-of-day tariff, and it requires the same metering devices. However, experiments are necessary to be sure that this type of tariff is well understood by small customers.

VI.4 News developments

Customers who invest in dual energy systems, take their decision according to the total operating cost of their equipment : fuel plus electricity. Moreover, the decision is very sensitive to unpredictable changes in the profitability of the investment. If the equipment was operated in the most economical way, that is to say by taking into account the inter-annual variations of marginal costs, the total operating cost of the equipment remains very stable. As a consequence, new tariff system are now under study, to make possible inter-annual variations of period of use of such systems, and also inform the customer in advance of the cost of its uses. A solution, currently under experiment, is to charge energy prices according to the equilibrium price between the two alternative energies, to deliver the power demand to the customer during a period of a random duration at impredictable dates and, as a counterpart, to give a pay-back (in F/kW) to the customer. The pay back is calculated as the difference between short run marginal cost of the supply and the equilibrium price.

CONCLUSION

Like the peak day withdrawal option, the modulatable option is a good illustration of the consistency of the marginal cost tariff system with a load management policy.

Some are proposing to confront load management techniques with marginal cost tariff systems to determine which system yields the best economic efficiency. In fact, as shown by the example of the peak day withdrawal option, this confrontation is not valid, since it is possible to define tariffs which can show the marginal cost of specific supplies which these techniques will be able to offer the customers. This correct indication of prices is vital, since this is the factor allowing for consistency between the means of control at the level of supply and the means of control at the level of demand.

BIBLIOGRAPHICAL REFERENCES

[1] M. BOITEUX, "La vente au coût marginal"
RFE, Paris, December 1956

[2] M. BOITEUX, P. STASI, "La tarification des demandes en pointe : application de la théorie de la vente au coût marginal", RGE, Paris 1949

[3] M. BOITEUX, "Sur la gestion des monopoles publics astreints à l'équilibre budgétaire", ECONOMETRICA pp 22-40, January 1956

[4] J. BERGOUGNOUX, "Marginal cost pricing and investment financing" EDF paper and 2nd Technical Summer Workshop Inter-american Development Bank, Washington, 1982

[5] Y. BALASKO, "A contribution to the history of the green tariff. Its impact and its prospect", Conference of the Institute of Public Utilities, Detroit, May 1975

[6] Y. PIOGER, "An analysis of the change in the daily load curve in the total load of the system and in individual or group consumption : an evaluation of the effect produced by a pricing policy", Contribution to the first technical seminar on marginal cost analysis and pricing, Inter-American Development Bank, Washington, October 22-23-24, 1980

[7] J.P. ACTON, D. McKAY, "Quantitative Aspect of Industrial Use of Electricity Under Time-of-use Rates in France, England and Wales". The Rand Corporation, Santa Monica, March 1983

[8] M. FRANCONY, B. LESCOEUR, Ph. PENZ, "Marginal cost pricing : updating the french electricity tariffs to reflect changes in the characteristics of the supply demand system" : 13 th Annual Conference, Institute of Public Utilities, Williamsburg, December 14-15-16, 1981

[9] S. BUSSON, "Panel BT Domestique 1982 - 1983 "Bulletin de la Direction des Etudes et Recherches, EDF - Série B n° 2 - Paris, 1985

A SPOT PRICE BASED ENERGY MARKETPLACE

Professor Fred C. Schweppe
Laboratory for Electromagnetic and Electronic Systems
Massachusetts Institute of Technology

This paper provides an overview of the concept of a spot price based energy marketplace.

Section 1 motivates and defines the basic concept. The three steps to an energy marketplace are presented in Section 2. Section 3 addresses customer response in an energy marketplace. Sections 4 and 5 discuss energy marketplace operation for developed and developing countries.

1. What is an Energy Marketplace?

The electric utility industry today is undergoing rapid and irreversible changes. Volatile fuel costs, less predictable load growth, a more complex regulatory environment and a deceleration in conventional technical progress are important examples of these changes. Yet the need for growth in productivity and efficiency, and for increased flexibility to handle future uncertainties is stronger and more challenging than ever. The utility industry, which comprises a substantial economic sector in all industrialized countries, must evolve to meet this challenge. New directions for the utility industry are being sought by many interested parties in the government, the private sector, and the universities. One such direction has been widespread interest in utility-customer cooperation through innovative rates characterized by broader options and better use of information on utility costs and customer needs.

The goal of this paper is to summarize a theoretically sound, yet practical foundation for the implementation of utility-customer transactions based on today's needs. The foundation meets four criteria:

o Freedom of Choice: Provide customers with options on the cost and reliability of supply and how they choose to use electric energy.

o Economic Efficiency: Motivate customers to adjust their own electric energy usage patterns to match utility marginal costs.

o Equity: Reduce customer cross subsidies; i.e. a customer's charges are based on the utility's costs to serve that customer.[1]

o Utility Control, Operation and Planning: Consider the engineering requirements for controlling, operating and

---

[1] There are, of course, other definitions for "equity".

A. T. De Almeida and A. H. Rosenfeld (eds.), Demand-Side Management and Electricity End-Use Efficiency, 207–226.
© 1988 by Kluwer Academic Publishers.

planning an electric power system.

Present Day Transactions

Most transactions between utilities and their customers fall today far short of meeting these four criteria. Flat and time-of-use rates are related to actual marginal costs in only a gross, average sense. The demand charge is an anachronism with at most a tenuous relationship to marginal costs; it usually encourages counter-productive customer behavior. Load management approaches often involve direct utility control (i.e., the big brother approach) and sometimes encourage wasteful customer behavior. Cross subsidization between customer classes and within a given class is rampant. Utility-customer transactions are typically specified without considering the engineering problems of controlling, operating, and planning an electric power system.

The failure of most present day utility-customer transactions to meet today's needs can be traced to their historical foundations. They were established by individuals unconcerned with power system control and operation; during times when communication and computation were very expensive; when there was less incentive to use electricity in an efficient fashion; and when cross subsidies were of limited concern to society.

The four basic criteria cannot be achieved by putting "bandaids" on the present day approaches. They cannot be achieved by adding modern microelectronics whose functions are based on present day approaches.

The four basic criteria can be achieved only by returning to the first principles of economics and engineering and by viewing the utility and its customers as a single integrated system. The result of this integration is the spot price based energy marketplace which is the subject of this paper.

Energy Marketplace Transactions

Students taking an introductory course in economics are taught that the best way to achieve

o Economic efficiency
o Equity
o Customer freedom of choice

is the use of marketplace where market clearing prices are determined by supply and demand. For example, when a lot of fresh produce (e.g., lettuce) is available, the prices go down and consumption goes up. During other times of the year, or after a hard freeze, the supply goes down and prices go up to reduce consumption.

Five ingredients for a successful marketplace are:

1) A supply side with varying supply costs that increase with demand

2) A demand side with varying demands which can adapt to price changes

3) A market mechanism for buying and selling

4) No monopsonistic behavior on the demand side

5) No monopolistic behavior on the supply side

Electric energy is an ideal text book type commodity relative to the first four ingredients, i.e.:

1) Figures 1.1 and 1.2 illustrate the effect of changing supply-demand conditions on supply costs. Figure 1.1 shows measured marginal fuel costs while Figure 1.2 shows total marginal costs (for a different utility) which also includes non-fuel effects such as quality of supply/reliability premiums and recovery of embedded capital costs (revenue reconciliation). These marginal cost variations are highly correlated with variations in demand levels.

2) Industrial processes and space conditioning are only two examples from a host of demands that are price responsive even to current rate structures.

3) The existing market mechanism (billing for electric power, etc.) can be adapted. Industrial customers who already have recording demand meters, needed no new hardware to implement a marketplace with prices varying each hour (assuming they also have a telephone).

4) Monopsonistic behavior is difficult on the demand side because the number of customers ranges from thousands to millions.

Relative to the fifth ingredient, electric energy is not an ideal text book example for a marketplace because the supply side is a monopoly that is either government regulated or government owned. Rate of return regulation is used today to limit the ability and incentives to behave monopolistically. This would continue with spot pricing.

We define

Energy Marketplace: The buying and selling of electric energy between independent customers and a regulated or government owned utility.

The energy marketplace is designed explicitly to include engineering issues associated with power system control and operation. Hence all four of the original criteria (efficiency, freedom of choice, equity, and control/operation) are met.

In the energy marketplace, all utility-customer transactions are based in a self consistent fashion on a single quantity, the hourly spot price. The hourly spot price is determined by the demand at that hour and the hourly varying costs and capabilities of the generation, transmission and distribution systems. The hourly spot price is defined in terms of marginal costs subject to revenue reconciliation.

In the energy marketplace, there is closed loop feedback between the utility and its customers. The whole electric power system (generation, transmission, distribution, and customers) is controlled and operated in an integrated fashion, without removing the customers' freedom of choice. This is made possible by the diversity in customers' characteristics, desires and needs.

The benefits of well designed, real time, utility-customer feedback are obvious. However, so are the metering and communications costs associated with conveying the necessary information. Therefore, the energy marketplace transactions are designed to match benefits to transactions costs. Some customers (e.g. large industrial) might see prices updated each hour while other customers (e.g., residential) might normally see prices updated each billing period. However, a residential customer who wants to exploit hourly price variation, has the option of seeing more frequent price updates, provided the customer pays the additional costs.

What is the Difference?

The initial reaction of many people to spot pricing is that the major difference between spot prices and present day transactions is that spot prices have complex time variations. Actually, present day prices also exhibit complex time variations so the major difference is in the nature of the price variations, not their presence. As one example, consider Figure 1.3 which plots the historical variation of prices ($/kWh) for one utility. The hourly cost (spot price) variations of Figures 1.1 and 1.2 simply exhibit finer time detail. As a second example, consider an industrial customer with a 5 ¢/kWh energy charge and a 5 $/kW demand charge based on the energy used during the customer's peak hour during the month. The corresponding energy rate paid by the industrial customer is plotted in Figure 1.4. It displays a dramatic time variation which bears little resemblance to either Figures 1.1 or 1.2. Many existing interruptible contracts and direct control schemes can also be interpreted in terms of rates that vary in real time.

A second difference lies in the nature of the relationship between the utility and its customers. In a spot price based energy marketplace, the utility and its customers are partners working together to achieve the maximum benefit from electric energy usage at minimum cost. The amount of such partnering found in present day utility-customer relationships is small at best.

Present day flat and time-of-use rates are specified by formulas based on expected costs, rates of return on equity, etc. An hourly spot price is based on the same principles; but the formula is solved every hour by computers instead of once a year or once per month (for fuel adjustment clauses implementation). Present day power system operation often involves a real time spot market for purchases and sales between utilities. The spot price based energy marketplace simply extends this spot market to include the customers. Thus it can be concluded that

o The spot price based energy marketplace is the logical evolution of present day rates and load management techniques; married with present day practices of power system operation, the concept of utility-customer partnering, and the availability of inexpensive communications and computation equipment.

2.   Three Steps to an Energy Marketplace

A spot price based energy marketplace which meets the four criteria of Section 1 can be achieved in three steps:

Step I: Define hourly spot prices and evaluate their behavior.

Step II: Specify an appropriate set of utility-customer transactions based on the hourly spot price and associated transactions costs.

Step III: Implement the energy marketplace considering the needs and capabilities of both the utility and the customers.

Step I: Define Hourly Spot Prices

The fundamental quantity underlying the energy marketplace is the hourly spot price.[2]

The hourly spot price is defined in terms of marginal costs subject to revenue reconciliation (i.e. recovery of operating and embedded capital costs).

The value of the spot price at any hour depends on the hourly variations of

o Generation fuel costs and capacities
o Transmission distribution network losses and capacities
o Aggregated customer demand patterns

The hourly spot price is a random process (e.g. see Figures 1.1 and 1.2). Its future value cannot be predicted perfectly because of random equipment outages and demand variations.

An hourly spot price can be determined for a utility which is buying energy from, as well as selling energy to its customers. The buy-back hourly spot price can be either greater or less than the selling hourly spot price (because of revenue reconciliation effects).

Step II: Specify Utility-Customer Transactions

---

[2] The use of an hour as the fundamental time unit is convenient, but not essential. Definitions could range from minutes to several hours.

All utility-customer transactions are based on the hourly spot price defined in Step I. Three general types of transactions are

o Price-only
o Price-Quantity
o Long Term Contracts

Examples of price-only transactions are

o One Hour Update: Customers see prices ($/kWh) varying each hour, predicted and communicated one hour in advance.

o 24 Hour Update: Customers see prices ($/kWh) varying every hour, predicted and communicated 24 hours in advance.

o Time of Use (TOU) and Flat Rates₃: Rates ($/kWh) are calculated using predictions of hourly spot price behavior one billing period in advance.

The choice of what type of price-only transactions to offer is based on a trade-off between the transactions costs (metering, communication, computation, etc.) and the benefits obtained from the transactions. For example, large industrial customers might see one hour update prices while small residential customers see flat rates.

Price-quantity transactions involve a short term utility-customer contract. For example, a customer may choose to receive a specified quantity of energy at a lower price with a contract to drop all or a part of such usage on a signal from the utility. Price-quantity transactions include as special cases present day interruptible contracts and direct load control. However the energy marketplace transactions are more robust and can better match the customer's needs and capabilities. Use of price-quantity transactions instead of price-only transaction can be motivated by:

o A need for fast acting, accurate load control (seconds to minutes rather than hours) to maintain power system security.

o A desire to reduce transactions costs below those of rapidly varying price-only transactions. Some price-quantity transactions are cheaper to implement than some price-only transactions.

Long term contracts are fixed price and fixed quantity. Thus they are like present day commodity contracts, and like present day long term contracts between utilities they can extend hours, days, months and

---

₃These energy marketplace TOU and flat rates differ from conventional ones because they vary each billing period and a demand charge is not used.

years into the future. As an example suppose that on January 1, an industrial customer contracts for 1000 kWh of energy to be delivered between 10 and 11 a.m. on July 1, for 10 ¢/kWh. If when July 1 finally comes, the price between 10 and 11 a.m. is actually 9 ¢/kWh, the actual cash flow is as follows:

If Customer Uses	The Customer Pays
1000 kWh	$100
2000 kWh	$100 + $90 = $190
0 kWh	$100 - $90 = $10

Thus the incremental cost of the customer's usage that hour is 9¢/kWh. Such transactions enable a customer to buy an insurance policy by locking in 1000 kWh at 10 ¢/kWh, even though the customer still sees the spot price for actual usage.

A variation on this long term contract is the option wherein the customer buys the right to buy up to a fixed amount of energy at a fixed price. The customer exercises the option if the hourly spot price turns out to be greater than the option price.

Step III: Implement Energy Marketplace

Implementation of the energy marketplace involves the utility and its customers operating as partners.

Utility implementation concerns include real time calculation/ prediction of hourly spot prices; metering - communications - billing; and system control center operation using the new control signal called price. The impact of the energy marketplace on utility long term investment decisions is also of concern.

Customers who choose to exploit the energy marketplace potentials must implement the appropriate response systems, which could range from simple manual response to sophisticated digital control. The utility can provide the control mechanism as a service to customers.

Sections 4 and 5 provide examples of energy marketplace implementations.

3. How Will Customers Respond?

A key question for a spot price based energy marketplace is: How will customers respond (change their usage patterns) to price changes? Two ways to try to answer this question are:

o Use direct observations of actual responses.
o Use models of response.

Unfortunately, not enough direct observations of the right type are available today. There is a lot of data (and studies) on customer response to time of use rates which are a special case of spot price based transaction. However, there is a large difference between a time of use rate behavior and that of say Figures 1.1 and 1.2. This difference combined with the nonlinear nature of customer response makes an extrapolation of response from time of use results a vague guideline at best. Interruptible rates and certain types of direct load control are also special cases at spot price based transactions but their method of implementation makes it very difficult to draw conclusions for an energy marketplace type operation. There are various implementations of 1 and 24 hour type spot prices in operation but the data coming from them is still too limited to provide a basis for any general conclusion.

When enough direct observations are not yet available, it is necessary to resort to the use of models. Models can range in sophistication from statistically based computer simulations or optimizations to simple mental pictures obtained by having an industrial plant manager spend a few hours showing one around the plant and explaining the processes and operating constraints.

We will now provide some discussion on what we know of customer response in the industrial and residential sectors.

Industrial Response

We have modeled (at various levels of detail) many industrial customers in different parts of the USA. This experience showed that five issues to be considered when talking to an industrial customer are:

o The negative initial reaction syndrome.
o The importance of non-physical and non-economic factors.
o The difference between a demonstration and the real world.
o The importance of advanced knowledge.
o The nonlinear character of the response.

The initial reaction of an industrial plant manager to the idea of facing a 1 hour or 24 hour update spot price can be expected to be negative. However, in most cases this negative reaction can be turned around. For example, a negative reaction to the possibility of very high energy rates at certain times can be countered by the existence of very low energy rates at other times and the death of the demand charge. As a second example, an initial negative reaction based on the concept of moving entire work shifts around can often be countered by discussing large electrical using devices whose operation can be rescheduled without effecting more than a few workers. As a third example, an initial negative reaction about uncertainty associated with rapidly varying rates (hours to days) can be replaced by an appreciation that a fluctuating market is a place where money can be made. One customer, who was on a one-half hour update spot price, complained that during some seasons of the year the price didn't vary enough for them to make any money.

When we started looking at industrial response, we naively assumed that response depended only on the physical character of the plant and the economics of its operation. However, in the real world other factors also play a key role. These factors include the nature of union contracts, company management (locally owned or part of a big corporation), and the character of the work force (social background) and the plant location (was it safe to work there at night). Two plants, with identical SIC codes and located within twenty miles of each other, had completely different response capabilities.

It became very clear that industrial response to a one year test or demonstration program would be much less than for a real world spot price based energy marketplace that won't go away in 12 months. Many customers could greatly improve their response capabilities by relatively minor additions of new control, process, etc., equipment. However, such modifications could not be cost justified for a mere demonstration or test.

Industrial response can be greatly increased if the customers know a few hours to a day in advance when exceptionally high or low prices will or are expected to occur. Such lead or warning time allows rescheduling of high energy consuming processes with much less cost. The use of a 24 hour update provides a "hard price" a day in advance. For a one hour spot price, it is important for the utility to provide a forecast of future prices (the same forecasts which formed the base for updating the 24 hour spot price). The importance of warning or lead time is another reason why it is difficult to extrapolate from time of use or interruptible response to energy marketplace response.

Industrial response is usually a very nonlinear function of price. Doubling the price from 5 to 10 ¢/kWh may create little response while doubling it from 10 to 20 ¢/kWh could cause a sizable reaction because of the threshold effects. Price differentials during a day can be equally if not more important than the absolute values of the prices.

Residential Response

Residential response is entirely different from industrial response. Our studies (ponderings) on residential response to a spot price based energy marketplace lead to the conclusion that one key issue is the character of the electronic support system that is provided. One cannot expect the typical residential customer to interact continuously with a spot price based energy marketplace without electronic support. If a residential customer has available a well designed microprocessor based information and control system, with user friendly interface, extensive response can be expected. Residential customers without such support available should, in general, see only very simple spot price type rates.

The cost of microelectronics is continuing to fall at a rapid rate. Therefore a key issue in residential response is how long will it take before the costs of such electronics becomes less than the benefits (to the customer and/or utility) of the resulting response. Some might argue that such a time has already arrived.

The importance of warning time and the nonlinearity of response apply to both residential and industrial customers.

Why No Numbers?

We have devoted a lot of time and effort to trying to understand how industrial and residential customers will respond. We firmly believe large response will exist. However, we have chosen not to present explicit numbers on customer response. One reason for this decision was to desire to keep the size of the paper under control. However the main reason is that our understandings have not reached the point where we can provide numerics that can be extrapolated to cover a utility service territory. We are like the classical seven blind men feeling different parts of an elephant. We have a lot of information but still are not certain what the overall elephant looks like. However, we do know that it is big.

4. Energy Marketplace Operation: A Developed Country

A simplified example of a sophisticated energy marketplace in operation in a developed country is used to illustrate its basic functions.

Figure 4.1 summarizes the main energy marketplace transactions and information flows. For simplicity, only one customer is shown. Each box in Figure 4.1 is discussed separately.

Generation Transmission Distribution System:

The energy marketplace has a long range impact on generation and network capacity requirements. In many situations, the result is a reduction in installed capacity (per unit of demand) and less reliance on generation peaking units.

Power System Operation

All of the basic present day power system operating functions continue in an energy marketplace, but some of the functions are modified. Short term demand forecasts now include the effects of price. Unit commitment logics incorporate the effect of price feedback. Operating reserve requirements are carried by the load or generation, whichever is least expensive. All of these changes reduce total system costs.

Meter

A recording meter measures and stores hourly energy usage for each customer. It is read once a month and used to compute the monthly bill.

Price-Only Transactions

At ten minutes before the hour, the 1 hour update spot price is computed based on a seventy minute forecast of system operation conditions (fuel costs, losses, generation reserve margin, and network capabilities) and demand (taking into account price effects). The one hour update spot price is automatically communicated in digital form over telephone lines to the computers of those customers who are on the 1 hour update. It holds for one hour.

At 3 PM of each afternoon, the hourly spot prices for the 24 hour period starting at 3:00 AM of the next morning are computed using forecasts of operating conditions. Customers on a 24 hour update price can receive the 24 numbers by telephoning the utility any time after

4 pm. Prices are available in verbal or digital format. Next day prices are also communicated using newspapers and TV.[4]

Price-Quantity Transactions

The price-quantity transactions are available only to those customers on a 1 hour update spot price. At ten minutes before each hour, an interruptible energy rate for the subsequent hour is computed, based on system operating conditions and forecasts of how much interruptible energy will be purchased by customers. It is communicated in digital form over telephone to those customers who request it. Customers who choose to buy interruptible energy do so by communicating the secure energy level they want, so that all usage above that level is at the interruptible price. If the utility experiences short term emergency operating conditions (such as a major generation forced outage), the reduction required from each customer is computed and communicated via telephone to the customer's computer.

Short Term Price Forecast

Forecasts of the hourly spot price for each hour of the next week are made available if requested by the customer via telephone, in either verbal or digital format. Naturally these are only forecasts and can be wrong.

Billing

Customer bills are computed each month by summing the metered and recorded hourly energy use, times the spot price that existed at the corresponding hour.

Long Term Price Forecasts

Monthly and yearly price forecasts are made by the utility itself or by independent information consultants who are not part of the electric utility.

Long Term Contracts

Long term fixed price, fixed quantity contracts are written through energy brokers who contract directly with the customers. These transactions have no effect on power system operation.

Now consider the customer boxes of Figure 4.1. Three different customers will be discussed.

- o A large industrial customer who sees one hour update spot prices, purchases interruptible energy, and has a long term contract.

---

[4] Of course, more sophisticated and expensive communication systems can also be used.

o A sophisticated residential customer seeing 24 hour
update spot prices.

o A simple residential customer seeing 24 hour update spot
prices.[5]

Large Industrial Customer (Foundry)

This industrial customer is a foundry with many metal working
machines (1 kW to 100 kW), electrical metal melting (10 MW to melt, 1 MW
to maintain molten), lighting, and space conditioning in the office.
This customer has a sophisticated computer system which is used for both
production scheduling (hourly for the next day and week) and direct
control of scheduled processes. There are two individuals responsible
for this customer's interaction with the energy marketplace; a production
scheduler and a production manager. The production manager is concerned
with long term (monthly to yearly) issues and makes long term contracts
by combining long term price forecasts with forecasts of this customer's
future electricity needs and cash flows. The production scheduler is
concerned with hourly and daily operation and provides the inputs into
this customer's control and scheduling computer.

The scheduling of plant operation is done by combining the short
term price forecasts with the specified product mix that the plant is to
produce.   Metal melting is always scheduled to occur at the time of
lowest forecast prices. When there is sufficient variation in the spot
price, some of the largest metal working machines (that require only a
few operators) are scheduled to times of low prices. The office space
conditioner is scheduled to make use of preheating and precooling the
building such that the space conditioning load is lowest during time of
high spot prices (a forecast of future outside temperature is an input to
the optimization logic). When prices get extremely low at night, some of
the metal curing processes that are normally done by natural gas are
switched to electric energy. The price-quantity transactions are
adjusted so that metal melting is always done on an interruptible price-
quantity basis.   During times when there is a large difference between
the secure and interruptible prices, the secure level of energy purchase
is reduced so that some of the other end uses are also on the
interruptible rate (unless the plant has a very tight production schedule
to meet).   All of those scheduling operations are done automatically by
the computer using input instructions and data provided by the production
scheduler.

The direct control function of the computer makes minor adjustments
on the actual hour by hour operations depending on how the hourly spot
prices actually behave (i.e., corrects for differences between actual
spot prices and the short term forecasts). When the utility calls for an
interruption, the computer drops the overall energy usage to the

---

[5] Many residential customers will see spot priced based flat or time-of-
use prices which are updated each billing period, but 24 hour update
residential customers make a more interesting example.

specified secure level by turning off the processes which had already been classified as being interruptible.

The plant management is happiest when the spot prices have large time variations, as it is these variations that offer an opportunity to save money.

A Sophisticated Residential Customer

This sophisticated residential customer under the 24 hour update spot price has a small, special purpose computer which automatically dials the utility once each day to get the 24 prices for the next day. The computer then controls the space conditioning and water heating to meet this customer's desires (as told to or learned by the computer) with minimum cost. If the prices rise above a critical level specified by this customer, the computer warns the customer so manual control of usage can be undertaken if desired. The computer acts as an expert system to help the customer diagnose his/her own needs/desires and make rational decisions for during normal and very high price variations days.

A Simple Residential Customer

This simple residential customer also sees a 24 hour spot price but has no computer. Most of the time, this simple customer ignores the price variations and operates as usual. However, during the few critical times of the year when the price gets very high, the customer exercises manual control. This customer learns of such times by reading the newspaper and/or listening to announcements made by the utility over the radio and TV.

5. Energy Marketplace Operation: A Developing Country

The example implementation of Section 4 emphasizes utility and customer computers talking to each other. This is reasonable for high technology parts of the world where both the electric power system and some of the customers are very sophisticated. Our second example considers a developing part of the world where the electric power system is being expanded as fast as possible to meet and fuel the needs of a rapidly changing, less sophisticated society.

Instead of repeating the box by box discussion of Figure 4.1 the main differences between this case and the sophisticated implementation of Section 4 are discussed.

In a developing country implementation, only large industrial and the largest commercial customers see 24 hour and 1 hour update spot prices. The 24 hour update is used for most cases. Residential customers (who have meters) see flat rates which may or may not be updated each billing period.₆

For the developing country, all long term contracts and price forecasts are provided by the utility itself.

The biggest difference is that for the developing country, new industrial and commercial facilities are built knowing, from the start, that they will be seeing energy marketplace spot prices. Thus their basic design incorporates control switches, storage capabilities, fuel switching capabilities, etc. that enable fast and large response to changing prices. With a large labor force, fancy digital hardware is not needed for control. This yields a much more price responsive industrial load than found in the developed country, where industries were built assuming they would always be furnished highly reliable power at low fixed rates. As a result the energy marketplace has a much faster and larger impact on the choice of generation mix and the amount of generation and transmission to be built.

To illustrate the impact on expansion planning, assume a policy decision is made to channel limited financial resources toward renewable generation and remote hydro sites with relatively weak transmission links to major load centers. By present day standards used in developed countries, the resulting system is unreliable in the sense that it cannot always meet the demand. However, having an industrial-commercial load that is very price responsive completely changes the meaning of reliability. Times of limited capacity are no longer handled by rotating blackouts. Instead industrial and commercial customers see high prices and reduce electrical usage in ways that do not have major financial impacts on their own operations. Industries that must run at a specific time, can have all the energy they need, albeit at the high price.₇

6. Discussion

This paper has provided an overview of a spot price based energy marketplace. We strongly believe that a such an energy marketplace is the logical and inevitable successor of the present day system. The key question is not if it will happen, but when, and to what degree.

₆ The choice of what types of residential metering systems to install is made keeping in mind the fact that, eventually, developing countries become developed and later, the energy marketplace will be extended to the residential sectors. Thus an electronic based system might be employed to avoid the "trap" developed countries now face of having a major investment sunk into already supplanted technology, in this case simple meters.

₇ A variation on this approach is to replace the use of prices to allocate a scarce resource by allocating each industry energy credits, where a kWh used during critical times requires the use of more credits. This leads to the possibility of bartering among customers.

REFERENCE

This paper is based on Chapter 1 of the book "Spot Prices of Electricity" by F. Schweppe, M. Caramanis, R.D. Tabors and R. Bohn to be published by Kulwer Press.

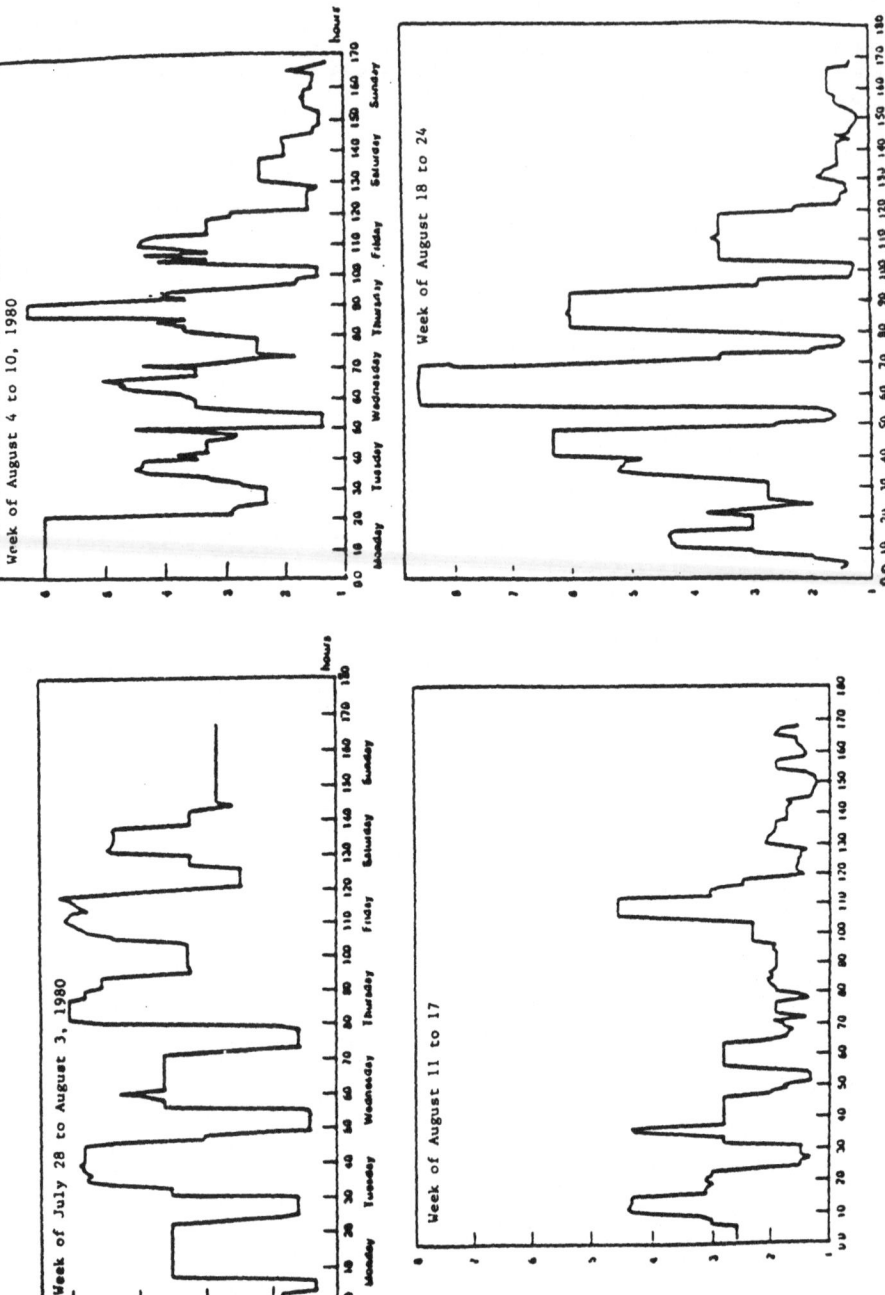

Figure 1.1

223

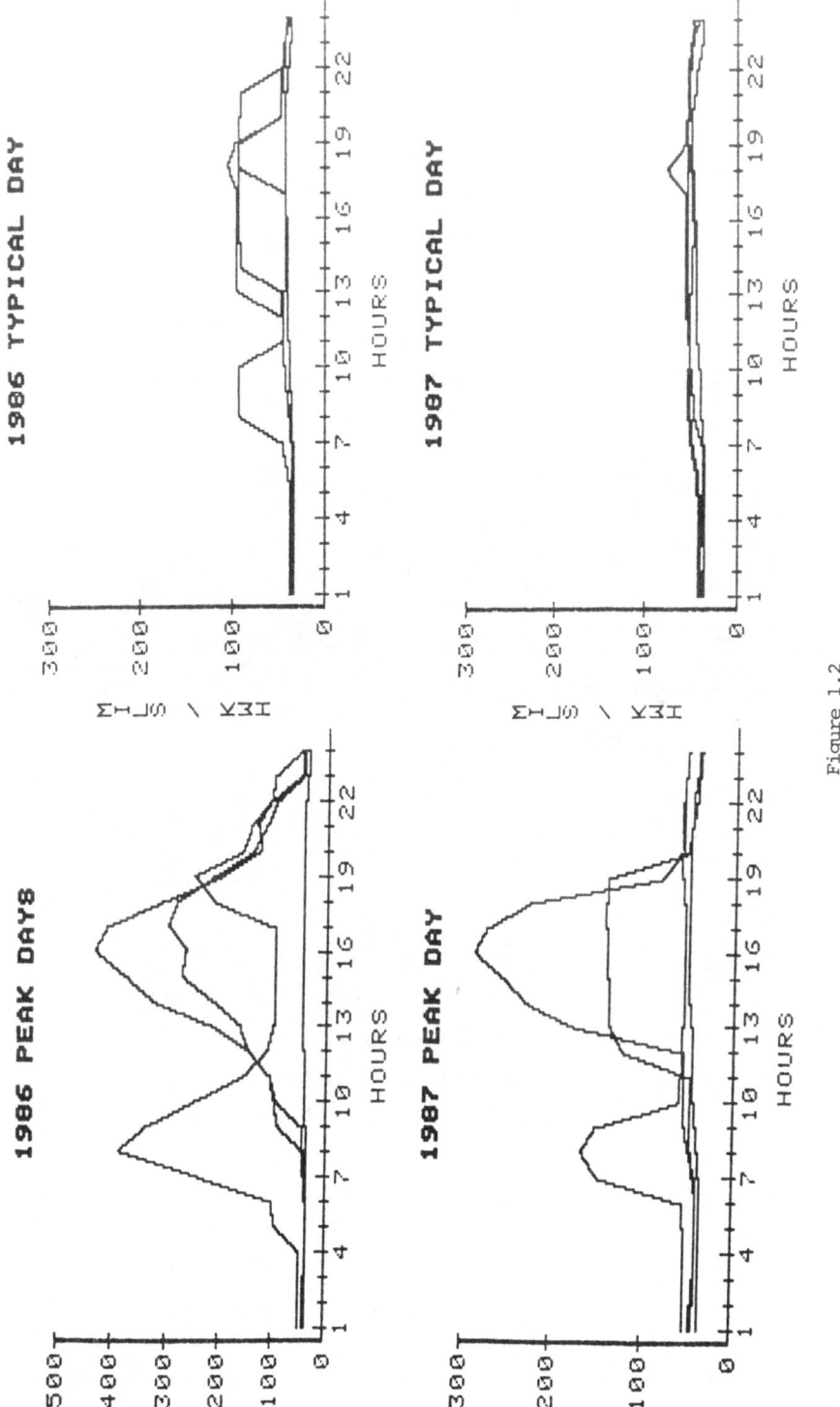

Figure 1.2

224

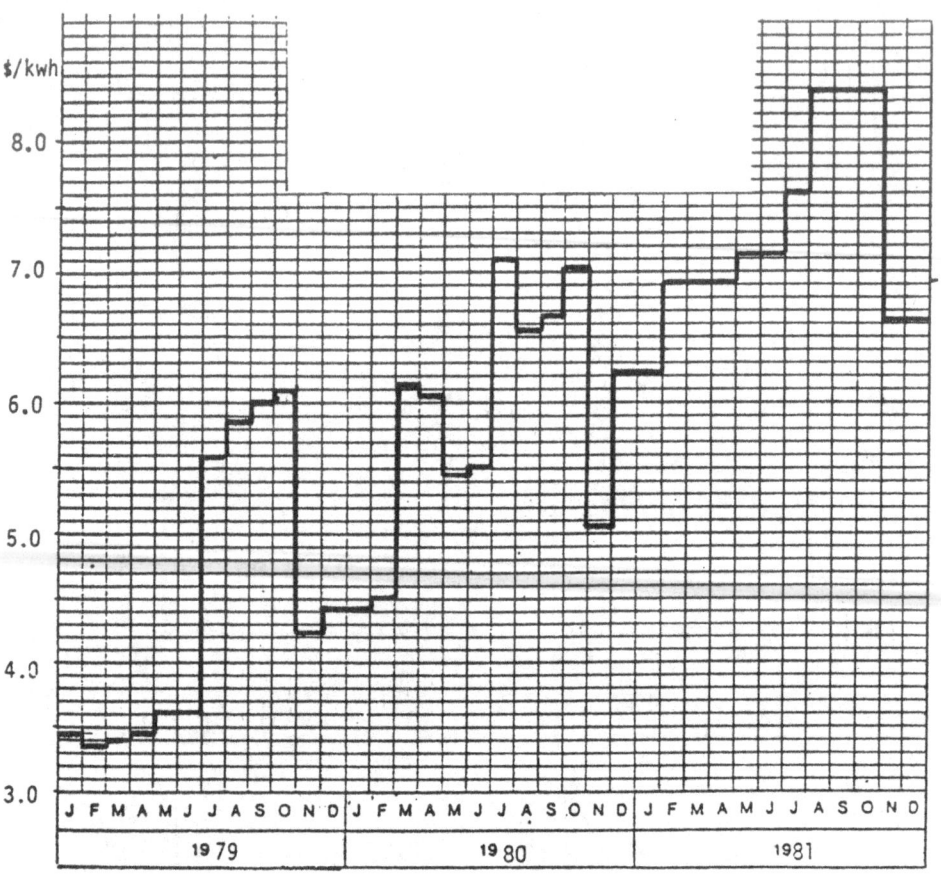

Figure   1.3

Examples of Monthly Variations in Residential Prices

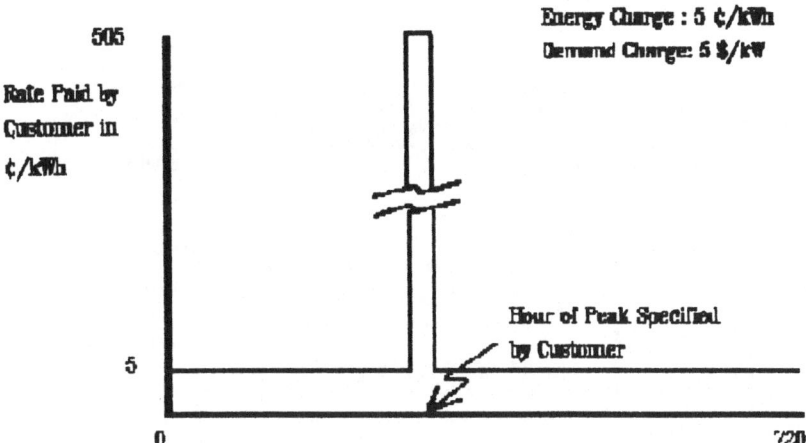

Figure 1.4

Effective Price Variations Resulting from Demand Charge

226

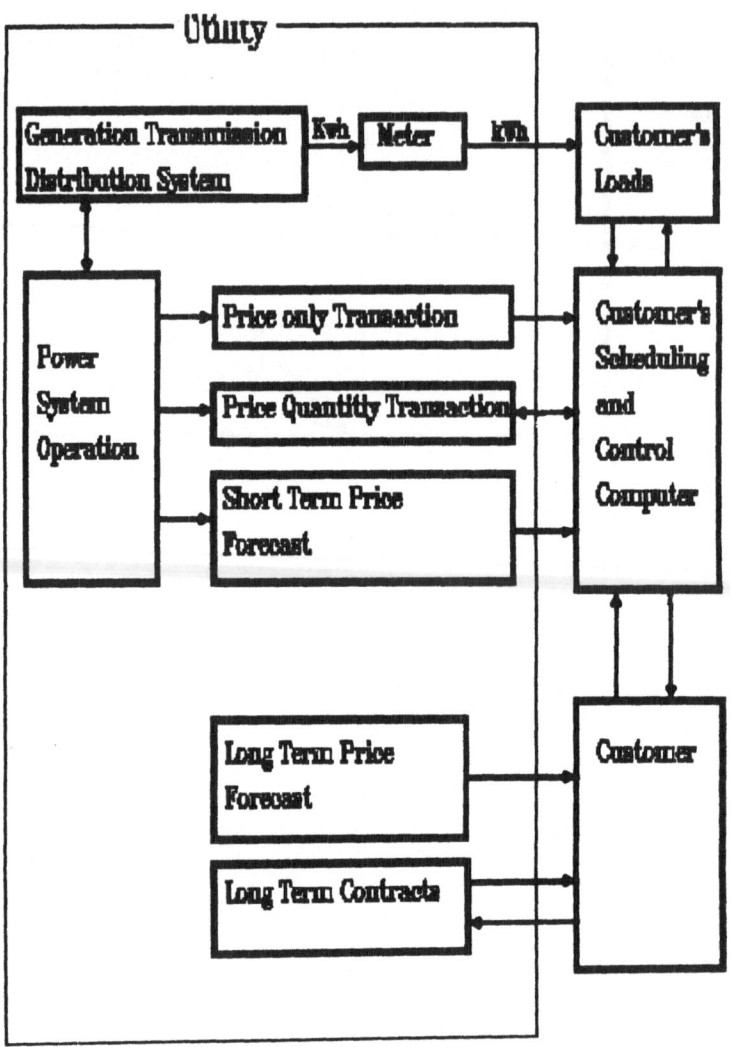

Figure 4.1

**Functions and Information Flow
in an Energy Marketplace**

DEREGULATION AND WHEELING

Professor Fred C. Schweppe
Laboratory for Electromagnetic and Electronic Systems
Massachusetts Institute of Technology

This paper addresses the concept of deregulation as it is being discussed in the USA where

> Deregulation: Replacement of a system of privately owned, state regulated electric utilities by a competitive system wherein at least the generators are privately owned and operated without state or federal regulation.

However we believe that at least some of the concepts are applicable to other types of relationships between electric utilities, customers and governments.

A companion paper "A Spot Price Based Energy Marketplace" outlines the principles underlying an energy marketplace which is designed to operate in a regulated environment (regulated private company, or government owned). However, its implementation opens the door to deregulation of all generation. This door is already being opened slowly in the United States at both the Federal and State level through regulations concerning cogeneration, small power production and wheeling. The purpose of this paper is to look behind that slowly opening door and try to see what might lie behind it.

This paper only presents a set of basic ideas; it does not analyze their impacts because such analyses have not yet been done. Since the advantages and disadvantages have not been quantified, we are not advocating deregulation (i.e. we do not know whether there is "a lady or a tiger" behind the door).

This paper shows how the establishment of a spot price based energy marketplace in a regulated environment (which we do advocate) can evolve towards or into a deregulated system. The reader may be surprised to learn that the trip from regulation to deregulation need not be very long (although it may be bumpy).

Many articles have been written on deregulation. In this paper we only consider the approach that evolves naturally from the regulated, spot price based energy marketplace.

The deregulated energy marketplace to be discussed here is summarized in Figure 1. It has three main participants:

o   A single regulated T and D company that controls the transmission and distribution system, and acts as a middleman in the energy marketplace.

o   Many independent, private generating companies which

A. T. De Almeida and A. H. Rosenfeld (eds.), Demand-Side Management and Electricity End-Use Efficiency, 227–244.
© 1988 by Kluwer Academic Publishers.

sell energy to the T and D company. The amount sold
at each moment is the decision of each generating
company.

o   The users of electricity who buy as much energy as
     they want from the T and D company.

Section 1 discusses the different elements of Figure 1. Sections 2
and 3 discuss how the deregulated marketplace of Figure 1 fulfills the
basic functions of an electric power system that are summarized in Table
1. Section 4 concludes with a scenario that might lead to Figure 1.

1. A Deregulated Energy Marketplace

The Regulated T and D Company

The transmission and distribution (T and D) company of Figure 1 acts
as an intermediary, both physically and financially. It buys all the
energy offered for sale at each moment, while selling all the energy
demanded by users.[1] It also periodically collects all payments by users
and pays all generating companies for whatever they produced. The
difference covers the T and D company's costs of maintaining and
operating the system and the marketplace. Its specific duties are to:

o   Build, maintain, and operate the transmission grid
     and distribution systems;

o   Determine the spot prices and communicate them to
     independent generators and customers.

o   Control the stability of the electric power system through
     a combination of pricing, direct management of certain
     physical attributes, and price-quantity contracts.

o   Collect money from users and pay money to generators.

The T and D company is regulated in a traditional rate of return
framework (or is owned and operated by some government agency). Revenue
reconciliation (to cover embedded capital costs, etc.) is done only with
respect to the T and D system by modifying the unreconciled spot prices
(up or down as needed) or alternatively using a revolving fund or
surcharge-refunds.

Generating Companies

Generation represents the bulk of the assets in most power systems.
In the deregulated marketplace of Figure 1, these assets are owned by a
number of private firms which

o   Build, maintain, and operate generation and storage

---

[1] Sales directly from generators to users are not forbidden. However,
when spot prices are properly calculated, they are irrelevant; customers
and generators always do at least as well by dealing only with the T and
D company.

units;

o   Sell electricity to the T and D at the current spot prices;

o   Meet zoning, environmental, and other restrictions, like any industrial firm;

o   Are not subject to regulation by public utility commissions;

o   Are barred by antitrust laws from explicitly cooperating with other generating companies in their areas, or owning too many units in one region;[2]

o   Are motivated by profit;

o   Are forced by competition to act in socially beneficial ways.

Today, ownership of generating units is based on geography. One utility owns almost all the units within its service territory. In contrast, ownership in the deregulated marketplace of Figure 1 is based on functional type. Different generating companies specialize in constructing and operating different plant types. They are then able to develop expertise, for example, in coal or nuclear power, instead of maintaining proficiency in all generating systems. This avoids the current problem of one utility owning and operating a single nuclear plant.

Today more than 200 local monopolies control generation. Even if ownership shifts to a smaller number of interspersed competitors, the electricity generation business will be less concentrated than many other U.S. industries.

Users

From a functional point of view, the electric energy users of Figure 1 behave in much the same way; independent of whether the energy marketplace is regulated or deregulated. Large industrial-commercial and sophisticated residential customers see 1 hour and 24 hour update spot prices while small residential customers see billing period updates. However, the deregulated energy marketplace employs a significant number of price-quantity transactions involving short term (seconds to minutes) transactions required for emergency state control (e.g. to carry operating reserve on the load). Such short term actions are also desirable in a regulated marketplace but are not as essential.

Although going from a regulated to a deregulated energy marketplace has little functional impact on the users, the hope of those who advocate

---

[2]Precisely defining how many is too many is a difficult task that depends on overall system characteristics.

deregulation is that the prices themselves will be lower (in the long run).

Information Consultants and Energy Brokers

Figure 1 shows two other participants; Information Consultants who forecast future spot prices (short and long term) and Energy Brokers who arrange side contracts to shift risks. Their roles are discussed further in Sections 2 and 3.

Decomposition of the Regulated T and D Company

The structure of Figure 1 combines the bulk transmission and the local distribution into one company. An alternative structure which has both advantages and disadvantages involves a single bulk transmission company and many individual local distribution companies; all of whom are regulated. We will only discuss the combined T and D company as it is somewhat simpler but almost all of the ideas apply equally to the decomposed structure.

2. Short-Term Operation and Control

This section discusses how the short term operating and control functions of Table 1 are fulfilled in the deregulated energy marketplace of Figure 1.

Using Spot Pricing to Coordinate Generation and Load

In the deregulated world of Figure 1, the generators are not centrally dispatched. Instead the Market Coordinator of Figure 1 sends each generator a spot price and each generator self dispatches itself by generating if the spot prices paid for electric energy exceeds the plant's marginal operating costs. A perceptive reader might say; "such generator self dispatch and central utility dispatch are theoretically equivalent." Such a reader would be right.

The Market Coordinator of Figure 1 keeps supply and demand in balance by continuously adjusting the spot price. On nights when demand is minimal, the spot price declines until only generating units with low operating costs remain on-line (e.g., nuclear or wind-powered units and perhaps some coal-fired units). As demand increases during the morning, the spot price may rise to the point where owners of electricity storage units, which had purchased and stored power during the night, begin selling back energy.

On the demand side of the market, users reschedule their electricity-intensive operations to times of low spot prices, and generally respond to spot prices to maximize their net benefits from electric power usage.

If a sudden temporary outage of large generating units occurs, the spot price immediately increases. Generating units already on-line increase their output to the maximum. Users' process control computers automatically delay the on cycles of air conditioners, furnaces, heaters, pumps and similar equipment. Other generators bring their units on-line. Owners of reservoir hydro units open their sluices to increase production. If the outage is severe, the regulated T and D company exercises some of its operating reserve price-quantity contracts for the first few minutes. These measures gradually reduce the spot price, which continues to decline as more units came on line. If the spot price remains quite high, some customers close down portions of their operations to save money. Eventually, equilibrium is restored at a higher spot price than before the outage.

Specification of Spot Prices

In a regulated energy marketplace, the utility's central control system knows the operating costs of all the generators, so the marginal fuel cost component of spot prices is computed by the formulas of this book. However, under a deregulated structure, the regulated T and D company's Market Coordinator does not necessarily know the precise operating cost characteristics of the private generators. Hence, the marginal operating cost component of spot prices could be determined empirically; by observing how the generation responds to different prices at various times. Alternately, the Market Coordinator could ask the private generating firms to furnish their operating cost data (confidentially of course) to facilitate system dispatch. We can see little reason for the private firms to say no or to try to play games and provide incorrect data.

Since the Market Coordinator knows the details of the transmission and distribution systems, the network components of the spot price are determined directly by the equations in the main text of this book.

o   Conclusion: Spot price evaluation is mostly, but not entirely the same for both the regulated energy marketplace and the deregulated system of Figure 1.

System Security

Security means avoiding major disruptions. The Market Coordinator of Figure 1 performs system security functions such as contingency evaluation and state estimation as is done today.

The Market Coordinator uses the available reactive control inherent in the network to maintain voltages within limits, as much as possible. However, when the voltage control capabilities of the network are exceeded, or when line flows threaten to exceed their thermal or dynamic limits, the participants' generation and usage patterns are changed by introducing transmission quality of supply components into the spot price for energy (reactive as well as real energy is charged/paid for).

The Market Coordinator maintains sufficient reserve to respond to major unexpected generation losses, dying tie-line support, and other dilemmas using price-quantity contracts which can be exercised as needed. Either the customer would be paid up front, or a very substantial payment would be made whenever the option is exercised. Analogous contracts with generators provide spinning reserves.

o   Conclusion: The security monitoring and control functions look a lot like those of a regulated energy marketplace.

System Dynamics and AGC

An interconnected system of generators can display oscillating or unstable behavior due to either small or large (e.g., faults) disturbances. Users and privately owned generators who aggravate such undesirable dynamic behavior are penalized (charged extra) by the T and D company while those whose characteristics improve overall dynamic behavior are rewarded (paid).

If the regulated T and D company is interconnected with other T and D companys; automatic generation control (AGC) systems are needed to keep frequency close to its desired value (60 or 50 Hz), to keep the sum of tie line flows near the net scheduled interchange levels during normal operation, and to provide emergency support when needed. The regulated T and D company pays private generators to accept and respond to AGC signals. The price level is determined by the marketplace. Of course, it is conceivable for the regulated T and D company to own some limited generation (pumped hydro would be nice) whose sole purpose is to accept AGC and emergency response signals.

o   Conclusion: At long last, the deregulated marketplace has resulted in something new; the need for pricing dynamic behavior. Research is needed before the details of how to do this are clear, but no major obstacles are foreseen.

o   Conclusion: Dynamic pricing might also be desirable in a regulated energy marketplace. For example, the dynamic effects of private generation in the USA operating under the PURPA (PL 95-617) might be best handled by pricing rather than trying to specify a set of standards which have to be met.

Relaying

Network relaying is done by the regulated T and D company following today's standards. The regulated T and D company establishes standards for the protective relaying at the generator-network interface. Internal generator relaying is specified by the private generation firms.

Exercise of Monopoly Power

A perceptive reader probably started to have second thoughts about deregulated short term operation when she/he began to worry that a single generating firm might have monopoly power in a region and hence be able to drive the spot price up artificially (beyond the marginal cost of generation). We have assumed away such a possibility in our formulation of Section 1. However, more analysis is needed before we are willing to believe the assumption is true. This is the first of many reasons why we do not advocate adoption of the deregulated marketplace of Figure 1 without further study.

3. Long Term Operation and Planning

Now consider the long term operation and planning functions of Table 1 and how they are done in the deregulated world of Figure 1.

Long term operations (hours to years) and planning (1 to 20 years) for any electric power utility requires forecasts of certain critical variables, such as fuel prices and demands for electricity. Decisions are made and implemented on the basis of these forecasts. When forecasts can be compared with actual events, new predictions are made, and the decisions are reevaluated, and possibly altered. Then the process is begun again.

This generic process is the same, whether done by a regulated utility, or in the deregulated energy marketplace of Figure 1. However, there are major differences in the nature of the variables, as well as the motives of the participants.

Under today's system, the regulated utility is expected to act in the best interests of its customers by minimizing their electricity costs. Under the deregulated structure of Figure 1 private generator owners try to maximize their own profits. But because of the market mechanism's "invisible hand", they also improve the welfare of their customers, assuming that a competitive market structure arises.

Investment by Private Generators in a Deregulated World

Since ownership and investment decisions are decentralized and deregulated, there is no need for central power studies or central decision making processes. Instead, private firms build plants if they think they will be profitable.

This is the approach followed in any industry that sells its product in a spot market. For example, wildcat oil wells, refineries, and new hotels are all built because investors expect the investment to be profitable. The basic investment rule is the same as in any other deregulated industry: if net revenues (appropriately discounted and adjusted for taxes) are expected to be appreciably larger than building costs, go ahead and do it. This rule is, in reality, quite similar in concept to the optimal capacity expansion process of a utility trying to minimize their customers' bills under a regulated energy marketplace.

NUMERICAL EXAMPLE: Assume a private firm is considering construction of a new coal-fired generating unit, which would have a marginal operating costs of 2.05 ¢/kWh. Thus the unit would generate at full capacity whenever the spot price is above 2.05 ¢/kWh, and shut down otherwise.[3] Whenever the spot price is above 2.05 ¢/kWh, the unit will more that

---

[3] This is oversimplified, as it ignores ramp rates, minimum downtimes, start-up costs, and the changes in a unit's heat rate at different output levels.

cover its operating expense, and it will have a positive net revenue equal to the difference between the current spot price and 2.05 ¢/kWh, times the unit's output. This net revenue makes up the positive cash flow which covers profits and the unit's capital cost.

Figure 3.1 shows how spot prices would have varied for a Midwestern utility for two weeks in 1980.[4] During the week of January 7 to 13, because the spot price rose above 2.05 ¢/kWh each day, the unit would generate daily. Its total net revenues for the week are the shaded area, times the unit's capacity in megawatts. During the week of October 6 to 12, on the other hand, the unit would generate for only a few hours on Wednesday, and about 15 hours on Saturday.[5] Looking at the data for the whole year, the unit's total net revenues would be $69,900 per megawatt of generating capacity, if the unit was available whenever needed. In fact, a coal unit would be down for maintenance some weeks. To keep profits as large as possible, all maintenance would be scheduled during weeks with low anticipated spot prices. Thus, the actual net revenues would be somewhat below $69,900 per megawatt for the year. If the private firm invested in a 500-megawatt coal unit, they can anticipate net revenues of roughly 500 megawatts times $70,000 or about $35 million per year. If the tax-adjusted value of this and the corresponding net revenues in later years are larger, when discounted to the present, than the construction cost of the unit, it is a profitable investment.

Economies of Scale: Generation

Economies of scale are a key issue in the argument of whether private firms will build generators that are socially desirable (economically efficient). Private firms will tend to build smaller plants that can be put on line fast, so they can start to get their money back sooner. If one lives in an ideal world where future costs, demands, technologies, environmental concerns, regulatory actions, etc. can be forecasted perfectly; big customized power plants that take a long time to build usually appear to be most efficient. In the real world, with all its uncertainty; smaller, modular power plants that can be constructed rapidly can generally be a much better investment. However, more analysis is needed before explicit answers are available. This is another reason we do not advocate immediate establishment of the deregulated marketplace of Figure 1.

Investment by Regulated T and D Company

In a regulated energy marketplace, the vertically integrated utility can coordinate its generating and network expansion plans so the network is not overloaded and desirable system dynamics are maintained. One of the potential weaknesses of the deregulated energy marketplace of Figure 1 is that such coordination will not be as close.

---

[4] A real planning study would of course use forecasted, not historical values.

[5] In fact, the unit might not be committed for this week, just as a central utility would not commit it.

When a private generation firm is considering building a new plant, it will pay the regulated T and D company to do load flow and system stability studies to determine whether the new generation can be expected to see unfavorable network and/or dynamic quality of supply components of the spot prices due to weak transmission in the area. If the network is too weak, the regulated T and D company will decide whether or not it is socially desirable (relative to the customer's bills) to strengthen the network. Alternatively, the regulated T and D company might offer to build new lines at the private firm's request and then charge for their use by using line by line decomposed method of revenue reconciliation so the private generation firms pay for the amount of the new lines they use.

A potential obstacle to overall economic efficiency of the deregulated marketplace is that the regulated T and D company will have to forecast what other types of private generation might be built in a given area in subsequent years. For example, for a given new generating plant, adding a 130 kV line might be optimal but if a second generator was to be added a few years later, building a 230 kV rather than 130 kV line might be better. This exemplifies the economy of scale issue for transmission. There is also the potential problem of the regulated T and D company building lines for a proposed new generation plant which is subsequently cancelled. These potential problems are still another reason we do not advocate deregulation as in Figure 1 until detailed analyses are done.

Forecasting in a Deregulated World

Operation and planning of any electric power system requires forecasts of future conditions.

In today's system (or a regulated energy marketplace), all the forecasting and decisions are done by the central utility. Demand and equipment availability are the critical variables influencing long term operation (unit commitment, plant maintenance, and fuel purchase) decisions. For investment planning, other variables such as future fuel prices, the cost and availability of capital, labor conditions, and the possible availability of alternative generation technologies are also forecast.

Under the deregulated energy marketplace of Figure 1, separate forecasts are made by each private generation firm (or purchased from Information Consultants) and by the regulated T and D company. Private generation firms base their long term operating decision on forecasts of future spot price patterns over the appropriate time span. These spot price forecasts replace the demand and unit availability forecasts used in the regulated case. These firms' investment decisions are also based on a similar spot price behavior forecast in addition to other variables used in the regulated case (fuel prices, capital prices, and so on).

One advantage of the deregulated case is that many different, independent forecasts and subsequent decisions are made. Some private generation companies may go bankrupt because of their forecasting errors, but others will realize healthy profits.[6] This is more desirable than the centralized case, where a single company's forecasting errors can influence all of the generators in a given region. Moreover, price feedback makes spot price patterns easier to predict.

Unfortunately, today we do not know whether the errors in long term forecasts of spot price behavior will have a larger or smaller impact on operating and investment decisions than the forecasting errors of the central regulated utility. This is still another reason we are not advocating establishment of the deregulated energy marketplace of Figure 1 at the present time.

Risk and Long Term Contracts

Generating electricity is a risky business. At present, regulatory action or inaction in response to changes in demands and costs increases these risks to a regulated utility. Under the deregulated marketplace of Figure 1, the total risks are the same. However, the risks borne by private generators and users are distributed differently.

Long term fixed price-fixed quantity contracts permit private generators to hedge some of the risks of power generation.[7] Such contracts are purely financial instruments, just like commodity futures contracts in agricultural and metals markets. They may be purchased by producers, customers, or speculators. The Energy Broker of Figure 1 can expedite such transactions by bringing buyers and sellers together. These contracts do not affect the efficient operation of the spot energy marketplace since they are based on fixed quantities of energy at fixed prices.

NUMERICAL EXAMPLE: Assume Firm A sells to Firm B a contract for 200 MWh of energy at $50 per MWh. Each hour, the two firms close out that hour's contract by comparing the actual spot price with the contracted forward price. If the spot price turns out to be $60 per MWh, Firm A pays Firm B $2,000 or (60 - 50) x 200. Owning generation equipment automatically puts any firm into a long position in electricity. Generators can hedge this long position by going short with futures contracts, i.e., taking the role of Firm A. Conversely, a user can hedge its implicit short position by going long, i.e., buying contracts. For example, if Firm A owns a 200 megawatt coal-fired power plant, it might want to lock in the $50/MWh price to satisfy lenders. When it signs the futures contract, its net revenues each hour are then composed of two parts. One is a financial gain or loss from closing out the long term/futures contract. This gain or loss is almost completely independent of how much the firm

---

[6] Bankruptcy will not in and of itself disrupt the operation of a generating unit. If the electricity it can produce is more valuable than the fuel it uses, it will continue to operated, albeit under new management.

[7] Another protection against this is geographic diversification. A single firm can build generating units in many regions, thus diversifying the risk of sustained regional price fluctuations.

generates.[8] The other component is its revenues earned by generating electricity, which depends on the actual spot price paid by the T and D company minus its variable operating costs. If the spot price for a given hour is too low, the firm will not generate, but still gains revenues from the futures contract. If the spot price for a given hour is only 10 $/MWh, B pays A $8,000, or (50 - 10) x 200, regardless of how much A generates or how much B consumes that hour. Generator A can only sell its output for 10 $/MWh. If this is less than its incremental generating cost, it will shut down for the hour. Therefore, although it earns nothing from generating electricity that hour, the $8,000 gain on its contract hedges A against the change in the electricity price. Conversely, B can buy energy for only 10 $/MWh for that hour, but it pays out $8,000 in addition to the value of its actual electricity use. If the spot price is high, firm A generates 200 megawatts and makes larger operating profits. However, it loses on the futures contract and in effect, gets to keep only 50 $/MWh generated. The net effect is that A's decision to generate and B's decision to use electricity depend only on the spot prices. The T and D company need not be concerned that Firms A and B have a futures contract between them. This is very different from fixed price, variable quantity long term contracts which can disrupt efficient operation of the power system.[9]

A perceptive reader has undoubtedly already observed that these long term/futures market transactions in the deregulated world of Figure 1 look like the long term market transactions for a regulated spot price based energy marketplace.

Regulation in a Deregulated World

The deregulated energy marketplace of Figure 1 still has the T and D company which is subject to regulation.[10] Thus regulation is done pretty much like in the present system except that only the rates the T and D company charges to transmit and distribute the energy are regulated (e.g., the revenue reconciliation multiplier is set to recover network capital costs). The regulatory commission does not directly set the rates the users pay or the generators receive.

---

[8] Under perfectly competitive assumptions, or if the generator is very small relative to the system, the gain or loss is completely independent. If the plant is large or the market is imperfect, the firm could affect the spot price.

[9] Utilities presently participate in a variety of long term contracts among themselves; either as independent utilities or as members of a pool. These existing long term contracts are much more like the contracts being discussed here than the fixed price, variable quantity often used for present day PURPA generators. This should be no surprise.

[10] The regulatory commission may also oversee the futures market and prevent any one generating company from obtaining monopoly power in any geographic region.

The regulatory commission maintains incentives to encourage the regulated T and D company to operate and plan efficiently. These management concerns are analogous to regulating today's vertically integrated companies.

One important issue remains. What would prevent regulators from reimposing price controls through the back door, by limiting how much the T and D utility is allowed to pay for purchases? Paradoxically, a property unique to electric power systems makes it very difficult to impose such limits. Electricity is unlike most other commodities because supply and demand must always be carefully balanced. A deviation of only a few percentage points for a very short time can cause a total blackout. Thus across-the-board price manipulation becomes extremely conspicuous because the T and D company is forced to perform involuntary power rationing, or do nothing and watch the whole power system black out. Either alternative leads to strong public pressure on regulators to back off.

Selective price discrimination, however, is not as easy to prevent. Suppose regulators forced the T and D company to offer a price lower than the spot price to a private nuclear unit. The unit still keeps generating as long as this price is above its marginal operating cost. This can be avoided by having the T and D company and each private generator sign contracts that rule out price discrimination. Such most-favored nation type contracts are found in other commodity markets. With this protection, the regulators are obliged to abide by the forces of supply and demand.

As in any industry, of course, legislators could override contracts and reimpose regulation. If potential investors thought such legislative action was a serious possibility, private investment would be retarded or even prevented.

4. A Scenario

Sections 1, 2, 3 addressed the deregulated energy marketplace of Figure 1. In this section, we hypothesize a scenario of events (happenings) wherein today's regulated structure can move toward and possibly eventually encompass the complete generation deregulation of Figure 1. Table 1 summarizes the four steps of this scenario.

Step I of the scenario is the topic of the companion paper (A Spot Price Based Energy Marketplace). We strongly advocate its implementation. We have highlighted our conclusions that there are many similarities between the regulated energy marketplace of the rest of the book and the deregulated system of Figure 1. However important differences exist such as:

o   In the regulated marketplace, users buy energy generated (or purchased) by their own utility and pay directly for the generation capital costs via revenue reconciliation.

o   In the deregulated marketplace of Figure 1, customers

effectively shop around to buy energy from the
cheapest available generator (with network costs
added) without explicitly worrying about generation
capital costs. The regulated T and D company acts
like a middle man in the transactions.

Steps II and III of Table 4.1 can remove these differences.

Step II: Mandatory Wheeling

Wheeling is the transmission of electric energy from a buyer to a
seller through transmission and/or distribution lines owned by the
wheeling utility. It is a common practice in the present system when the
buyer and seller are other utilities. Step II makes wheeling mandatory
and allows private generators and users to play buyer and seller roles.
Step II is a major change from present day wheeling because it involves
wheeling rates based on the spatial variation of the spot prices. The
wheeling rate from point A to point B is the difference between the spot
prices at points A and B. It turns out that the impact of including the
generation revenue reconciliation component into the spot prices that
define the wheeling rates can be very dramatic. Generation revenue
reconciliation can dominate the wheeling rates most of the time.

At the present time we are not quite ready to advocate the
implementation of Step II. More study is needed to make sure that
mandatory wheeling based on spot prices does not lead to horrendous
regulatory problems.

Step III Renouncement of Obligation to be Served

The wheeling rates of Step II use spot prices which incorporate
revenue reconciliation for both generation and the network. As a result,
a private user, for example, would not find it advantageous to buy from a
private generator or other utility solely to escape large generation
capital costs of his/her own utility (say resulting from a new nuclear
plant entering the rate base). The underlying logic is that as long as
the user's own utility has an obligation to serve the user, the user has
an obligation to help pay for generation capital costs. A similar logic
applies to private generators from which the utility has an obligation to
buy.

Step III of Table 4.1 establishes procedures which allow a user or
private generator to renounce their obligation to be served. Such
renouncing would probably be accompanied by payments from the user to the
utility or vice versa depending on whether the local utility has over or
under generation capacity. Alternately, a long advance warning (say 5 to
10 years) of the intention to renounce could be required. When a utility
no longer has an obligation to serve a given user or private generator,
such users or generators become free agents who can buy/sell from other
free agents or regulated utilities at wheeling rates that do not
incorporate generation capital costs are not included. They can buy from
where ever the operating costs are lowest (subject of course to the
network costs due to losses, finite network capability and network
capital costs).

This concept of renouncing an obligation to be served has been neither worked out in any detail nor analyzed. Thus at the present time, we advocate further study, not immediate adoption. Renouncing of obligation to be served raises a host of interesting questions which may not have nice answers. However, it is much easier to deal with than the fully deregulated marketplace of Figure 1.

Step IV: Wait and See What Happens

The result of Step III is a mongrel system which combines regulated utilities (with their own customers whom they have to serve) and free agent, users and private generators. Step IV is to wait and see how many users and private generators choose to enter the free market and what effect such actions have on overall costs, system behavior, etc.

Sometime during this wait and see period it could become necessary to introduce the dynamic pricing discussed in Section 2. However, as discussed earlier, it could also come into existence during Step I; i.e., as part of the regulated energy marketplace.

If free agent status becomes popular and has desirable effects, the regulatory commission (or legislature) might take further action such as discouraging (or preventing) the regulated utility from building more generation. This would eventually lead to the deregulated marketplace of Figure 1. The regulatory commission might even go all the way and have the regulated utility sell off its existing generation to the highest bidders (which introduces another host of questions for which answers do not exist today).

The main advantage of the four step scenario of Table 4.1 is that one is able to play court to deregulation for a long time before committing to a binding marriage. It is also possible to achieve many of the potential advantages of deregulation without full implementation, i.e., living with a mixed marriage of both regulated and deregulated participants.

5. Discussion

As stated many times in this paper, we do not advocate establishment of the deregulated energy marketplace of Figure 1 at the present time. There are simply too many unanswered questions. We wrote this paper because the ideas are fascinating and further analysis may prove them to be desirable.

Going all the way from today's system to the deregulated energy marketplace of Figure 1 in one step is a mind boggling concept. Fortunately for those who favor deregulation, there is a scenario which leads towards deregulation in a reasonable step by step process.

REFERENCE

This paper is based on Chapter 5 of the book "Spot Prices of Electricity" by F.C. Schweppe, M. Caramanis, R.D. Tabors and R. Bohn to be published by Kulwer Press.

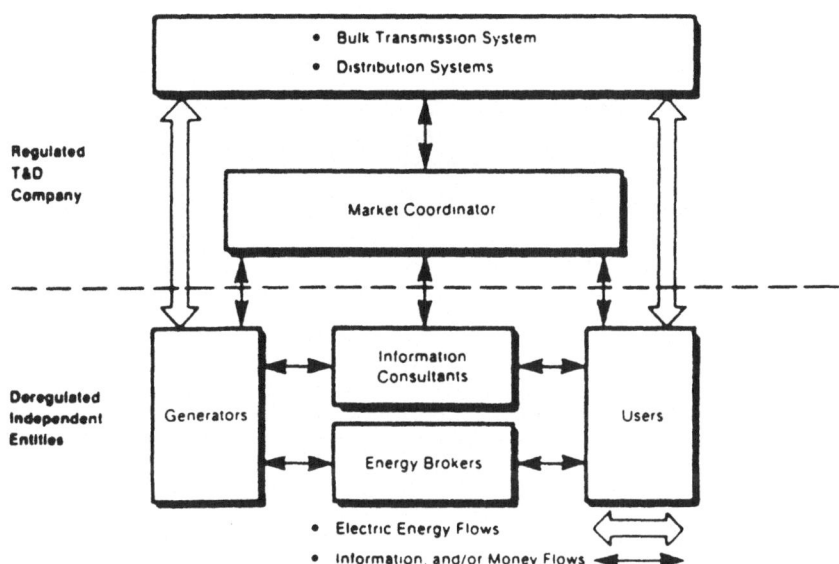

Figure 1

A Deregulated Energy Marketplace

Short Term Operation and Control

o   Dispatching generation minute by minute.

o   Operating the transmission system during normal conditions; system security, system dynamics, relaying.

o   Controlling the overall system during emergency state conditions such as sudden generation or transmission outages.

o   Setting prices to customers; choosing which customers will receive electricity during system emergencies.

Long Term Operation and Planning

o   Unit commitment, maintenance scheduling, and fuel purchasing.

o   Investment planning. Choosing what kind, where, how large, and when to build:

Generating units,
Transmission lines,
Local distribution systems.

o   Forecasting future conditions.

Table 1

Basic Functions of an Electric Power System

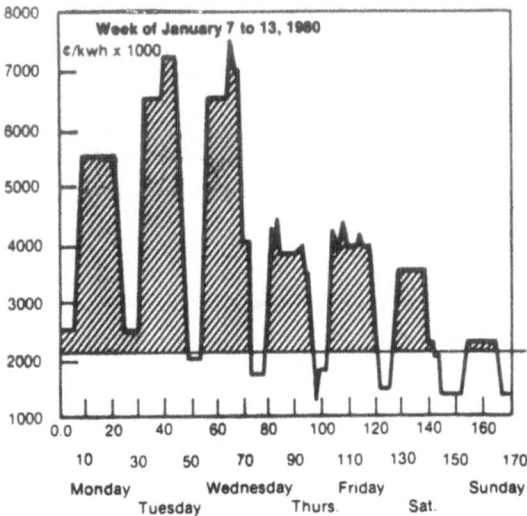

Shows net revenue under the spot price curve for the week January 7 to 13. Only one
unit's net revenues (based on a marginal cost of 2.05¢) is shown for sake of clarity

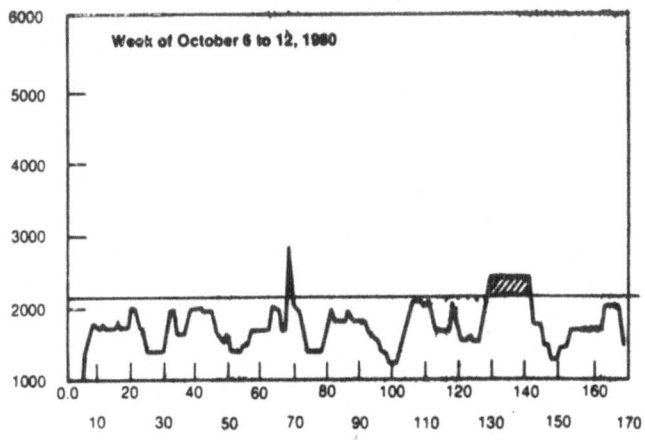

Shows net revenue under the spot price curve for the week October 6 to 12. Only one
unit's net revenues (based on a marginal cost of 2.05¢) is shown for sake of clarity. Total
revenue for a year is the sum of weekly revenues for each week the unit is available

Figure 2

Spot Prices and Net Revenues: An Example

Step I: Establish a regulated energy marketplace where users and private generators see spot price based rates which incorporate both generation and network revenue reconciliation.

Step II: Establish mandatory wheeling based on the spot price based wheeling rates which incorporate both generation and network revenue reconciliation.

Step III: Allow users and private generators to renounce their obligation to be served.

Step IV: Wait and see what happens.

Table 2

A Scenario that Might Lead to Deregulation

IV. Cogeneration

'COGENERATION PRINCIPLES AND PRACTICES AND A EUROPEAN OVERVIEW'

B W GAINEY
HEAD ENERGY MANAGEMENT MARKET DEVELOPMENT
Shell International
Petroleum Company Limited
Shell Centre
London SE1 7NA

1. INTRODUCTION

Cogeneration is the simultaneous conversion of fuel into heat and power; into a combination of steam and electricity.

The technology has merit because it is more efficient to produce steam and electricity simultaneously rather than produce them separately (see Figure 1).

Cogeneration is not new; it was extremely commonplace in the 1950s and before the advent of huge conventional power stations supplied over 50 per cent of the USA power needs. However the declining price difference between electricity and other energy sources led to a gradual decline in cogeneration practice (see Figure 2) [1].

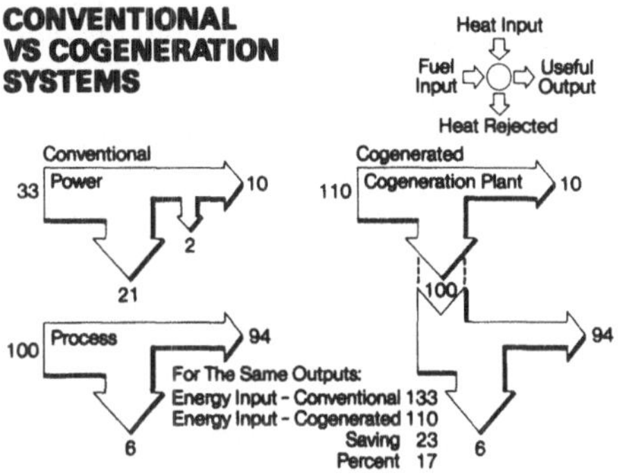

CONVENTIONAL VS COGENERATION SYSTEMS

Price Difference Between Electricity and Other Energy Sources in USA (1960–1985)

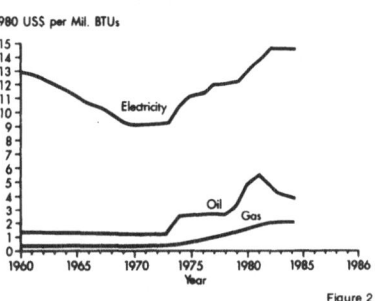

Figure 1

Figure 2

Economics of scale of the large but inefficient thermal power stations provided cheap electricity - economics denied to the small cogenerator with his tiny electrical loads. So despite his higher thermodynamic efficiency the cogenerator savings became increasingly insufficient to offset higher generating costs. As a result self generation of electricity declined.

However the oil price shocks of 1973 and 1979 removed the cheap fuel option and methods that generated electricity more efficiently, more effectively and more reliably were sought.

Cogeneration made a quick comeback in the USA, helped by a number of factors:

■ The Government introduced the <u>Public Utilities Regulatory Policy Act</u> of 1978 (PURPA)-legislation that required utilities to accept and pay for self-generated electricity at an appropriate 'avoided cost' i.e. that cost that the utility would have had to invest in order to produce that power at that time.

247

A. T. De Almeida and A. H. Rosenfeld (eds.), Demand-Side Management and Electricity End-Use Efficiency, 247–265.
© *1988 by Kluwer Academic Publishers.*

- The inability of many utilities to satisfy Nuclear Regulatory requirements. This caused delays or cancellations of nuclear power plants, cost overruns and increased electricity charges.
- The increasing impact of environmental regulations on emissions. Cogeneration with gas provided clean, efficient energy when NOx abatement procedures were employed.

The increase in energy prices had served one very valuable purpose. The search for higher-efficiency power-producing machinery was stimulated especially in the commercial aircraft business. The result was a rapid leap forward in both aeroderivative gas-turbine technology and the industrial versions. For example, the current largest single gas-turbine size is 140 MWe with a 200 MWe size machine announced for service in 1988 (the GE MS 9001 F) [2]. This turbine uses technology proven in aeroderivative machines extensively . Its thermal efficiency will be 34 per cent.

Large combined-cycle gas power plants are under construction in several parts of the world. For example, in Japan a 1 090 MWe station has been built [3] and a 2 000 MWe plant is near to completion [4]. In Turkey, half of a 1 200 MWe station [5] is operating with completion set for 1989.

In many countries the energy-intensive process industries are turning to cogeneration of heat and power to solve their escalating energy bills. In particular, refineries, paper and pulp mills, food processing and chemical industries have switched to cogeneration. In the commercial sector hotels, universities, hospitals, schools, supermarkets and laundries are beginning to realise the economic incentive that cogeneration provides.

2. THE TECHNOLOGY

Any system that supplies both the steam and electricity needs of a customer qualifies as cogeneration. Many systems are available involving both reciprocating engines, like diesels, and rotary engines, like steam turbines and gas turbines. All have their appropriate heat/power range over which they are most effective for a particular application.

This paper considers only gas-fired cogeneration systems that employ gas turbines because, under normal circumstances, their use is usually the most economically attractive. Therefore specifically excluded are all cogeneration plants that utilize any fuel in conjunction with a steam turbine alone.

There are numerous applications of cogeneration which depend upon whether a particular heat application or electricity application is required. Some typical heat opportunities are:

- Steam to industry.
- District heating.
- Industrial heat.

- Agriculture/Aquaculture.
- Desalination.
- Absorption cooling.
- Enhanced oil recovery.

Power opportunities are:
- Repowering existing power stations, boilers or furnaces.
- Power generation.
- Heat and power supply to industry.
- Small-scale cogeneration for hospitals, schools, buildings, hotels, supermarkets and laundries.
- Use of cheap fuels in power generation.
- Load management/energy storage.
- Fluidized bed air-turbine combined cycle.
- Coal-gasification combined cycle.

The available options are therefore:
■ GT (simple cycle) + Heat Recovery Steam Generator (HRSG) + supplementary firing (Figure 3).

Simple Cycle Gas Turbine System

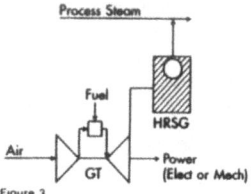

Figure 3

■ GT - combined cycle. (Figure 4.1,4.2).

Combined Cycle Steam

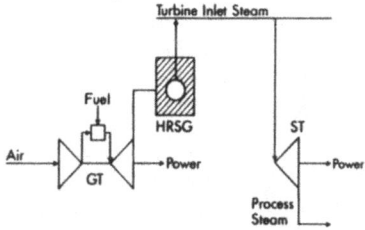

Figure 4.1

Combined Cycle with Dual Pressure

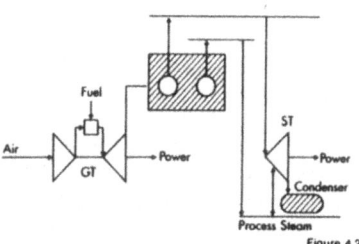

Figure 4.2

■ Fluidized bed/GT systems. (Figure 5) (AFB/PFB).
■ Coal gasification/combined cycle. (Figure 6).

PRESSURIZED FLUIDIZED BED (PFB) WITH COMBINED CYCLE GAS TURBINE

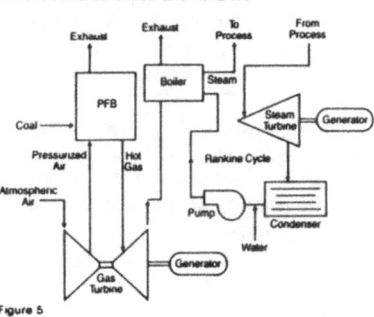

Figure 5

Gasification Combined Cycle

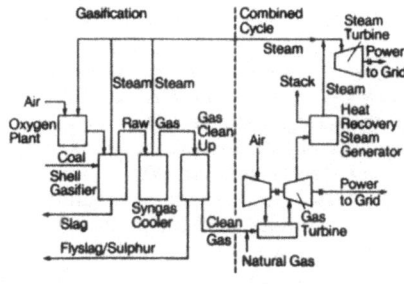

Figure 6

■ Applications that repower other systems with gas turbines. (Figure 7.1, 7.2).

Repowering a Power Station

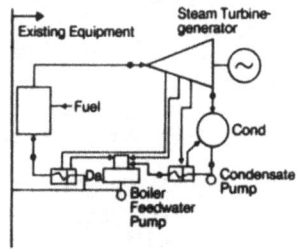

Figure 7.1

Repowering a Power Station

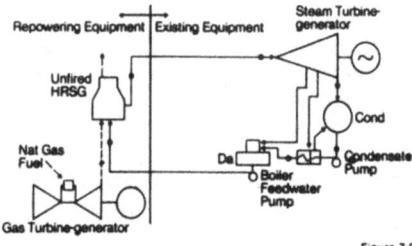

Figure 7.2

The technology involved in gas-fired cogeneration can perhaps best be explained as a sequence of operations designed to increase, progressively, power and heat generation efficiency.

A single gas turbine will generate electricity (say 100 MWe rated output) at an efficiency of 32 per cent, which is somewhat less than that of a modern conventional power station using oil, coal or nuclear fuel. However this gas turbine also generates a considerable amount of heat that, in simple-cycle power-only mode, is wasted. A waste-heat boiler can be used to convert this waste energy into steam or hot water for process use. Supplementary gas-fired duct burners can be installed that uses the oxygen present in the exhaust gases to produce additional heat which converts into more steam. It is therefore possible to design a system that can supply the varying instantaneous process plant requirements for heat and power. This is called combined heat and power (CHP).

Figure 1 shows an example of the efficiency advantage to be gained from producing both process steam and electricity simultaneously. A conventional power station will generate electricity with an efficiency of about 30 to 35 per cent. A process boiler can produce steam with an efficiency of about 94 per cent. Therefore to produce ten units of power and 94 units of steam, 133 fuel units are required. Cogeneration of this same steam and power demand requires only 110 fuel units, a thermodynamic saving of 17 per cent.

However in some cases only electricity and no steam is required. In this case the steam generated in the waste-heat boiler can be expanded through a steam turbine to create more electricity. This is called a combined cycle because it involves both a gas-turbine cycle and a steam-turbine cycle for electricity production. Where excess gas supplies are available in a country that needs additional electricity, gas fired-combined-cycle plants can offer an attractive, efficient, profitable alternative to conventional power generation. The latest large scale combined cycle plants have electrical efficiencies close to 50 per cent [6] which, in comparison with 30 to 35 per cent for conventional stations, shows a considerable fuel saving. Further gas combined-cycle plants can be installed much quicker than conventional plants and their resultant installed capital costs are significantly lower.

Environmental issues have become extremely important for the future economics of conventional power stations. 'Acid rain' issues and the need to restrict emissions of SOx, NOx, dust, smoke and unburnt hydrocarbons has led to the need to use 'cleaned' coal or the proposed introduction of flue gas-desulphurization units and in some cases baghouse filters. Such anti-pollution measures are expensive. Therefore the main challenge that coal-gasification/combined-cycle plants face is to provide environmentally acceptable economic electricity from coal.

Nuclear stations are suffering considerable project delays due to enhanced safety considerations which likewise have increased their capital cost. Gas, on the other hand, is low in sulphur and nitrogen oxide emissions are low. NOx can be reduced to levels of only 25 ppm by a steam-injection technique and solids emissions are negligible. Selective catalytic reduction (SCR) units can reduce the NOx levels still further where necessary [7]. Therefore where gas is plentiful and inexpensive, and where electricity is required and relatively expensive, gas-turbine combined cycle plants can offer an extremely attractive power generation opportunity.

Obviously plants can be designed that are a hybrid of the CHP or the combined-cycle plant. In these plants, power from gas turbines and steam turbines and steam for process use can be produced to requirements.

A sizeable market is predicted in the future for the repowering of existing boilers, furnaces and conventional power stations. Essentially a gas turbine, or a series of gas turbines is installed in front of existing equipment. The gas turbine provides electricity and exhaust heat. This heat is used as supplementary air to a furnace; to provide steam either to a boiler or to a conventional steam-turbine power plant. While these are often not optimised energy solutions the refitting leads to savings from the higher overall energy efficiency of the unit. As a measure of the profitability of such ventures one can take the 511 MWe Halfweg power station in Amsterdam [8] which has been repowered with a 140 MWe BBC type 13E gas turbine. The real term before tax-earning power of this refitment is considered to be in excess of 40 per cent. This is not an optimised plant - in such a plant about two-thirds of the produced power comes from the gas turbine while one-third derives from the steam turbine. Some of these custom-made combined-cycle plants have expected earning powers greater than this Amsterdam example.

3. MARKET-PLACE AND INFLUENTIAL FACTORS

The decision to install cogeneration may be made in various situations: to supply a plant's internal heat/power requirements only; to provide power to export to the local utility; or for a plant built solely for private power production, all of which will be sold to the power utility. Whenever power is sold to the utility from a planned new cogeneration facility, many interested parties become involved, e.g. the power utility, the gas supplier, the industrialist, the financier, the system contractor, vendors, the designer, and often the government. Success in such ventures depends upon being able to put together a mutually acceptable package that meets everyone's objectives. Therefore, in order for a market in cogeneration to flourish in a particular country, the following positive influences need to be present.

3.1 Government

Without the support of the government it is extremely unlikely that a flourishing industrial, commercial or private utility market will develop in cogeneration.

Support can take many forms but suitable legislation for the purchase of self-generated electricity, and a suitable transparent price-setting mechanism are considered essential. The government acts as the referee between the gas seller, the electricity self generator and the electricity purchaser. Cogeneration plants are long-term investments (20 years) and suitable arrangements that also take account of economic changes over this time period are required.

3.2 Electric utility

If an electric utility is going to buy electricity from a cogenerator, he has to need the power and it must be cheaper and more cost effective for him to purchase rather than to construct a new power station himself. High industrial electricity rates arise usually because of high fuel costs, low-efficiency generating plant or electricity shortages. In such circumstances the utility is often keen to purchase reliable cogenerated power that it can sell at a reasonable profit. High industrial electricity rates encourage energy-intensive industries to consider self generation for their own use as well as export.

Tariffs must be available, and their escalation formulae transparent, for power buyback from industrialists. In addition, the cost of reserve power and a long-term electricity contract needs to be available.

3.3 Gas retailer

In order for cogeneration to flourish, a long-term supply of gas of the appropriate quality, at the correct pressure and available at a known price, is required. The possibility of a long-term contract for gas supply considerably reduces the risk of projects of this type.

3.4 Host industrial

The industrial or commercial company interested in cogeneration is seeking cheaper, more reliable power. Therefore the current and future interfuel economics for gas use for steam and electricity production must be favourable.

In addition it must make good business sense for a change in the energy production method to be made. Projects must be commercially viable and attractive, and finally, the industrialist must be willing to convert to cogeneration. Some concerns expressed by managing directors in industry are:

- Unfamiliarity with new technology.
- Lack of finance to implement scheme.
- Fuel supply security and price uncertainty.
- Process uncertainties - effect of product mix - new technology conservation measures.
- Disruption of production.
- Environmental impact.
- Reluctance to self-generate electricity.

As a result in the USA many companies have emerged since about 1980, who are prepared to offer a complete installation service. These companies take responsibility for design, construction, commissioning, operation and maintenance. In addition they will undertake to provide the project financing package off the industrialist's balance sheet.

The rationale for this type of business is very simple - industrialists do best when they concentrate on their main-line activity. Energy production, use and control is not their main activity. Therefore, under the right terms and conditions, industrial and commercial clients are prepared to give responsibility for their energy needs to specialist energy companies. Such companies provide a service, delivering what the industrialist actually needs ... thermal comfort or steam, heat or electricity, in the right amount, and at the time and place required.

Such businesses are still in their infancy here in Europe. One such company is Shell Industrial Energy Management (SIEM) [9] that operates in the Netherlands and is based in Rotterdam. It is to be expected that if cogeneration becomes widespread in Europe other companies will offer this service.

4. PROSPECTS IN EUROPE

Cogeneration will only develop in a particular country if the economics for its introduction are favourable. A long-term gas supply at an appropriate pressure is required with a favourable tariff. Demand for electricity should be outpacing supply and be relatively expensive. Therefore in order to assess the prospects for cogeneration a knowledge of the gas supply and the electricity production both now and in the future is required.

4.1 The European gas market

Deregulation of the USA gas market, coupled with a declining demand for gas, has resulted in falling prices and underutilized production and transportation capacity. This has led to a spot market and a contract transportation business. It has also aided cogeneration because the

availability of low-priced copious quantities of gas is a favourable
factor that promotes gas-fired combined heat and power and combined-cycle
power generation. In addition, multifuel options in commerce and industry
lead to a volatile unpredictable market. In Europe the gas contracts tend
to be longer term and therefore less susceptible to spot trading but the
USA experience may be a trend for Europe in the future. However it is
expected that the European gas market will develop in a stable, gradual
manner.

The extent of development of gas markets in Europe varies considerably.
Some countries have a well-developed infrastructure for utility,
industrial, commercial and residential customers. Gas sales in the mature
markets of the Netherlands, West Germany, France and the UK are very
competitive. Developing markets are Denmark, Greece, Sweden and Ireland,
while further maturity is sought by Italy and Switzerland. The main
restriction to gas use in power generation comes from certain EEC and
country legislation that categorizes gas as a valuable premium fuel
designated for other uses. On the other hand, gas is the most appropriate
'environmentally-acceptable' fuel.

Expectations are that gas demand will grow steadily [10], with the
residential/commercial sector gradually taking a larger share from
conventional power generation. This is the opportunity for combined heat
and power in industry. Large scale gas-fired combined-cycle plants may
well become important in the future where strategic gas supplies are
available and electricity required. (Such assumptions are not included in
Figure 8.)

An important aspect of the European gas market is the complex network of
transmission piping that now links Algeria and the USSR to many European
countries (see Figure 9).

**Natural gas consumption Western Europe
market share per sector**

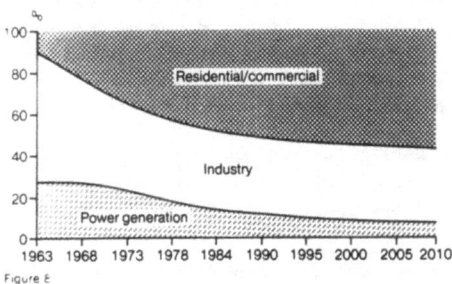

Figure 8

The European Gas Network

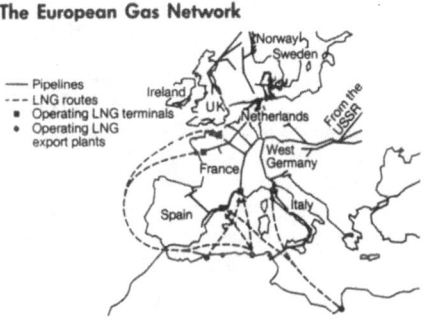

Figure 9

West Germany is particularly well placed strategically at the hub of
this developing network which allows considerable export from USSR,
Algeria, Norway and the Netherlands (Figure 10).

Figure 11 shows the export trade in 1984 to the five major European
countries [11] which use the bulk of gas consumed in Europe. This
leads to a high transportation and distribution cost element in the gas
price. The recent drop in energy prices has made it difficult for these
costs to be absorbed.

One method of combating these high unit costs is to develop a
significant power generation market. Added value is earned by converting
to electricity, and, by using the highly-efficient combined-cycle
configuration, profitability is optimised. Such approaches may develop in
Norway, UK and Sweden.

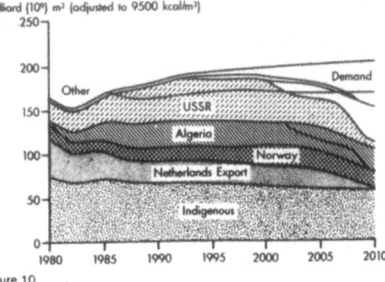

Continental Europe – natural gas demand/supply
(including Troll and extended Algeria and USSR)

Figure 10

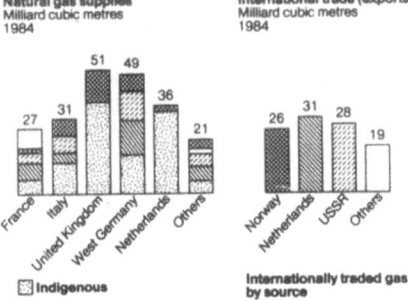

Figure 11

4.2 The European electricity market

In assessing the prospects for the European cogeneration market one must take account of the electricity supply and demand picture. In each country, and even within regions of each country, the electricity business is quite different in supply/demand balance, cost structure and fuels used to generate that power. These factors will determine whether it is desirable to self-generate electricity for own use and whether it is profitable and possible to export to the power utility.

The European electricity market is extremely complex [12]. Nuclear power plays a dominant role in some countries (France), hydroelectricity in others (Norway). Some are coal based (Germany), and some gas based (Netherlands). Significant electricity exchange occurs between countries as the market begins to integrate. Environmental requirements have become more stringent, especially from the EEC. Government control, private utilities and local government control influences differ from country to country. Some utilities are highly geared, others are unprofitable. Cogeneration and private power generation via combined cycle could begin to emerge in these countries. Others are short of power and might welcome cogeneration. Some countries, post Chernobyl, have decided against nuclear power already planned. Industrial cogeneration and gas combined-cycle power plants might emerge as the logical power replacement, especially since they can be built more quickly than conventional power plant.

In order to determine whether cogeneration may be an option in any particular country it is necessary to take a micro-economic look at the electricity supply/balance picture both today and in the future.

4.3 Cogeneration market

Figure 12 illustrates a range of markets where cogeneration could be used. Small-scale cogeneration with gas engines or modified reciprocating diesel engines aimed at the residential/commercial/industrial sector are outside the scope of this paper [13]. Here we consider gas turbines in sizes above about 3 MWe. The main areas of use of this technology are industry, district heating and power generation. Considerable use of cogeneration is made by some oil companies [14] involved with chemicals, refineries, enhanced oil recovery and offshore facilities. In the industrial sector the food processing, pulp and paper business have made most use of cogeneration. Several studies have been made on the expected USA market size [15]. One such prediction by use (Figure 13) and by technology (Figure 14) are shown overleaf.

Markets

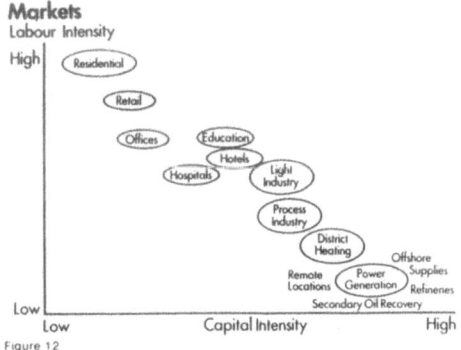

Figure 12

The Expected USA Cogeneration Market
(1985-2000)

Commercial
(No of sites 302,300) Av. Size (kw)

Figure 13

The Expected USA Cogeneration Market
(1985-2000)

Industry
39,348 MW (Plants 3644)

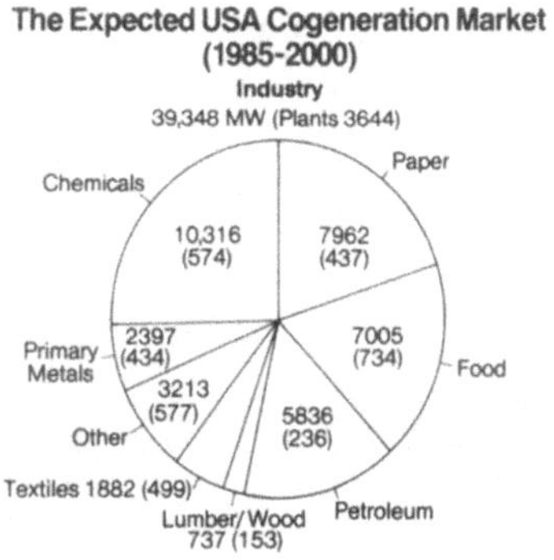

Figure 14

Likewise in the Netherlands studies have defined the current and the likely market size (Figures 15 and 16) [16] [17]. However the European market size is much more difficult to determine. Each country is at a different stage of development and, apart from in the Netherlands, the technology is being gradually accepted.

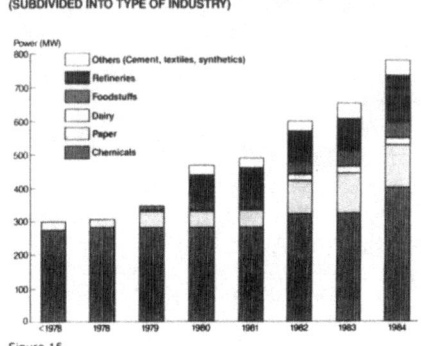

INDUSTRIAL USE OF GASTURBINE POWER IN THE NETHERLANDS
(SUBDIVIDED INTO TYPE OF INDUSTRY)

Figure 15

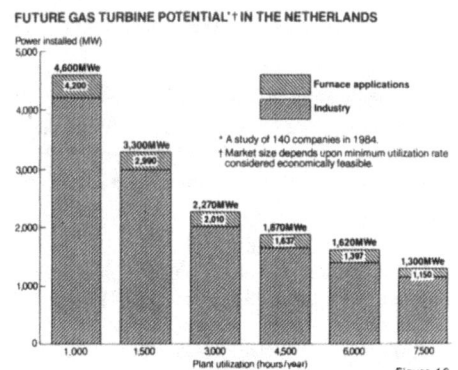

FUTURE GAS TURBINE POTENTIAL*† IN THE NETHERLANDS

Figure 16

One method of predicting the likely market size for cogeneration often used by gas-turbine manufacturers is to use historical data as a basis for future predictions [18].

Figure 17 shows the historical trends in the world market for power generation by gas turbines in all sectors.

The annual market gradually dropped between 1980 and 1986 from 10 000 MWe to around 7 000 MWe (corresponding respectively to between 600 and 400 units). In the five years between 1980 and 1985 Western Europe claimed about 13 per cent of this market, selling 600 units [18].

Figure 18 shows that, in size categories, the decrease was most rapid in the 0 to 10 MWe range, whereas the 10 to 20 MWe range remained approximately constant [18].

Gas turbines for power generation (World)

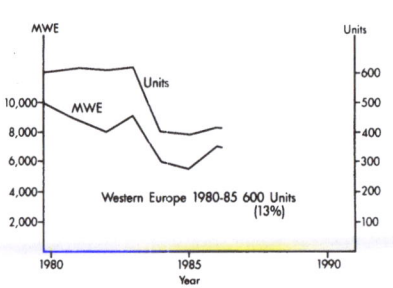

Figure 17

Gas turbine shipments by MW range (World)

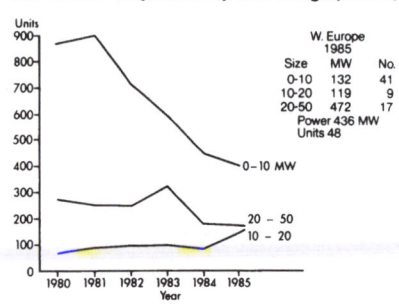

Figure 18

In 1985, 48 units delivering 436 MWe were sold in Western Europe, most being in the size range 0 to 10 MWe. The expectation from a study by PRS Consultancy Group [18] is that this market will increase in the next few years. The detailed analysis of the market in 1985 (see Table 1) reveals that most units were sold into the Netherlands, followed by Norway, UK, West Germany, Greece and Italy [18]. Most of the gas turbines were in the range 20 to 0 MWe, followed by 10 to 20 MWe, and those larger than 5 MWe.

This study [18] anticipates a gradual market growth by some 3 000 MWe between 1986 to 1990. This is comparable with Japan, larger than expected for China and India, but less than expected for the Middle East. With the partial collapse of oil prices, it remains to be seen whether the Middle East is still the high-growth area for gas turbines that this survey predicts.

5. SOME OPPORTUNITIES EXPLOITED IN EUROPE

A wide variety of opportunities has been exploited in Europe in several markets with different technology. Most of the applications have been in the Netherlands. An example of the range of applications is given below:

5.1 Paper industry

The use of aeroderivative gas turbines like the Allison KB5-501 or 571 (see Figure 19), or the Ruston Tornado for heat and power requirements. Examples are Smith, Stone and Knight (UK) [19] and KNP (Netherlands). However other examples have been documented [20].

5.2 Food processing industry

The use of aeroderivative gas turbines for heat and power applications (Allison, Ruston, Solar). Some examples are:

- Heineken Brewery, Netherlands.
- Whiskey Distillers, UK. [21]
- Cacao Fabriek, De Zaan, Zaandam, Netherlands.

New Installed GT Power Units 1985
Western Europe

Size	Netherlands	Norway	UK	W. Germany	Greece	Italy	Spain	France	Total
< 5 MWe	23	5	22	10			7	4	71
5-10 MWe		6		12	12				30
10-20	30	46	10						86
20-50	180	21	20	28					249
Total	233	72	48	48	12	12	7	4	436

Table 1

Typical Cogeneration Scheme

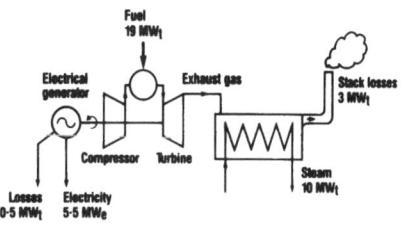

Electrical efficiency = 29% (i.e. 5·5/19)
Combined thermal efficiency = 81·5% (i.e. (10+5·5)/19)

Figure 19

5.3 Chemicals and refinery industries

Larger units are used in these industries in view of their immense energy requirements. The technology used is either larger industrial gas turbines (GE FRAME 3,5,6 or BBC 8,9) or larger aeroderivatives (e.g. GE LM 2500, LM 5000). An example is the Pernis Refinery [14] in Rotterdam which uses a BBC 8, 9 and a GE LM 2500. The BBC units are used in combined heat and power units, and the LM 2500 is used to repower a furnace in the refinery.

Another example is the CHP scheme at ICI Runcorn [22] which uses a Sulzer 10 gas turbine. At Gist-Brocades, a pharmaceutical company in the Netherlands, three Allison 501 gas turbines are used in a CHP scheme, which also has provision for electricity production from an existing steam turbine.

5.4 Hospitals/universities

In the institutional sector the use of gas-turbine CHP and district heating schemes is gaining popularity. Two examples are:
- Liverpool University (1 x 501 Allison).
- Stuggart University (2 x 571 Allison).

5.5 District heating schemes

Several authorities in the Netherlands have installed gas turbines for power generation. Examples are the Rolls Royce Olympus engines (2 x 18 MWe) installed by Energie Bedrift Rijnland, Leiden and NV Mesa, in Alsmere which uses a BBC-type 8 machine (45 MWe). The waste heat is used in the district heating system [23]. In the UK a proposed district heating scheme for the city of Leicester [24] would use two gas turbines purchased from the CEGB which were formerly used for peaking purposes (Figure 20).

5.6 Power utilities

Ten power utilities in the Netherlands have decided to improve the efficiency of some of their gas and oil-fired conventional plant [23] by repowering (see Table 2). This involves placing a gas turbine(s) in front of the conventional station. The turbine exhaust gases are routed to the conventional boiler, where the steam generated produces electricity in the conventional manner from a steam turbine. The GEB, in Amsterdam at its 511 MWe Hemweg plant, has installed the prototype 140 MWe BBC 13E machine [8].

Norway recently announced that it has several large combined-cycle plants under consideration. Gas is available from Statfjord and the Troll Sleipner fields, and with increasing resistance to more hydro stations on environmental grounds, several thousand megawatts of gas combined cycle could appear in the future. Planning continues at Rogaland, Monstadfeldt and Nord Tronland. In Japan the Niigata gas combined-cycle plant [3] (1 090 MWe) has operated since 1985. Based upon Westinghouse/Mitsubishi

Combined Heat and Power For A City District Heating Scheme

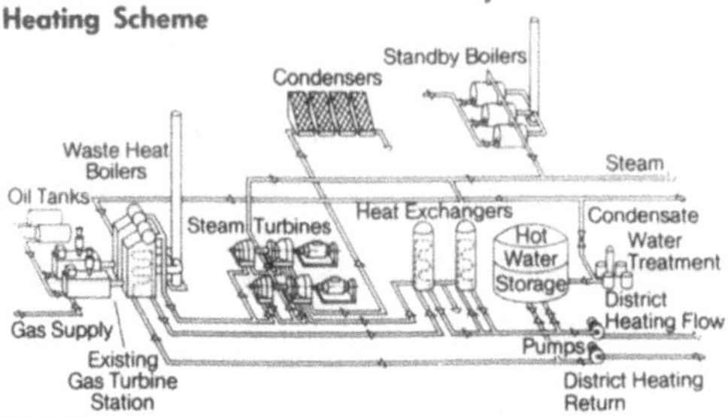

Figure 20

Repowering Dutch Power Stations

Utility	Steam Plant	Fuel	Year	Rating	Gas Turbine Model	GT Rating	New Plant Rating
EGD	Groningen-2	G	1977	611 MW	Kraftwerk Union V94	128 MW	645
GEB Amsterdam	Hemweg-7	G/O	1978	511 MW	Brown Boveri Type 13	135 MW	622
PEB Friesland	Bergum-1	G	1974	322 MW	Thomassen 2 x MS6001	2x37 MW	336
PEB Friesland	Bergum-2	G	1975	322 MW	Thomassen 2 x MS6001	2x37 MW	336
EPON	Harculo-5	G/O	1972	328 MW	Thomassen 2 x MS6001	2x37 MW	328
EPON	Flevo-3	G/O	1974	465 MW	·	100 MW	451
Pegus	Lage Weide-5	G/O	1976	255 MW	Brown Boveri Type 11	70 MW	273
EZH Rotterdam	Woalhaven-4	G/O	1971	340 MW	- -	80 MW	313
EZH Rotterdam	Woalhaven-5	G/O	1972	340 MW	- -	80 MW	313
EZH Dordrecht	Dordrecht-6	G/O	1968	150 MW	Thomassen MS 6001	37 Mw	159

Table 2

gas-turbine technology, the plant has two energy trains which each involve three gas turbines connected to one steam turbine. The Japan Electric Power Futtsu plant [4] is 2 000 MWe, of which 1 000 MWe is currently operating. This plant uses 14 gas turbines and 14 steam turbines in a 1:1 single-shaft configuration, based upon the GE FRAME 9E gas turbine.

In fact Japan [25] will have more than 7 200 MWe of combined-cycle power using gas installed by 1994 (see Table 3).

Japanese Combined Cycle Plants in Service and Planned

Operator	Plant Site	Output	GT Type	Supplier	Service
Japan Natl Railway	Kawasaki	1 x 125 Mw	Frame 9B	Hitachi	04/1981
Tohuku Electric	Higashi Niigata-1	1 x 545 Mw	3xMW 701D	Mitsubishi	04/1985
Tohuku Electric	Higashi Niigata-2	1 x 545 Mw	3xMW 701D	Mitsubishi	10/1985
Tokyo Electric	Futtsu-1	7 x 143 Mw	Frame 9E	GE and MA	11/1986
Chubu Electric	Yokkaichi	5 x 112 Mw	Frame 7E	–	03/1988
Tokyo Electric	Futtsu-2	7 x 143 Mw	Frame 9E	GE and MA	11/1988
Hokuriku Electric	Nana Ohta	1 x 550 Mw	5xFrame 7E	–	03/1990
Chugoku Electric	Yanai-1	6 x 115 Mw	Frame 7E	–	07/1990
Kyushu Electric	Shinohita-1	6 x 115 Mw	Frame 7E	–	07/1990
Kyushu Electric	Shinohita-2	6 x 145 Mw	MW 501D	Mitsubishi	07/1991
Chugoku Electric	Yanai-2	6 x 115 Mw	Frame 7E	–	03/1993

Table 3

In Turkey a 1 200 MWe combined-cycle plant is being built by BBC Brown Boveri - ENKA for the Turkish Electrical Authority (TEK). The plant based upon the BBC-type 13D gas turbine will be fully operational in 1989. Its special feature is its massive dry cooling towers. [5]

In the USA the abandoned Midland nuclear plant will be converted to a gas-fired 1 330 MWe combined-cycle facility by the addition of gas turbines, waste-heat boilers and modification of the existing steam turbines. [25]

5.7 Landfill gas

Methane generated from mature refuse landfill sites represents a cheap and valuable source of fuel that can be exploited in gas turbines to produce electricity. One example of this is the entrepreneurial Packington Estate Enterprises in Meridian, UK, where an Allison 501 packaged by CENTRAX Ltd is used to produce 3.6 MWe. Another example is the Purfleet Board Mills combined cycle CHP unit at Aveley on the Thames [26]. In Europe there are in excess of ten thousand such sites (over 600 in the UK), which when commercially analysed, might produce profitable electricity as a by-product.

6. FINANCING AND PROJECT ECONOMICS

In the USA, since PURPA, it has been commercially attractive to consider in-house production of heat and power, and selling off excess electricity at a competitive price back to the utility. Changes in Federal, State and local regulations, and tax laws aimed at encouraging cogeneration, have led to the emergence of new innovative financing methods to maximise these benefits.

A growing business has developed where financial and technological packages are offered to industrial clients to provide them with their steam and electricity needs, using cogeneration. Large units, centrally situated, can supply a number of industrial customers and export excess electricity to a utility. Many utilities have become partners themselves in such schemes.

Industrial customers are faced with a choice - 'make or buy'. That is, build and finance their own plant and make the required energy, or buy the energy from an energy management company who invests in the required equipment on the client's site and supplies the required heat and power.

The advantage to the industrialist is that he incurs no risk, no capital investment, no balance sheet exposure and has no need to operate energy equipment. He receives either cheaper energy or a rebate on his current energy bill. This allows the industrialist to invest his money and devote his attention to process upgrading, automation or any other area where returns are either favourable, or more predictable, or both.

The energy management company that supplies heat and power has specialized technical, financial and commercial expertise in the business as this is their dedicated job. Therefore such projects can be assembled and progressed more rapidly and efficiently than by the industrialist venturing beyond his established business into a new area on a one-off project.

The approach of the energy management company is to make a commercial proposal to the client, based on its own feasibility study to install a cogeneration plant. In the package to the industrial customer is the design of the plant, construction, training, monitoring and plant optimisation, as well as project financing.

The energy management company optimises the following variables:
- Technology/system chosen.
- Type of fuel.
- Project economics.
- Financing arrangements.
- Compliance with all regulations.
- Contractual conditions.

The technology chosen depends upon the specific site requirements. Some considerations are:
- Size of energy plant.
- Range of heat to power required.
- Average heat to power needs.
- Annual plant utilization.
- Existing equipment available.
- Available space.
- Operator experience.
- Possible future demand modifications due to process restructuring.
- Electrical tie-in possibilities.
- Fuel supply constraints.

260

Since 1983, Shell has offered this service in Europe, installing cogeneration plants for industrial customers on an own-operate basis [9]. Called Shell Industrial Energy Management (SIEM), the company, based in Rotterdam, has four sites operational using eight gas turbines providing 30 MWe. The investment in such plants exceeds £30 million and these turbines have run for 85 000 hours.

Current projects under construction will increase the power supplied to 47 MWe.

SIEM's operations do not end when a plant is built and commissioned; SIEM remains involved in operational optimisation, high-technology maintenance and trouble shooting. High availability of these plants is essential because Shell's profit derives from their efficient reliable operation. To help ensure this, SIEM carries spares and a spare gas-turbine engine which can be installed by its own skilled staff.

This small step-out company is backed by considerable cogeneration experience gained in oil company operations. For example, in Brunei, a joint venture with the Government supplies 50 per cent of the country's electricity needs. In Pernis refinery, Rotterdam, about 125 MWe of electricity is generated by a BBC-type 8, BBC-type 9 and a GE type LM 2500. Steam injection to minimise NOx emission leads to a higher electrical output from the BBC machines.

Financing or co-financing of these projects can occur in a number of different ways:

- Third-party ownership with shared savings
- Joint venture
- Lease
- Loan

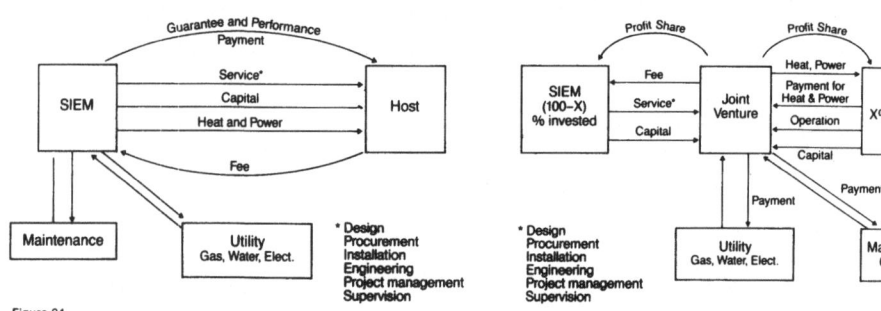

Figure 21

Figure 22

The particular method chosen depends upon the client's requirements. The economics of the project are determined as illustrated in Figure 23.

This illustrates the method for evaluating cogeneration economics. The basis for comparisons is the future expected cost of energy from the existing facilities, assuming no changes are made, compared with the installation of an optimised combined heat and power plant. This comparison considers both current and future energy cost scenarios in establishing the likely annual savings. In arriving at the best technical option, consideration is given to the technical criteria for the plant, the plant capital cost and the method of project financing. From these perceived annual savings the plant capital cost is repaid, and the industrial host and the energy management company share the remaining profit in an agreed negotiated manner.

7. FUTURE TECHNOLOGY DEVELOPMENTS

Gas-turbine performance and size range continues to improve at an impressive pace. Better fuel efficiency, lower emissions, better reliability [27] and lower unit costs help to establish the technology in the power generation arena. Developments from the aero-industry feed across to the industrial aeroderivative designs which especially permit the construction of larger industrial gas-turbines. Today, the largest available gas turbine delivers 140 MWe, but by the early 1990s, 200 MWe machines should be commonplace.

Businesswise, the gas-turbine industry has had a difficult time; mergers and joint marketing agreements continue to be signed as the industry restructures for the future. However fuller integration of combined heat and power plants with the actual industrial process can be expected, as for example, has occurred in Stade, West Germany (Figure 24).

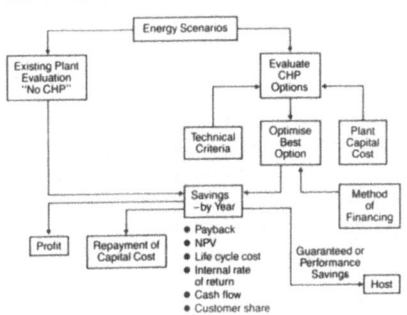

Figure 23

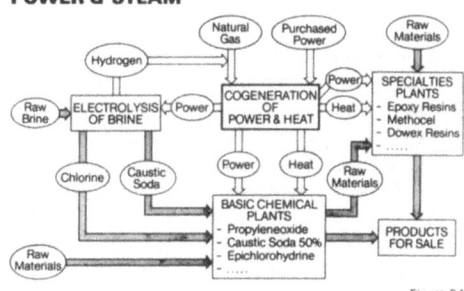

Figure 24

The NOx emission levels from gas turbines continue to drop due to design changes, steam injection or the use of selective catalytic reduction (SCR).

Research into the use of solid fuels both indirectly and directly in the gas turbines continues, particularly on coal and wood waste. Also under investigation is the efficient combustion of low heat content gases, like biogas or hydrogen.

Better heat utilisation will lead to further, perhaps dramatic increases in power output. Work continues on recuperation and the use of regenerative-cycle gas turbines [28]. These cycles use steam injection to provide an evaporative cooling cycle, followed by regenerative heating of compressor discharge air, which can boost simple gas-turbine power output by up to 50 per cent. Called intercooling, it is expected to revolutionise the designs, cost efficiency and performance of gas turbines in the future.

In the smaller size ranges, packaged units which simplify installation and reduce unit cost are emerging. Several new gas-turbine designs are now offered in the less than 5 MWe range from companies like Daihatsu, Centrax, Fiat, Yanmar, Ruston and Radial Turbines International.

Pressured fluidized bed technology burning coal can be integrated into a combined-cycle power plant to generate electricity via a gas turbine from coal (See Figure 5). Two coal burning combined-cycle conversions to power stations in the USA and Spain have been announced using ASEA - STAL's concept (GT 35P) [29].

Higher efficiencies of gas-turbine power plant are possible when the Kalina cycle [30] is fully developed. A small prototype project is in the planning stage.

Finally, Columbia Energy Storage Inc. of Florida has developed a programmed air-compression energy recovery (PACER) [31] peaking system designed around the Allison 501 gas turbine. This unit compresses air with cheap off-peak electricity and expands it through the turbine to produce peak-value electricity when required. This is not cogeneration, but it demonstrates how gas-turbine technology can be used in a different form for power generation.

Lastly, several prototype coal-gasification schemes [32] are underway and the first commercial plant cannot be far away. Although the economics for gas production from coal are not sufficiently attractive at the moment, the technology is available for immediate use once natural gas supplies become either scarce or expensive. In the Netherlands the far-sighted government is offering special subsidies to allow the first of a new kind of coal-gasification/combined-cycle power generation plant to be constructed.

8. CONCLUSIONS

Energy prices will escalate in real terms in the future: the problem is that when is uncertain. Environmental regulations will continue to be more rigidly, and more severely applied which will adversely affect the economics of conventional power generation.

Gas combined-cycle power plants and repowered large conventional plants may be given a major opportunity if new nuclear plant is delayed or cancelled as a result of Chernobyl. It is expected that combined-cycle plants will be built where long-term gas, liquified petroleum gas or gas oil is available, while electricity is expensive and in short supply.

The 7000 MWe of combined-cycle power being built in Japan may be exceptional, but this may encourage other Pacific Basin governments to consider gas for power generation. It is quite possible that electricity privatisation initiatives may develop in some countries providing the necessary impetus for cogeneration where the economics are attractive. However, cogeneration is unlikely to make a significant impact in any country unless favourable government support and legislation regulating interaction with gas and utility companies is introduced. The technology will be steadily accepted by industry and further integration between energy plant and process will occur. Large industrial energy parks could develop, with energy supplied from a central-supply facility.

Enhanced oil-recovery techniques [33] that use steam will use gas-turbine technology for electricity/steam requirements once the oil price improves. Already extensively used in the vicinity of Bakersfield, California, the technology at one Kern River site produces 1.7 million lb/hr steam and 300 MWe electricity, saving 7000 barrels per day of crude oil.

Steam injection into gas turbines not only decreases NOx levels but boosts electrical power output. More use will be made of this technology for power boosting in future. Intercooling depends upon water-injection technology to improve the efficiency and power output of turbines using an evaporative regenerative cycle. Power output of modified existing turbines could be improved by as much as 50 per cent [28].

Emissions, especially NOx, will become as important world-wide as they are now in California, where the turbine manufacturers have met the stringent requirements set [34].

REFERENCES

1 Alison Tucker, Cogeneration in the United States, past trends, present status and future potential EPR1 October 1985.
but see also:
  Power 131, 6, c1 (1987)
  Power 130, 5, s1 (1986)

2 New Frame 9 gas turbine ups output by 80 per cent, Modern Power Systems 7, 5, 49 (1987).

3 R Farmer, 1090 Mw Higashi-Niigata sets efficiency and reliability record. Gas Turbine World 16, 3, 32 (1986).

4 D Hayes and J Roderick, Futtsu - the first 1000 Mw. Modern Power Systems 7, 6, 27 (1987).

5 Mustafa Geçek and Joseph Reisner, Turkey installs major combined cycle plant. Modern Power Systems 7, 6, 33 (1987)

6 See for example:
  D E Brandt, The design and development of an advanced heavy duty gas turbine.
  ASME Gas Turbine Conference, Anahein California June (1987). Paper no 87-GT-14 and Gas Turbine World 15, 7, 15 (1985)
and, R Kehlhofer and A Plancherd, Brown Boveri Combined Cycle Power Plants. Publication CH-TO40 183E.

7 Earl Dunckel, Selective catalytic reduction: cogen bane, ban or boom? Cogeneration 2, 1, 10 (1985).

8 Robert Chellini, Repowering a steam powerplant with a gas turbine. Diesel and Gas turbine Worldwide XVIII, 8, 28 (1986)

9 B W Gainey and P A Ward, Shell experience in cogeneration, Institute of Mechanical Engineers Conference, May (1986) London Technical and economic impact of cogeneration p 35.
  Gas Turbine World 17, 3, 42 (1987)

10 Dick de Jong, Europe's gas industry - an insider's overview.
  Preserving gas competitiveness in Europe, ESC Conference, Geneva, November 1986.

11 Regional outlook for natural gas.
Shell Briefing Service, 5 1985 available for SIPC, London (PAG/331).

12 Andrew Holmes, Electricity in Europe - present status and prospects for the 1990's. Financial Times Business Information (1986).

13 K G Davidson, Packaged cogeneration systems for commercial buildings.
  The Cogeneration Journal 1, 2, 65, (1986)

14 A D Benz, B D Degen, J R McKibbin, Cogeneration in the Petroleum
Refinery, Chen Eng. Progress, 10, 21, (1986).
   Gas Turbine World 15, 6, 12 (1985)
                     16, 6, 12 (1986)
                     16, 5, 10 (1986)
   Cogeneration   2, 3, 28 (1986)

15 Dun and Bradstreet Technical Economic Services and TRW Energy Development
Group, Industrial Cogeneration Potential (1980 - 2000) for applications of
four commercially available prime movers at the plant site, Volume 1 DOE/CS
140403-1, August 1984.
   Frost and Sullivan Inc, The packaged cogeneration systems market in the
United States, April 1984.
   Resource Planning Associates, Inc., The potential for industrial
cogeneration development by 1990, July 1981.
   The potential for industrial cogeneration development to defer new utility
electric power capacity over the next decade, June 1983.
   US Office of Technology assessment, Industrial and commercial
cogeneration, February 1983
   Industrial and use, June 1983.
   Hagler, Bailly and Co, USA industrial cogeneration, February 1982.
   General Energy Associates, Industrial cogeneration potential targeting
opportunities at the plant site, December 1981.
   SRI International, Industrial cogeneration, February 1981.

16 W A van der Lugt, The importance of gas fired turbines for energy
production and the environment in the Netherlands.
Energy Economy 1984 Symposium Amsterdam, December 1984.

17 J G Hellemans, A study of the potential of CHP in Netherlands industry
(1984) made on behalf of NEOM by Vereniging Kraftwerktuigen, Amersfoort.
Published September 30th 1985.

18 Planning Research Systems PLC, The world market for industrial gas
turbines, 1981-90. A production and market forecast, London, January 1987.

19 UK DOE Energy Efficiency Demonstration Scheme.   ETSU report ED/93/188
(1986).
   Cogeneration 3, 3, 24 (1986)
   Gas Turbine World 16, 5, 10 (1986).

20 ASME Gas Turbine Conference.
   Paper No 83-GT-102.
   Gas Turbine World 15, 5, 18 (1985).

21 Gas Turbine World 16, 1, 26 (1986).

22 Gas Turbine World 17, 1, 10 (1987).

23 M J J Linnemeijer, J P van Buijtenen and A U van Loon, Boosting steam
plant thermal efficiency and power output through the addition of gas
turbines.

24 ASME Gas Turbine Conference Anahein Calif, June 1987, Paper no. 87-GT-4.
J J Veenema, H Brueckner, H H Finckh, Topping the Groningen Steam Turbine plant with a Gas Turbine.
   ASME Gas Turbine Conference, Anaheim Calif, June 1987, Paper no. 87-GT-38.
   Diesel and Gas Turbine Worldwide XVIII, 9, 40 (1986).
   Cogeneration 3, 3, 20 (1986).
   Gas Turbine World 17, 3, 27 (1987).
   Gas Turbine World 15, 6, 16 (1985).
   Gas Turbine World 16, 2, 4 (1986).

25 Gas Turbine World 17, 1, 16 (1987).

26 See for example Gas Turbine World 16, 5, 7, 86 (1986).

27 Diesel and Gas Turbine Worldwide XIX, 5, 16, (1987).

28 Gas Turbine World 17, 3, 34, (1987).

29 Gas Turbine World 17, 3, 20 (1987).
                     15, 4, 22 (1985).
   Cogeneration 2, 5, 8 (1985).

30 A L Kalina and H M Liebowitz, Applying Kaline technology to a bottoming cycle for utility combined cycles.
   ASME Gas Turbine Conference, Anaheim, Calif, June 1987, Paper No. 87-GT-34.
   Cogeneration 3, 1, 8, (1986).

31 Victor Biasi, Modified 501 powers 10 MW and 25 MW storage peakers, Gas Turbine World 17, 3, 50 1987.

32 J R Joiner, W G Snyder, Coal gasification systems for power generation.

   RETSIE/IPEC Conference, Anaheim Calif, (1987).
   Gas Turbine World 15, 4, 16 (1985)
                     16, 4, 10 (1986)
   Modern Power Systems 6, 10, 31 (1986)
                        6, 10, 37 (1986).

33 Cogeneration 3, 2, 30 (1986)
   Gas Turbine World 15 , 6, 20 (1985)
   Turbomachinery 26, 9, 23 (1985).

34 J C Solt, Coping with Gas Turbine Emmission Regulations.
   ASME Gas Turbine Conference, Anaheim, June 1987.
   Paper No 87-GT-239.
   B Becker et al, Premixing gas and air to reduce NOx emissions with existing proven gas turbines combustion chambers.
   ASME Gas Turbine Conference, Dusseldorf, W Germany, June 1986.
   Paper No 86-GT-157.

V. End-Use Technologies

ELECTROMECHANICAL ENERGY CONVERSION BY CONTROLLED ELECTRICAL
DRIVES

W. LEONHARD

INSTITUT FÜR REGELUNGSTECHNIK
TECHNICAL UNIVERSITY BRAUNSCHWEIG, GERMANY

1. INTRODUCTION

An electrical grid serves many different tasks, most
important the transmission of electrical power from the points
of origin to that of consumption and of pooling the statistically varying demands of innumerable consumers with the power
generated from the primary energy sources. The storage
medium, buffering the unavoidable short term mismatch of the
generated and consumed power is the kinetic energy of the
rotating machines, reducing the temporary differences of power
to minor variations of frequency and phase angles.

Another important aspect of an electrical power grid is that
it links any source of primary energy, be it fossil, nuclear
or regenerative, to all sorts of consumers, whether thermal,
chemical or mechanical, found in industry, transportation and
households. Electricity as an intermediate form of
energy thus makes it possible to exploit primary
energy that could not be used directly due to distance,
cost, radiation or environmental effects. If renewable energy
from wind or sun should some day be of practical importance,
it will undoubtedly also pass through electrical form.

Electricity is clearly the most versatile energy that can,
with low transmission and distribution losses, be readily
converted to any other form. It is most closely related to
mechanical energy because practically all electrical energy is
today produced by rotating generators acting as mechano-
electrical converters. The same is true at the end of
delivery, where more than half the electrical energy
produced in an industrialized country is converted back to
mechanical form, used wherever physical activities take place,
such as production processes of all kinds or transportation of
goods and people. The electro-mechanical energy conversion is
performed by electrical drives, which are the subject of this
review.

As sources of mechanical power they excel for a variety of
reasons:
- They are built for a wide power range, from $\ll 1$ W in
  electric watches to $> 10^8$ W for driving pumps in hydro-
  storage plants; high torque, $> 10^6$ Nm, for rolling mills
  and high speed, $> 10^5$ min^{-1}, for centrifuge drives are
  realizable.

269

A. T. De Almeida and A. H. Rosenfeld (eds.), Demand-Side Management and Electricity End-Use Efficiency, 269–297.
© 1988 by Kluwer Academic Publishers.

- Electric drives can operate in all four quadrants of the torque-speed plane, which makes them suitable for active loads, such as mine hoists without brakes or reversing gear. In the braking quadrants, they are regenerating, feeding electrical power back to the supply line.

- Electrical drives exhibit high efficiency, > 0.9 at larger ratings; hence the input power is determined by the load.

- Full power is immediately available upon turn-on, no idling or warm-up is required; the drive can be designed for high short time overload, limited only by thermal considerations.

- Electrical drives are characterized by smooth torque, low vibrations, relatively low temperatures and long life time; they can operate under any conditions, even submerged in hazardous or radioactive liquids.

- They have no environmental effects at the point of application which makes it possible to integrate the drives into the machines; this is most prominent in machine tools and robots employing distributed electrical drive systems; controllable mechanical power is applied wherever and in what form it is needed, thus avoiding complicated mechanical transmissions.

- Most important, electrical drives can be rapidly and precisely controlled, which renders them adaptable to almost all situations, where controlled mechanical power is needed. For example, an elevator drive must alternately operate with controlled torque, acceleration, velocity or position in order to produce the desired smooth and swift motion of the elevator cabin.

- The steady state characteristics can be shaped by the control to suit particular needs, such as constant power over a wide speed range for use on traction drives without changing gears.

As a consequence of this unique combination of desirable properties, electrical drives are the natural choice in most stationary applications, where a continuous electrical power supply is available. Some limitations exist, however.

- Without continuous power supply, for instance on railways without catenary and power rail or with road vehicles, the electrical energy source, which is usually bulky, heavy and expensive must be carried on board (storage battery, rotating generator with internal combustion engine or turbine, solar cells). This has so far prevented the general use of electric drives on road vehicles; the weight of a present day lead-acid battery is about 50 times that of a liquid fuel tank storing an equal amount of energy, even when taking the low efficiency of an internal combustion engine into account.

- The tangential force per unit surface of the rotor of an electrical machine is limited by saturation of the iron and cooling problems; this has the effect that a high power density can only be achieved by resorting to high speeds. Where large force, short stroke actuators are needed, such as for positioning control surfaces on aircraft, gearless high pressure hydraulic drives have a higher power to weight ratio.

- Electrical drives can be designed for linear motion, such as needed for magnetically levitated trains, but their properties are normally inferior to those of rotating machines where all parts of the stator and rotor structure can be utilized.

Most of the electrical drives in use today are of the (approximately) constant speed type, as exemplified by the line-fed induction motor with cage rotor, which is found in large numbers on every farm. Only a small fraction of the drives , perhaps 20%, are variable speed drives, requiring some sort of control; this is the most interesting part of the field, which is likely to expand due to advancing automation in industry. Also, there is a considerable potential of energy conservation at partial load if the speed can be varied; this is evident with centrifugal pumps and ventilators, where control of pressure and flow by varying the speed consumes far less energy than control by throttle. Naturally the cost of a variable speed drive is higher than that of a constant speed drive.
At present, the field of adjustable speed drives is dominated by direct current (dc) drives but there is a strong tendency towards development of alternating current (ac) drives, accelerated by the progress of semiconductor technology. It is the aim of the paper to describe the present state and the foreseeable future development.

2. ELECTROMAGNETIC TORQUE GENERATION
The most common principle of generating torque in an electrical machine is described by the simplified model shown in Fig. 1, where a cylindrical core (rotor) revolves concentrically in a hollow ring (stator), both of which are made of magnetic material. The cylindrical surfaces bounding the airgap are assumed to be circular, even though they normally have axial slots, possibly skewed, containing current conductors and may also be shaped to form salient poles protruding into the space between stator and rotor; both parts of the machine have the axial length L.

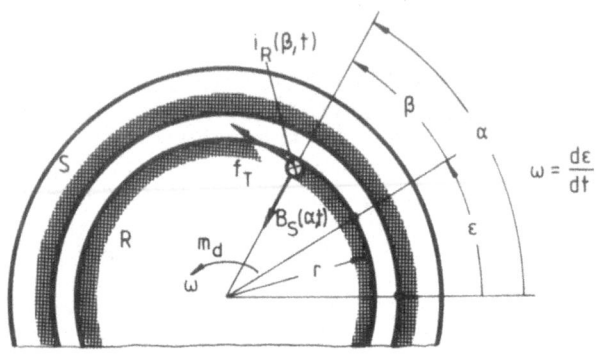

FIGURE 1. Simplified cross section of electrical machine

α,β are the angles of circumference in the stator and rotor respectively; ε is the angle of rotation, hence ω=dε/dt the angular velocity of the rotor.

It is now assumed, that the stator contains windings consisting of axial conductors, connected at the ends; when carrying currents, they produce a magnetic field distribution B_S (α,t); the field pattern is periodic with $2\pi/p$, where p is the number of pole pairs. Due to the permeability of the iron, the field lines end perpendicularly on the iron surfaces. If the stator currents are constant (dc), the stator field is stationary, $B_S(α,t)=B(α)$, but when the stator currents form a symmetrical polyphase system of frequency $f_1=\omega_1/2\pi$ and flow in a corresponding polyphase stator winding, the field pattern B(α,t) moves with the angular velocity ω_1/p while approximately maintaining its shape.

A similar situation exists on the rotor, where a winding with axial conductors is assumed to be attached to the rotor surface. $i_R(β,t)$ is defined as the current in a conductor at the rotor-based circumferential angle β. By again assuming a polyphase winding carrying alternating currents of frequency $f_2=\omega_2/2\pi$, $i_R(β,t)$ represents a current density distribution, which is moving across the rotor with the angular velocity ω_2/p, while the rotor itself revolves at the angular velocity ω.

Each conductor on the rotor surface which is exposed to the radial field experiences a tangential Lorentz-force

$$f_T(β,t)=L\ B_S(α,t)i_R(β,t)=L\ B_S(β+ε,t)i_R(β,t);\qquad(1)$$

summing over all N_R conductors on the rotor surface results in the electrical driving torque exerted on the rotor

$$m_d(t) = rL \sum_{N_R} B_s(\beta+\epsilon,t)i_R(\beta,t). \tag{2}$$

Clearly, constant electrical torque can only by produced, if the field and the current density patterns maintain their shapes as the rotor moves and remain "in synchronism", i.e.

$$\omega_1 = p\omega \pm \omega_2 ; \tag{3}$$

the ± sign depends on the sequence in which the rotor windings are connected.

This simplified principle of generating torque in an electrical machine may be modified in many ways. For example, the field distribution can be produced by permanent magnets or the role of stator and rotor may be interchanged by making the rotor the primary part. With a reluctance motor, the stator winding takes care of flux and currents, while the geometrical shape of the rotor carrying no windings defines the flux distribution .Further, the machine may act as motor or generator, depending on the direction of torque and rotational speed. This is discussed with the example of a dc-machine, which is the variable speed drive most widely used today.

3. DC-MACHINE FOR USE IN ADJUSTABLE SPEED DRIVE
In a dc-machine, the stator field is stationary, being produced by permanent magnets or a constant stator current, $\omega_1=0$; the shape of the field distribution is an approximately trapezoidal function of the angle α. The rotor winding (armature) is fed through a commutator, a mechanical switching device attached to one end of the rotor, where continuous armature current is injected through stationary carbon brushes sliding on the commutator bars. This has the effect that the current is switched to subsequent conductors as the rotor moves; as a result, the current pattern of the rotor also remains stationary, being fixed with respect to the stator. This is shown in Fig. 2a with the example of a dc-machine having 4 salient poles and concentrated stator windings. The field distribution $B_s(\alpha)$ is mainly determined by the shape of the poles.

While the current fed to the brushes is continuous, the currents in the conductors of the revolving rotor winding alternate, having an angular frequency $\omega_2=p\omega$; this follows from equ. (3) with $\omega_1=0$. Hence the commutator acts as a synchronous inverter clocked by the angle of rotation.

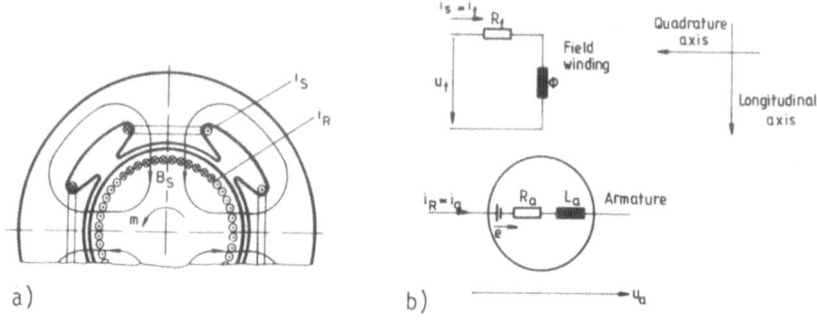

FIGURE 2. Four pole dc-machine
a) Simplified cross section. b) Equivalent circuit.

This model is of course greatly simplified; due to the inductance of the armature winding the current in individual coils
cannot be commutated instantaneously. In order to accelerate
the commutation transient, special auxiliary poles, not shown
in Fig. 2a, are placed in the neutral zones between the main
poles. Still, commutation tends to limit the speed and the
armature current, at which the motor can be operated. Also,
the quadrature magnetic field caused by the armature winding
as a whole tends to distort the flux wave B_s (α) produced by
the stator poles and has further detrimental effects on
commutation. In order to cancel the quadrature field a compensating winding may be placed in the pole shoes, close to the
armature; this also reduces the armature inductance, allowing
a more rapid change of the armature current and torque.

Generation of torque is of course only one aspect of a motor,
another is speed of rotation. Clearly, as the armature ro-
tates, flux change is experienced and there are alternating
voltages induced in the windings; they are rectified by the
commutator and become effective as direct voltage u_a between
adjacent brushes. An external voltage source must be applied
to the armature circuit in order to maintain the desired
armature current and torque.

An equivalent circuit of a dc-machine is drawn in Fig. 2b
showing the two separate circuits; the stator field winding
is passive, serving only for excitation. The electric power
u_a i_a fed to the armature is the main input, which after
subtracting the unavoidable losses represents the equivalent
of the mechanical power m_d ω, delivered at the shaft.

As mentioned before, dc-motors are the usual choice today
where variable speed electrical drives are needed; most of
these drives require closed loop control for armature current,
speed and, possibly, angular position. A block diagram of a
separately excited dc-machine is drawn in Fig. 3, showing the
dynamic interactions between the input variables u_a, u_f and
the feedback variables i_a, m_d, ω, ε. There are nonlinearities due to the formation of torque (Lorentz) and induced
voltage (Faraday); the nonlinear curve describes the effect of
magnetic saturation in the iron core.

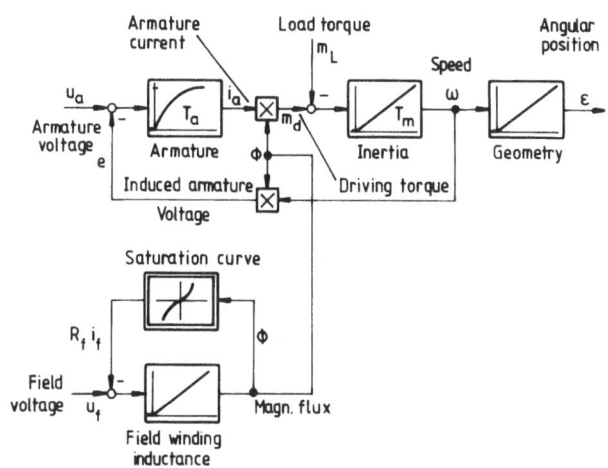

FIGURE 3. Block diagram of dc-motor with armature and field
           supply.

The main control input is the armature voltage u_a, while
the field voltage u_f only serves as an auxiliary input which
is normally kept at rated value; reducing the flux has
the effect of increasing the speed at a given armature
voltage, at the same time lowering the torque; it may be
likened to a continuously changeable gear ratio.

In order to operate a dc-machine at variable speed and torque
it is necessary to supply the armature with variable direct
voltage and current. The most common four quadrant controllable dc-power supply in use today is a bidirectional line-
commutated converter as seen in Fig. 4a in the form of a
six-pulse reversible thyristor circuit. Converters of
this type are available in a wide power range, reaching - in
multiple connections - up to the GW level for high voltage dc
transmissions, far beyond the 10-20 MW, needed for the largest
dc-drives.

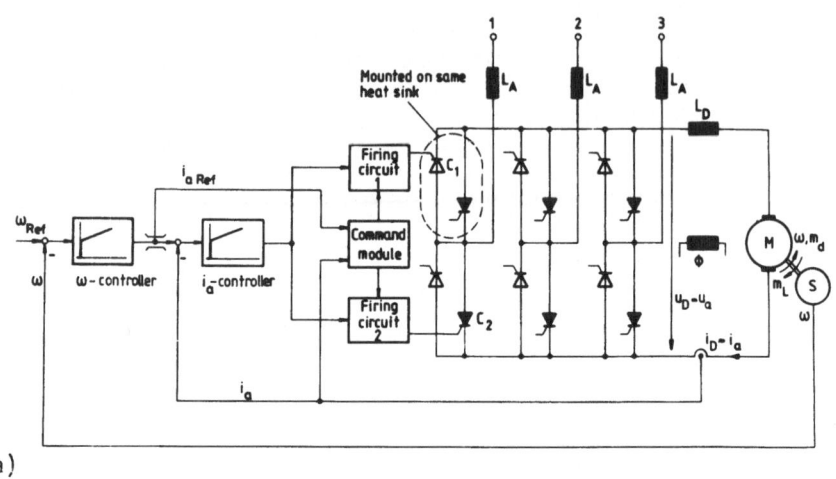

a)

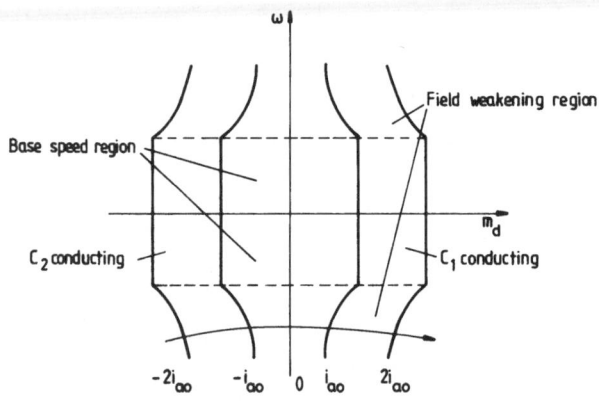

b)

FIGURE 4. Dc-drive with bidirectional 6-pulse thyristor converter.
a) Circuit. b) Operating range.

The control function is achieved by phase shifted firing
pulses that are supplied to converter C_1 or C_2, depending on
which direction of armature current and torque is necessary
for maintaining the commanded speed. The operating regions of
the drive in the four quadrants of the torque-speed plane are
seen in Fig. 4b.

The control scheme in Fig. 4a is typical of controlled
drives, be it dc or ac; it contains an inner current loop
serving for fast build-up of armature current and torque and a
superimposed speed controller generating the reference value
for the current controller. By electronically limiting the

commanded current reference, the converter, the drive and the mechanical load are protected against accidental overcurrent in the low impedance armature circuit. In view of the limited thermal storage capacity of power electronic components, the inner current loop is an important protective feature for all types of converters and drives.

The dynamic response of a line-commutated 6-pulse converter is excellent, allowing 6 regular firing intervals per period of line frequency. With a 60 Hz power supply this means that the firing instants, at with a changed control signal can become effective, are only 2.77 ms apart in steady state. For large drives, several 6-pulse converters with out-of-phase supply voltages are used so that the firing intervals are further reduced. Since the response of the current control loop is ultimately determined by the discontinuous action of the converter, this control delay results in an equivalent lag of the current control loop of less than 10 ms which is adequate for most drives. With small servo motors for feed drives requiring even faster response, the thyristor converter is replaced by a dc-supplied transistor chopper having a switching frequency in the kHz range; the equivalent lag of the current control loop can then be shortened to < 1 ms, equivalent to nearly impressed armature current. The response of the speed control loop depends of course on the inertia of the drive; it is necessarily slower than that of the current loop because the speed controller operates through the current control loop. The same is true for a possible angle control loop.

From the preceding section it is apparent that a dc-machine is ideally suited for adjustable speed drives, except for the mechanical commutator which

- limits power rating and speed of the machines that can be built, because of the flash-over voltage between commutator bars and the time interval required for commutation,

- imposes restrictions on the ambient conditions, at which the motor can be operated, because there is always the possibility of sparking,

- is subject to wear, calling for periodic maintenance and

- increases the length, weight and inertia of the motor, which is a disadvantage for traction and servo drives.

- Furthermore, dc-machines are of complicated mechanical construction requiring considerable manual labour.

It is for these reasons, that enginers have for a long time been investigating the possibilities of variable speed drives without sliding contacts, i.e. adjustable speed ac-drives.

4. ADJUSTABLE SPEED AC-DRIVES

From what was said in the introduction, an ac-drive requires an external power converter capable of producing alternating voltages and currents with controllable amplitude and frequency. The conversion from the constant voltage, constant-frequency supply voltage is often performed in two steps employing an intermediate dc-link, as seen in Fig. 5.

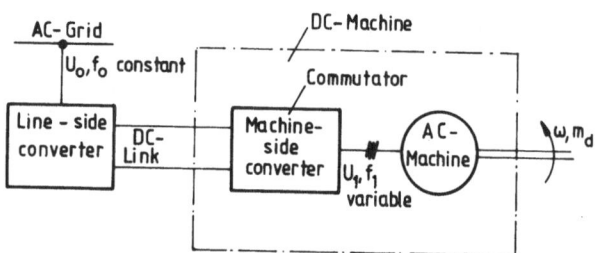

FIGURE 5. Adjustable speed drive with dc-link

The line side converter feeding the dc-link with variable voltage and current could be of the line commutated type as shown in Fig. 4a; in the simplest case, assuming a constant voltage dc-link and excluding regeneration, it could even be a diode rectifier.

A machine side converter (inverter) is needed for generating alternating voltages of variable amplitude and frequency; there is a wide choice of inverter schemes to be discussed later. With a dc-drive, the mechanical commutator represents the machine-side converter. As this is part of the motor, the power circuits of todays dc-drives are quite uniform. There are only few variations such as excitation by field winding or permanent magnets; if the load calls for unidirectional torque only, the converter in Fig. 4a can be simplified by omitting part C_2. When the drive is fed from a dc power supply, for instance a battery on board a ship or a dc power rail on a rapid transit vehicle, the line-side converter is replaced by a power electronic chopper for speed control.

With an adjustable speed ac-drive, the situation is much more involved because many different types of ac-machines and converters are now applicable. This is shown in a synopsis (table 1) containing the most important types of ac-machines and converters that may be combined to adjustable speed ac-drives. There is a considerable variety but only the combinations where comments are entered are presently of importance. It is quite possible that shifts of emphasis will occur, such that some combinations will lose and others may gain interest, as the field of adjustable speed ac-drives matures; however it is unlikely that it will converge to only one or two combinations as with dc-drives, because each solution in table 1 has its particular merits. The matrix in table 1 is arranged such that the power rating begins at lower values on the upper left and increases towards the lower right. The numbers indicate typical values, not the maximum possible power ratings.

POWER CONVERTER

AC-MACHINES	CONVERSION WITH DC-LINK				DIRECT CONVERSION
	Voltage source transistor inverter A	Voltage source thyristor (GTO) inverter B	Current source thyristor (GTO) inverter (forced commutation) C	Current source thyristor inverter (natural commutation) D	Cycloconverter E
Synchronous machine with permanent magnet excitation 1	Low power, high dynamic performance (servo drive) 10 kW	Medium power, high power density 1 MW			
Reluctance machine 2		Low to medium power 100 kW			
Induction machine with cage rotor 3	Low power, high speed high dyn.performance (servo or spindle drive) 100 kW	Medium to high power high dynamic performance 2 MW	Medium to high power, high speed 4 MW		High power, low speed high dynamic performance 7,5 MW
Doubly-fed induction machine with wound rotor 4				High power, subsynchr. operation (Scherbius) 20 MW	High power, restricted speed range 3 MW
Synchronous machine with field and damper windings 5				High power, high speed 20 MW	High power, low speed high dynamic performance 10 MW

TABLE 1. Synopsis of adjustable speed ac-drives

4.1. Choice of ac-machine

All the motors in table 1 are inverted, when compared to a dc-machine (Fig. 2b), because the main power input is now at the stator side, thus avoiding the problems associated with sliding contacts at high power and speed.

4.1.1. Synchronous machines

Three of the five motors (1, 2, 5) are of the synchronous type, where the flux wave revolves in synchronism with the rotor ($\omega_2 = 0$) and the frequency of the stator currents is proportional to speed, $\omega_1 = p\omega$. The rotor excitation is produced by permanent magnets (1) or a field winding (5) which is supplied with continuous current through slip-rings and brushes, or by a rotating rectifier and an ac-exciter (an auxiliary ac-generator), as is common with large turbo-generators. The reluctance motor (2) is a special case where the field and the ampere turns are produced by the stator windings, while the toothed rotor, carrying no windings, locks in with the field, similar to a stepping motor.

Motors with permanent magnet excitation which benefit from low rotor losses, are a frequent choice with servo drives for high dynamic performance; they may be designed for sinusoidal voltages and currents or for trapezoidal voltages and intermittent phase currents (brushless dc-motor). The magnets are usually ferrites requiring flux concentration and protection against demagnetization by high stator currents; more recently rare earth magnets are preferred because of their high energy product (remanence flux density>1T, coercive force > 7 kA/cm), which results in much reduced volume of the (expensive) magnet material, permits surface mounting without flux concentration and offers ample safety against accidental demagnetization. Permanent magnet synchronous motors can be either of the conventional drum type with radial field or of a disk motor design with axial field, where the magnets embedded in a nonconducting rotor disk move between stator plates carrying radial windings. Low inertia, which is important with servomotors where high dynamic performance is specified, can be achieved with both designs. Very recently, permanent magnets are also used in large motors, for example a gearless 1 MW drive with high power density for ship propulsion (Fürsich 1986).

Synchronous motors operating off the constant voltage/frequency line are characterized by oscillatory response, which makes it necessary to install short circuited damper windings in the rotor; they also serve for starting as an induction motor. With a controllable power supply both effects can be achieved through the converter making the damper windings on motors of type 5, in principle, superfluous. However, a damper winding may still be desirable for reducing the subtransient reactance of the machine, thus facilitating commutation of the inverter. As with generators, large synchronous motors have either solid cylindrical rotors at high speeds or salient poles at low speed, they are combined with different types of converters (5D, 5E).

In steady state, the stator current ampere turn wave of a synchronous maschine is locked in synchronism with the rotor, hence it has advantages to control the stator currents in rotor coordinates, using a coordinate-transformation based on the rotor angle. This is explained in more detail later.

4.1.2.Asynchronous machines.
As mentioned earlier, induction motors with cage rotor (3) are the most widely used electrical machines at constant speed; their main attractions are mechanical simplicity, ruggedness, and low cost, particularly at lower power where the rotor windings are produced by die casting. With power electronics, induction motors can be used also for adjustable speed drives. Refering to Fig. 1, the rotor currents are now induced in the short circuited winding by the stator flux wave which moves across the rotor surface at slip speed; this is only a few percent of rated speed, otherwise the power factor would decrease and the rotor losses would be too large.

The choice of the sign in equ (3) depends on whether the motor operates below or above synchronism, i.e. motoring or regenerating; synchronous speed is determined by the stator frequency, $\omega_0 = \omega_1/p$. The induction motor is an ideal solution for many applications of ac-drives, especially when high speed is desired. Of course, the inverter must deliver not only the active input power but also the reactive power for magnetization, as no external excitation is present.
Control of an induction motor is not difficult as long as high dynamic performance is not the main objective; however, it becomes quite involved if performance similar or even superior to that achieved with a converter-fed dc-motor is specified. The reason is that the position of the flux wave, being produced by the stator as well as the not measurable rotor currents, is unknown. This calls for flux modelling, the accuracy of which may be affected by parameter changes of the motor. Most of these problems have been solved in recent years with elaborate digital control schemes employing microelectronics.

Some applications of drives require only a limited speed range, such as 20% variation around nominal speed. This is the area of the doubly-fed induction machine (4); typical examples are fan drives and boiler feed pumps in power stations, where the load torque varies with the square of the speed; drives with fluctuating load, that are buffered by a fly wheel also belong to this class. More recent applications are variable speed wind power generators exposed to gusty wind conditions.
By connecting the stator winding of a wound rotor induction motor to the constant frequency line and extracting or

injecting power at slip frequency into the rotor windings, the rating of the rotor-side electronic conversion equipment is determined on the basis of slip power only, i.e. by the specified slip speed. Various options exist, ranging from a very flexible solution with nearly sinusoidal low frequency rotor supply by cycloconverter to simplified versions with diode rectifier and line commutated inverter for slip power recovery (Scherbius- and Krämer-drives).

4.2. Choice of power electronic switching devices and converter circuits

4.2.1. Switching devices

Inserting an electronic device with variable resistance in an electrical circuit offers the means of controlling the current. However, in view of the otherwise high power losses, it is important that either the voltage across the device or its current are as small as possible, i.e. that it acts like a switch that is opened and closed with varying duty-cycle. This has the consequence that the voltages produced by all power electronic converters are highly distorted; the waveforms of the currents are usually smoother having been filtered by the circuit inductances.

There are many electronic devices offering variable resistance to current flow. Those presently applied in power electronics are, in order of power rating,

- Field effect transistors (Power-MOSFET) for lower power, up to 50 A or 1000 V, solid state switching devices with characteristics resembling those of vacuum tubes. Their main advantage is short switching time ($<1\mu s$) making it possible to design inverters with switching frequencies beyond 20 kHz, where the machines and associated magnetic circuits cease to generate audible noise. A disadvantage of Power-MOSFETs is the relatively high on-state resistance. Like all transistors they can be switched on and off by low power gate signals, which makes them readily suited for inverters with dc-supply.

- Bipolar transistors for intermediate power, to several 100 A, 1200 V. They are available in modules containing up to six-pulse bridge connections and are widely used in ac-drives to about 100 kW. The usual switching frequency is several kHz.

- Thyristors are the typical electronic switches for high power applications. They are available in a wide range of voltages, currents and switching speeds, characterized by the time for recovering from conducting to forward blocking condition. Line grade thyristors are available up to about 5 kA forward current and 5 kV forward and reverse blocking voltage, the recovery time may be as high as 500 μs, making these devices mainly applicable to line commutated converters with 50/60 Hz supply frequency. They are also used for reactive power compensation and very high power HVDC-transmission.

Thyristors for use in inverters have much shorter recovery time (< 50 µs) but show other parameters that are less favourable such as higher voltage drop in on state; also the available ratings are somewhat lower. Thyristors in general have the property that they can only be made conducting by a gate signal; returning a thyristor to the blocking state requires a brief interruption of the main current with the help of external components. With inverters fed by a dc-supply, this "forced commutation" calls for additional commutation circuits.
This drawback is eliminated with the more recently developed

- Gate-turn-off thyristors (GTO) where turn-off is also effected by a short time gate signal, though at a low current gain of 3 to 5.The elimination of commutation circuits greatly simplifies the circuitry of GTO-based inverters. GTO-thyristors are available in ratings up to 4,5 kV, 2,5 kA, which makes them suitable for large drives. The switching frequency is usually less than 2 kHz.

After introducing the power electronic switching devices, the converter circuits listed in table 1 will be briefly discussed.

4.2.2. Converters with dc-link
With the converter schemes (A-D) in table 1 the conversion from constant voltage, constant frequency (CVCF) to variable voltage, variable frequency (VVVF) is performed in two steps which provides decoupling and freedom from undesired interactions.
The basic circuits are shown in Fig.6.

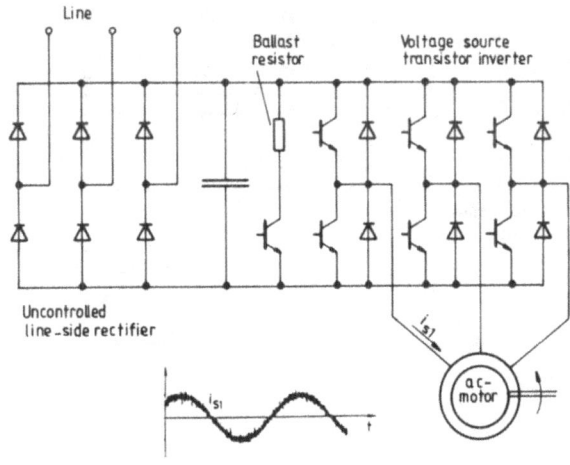

FIGURE 6a)

284

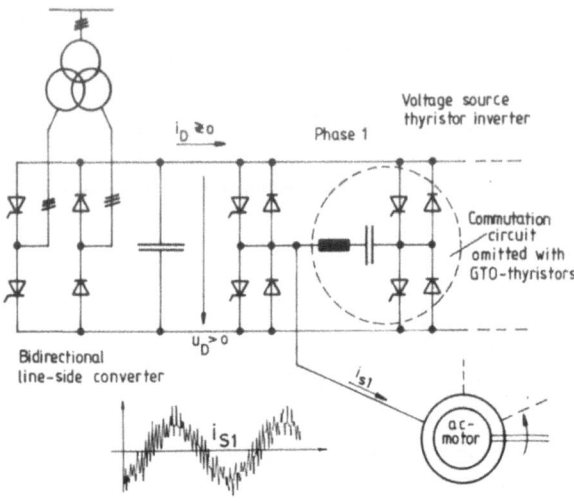

FIGURE 6b)

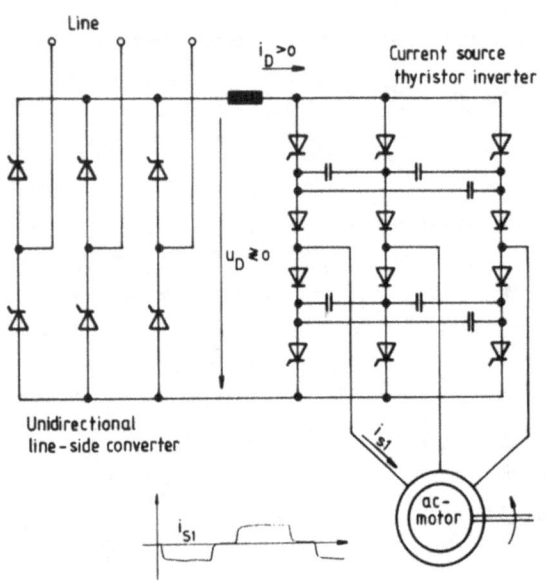

FIGURE 6c)

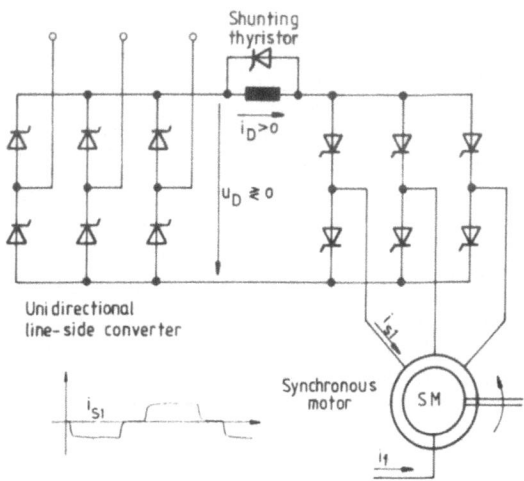

FIGURE 6d)

FIGURE 6 Converters with dc-link
- a) Diode rectifier, voltage source transistor inverter (A),
- b) Bidirectional line-side converter, voltage source thyristor inverter (B),
- c) Unidirectional line-side converter, current source thyristor inverter with forced commutation (C),
- d) Unidirectional line-side converter, current source thyristor inverter with natural commutation (D).

The converters shown in Fig. 6a, 6b possess a low impedance dc-link presenting a nearly impressed constant voltage to the machine-side inverter (voltage source converter) which operates at bidirectional link current i_D. The control of the inverter is performed by pulse-width modulated switching, connecting the motor terminals alternatively to the upper or lower dc-bus, so that approximately sinusoidal motor currents are achieved, at least at low stator frequency. At high speed, the pulse-width modulation ceases so that the inverter generates square wave voltages exhibiting the maximum fundamental component. If the thyristors in Fig. 6b are replaced by GTO-thyristors, the commutation circuit (two thyristors, resonant circuit in each phase) are omitted. The main inverter circuit then resembles Fig. 6a with GTO-thyristors instead of transistors.

The schemes in Fig. 6c, 6d employ a high impedance dc-link with a series reactor carrying the unidirectional link current. The reactor serves to supply the machine-side inverter with approximately impressed current, hence they are called current source converter. The motor currents are intermittent, exhibiting a 60° blank during each half cycle. Converters with current link are usually more economical than those with voltage link because regeneration is achieved by inverting link voltage rather than current, hence a unidirectional line-side converter suffices. The link reactor may be bulky and heavy but it simplifies protection in case of inverter failure by limiting the rise of the link current. At low speed the current-source inverter can also be pulse-width-modulated in order to reduce the pulsations of the motor torque. The thyristors can again be replaced by GTO-thyristors with the ensuing simplifications of the circuitry.

The current source inverter in Fig. 6c is force-commutated which makes it suitable for any ac-motor, while the inverter shown in Fig. 6d employs natural commutation; this is only feasible with synchronous motors above a minimum starting speed, where the internal induced voltages are of sufficient magnitude to perform the commutation. Below this minimum speed the machine-side converter is commutated by temporarily interrupting the link current through control of the line side converter; shorting the link reactor by another thyristor helps to accelerate commutation. Converters of this type are used mainly for high power, high speed drives, such as 20 MW, 5000 min^{-1}, where presently no alternative, certainly no dc-drive, exists (5D).

4.2.3. Converter without dc-link (E)
Converting from constant voltage/frequency to variable voltage/ frequency is also possible without dc-link, again with forced or natural commutation. So far only the second option is being applied in practice. The idea is that a bidirectional, line-commutated converter as used with dc-drives, Fig. 4a, represents a power supply that can operate in all four quadrants of the u_D, i_D-plane. Hence, it can be used to supply one of the stator phases of an ac-motor with voltages and currents of low frequency. By using separate power supplies for each of the stator windings, a symmetrical polyphase voltage system is obtained. Converters of this type are called cycloconverters because the output voltages are made up of sections of the input line voltages. Hence the frequency of the output voltages is restricted to a fraction, such as 40%, of the line frequency. With a 50/60 Hz grid, this limits its use to low speed drives, for example gearless drives for cement kilns or rolling mills. A cycloconverter with three phases at the input and output requires a minimum of 36 thyristors which indicates that its application is mainly in the field of large drives. However as its operation involves only natural commutation, the thyristors can be of the line-grade variety, where standard equipment up to the highest power ratings is available.

5. HOW TO CONTROL AN AC-MACHINE?
Dc-machines are of complicated mechanical construction but, due to the stationary orthogonal field axes, have simple control dynamics. With ac-machines the situation is inverted; since commutation is performed outside the machine, its design is simplified but, owing to the revolving fields, the control

structure is much more complex. This is most pronounced with
the simple cage rotor induction motor which represents a
nonlinear multivariable control plant, where important variables (the rotor currents) cannot be measured, unless a sensor
is built into the rotor. The difference to a dc-machine becomes obvious by comparing the block diagram in Fig. 3 with
that of an induction motor (Fig. 7), valid for arbitrary wave
form of stator voltages.

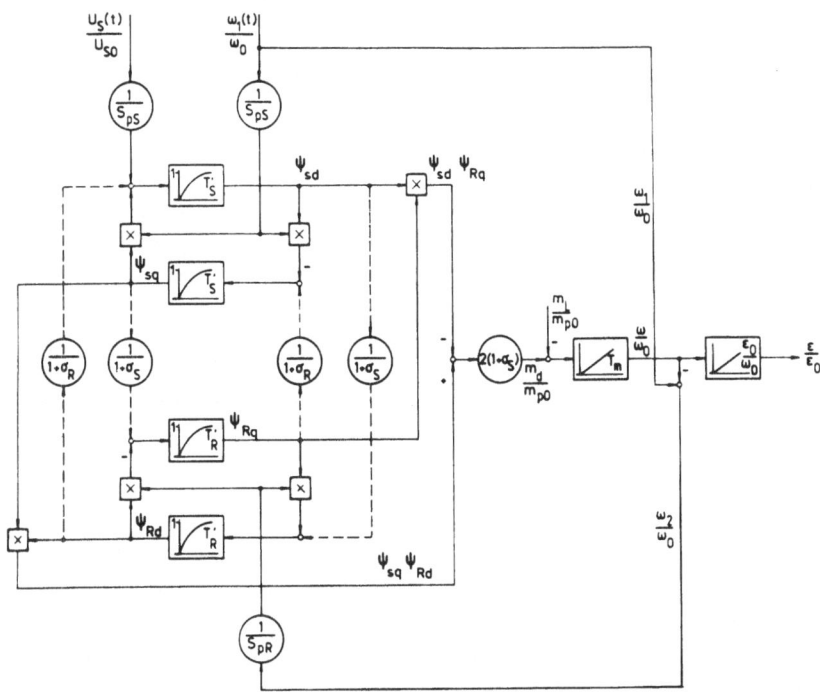

FIGURE 7. Block diagram of a two pole induction motor with
cage rotor

$u_s(t)$, $\omega_1(t)$ are the magnitude and instantaneous angular
velocity of a voltage vector $\underline{u}_s(t)$, which is a combination of
the terminal voltages; ψ_{sd}, ψ_{sq} and ψ_{Rd}, ψ_{Rq} are orthogonal flux components in stator and rotor, defined in a moving
coordinate system determined by the voltage vector. Clearly
this complicated block diagram is not suitable for developing
a transparent control strategy for a high dynamic performance
ac-drive. It has taken engineers considerable time until a
satisfactory solution was found in the form of field oriented
control, which can be adapted to any type of ac-machine and
inverter (Blaschke, 1972).

The basic idea of field orientation consists of transforming
the machine equations into a moving frame of reference,

defined (in the case of the induction motor) by the rotor flux wave. This provides the information needed to determine the effect of stator currents on torque or flux; hence the control structure becomes decoupled similar to that of a dc-machine. The problem is thus reduced to detecting the magnitude and position of the rotor flux wave; various schemes involving flux measuring devices and sensing coils in the stator have been proposed but eventually abandoned because of practical difficulties. The method usually adopted now is to use a flux model based on machine equations and measured input signals that are readily available, such as stator currents, terminal voltages, speed or angular position. With the help of redundant measurements it also is possible to detect parameter variations of the machine caused by temperature changes or saturation.

A high-performance ac-drive control scheme in field coordinates requires complicated signal processing involving many nonlinearities such as trigonometric functions or multiplications, which are difficult and expensive to accurately realize with analogue components. This is one of the reasons why control in field coordinates was not immediately accepted and applied to a wide range of drives. However, towards the end of the 70's digital microcomputers had become sufficiently powerful to handle the necessary signal processing tasks by software on-line. Beginning with 8 bit-processors, then converting to 16 bit and later to signal processors, the design of digital ac-drive controls has since become a major international research activity, where many laboratories in universities and industry have produced interesting results; it is not possible to adequately describe this evolution in a paper of limited length (Leonhard, 1986). In the meanwhile many companies are offering digitally controlled ac-drives of various types, as indicated in the synopsis in table 1. It may suffice to discuss only one example of a totally digital control scheme for a servo type ac-drive in some detail here (Lessmeier, 1986).

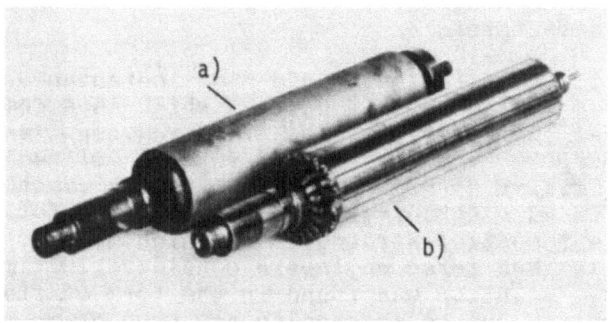

FIGURE 8. Rotors of 1.2 kW servo motor
a) Synchronous, b) Induction motor.

The stator of an experimental four pole, three phase ac-motor can alternatively be fitted with the rotors shown in Fig. 8. One of them is a permanently excited synchronous rotor with rare earth magnets attached to the circumference, the other is a cage rotor. Speed and position of the motor are measured with an optical incremental sensor, the resolution of which is increased by interpolation.

The stator is fed from a pulse-width-modulated voltage source transistor inverter, operating at a switching frequency of 4 kHz; the modulation is performed digitally by a microcomputer used for all drive control tasks including current control. Assuming a maximum speed of 6000 min^{-1} corresponding to 200 Hz stator frequency, approximately sinusoidal stator currents are obtained over most of the speed range.

The synchronous motor and the induction motor are controlled in rotor or field coordinates, respectively. This is indicated in Fig. 9 showing the reference frames for controlling the two motors.

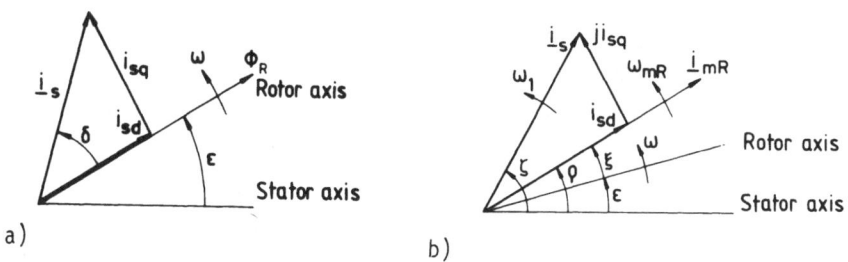

a)   b)

FIGURE 9. Moving reference frames for control of ac-motors
           a) Synchronous motor. b) Induction motor

The position of the flux vector in case of the induction motor is computed on-line by the microcomputer on the basis of measured stator currents and speed. The block diagram of the flux model is shown in Fig. 10. It includes the in-bound transformation of the stator currents from stator coordinates to field coordinates, thus immediately producing the orthogonal components of the stator current vector in field coordinates, i_{sd} and i_{sq}.

290

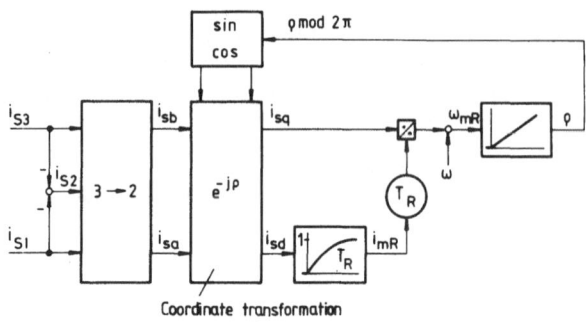

Coordinate transformation

FIGURE 10. Flux model of induction motor in field coordinates

The complete block diagram of the digital control scheme for
the two motors is drawn in Fig. 11, where all the control
functions are executed in software. For example, there are no
external current control loops; rather, the currents are controlled in rotor or field coordinates; thus the current controllers are processing dc-signals in steady state. The only
difference between the two control schemes is the flux model
which is superfluous in case of the synchronous motor.

The upper control levels take care of flux, terminal voltage
(restriction by the inverter), speed and angular position. All
control tasks are executed by a microprocessor/signalprocessor
combination at the sampling frequency of 4 kHz, synchronously
with the switching of the inverter.

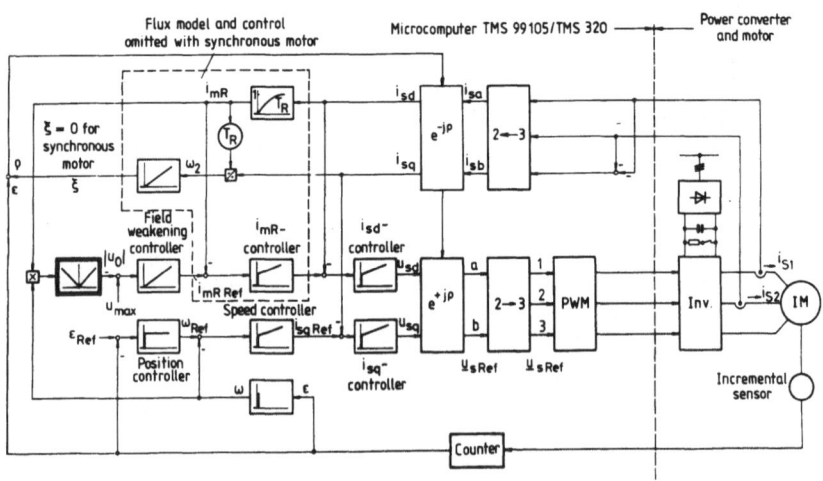

FIGURE 11. Block diagram of digital ac-motor control in
moving coordinates.

Measured step responses of the two drives at no load, with the position control loops closed, are seen in Fig. 12; they show very high dynamic performance corresponding to nearly ideal time-optimal behaviour. With the stator current limited to five times thermally rated current, the maximum acceleration is 35000 Rad/s^2. Following a change of reference, the torque changes within 250 µs, the transient is completed in less than 1 ms.

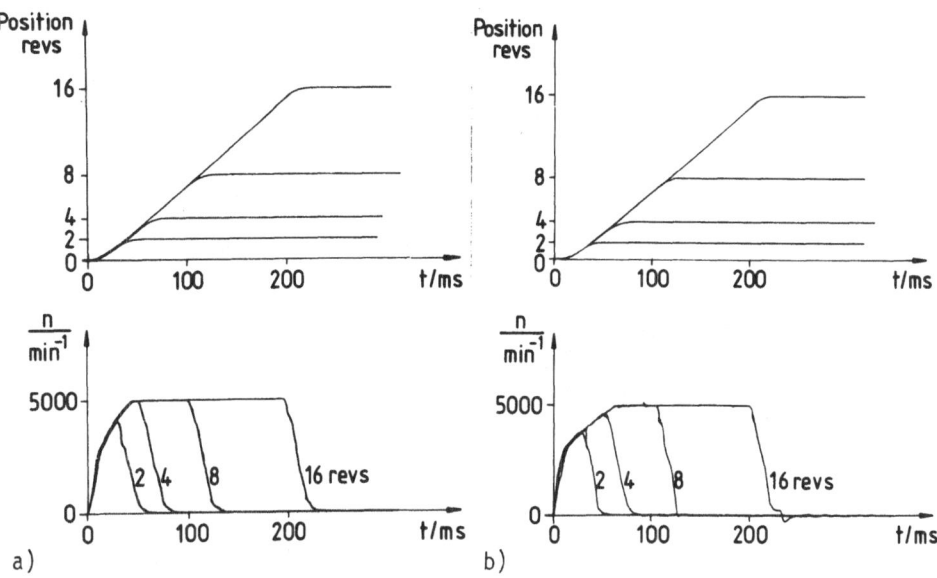

FIGURE 12. Step response of position controlled ac-servo drives
a) Synchronous, b) Induction motor.

Of course, this is just one example of a digital drive control system. Digital control schemes exist for the various types of ac-drives; their performance is continually improved as made possible by the progress of microelectronics. Some of the repetitive control functions will eventually be embedded in special chips in order to make the controller even more com-pact and to increase its computational speed. An example of this tendency is pulse-width-modulation, which today is often executed by means of pre-optimized pulse patterns, eliminating certain harmonics in steady state (Patel, Hoft 1973); in future it may be done by on-line optimization in order to take dynamic conditions of the drive into account (Holtz, 1983; Pfaff, 1983).

6. DIGITAL CONTROL OF DRIVE SYSTEMS
The flexibility offered by microelectronics can of course, be put to more extended uses than just on-line control. Depending on the required bandwidth and the complexity of the tasks, there will be a place for custom designed hardware as well as software, preferably written in high-level language. This

perspective is indicated in Fig. 13, showing the likely structure of future digital drive control systems.

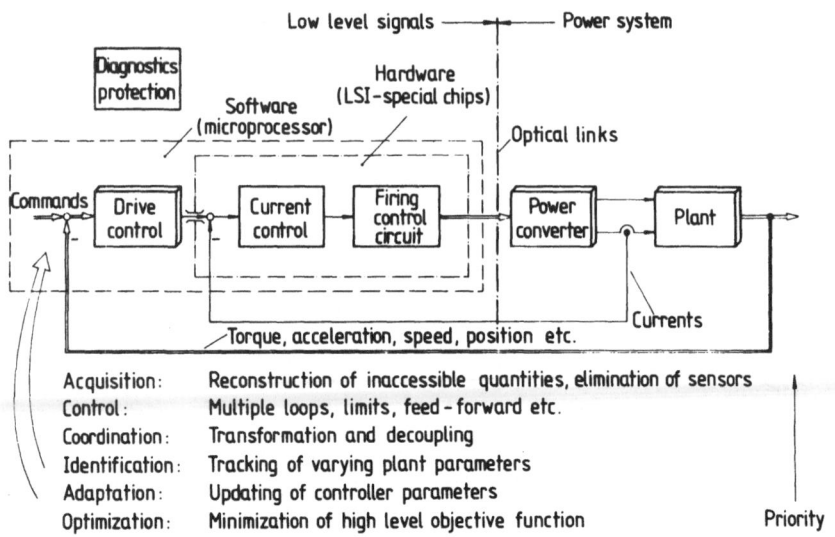

Acquisition: Reconstruction of inaccessible quantities, elimination of sensors
Control: Multiple loops, limits, feed-forward etc.
Coordination: Transformation and decoupling
Identification: Tracking of varying plant parameters
Adaptation: Updating of controller parameters
Optimization: Minimization of high level objective function   Priority

FIGURE 13. Digital control of electrical drives

The tasks close to the power electronic converters are inherently repetitive, requiring no high level decisions, but very fast response, possibly with a resolution of less than 1µs, for time-critical tasks such as pulse-width-modulation. If the control function is of a sufficiently uniform nature, so that adequate demand exists, this may be an ideal case for a single chip computer, a semiconductor gate array or a custom chip. In view of the necessary protection of the power electronic converter, the current controllers are also likely candidates for realization with special hardware. This would have the additional benefit that the current control becomes an integral part of the high power converter, even in the testing phase, and cannot be affected by software errors occurring in the higher levels of control.

While the control of the power converter may be complex, it is still a repetitive and relatively uniform task. This changes profoundly, as we approach the higher levels of control, such as speed or position control of single drives, or trajectory control of multidrive machine tools and robots. There is an enormous variety of different designs, sets of parameters and operational requirements that can only by software be efficiently handled. Considering the complexity of a digital real time program, when compared to the transparent structure of an analogue control scheme, there is a strong incentive to implement selfadjusting control and diagnostic control functions. The aim is to simplify the work with these systems during tests and in the field, possibly by personnel, who are not familiar with the details of the programming structure.

Suitable computer-based tools are also needed for an efficient design of these control systems, in order to assemble the programs from prefabricated software components and to assure testability and complete documentation despite future changes in hardware.

Clearly, all these are radical deviations from well established procedures in the design, production and test phases of the equipment, as well as training of customer staff. Hence, these transitions will take years to be completed, just like the substitution of dc by ac-drives. Naturally, this development must also be considered in the education of engineers at the universities, where topics like electrical machines, power electronics and control by microcomputer must be treated from a systems point of view.

7. ENERGY CONSERVATION MADE POSSIBLE WITH CONTROLLED DRIVES
The high efficiency of electrical drives, combined with the precision and fast response of the associated control offers numerous oportunities for energy conservation. This may be briefly discussed with the help of a few examples.

The advantages of driving centrifugal compressors, pumps and fans at variable speed as compared to a constant speed drive and control by throttle are well known. Variable speed offers large potential savings from very high power drives used in the process industry all the way down to home size air conditioners. With synchronous motors, fed through a dc-link (combination 5D in table 1) the range of high speed a n d high power is now for the first time accessible by variable speed electrical drives; at low power, transistor-fed induction motors offer an attractive solution.

The energy savings made possible in the steel industry by continuous casting and subsequent rolling in "one heat" have also been recognized for a long time. They are based on the accuracy and fast response of controlled electrical drives of all sizes.

A further example is electric ship propulsion. Recently the 90 MW steam turbines of the liner "Queen Elizabeth II" were replaced by a diesel-electric propulsion system with two 44 MW ac-drives, resulting in 50% fuel saving at greatly improved manoevrability.
The new standard locomotive of the German Bundesbahn with four ac-induction motors fed by voltage source thyristor inverters has a total power of 5.6 MW with the locomotives weighing only 64 to. Due to the continuous fast torque control the tractive force under slippery rail conditions is equivalent to that of the former generation of locomotives with six single phase ac-commutator motors and 50% more weight. In addition,the new

locomotives draw a nearly sinusoidal line current at unity or even leading power factor and permit regeneration. These locomotives will permit huge energy savings with goods trains as well as high speed Intercity-trains.

Finally, in production engineering, where industrial robots are employed in increasing numbers, the emphasis is, of course, on light weight design in order to reduce the necessary power of the servo drives and still achieve a high degree of mobility. Light-weight design invariably leads to resilience of the mechanical linkages and oscillatory response, which in turn can only be remedied by sophisticated control methods using highly responsive electrical drives for the joints of the robots.

These examples may suffice to show how advanced electrical drives can help in improving the energy balance of processes in industry and transportation. More details are likely to be discussed in other lectures.

8. CONCLUSION

Electromechanical energy conversion has been regarded by many as a classical subject, where few, if any, innovations take place; nothing could be further from the truth. It has been the aim of this survey to show that in the last 20 years, since power electronic switching devices on the basis of solid state semiconductors have become available in ever increasing variety, there have been enormous changes, particularly in the field of variable speed drives. With these devices it is now possible to design efficient, compact,very powerful and cost effective static converters required for variable speed ac-drives having no mechanical commutator.

In order to apply these drives to the many different loads, they need fast, precise and, above all, flexible control systems of considerable complexitiy; they can best be realized by microelectronics, using hardware as well as software. While power electronics may be called the muscle of modern drive technology, microelectronics would correspond to the brain; all three parts, the electrical machine with the associated mechanical load, the power converter and the electronic control constitute an inseparable drive system.

Due to the choice of motors and power converters that may be combined, ac-drives exhibit a large variety which, since each has its special merits, is unlikely to be consolidated in the near future, as was the case with dc-drives.

Research and development of controlled ac-drives is far from being completed. Important topics, where further progress may be expected in the coming years, are

- improvements and further cost reduction of power electronic converters,
- complete digitalization of control by microelectronics, use of custom chips for time critical tasks and of software for the higher level control problems,
- increased use of optical transmission of control signals in

order to achieve electrical insulation and reduce
interference by the high power converters,
- selfadjusting and adaptive control functions for drives with
variable loads,
- optimizing control strategies for high level objective
functions, such as energy consumption or line interactions.

These problems make research in electrical drive technology a
worthwile and demanding subject. The combination of power- and
micro-electronics including high speed signal processing and
software proves to present a challenge for the top group of
our students, some of whom may have grown a little tired of
nothing but computer science, sitting in front of terminals
with few contacts to the real world.

REFERENCES

Abbondanti, A. (1977). Method of flux control in induction motors driven by variable frequency, variable voltage supplies. Proc. IEEE/IAS Int. Semicond. Power Conv. Conf., 177.

Appun, P.; Runge, W. (1986). Three-phase ac-traction drives. Proc. ICEM 86, München p. 10.

Baitch, Th.L. (1986). Electricity consumption minimisation through the use of electric variable speed systems. Proc. ICEM 86, München, p. 407.

Blaschke, F. (1972). The principle of field orientation as applied to the new TRANSVECTOR closed-loop control system for rotating field machines. Siemens Rev., 217.

Bose, B. K. (1982). Adjustable speed AC drives, a technology status review. Proc. IEEE. p. 116.

Bühler, H. (1977). Einführung in die Theorie geregelter Drehstromantriebe, Vols. 1 and 2. Basel, Birkhäuser.

Dirr, R., I. Neuffer, W. Schlüter and H. Waldmann (1971). Neuartige elektronische Regeleinrichtungen für doppeltgespeiste Asynchronmaschinen großer Leistung. Siemens-Zeitschr., 45, 362.

Eto, J.H., de Almeida, A. (1987). Saving electricity in commercial buildings with adjustable speed drives. Proc. IEEE Ind. and Comm. Power Systems Conf.

Fürsich, H., E. Thum, J. Hamann, D. Köllensberger (1986). New converter-fed permanent-field motor. Proc. ICEM 86, München, p. 50.

Gabriel, R., W. Leonhard and C. Nordby (1980). Field-oriented control of a standard AC-motor using microprocessors. IEEE-Trans. Ind. appl.IA 16, 186.

Gleissner, G., H. Wokusch (1987). Umrichtergespeiste Drehstrommotoren verdrängen den Gleichstromantrieb im oberen Leistungsbereich. Siemens-Zeitschr., Energie u.Automation p.4.

Gölz, G. and P. Gumbrecht (1973). Umrichtergespeiste Synchronmaschinen. AEG Mitteilungen, p. 141.

Haböck, A. and D. Köllensperger (1971). Stand und Entwicklung des Stromrichtermotors. Siemens-Zeitschr., 45, 177.

Henneberger, G. (1986). Servo drives for machine tools and robotics. Proc. ICEM 86. München, p. 21.

Holtz, J., S. Stadtfeld (1983). A predictive controller for the stator current vector of ac-machines fed from a switched voltage source. Proc. Int. Power Electronics Conf. Tokyo, pg. 1665.

Ishihara, K., T. Kokubu et al. (1984). Application of new drive systems for plate mill drives. Proc. IEEE-IAS, p. 43.

Leonhard, W. (1985). Control of Electrical Drives. Springer, Berlin.

Leonhard, W. (1986). Microcomputer control of high dynamic performance ac-drives; a survey. Automatica 22, p. 1.

Lessmeier, R., W. Schumacher and W. Leonhard (1986). Microprocessor-controlled ac-servo drives with synchronous or induction motors: Which is preferable? IEEE-Transactions on Ind. Appl. IA 22, p. 812.

Meyer, A., H. Rohrer (1986). High-speed electrical motors for variable speed drives. Proc. ICEM 86 München, p. 880.

Mokrytzki, B. (1968). Pulse width modulated inverters for ac motor drives. IEEE Trans. Ind. Appl., p. 312.

Musil, R.v. (1986). Survey of large inverter fed motor designs for variable speed drives. Proc. ICEM 86, München, p. 32.

Nabae, A., K. Otsuka, H. Uchino and R. Kurosawa (1980). An approach to flux control of induction motors operated with variable-frequency power supply. IEEE Trans. Ind. Appl., p. 342.

Patel, H.S., R.G. Hoft (1973). Generalized techniques of harmonic elimination and voltage control in thyristor inverters. IEEE-Trans. Ind. Appl.S. Part I, 1973, p. 310; Part II, 1974 p. 666.

Pfaff, G., A. Wick (1983). Direkte Stromregelung bei Drehstromantrieben mit Pulswechselrichter. Regelungstechnische Praxis 24, p. 472.

Plunkett, A.B. (1977). Direct flux and torque regulation in a PWM-inverter induction motor drive. IEEE Trans. Ind. Appl., IA-13.

Saito, K., K. Kamiyama et. al. (1986). A multiprocessor-based, fully digital ac-drive system for rolling mills. Proc. IAS-Conf.

Salzmann, Th. and H. Wokusch. (1980). Direktumrichterantrieb für große Leistungen und hohe dynamische Anforderungen. Siemens-Energietechnik, p. 409.

Schönung, A. and H. Stemmler (1964). Static frequency changers with subharmonic control in conjunction with reversible speed drives. BBC-Review 1964, p. 555.

Schörner, J. (1986). Speed controlled industrial drives. Proc. ICEM 86, München, p. 40.

Schumacher, W. and W. Leonhard (1983). Transistor-fed AC servo drive with microprocessor control. Proc. Int. Power Electr. Conf., Tokyo, p. 1465.

Stephenson, J. M., P. J. Lawrenson and N. N. Fulton (1986). High power switched reluctance drives. Proc. ICEM 86, München, p. 1052.

Tanaka, H., T. Kuga (1985). 2000 kW - 8000 RPM very high speed induction motor drive system. Proc. IEEE-IAS, p. 676.

Terens, L., J. Bommeli (1982). Der Direktumrichter-Synchronmotor. BBC-Mitt., p. 122.

Walker, L.H. and P.M. Espelage (1980). A high performance controlled-current inverter drive. IEEE Trans. Ind. Appl., p. 193.

DEVELOPMENTS, APPLICATIONS AND COST-EFFECTIVENESS OF VARIABLE SPEED AC DRIVES

A.F.WICK

SIEMENS AG SYSTEMS ENGINEERING DEVELOPMENT
ERLANGEN, GERMANY

1. INTRODUCTION

Variable speed drives are nowadays applied across a wide field of application in industry, in power generation and in vehicles and are becoming ever more important. The rate of growth of electric drives is at present approximately 4.5 %. This growth may be attributed to a number of differing factors. Thus mechanical drives are being replaced by electric drives to an increasing extent since the latter are easier to control, have higher efficiency and require less maintenance. In the process industry in particular, there exists a growing demand for variable speed drives, either as a result of replacement of fixed speed drives or due to the need for variable speed drives employed in new production and manufacturing equipment. The reasons for this trend are: rationalization measures, increase in productivity, quality improvement and in particular measures to reduce energy consumption. Variable speed drives are nowadays a standard component in automated processes for energy production (e.g. boiler feed pumps, circulation pumps, flue gas blowers), in primary industry (e.g. rolling mill drives, ring-motor mill drives) and in manufacturing industry (e.g. drives for machine tools and robots).

Of all types of variable speed drives, the DC drive is most widely employed due to the simple control strategy. Its share of the world market is at present in excess of 80 %.

On the other hand, DC drives are at a considerable disadvantage as compared to AC drives due to the problems associated with mechanical commutation. A comparison of these drive motors reveals that the AC motor better fulfils the user requirements. Brushless AC drives allow employment of a higher degree of protection, which is of great significance under dirty or explosive environmental conditions (e.g. in the chemical industry). In the case of very high speed applications (e.g. for machine tools) and for very high drive ratings (e.g. for rolling mill drives and boiler feed pumps) the DC motor reaches its capability limit. Due to its more compact design, the AC motor has a higher space factor. In the case of drives subject to extreme dynamic loading the lower moment of inertia at equivalent output is a marked advantage. Moreover, the AC motor generally exhibits a higher efficiency and requires practically no maintenance.

On account of these reasons, research has been done for a number of years to develop technically adequate and economic variable speed AC drives and suitable control strategies. The control equipment and in particular the power section are, however, markedly more complex and more expensive than those of DC drives. For this reason, decisive innovation in the field of micro and power electronics provided the basis for the considerabely increased employment and high rate of growth of variable speed AC drives over the whole power range. A number of widely differing

A. T. De Almeida and A. H. Rosenfeld (eds.), Demand-Side Management and Electricity End-Use Efficiency, 299–320.
© *1988 by Kluwer Academic Publishers.*

drive concepts are employed in practise. Both induction and synchronous motors of various types are employed as variable speed AC drives.

Marked differences exist in the power supply concept and in the type of semiconductor components employed, not to mention the control strategies. The user is thus in a position to select the most favorable solution from a range of drive solutions on the basis of a number of different criteria such as control characteristics, mains pollution, efficiency, maintenance, operating and initial cost.

The major drive concepts and those characteristic features of major concern to the user will be described on the basis of a number of examples for applications in industry, traction and the power industry /1/, /2/.

2. LARGE VARIABLE SPEED AC DRIVES

In the power range above 1 to 2 MW, two drive variants are predominant, namely the cycloconverter drive and the load-commutated inverter (LCI) drive for brushless synchronous motors. Both concepts employ standard thyristors. Proved components, which have been employed with great success for more than twenty years for DC drives, are employed in the power section and for the control equipment /3/.

2.1 Cycloconverter drive

The cycloconverter is a member of the class of line-commutated converters employing a thyristorized three-phase bridge (Fig. 1). Two anti-parallel bridge circuits form a functional unit for generation of a controlled single-phase output voltage. The output voltage of such a converter is made up of chopped elements of the line-voltage. These elements are combined by suitable control of the thyristors in such a manner that the required approximately sine-wave voltage is obtained at the output of the circuit (Fig. 1b).

A three-phase output voltage system required for power supply of a AC machine is obtained on connection of 3 such functional units.

For optimal control of a AC motor, the amplitude and frequency of the motor voltage must be altered as a function of speed. As a result of this fact, a limit is set for effective use of the cycloconverter. Since the controllability of the cyclo-converter deteriorates as the output frequency approaches the input frequency, its application is limited to low-speed drives. Its primary field of application is thus for extremely low-speed ring-motor mill drives in the ore and cement industry, for winders in mines and in particular for main drives and roller table drives with extreme dynamic loading in the steel industry.

Up to a few years ago only variable speed DC drives could be employed as rolling mill drives on account of the stringent control requirements respecting speed accuracy and reversing characteristics. On the basis of intensive investigation carried out in development laboratories it could be proved that the control characteristics of an AC drive can be better than those of a DC drive. To achieve this aim, it is necessary to employ a technique which allows direct influencing of torque of the motor. The characteristics of DC drives are relatively good in this respect. As a result of the use of a commutator a fixed spatial relationship exists between the excitation field and the distribution of armature current on the rotor surface. Torque is thus proportional to the product of excitation field and armature current and is directly controllable by means of the motor current.

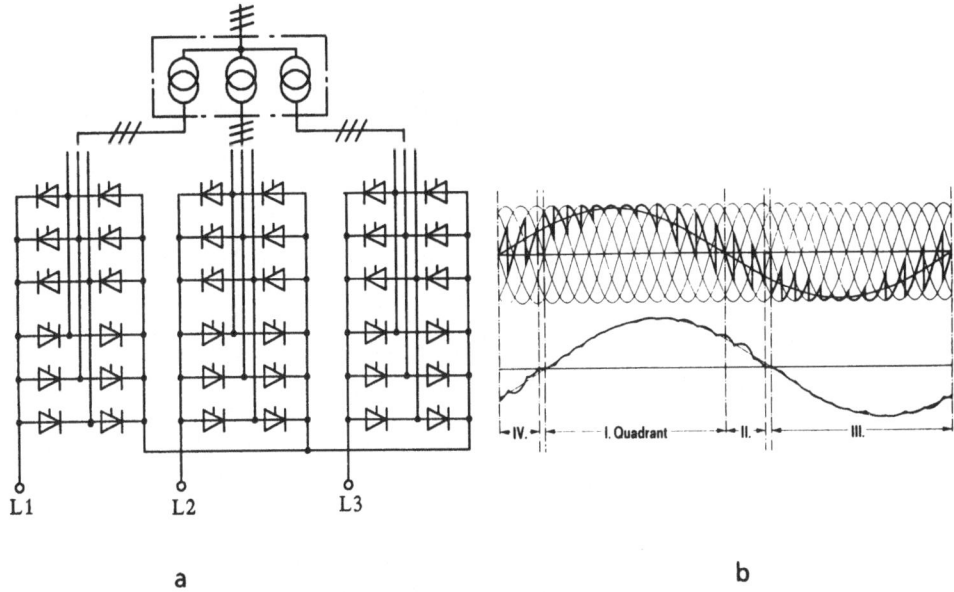

Figure 1.   a) Cycloconverter with three phase output
             b) Voltage and motor current of cycloconverter

In the case of an AC motor no fixed relationship exists between motor current and machine flux. This interrelationship must be achieved by the control unit. This requirement is satisfied by field-oriented control, whose basic functions and implementation in practice will be discussed on the basis of a cycloconverter drive (Fig. 2).

The basic principle of field-oriented control is that the motor currents are controlled in such a manner as to obtain a defined phase angle relative to the machine flux. Implementation of field-oriented control encompasses two tasks. First of all, it must be ensured, that the motor currents can be impressed by current controllers with high dynamic performance and control accuracy. The second task comprises identification of the machine flux. Thus, when the spatial position of the flux is known, it is possible to generate the reference values for the motor currents on the basis of a flux-dependent component and a torque-dependent component employing simple arithmetic functions. Now it is possible to directly influence the major physical variables, i.e. flux and torque, required for optimum control of an electric motor. The structure thus obtained is comparable to that of a DC drive.

Equipment for flux measurement forms the basis of such a control unit. Initially attempts were made to determine the magnetic flux by means of special measuring devices such as measuring coils and Hall probes, which were installed within the machine on the stator. This technique did not gain acceptance, because on the one hand standard machines could not be used and on the other hand reliability and availability of the drive system was markedly reduced by employment of sensitive measuring equipment which was difficult to repair in the event of fault. It is nowadays possible to determine the machine flux extremely accurately by means of

302

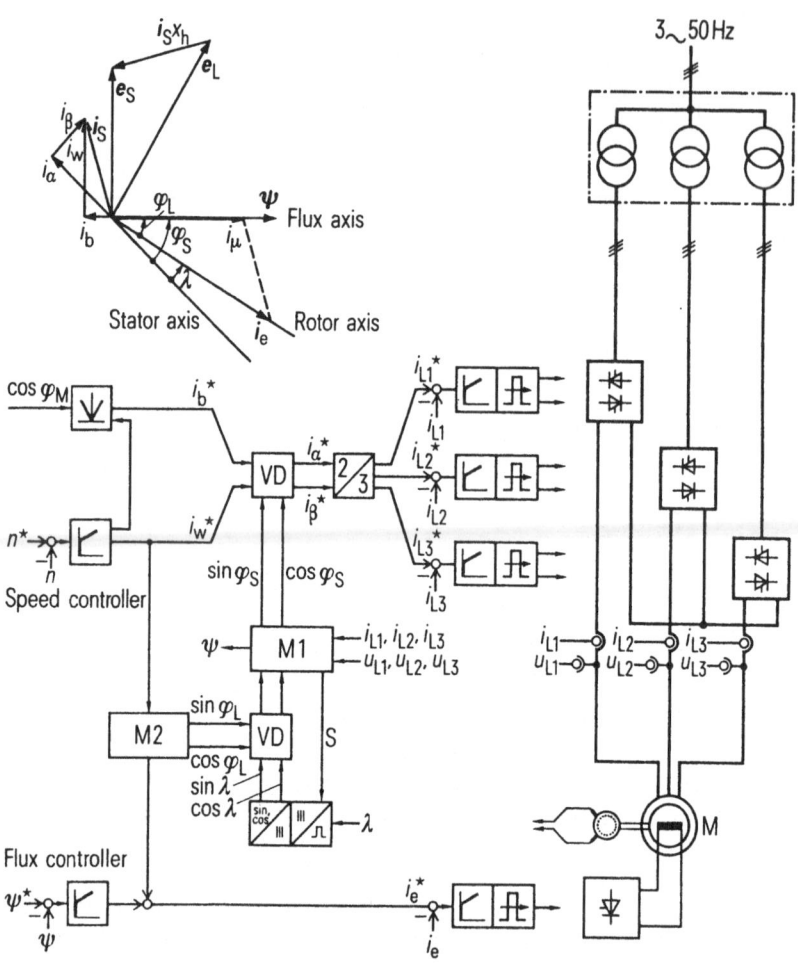

Explanation of symbols in Fig. 2

e_s	Air gap e.m.f.	n	Speed	$\cos\varphi_M$	Displacement facor at
e_L	Field e.m.f.	*	Setpoint values		syncronous-machine
i_s	Stator current	Ψ	Magnetic flux		terminals
i_e	Excitation current	i_μ	Magnetizing	I_{L1-3}	Line currents
			current	u_{L1-3}	Star voltage
i_w	Internal active current	φ_L	Rotor displacement	M1	Voltage-based simulator
	(torque-producing		angle	M2	Current-based simulator
	component)	φ_s	Flux axis angle	S	Rotor position setting input
i_b	International reaktive	λ	Rotor axis angle	VD	Vector rotator
	current (magnetizing				
	component)				

Figure 2. Control concept and vector diagram for field-oriented operation of a
synchronous machine fed by a cycloconverter.

electronic computation of circuits on the basis of easily measurable variables such as voltage, current and speed, which are always required for drive control. The electronic circuits used for flux estimation are in principle partial models of the rotating field machine, which employ two differing methods, the voltage model and the current model. The voltage model determines the machine flux employing Faraday's law. Calculation basically involves integration of the electromotive force, for which reason acceptable accuracy can be obtained only above a minimum speed. In cases where high dynamic control requirements exist for the drive at lower speeds, the flux is determined additionally or alternatively using the current model. This technique is based on Ampere's law. However, it requires precise measurement of speed and knowledge of various machine parameters.

The now standardized models are the result of extensive experience with converter-fed AC drives. These models provide the basis for employment of induction and synchronous motors in applications involving stringent control requirements.

As the drive rating increases, the efficiency of power conversion becomes a determining factor in addition to the control behaviour. Field-oriented control techniques also provide a good solution, since they allow tight control of the motor current and of the machine flux at every operating point so that the momentary torque requirement of the driven machine is satisfied at low losses.

Rolling mill drives. The suitability of variable speed AC drives for applications involving stringent control requirements could be proved under operating conditions on the basis of the main drives of a hot-rolling mill seven years ago. The significant advantages for the user are the freedom of maintenance of the AC motor and the compact design resulting in a low moment of inertia for the rotor. Significant advantages become apparent at higher ratings, where, in the case of DC drives, multi-motor drives would have to be provided on account of the maximum rating of DC machines.

Figure 3. Twin-drive 2 x 10 920 kW, 58.5 to 112.5 1/min with synchronous motors

Figure 4.
Working roller table with
asynchronus motors

There is nowadays practically no limit to the rating of AC drives particularly at low speeds. Let us look at an example. The twin drive depicted in Fig. 3 comprises the two drive motors of a rolling stand for a 5.5 m heavy plate mill. The rating per drive is approximately 11 MW within a speed range from 58.5 to 112.5 1/min. During the reversing phase the current is momentarily increased to 2.5 times its rated value, resulting in a peak load of up to 26.5 MW for each drive. The motor ratings are the maximum values of DC machines /4/.

Table 1. Comparitive data for drives of 10 920 kW; 58.5 to 112.5 min⁻¹

	Unit	D.C. motor drive	cycloconverter drive with synchronous motor	asynchronous motor
Maximum permissible speed for 250% current	min⁻¹	81	112.5	100
Rotor mass moment of inertia	tm²	107	79	99
Power loss for motor, excitation, motor ventilation, converter and cables	kW	1345	933.	1277
Power factor on supply side at basic speed		0.85	0.80	0.53
Inspection and maintenance requirements	h/year	approx. 145	approx. 36	approx. 30

In Table 1 results of various drive designs for this field of application have been compared. It can be seen that the cycloconverter drive employing a synchronous motor has the best characteristics. As compared to a DC drive, it exhibits no limitations regarding overload and the rotor moment of inertia is approximately 30 % lower. These two factors result in markedly improved reversing characteristics. A further advantage is revealed on comparison of the power losses. The somewhat lower power factor in the power supply system is insignificant compared to the major advantage for the user resulting from lower maintenance costs for the AC drive. This comparison also shows that synchronous motor drives offer a better solution in the upper power range.

Employment of AC motors which are compact and free of maintenance for rolling mill auxiliary drives is of particular economic significance in the case of directly driven work roller tables.

In the case of the example depicted in Fig. 4, it was possible to install the drive motors directly one beside the other with short couplings to the rollers without having to alter the distance between roller centres. In this application, the requirements are also characterized by a high reversing duty cycle and by high overloading capability. A significant advantage as compared to a DC concept is the fact that the induction motors can be designed with surface self-cooling. As a result, no cooling plant is required and it was possible to employ simpler foundations.

Low-speed motors with a maximum frequency of up to 20 Hz are required for directly driven work roller tables. The cycloconverter is also suitable for this application since in practice up to 30 motors can be operated as a group and can be fed from a single cycloconverter rated at up to 1.5 MW. The control requirements are satisfied employing field-oriented control and the overall efficiency of the drive system is optimized.

Important comparative data from the point of view of the user have been compared in Table 2 on the basis of an actual plant. It can be seen that the efficiency of the AC variant is better and that considerable economic advantages are gained by the user as a result of the lower maintenance requirements.

Table 2. Comparative data of roller table group drives (the efficiency and power factor refer to one motor with an operating speed of 235 min⁻¹ and a maximum torque of 14 kNm)

	Unit	D.C. motor with static converter supply	Asynchronous motor with cycloconverter
Efficiency motor + converter		0.80	0.87
Power factor on supply side at maximum speed		0.78	0.64
Motor cooling		force-ventilated	self-ventilated
Inspection and maintenance requirements for 97 motors including cooling system	h/year	approx. 1550	approx. 300

Although variable speed AC drives have only recently been employed in the steel industry, two further examples will be given for successful application of cycloconverter drives. The examples given are shaft winder drives and conveyor belt drives of high rating in mines and mill drives in the cement and ore industry.

Mill drives. Tube mills are used for size reduction of minerals in the ore and cement industry. In the case of cement mills they have a diameter of up to 6 m and of up to 12 m in the case of ore grinding. The tendency is to employ mills of ever larger diameter since the mill capacity and thus the process efficiency is improved on increase of the drop height of the material. Increase of the drive rating is consequential on the increase in diameter of the mill.

The highest drive rating manufactured nowadays is 12 MW at a speed of 9 1/min. Such designs were made possible by means of shaftless and gearless transmission of the torque by building the low-speed synchronous motor around the mill tube. The excitation equipment of the motor was mounted on the mill body. The synchronous motor is supplied from a cycloconverter whose particular advantages come into play in the case of extremely low-speed drives. Running up from standstill with high breakaway torque is possible. The drive can operate at low power losses. The speed can be optimally matched to the process requirements in conjunction with the control techniques described. Variable speed AC drives allow dispensing with gears and couplings in this case. As a result the following major advantages accrued:

- The size of the mill is no longer limited by the availability of gears.
- Relatively large overload capability of the motor
- Sharply reduced maintenance and spare part costs
- Better availability and reliability
- Inherent speed control by electronic means only

Mine winders. The rating of mine winders as employed for transport of persons and material in underground mines is in the range of 2 - 8 MW and requires a speed control range of 0 - 60 1/min. In the case of so-called rope shortening these drives must produce a torque many times the rated torque at standstill or at very low speeds. The cycloconverter drive also provides an optimum solution in this case. In addition to advantages of variable speed AC drives already discussed, considerable economic advantages become apparent, in particular at higher drive ratings. With AC drives it is possible to dispense with multiple motor drives. This leads to considerable savings of space, weight and costs particularly for use under ground or in winding towers.

2.2 Load commutated converter

Medium and high speed drives require a converter system whose output voltage is not limited by the frequency of the power supply system. This requirement is satisfied by the load commutated converter (LCI) drive. This is a combination of a synchronous machine and of a current-source converter as depicted in Fig. 5.

The converter comprises two simple identical functional units, which are connected by means of a DC link reactor. Power transmission is effected in two steps. The line-side inverter generates a DC current. This current is controlled by means of a current controller so as to correspond to the actual current requirement of the synchronous machine. The load-side inverter is assigned the task to switch the DC link current cyclically to the three-phase winding as a function of the motor speed. On suitable selection of the firing angles for the thyristor, it is possible to control not

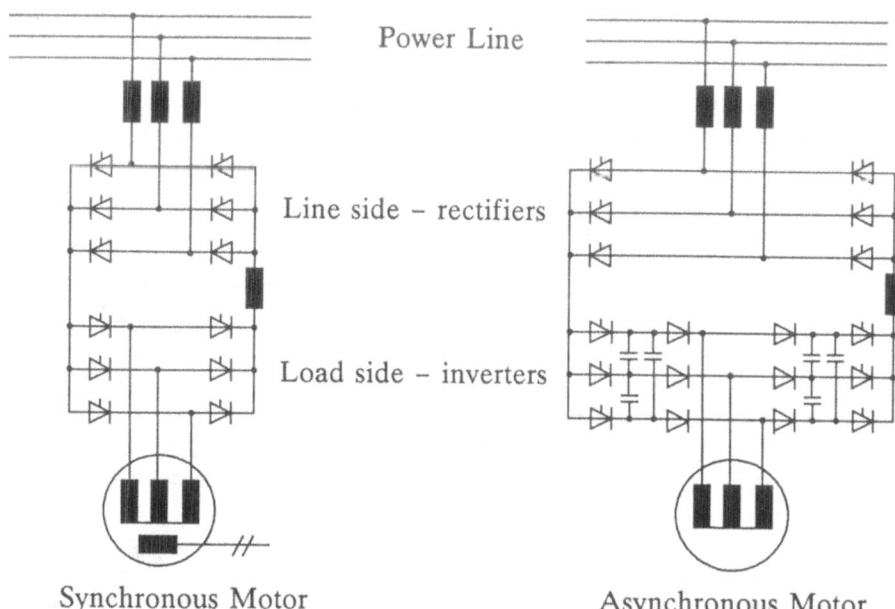

Power Line

Line side – rectifiers

Load side – inverters

Synchronous Motor

Asynchronous Motor

Figure 5.
Current-source converter with
load-commutated inverter

Figure 6.
Current-source converter
with self-commutated inverter

only the frequency of the AC current but also the precise angular position e.g. relative to the phase angle of the flux. The advantages of field-orented control as previously described are thus obtained employing this drive concept.

The load-side inverter receives the voltage required for commutation from the machine and thus operates always with leading reactive power. This mode is possible with a synchronous machine only and assumes the existence of a minimum speed. However, as a result of decoupling by means of the DC link reactor, the maximum obtainable operating frequency is no longer related to the frequency of the power supply system. As a result, converter-fed drives can attain speeds up to 7000 1/min or higher which are limited only by the maximum permissible circumferential speed of the rotor.

Coolant circulation pumps in nuclear power plants, starting converter and large industrial drives were among the first to employ converter-fed drives. As a result of the high efficiency of these drives, the absence of maintenance and the proven high reliability, converter-fed drives are suited for applications for medium and high speed drives in the upper power range, e.g. for boiler feedwater pumps and flue gas fans in power plants and compressors and pumps in the process industry. Such drives are employed not only for replacement of DC drives but are employed to an increasing extent for applications which were previously reserved to turbines and hydraulically controlled drives by reason of design problems. This trend is based on the simplicity of control of variable speed AC drives, on the higher efficiency particularly under partial load conditions and on the reduced maintenance requirement.

Employment of variable speed drives of high rating in the chemical and petrochemical industry in particular for continuous flow machines, such as turbo-compressors, rotary pumps and fans offers considerable economic advantages.

In cases where process variables such as pressure, flow or volume are to be controlled, variable speed drives can be used to replace fixed speed drives with associated mechanical control equipment. Replacement of dampers, control valves and control vanes in gas or fluid system results in a markedly improved overall efficiency and requires less maintenance (cf example in section 5).

3. INDUSTRIAL DRIVES IN THE LOW AND MEDIUM POWER RANGE

As a result of the rapid progress made in development of power electronics, the technology of AC drives is undergoing continuous development particularly in the lower and medium power range. It is no longer possible to distinguish the drive concept on the basis of power and speed. There is often a choice between a number of different concepts which are distinguished by the use of different drive components, for example the motor, the converter, the power electronic components and the control concept.

3.1 Drives with current-source inverter

The current-source inverter in conjunction with an induction motor (Fig. 6) still dominates in the power range from about 100 kW to the MW range. The current source inverter also comprises a line-commutated inverter employing a three-phase bridge circuit, which is responsible for the DC link current. The load-side inverter also employs thyristors. Additional components such as diodes and capacitors are required for safe commutation in every operating point. With this arrangement it is possible to distribute the DC link current to the relevant machine winding by means of suitable firing of the thyristors independent of the machine voltage. This type is designated a self-commutated inverter.

The robust, compact and maintenance-free induction motor in conjunction with such a converter, which is of comparatively simple design and easy to control, provides an economic solution in particular for individual drives. The converter allows four-quadrant operation motoring and regenerative braking in both direction of rotation similar to that provided by the previously discussed solution. Numerous drive tasks, e.g. operation of pumps, fans, centrifuges and material transportation systems, can be implemented employing simple control concepts. Field-oriented control and current source inverters, make possible to achieve very high dynamic performance, which is required for example by the motor-car industry in the case of test stands for testing combustion engines.

3.2 Drives with voltage-source inverters

Interests in new developments in the field of drive technology are nowadays focussed on variable speed AC drives employing a pulse-controlled voltage-source inverter (Fig. 7).

The inverter is supplied by a DC link voltage which can be simply derived from the three-phase AC power supply system by means of a diode bridge. A capacitor is employed to store energy in order to maintain the DC link voltage at a constant value.

The function of the inverter can best be explained by assuming that the power semiconductors are ideal switches. At no moment, both of the series switches of a

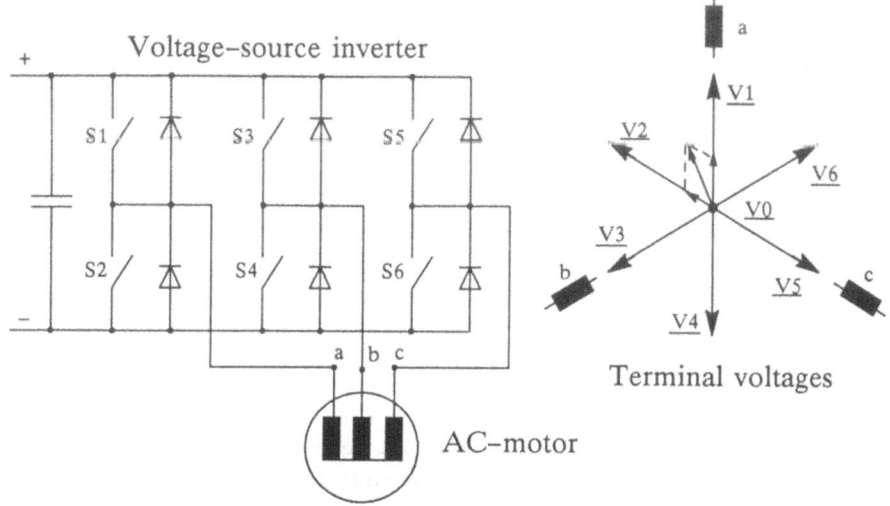

Figure 7.   Pulse-controlled voltage-source inverter (PWM) and space vectors of terminal voltages defined by the switching states of the inverter

phase leg (e.g. S1 and S2) may be closed simultaneously as otherwise an internal short-circuit would occur. Either the positive or negative potential of the DC link voltage can be connected to the appropriate motor terminal by control of the other switches. On the basis of these 8 possible switch combinations, 7 discrete voltage vectors can affect the three-phase winding of the motor. 6 of these resulting voltages, depicted in Fig. 7b, differ only in the phase angle, the amplitude being determined by the DC link voltage. The 7th voltage vector corresponds to the value 0. This voltage voltage occurs only when all switches connected to the same DC link bus are in the same switching state.

A three-phase voltage system of varying frequency can be generated as a result of cyclic sequential switching of the 6 discrete motor voltages as was the case with the current of the current-source inverter. However, for optimum control of a AC motor, the frequency and amplitude of the motor voltage must be varied as a function of speed. Two solutions may be implemented in conjunction with this converter design of which only the far more important one will be discussed.

Pulse width modulated (PWM) inverter

The PWM inverter is operated in conjunction with a fixed DC link voltage. Its output voltage is controlled by means of pulse width modulation. In addition to sequential switching of the 6 discrete voltage vectors, the voltage value 0 is used. The duration of voltage 0 and of one or two discrete neighbouring voltage vectors within a pulse-cycle determines the value of the average motor voltage. It is obvious that the closer the machine voltage and current approach the ideal sinusoidal wave form the larger must be the ratio between the pulse frequency of the inverter and the operating frequency of the motor.

The following objectives are achieved on increasing the pulse frequency:

- Reduction of additional losses in the motor
- Noise reduction in the vicinity of workplaces or public areas
- Reduction of the dead time to improve the dynamic control characteristic.

In order to build a PWM inverter, switches which are capable of being "opened" or "closed" by means of a control signal without delay and loss are required.

Standard thyristors are relatively unsuitable for this purpose. Although such semiconductors can nowadays be manufactured with a blocking voltage of up to 6000 V and for currents in excess of 3000 A and thus satisfy the maximum rating of drives , the thyristor itself does not satisfy the requirements as a switch. The thyristor may be switched on by means of a gate pulse, however, additional circuits with active and passive components such as reactors, capacitors, diodes and auxiliary thyristors are required in order to switch it off. The use of thyristors for PWM inverters results in an uneconomic solution with regard to cost, efficiency, size and weight as compared to other converter technologies.

Consequently, high power inverters employing thyristors are used only in cases where drive characteristics such as an unlimited control range from zero speed to the higher speed together with smooth running is of primary importance.

The use of PWM inverters on a wide scale became possible as power semi-conductors capable of being switched off became available . The first components to come from the manufacturers' laboratories were power transistors employing bipolar technology. They were later supplemented by MOS field effect transistors. Thyristors capable of being switched off, designated gate turn-off thyristors (GTO), became applicable for inverters at the end of the 1970s. Technology has since then advanced rapidly with increased ratings and a wide spectrum of power devices. Development is still in progress. This is illustrated in Fig. 8 which lists the limits of voltage, current and puls frequency of the components in addition to the output rating of three-phase AC PWM inverters. It is important to note that in view of the possibility of connecting components in series and in parallel there are no practical limits to equipment rating /5/.

MOS field effect transistors exhibit the best switching characteristics. It is possible to design equipment employing pulse frequencies in excess of 10 kHz or above the audible range. The cost of snubber circuits and control is very low. These devices are employed to an ever increasing extent in equipment rated at up to approx. 10 kVA. The voltage for which field effect transistors can be manufactured is limited. Thus for the higher power range bipolar transistors are employed. The forward characteristics of bipolar transistors and voltages up to 1200 V are good, but, the base drive has a higher power consumption. The control electronics including the protection functions are more complicated and the pulse frequency of present-day equipment is limited as a rule to a few kHz. However, this is generally sufficient for
drive requirements of today. The PWM inverter has become the standard inverter for AC drives within the power range covered by transistors, that is at present up to approx. 200 kW. Such drives employing either induction or a synchronous motors successfully compete with DC drives.

This trend is also apparent in the case of machine tool drives.This field of application is characterized by the stringent requirements regarding dynamic performance. Positioning drives are required to produce a peak torque of up to 10

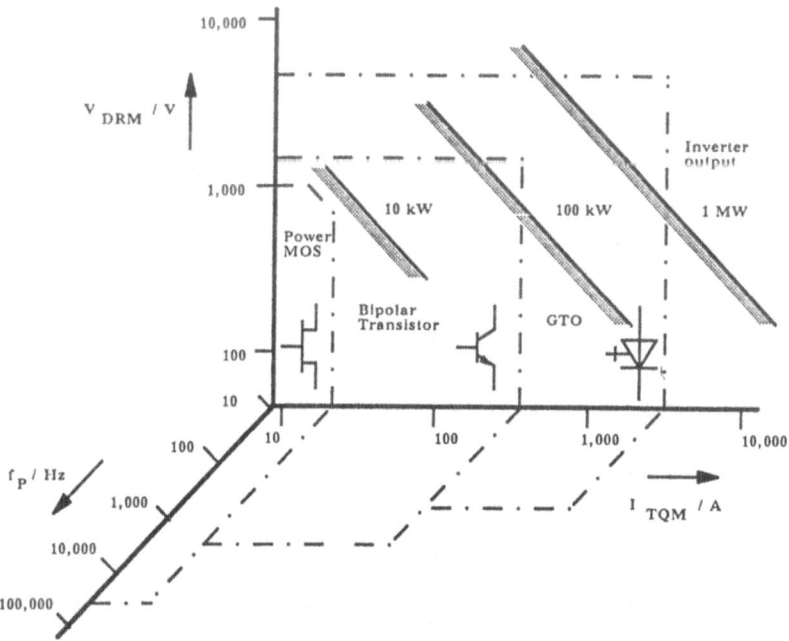

Figure 8.   Power semiconductor devices today

times the rated torque, and it must be possible to adjust a speed down to extremely low values of less than 0.1 1/min. High speed control accuracy is also required in the case of main spindle drives. Speeds of up to 14000 1/min or higher must be achieved in conjunction with a field weakening range larger than 1:8. Such extreme requirements can no longer be fulfilled even by advanced DC drives. More extensive requirements such as freedom from maintenance, high degree of protection and even higher speeds for main spindle drives at constant power can in theory be better satisfied by the robust AC motor. Economic solutions employing AC technology have been made possible as a result of the advances made in the field of power electronics (bipolar transistor) and data processing technology (powerful microprocessors). Nowadays most manufacturers quote sychronous motor drives employing permanent magnet excitation for positioning drives and induction motors for main spindle drives /6/.

The spectrum of switchable components is rounded off by gate turn of thyristors (GTO). These components allow use of considerably higher voltages and currents than is the case with transistors. However, their use is coupled with greater expenditure for the control electronics and for protection. The components are considerably more expensive than comparable normal thyristors. Only a limited number of manufacturers can supply such devices for the megawatt range but development is continuing in this area.

As a rsult of ongoing development the cost structure will change to the advance of AC drives so that the rate of replacement of DC drives by AC drives will increase in the next few years. In the near future there appears little likelihood that the PWM inverter will be able to compete with the converter technology described in the

312

medium to high power range. Each user must decide for himself which drive concept best satisfies the criteria relevant to his field of application.

4. THREE-PHASE AC DRIVES FOR TRACTION

Railway traction made the transition to three-phase AC drives for the major applications much earlier than industry. This development was triggered by the technical advantages of the AC motor, not to forget the considerable economic advantages.

The main reasons for implementation of three-phase AC drives for traction were:

- The high power density of high-speed induction motors allowed increase of the power rating within the limited space available on the bogie.
- The high overload capability offered advantages especially for rapid transit with high starting and braking torques.
- As a result of the extended speed control range at maximum power, it was possible to design universal locomotives for freight and passenger traffic.
- The improved utilization of energy as a result of regenerative braking and reduced interference to the power supply system.

A number of different drive designs are used on railway systems due to the different features of railway vehicles and the various methods employed for supply of electrical energy.

single-phase   PWM inverter
PWM inverter

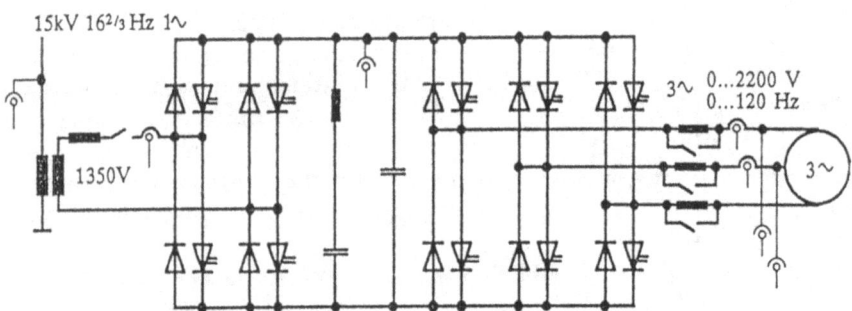

Figure 9. Drive System of the locomotive BR 120

In the case of AC railways, the drive system of the locomotive BR 120 (Fig. 9) is nowadays regarded as standard. A robust maintenance-free induction squirrel-cage motor supplied by a PWM inverter is used.Railway power supply systems are single-phase systems,wich are generally weak.Therefore, energy storage causes a problem. Capacitive energy storage offers a favourable solution as regards weight. This favours the use of the PWM voltage source inverter. The line-side inverter also comprises a single-phase PWM inverter, which compensates power supply voltage fluctuations and ensures a constant value of DC link voltage for operation of the pulse-controlled inverter. In addition, the line-side inverter ensures that the current taken from the power system approximates a sinusoidal wave form at unity power fator. Energy flow in both directions is possible /7/.

An alternative concept favoured by the French railways is a solution employing the LCI drive. The advantage here is the simplified load-commutated inverter. On the other hand, a synchronous motor must be used, whose rotor design is more complicated and whose excitation power must be supplied by an additional converter and fed to the rotor via sliprings /8/.

The advantages of the synchronous system is due to the simpler design of the inverter. However,it will soon disappear when GTO thyristors become available in this power range. The construction of suitable PWM inverters will be considerably simplified and the more complicated control unit will not necessarily be more expensive in this age of microprocessor technology.

Up to now AC drive technology has been employed for DC railways only for rapid transit systems. In this case, current-source inverters with induction motors offer the most favourable solution (Fig. 10). The input chopper allows continuous transition between driving and braking including regenerative braking. Field-oriented control of the motors which are generally connected in parallel is employed to optimize energy recovery on braking. Vehicles fitted with such drives can during braking feed back up to 40 % of the energy taken from the power system.As a result of this measure,operating costs, tunnel heating and energy consumption are considerably reduced.

Figure 10.
Drive system for
DC railways:
Input chopper, current-
source inverter, asynchronous
motors connected in parallel

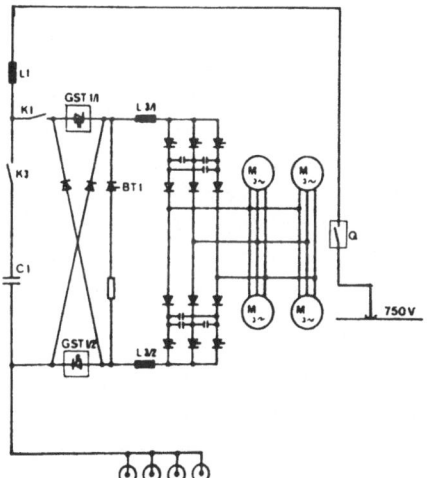

5. COST AND ENERGY SAVING THROUGH NEW DRIVE CONCEPTS

Power supply to an electrical machine from a converter initially entails some disadvantages. Converters generate harmonics and thus increase the machine losses. Energy conversion in the converter entails losses and converters cause interference to the power supply system, e.g. as a result of current harmonics or through additional reactive power demand. All these factors must be taken into account when assessing the suitability of a drive concept with respect to power demand. The overall efficiency of a variable speed electric drive is comparatively good. Efficiencies of the order of 95 % and more can be achieved for example for drives in the megawatt range.

Cost advantages are achieved when the technology of the driven machine can be optimized as a result of the unrestricted speed control capability of the drive. The variable speed electric drive satisfies these requirements:

- The speed, torque and power output can be varied without loss and can thus be adapted to the work process.
- The efficiency is high even in the partial load range.
- It facilitates regenerativ braking.

The AC solution brings further advantages. The speed limit can be further increased and there is practically no limit of the output rating. As a result AC drives are nowadays being employed to an increasing extent for fields of application which were hitherto reserved for turbine and hydraulic drives, e.g. in power plants and in the chemical industry. In the case of mill drives it has already been shown that the technology was optimized as a result of increase of the drive rating and that the total power consumption could be reduced. As a result of removal of limitations due to speed and power ratings, it was possible to introduce new drive concepts employing directly coupled variable speed drives with smooth starting characteristics instead of fixed-speed drives employing gearing or hydraulic couplings which entailed losses. Maintenance-free units, reliability and availability are frequently decisive factors for the user since they lower the operating costs and often justify higher investment costs. This is demonstrated by a number of examples.

Example 1: Cost advantages for railway traction drives employing three-phase drive systems

Although the capital investment for a vehicle of AC design is considerably higher than that for a cam-shaft controlled vehicle employing series wound traction motors, an overall cost analysis reveals substantial economic advantages mainly due to the maintenance-free drive motor and to energy recovery in conjunction with regenerative braking. For example, the following figures are available for the new Class 120 universal locomotives of Federal German Railway (Bundesbahn):

- Reduction in maintenance costs   approximately 30 - 40 %
- Reduction of energy consumption as a
  result of regenerative braking and reduction
  of power transmission losses   approximately 20 %

Further economic advantages accrue as a result of the universal capabilities of the locomotive allowing use both for high-speed passenger trains and for heavy freight traffic. Because of the resulting reduced need for standby capacity, capital charges are reduced since fewer locomotives are required.

The economic advantages are demonstrated by a second example of an underground railway double railcar unit (Table 3). The price of the AC vehicle is approx. 10% higher than that of a camshaft controlled vehicle. Reduction of the maintenance costs for the maintenance-free AC drives and as a result of elimination of heavily stressed switch contacts in the power circuits and the considerable savings due to the reduction in energy consumption with regenerative braking more than compensate for the higher capital charges after one year's operation.

On the basis of a yearly increase of maintenance costs by 5% and of energy costs by 4%, a capital gain of 3.5 times the additional investment is obtained for a vehicle life of 25 years (exclusive of interest on the capital gain).

	after one year's operation	vehicle life of 25 years
Capital investment of the three–phase drive system	300 000,–	
Higher capital charges (interest rate 9%/a)	30 500,–	762 500,–
Reduction of the maintenance costs (40%)	13 000,–	620 000,–
Reduction in energy consumption (energy costs 0.20DM/KWh)	29 000,–	1 207 500,–
Economic advantages	11 500,–	1 065 000,–

Table 3.   Economic advantages by example of an underground railway

Example 2: Energy balance for a pump

The discharge of a pump may be controlled by mechanical throttling at constant pump speed or by alteration of the pump speed. These methods differ considerably as regards the power consumption for the drive system, as can be seen from the characteristics depicted in Fig. 11.

If the performance of the pump is known, e.g. over a yearly cycle, the energy saving on employment of a variable speed drive can be calculated.

Example 3: Energy saving by matching belt speed to material flow and by no-slip starting

In 1986 the world's first conveyor drive with cycloconverter-fed synchronous motors was put into operation (Fig 12).

In this example coal is transported from a depth of 786 m by means of a conveyor belt which is approximately 4 km long. The drive rating is 6.2 MW. Previously such drives were designed as fixed-speed drives employing special starting equipment.

316

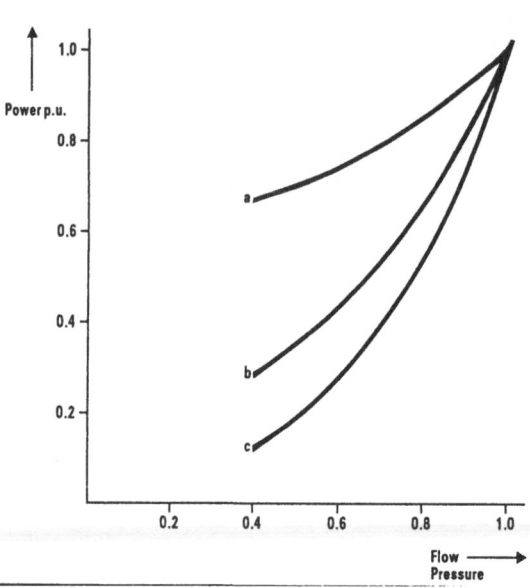

a = fixed speed drive with throttling

b = wound rotor motor with speed variation by variable rotor resistors

c = LCI drive

Figure 11.
Power requirement of fixed and variable speed drives.
Example: Circulation pump

Figure 12.
Inclined conveyor
of a underground mine
Rated power: 2 x 3100 kW
Length: 3743 m
conveying capacity: 1800 th⁻¹

Compared to a fixed-speed drive the following advantages with regard to energy consumption were achieved:

- Loss-free startup involving approximately 1-2 % of the rated power taken over the year.
- Absence of the losses due to the slip characteristics of fluid and eddy-current couplings

- Matching of speed to the material flow.
  The last point is illustrated in Fig. 13.

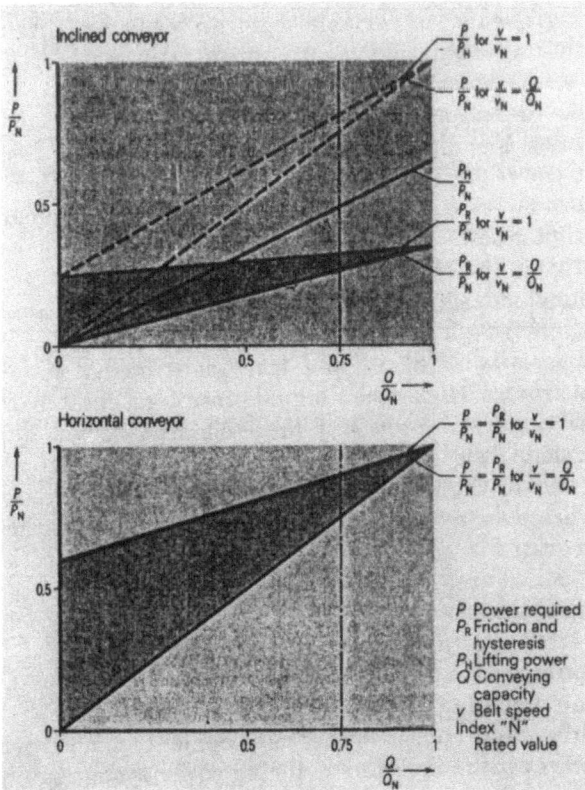

Figure 13.
Energie savings
by matching belt
speed to material flow

The power consumption of the drive system is depicted as a function of material flow for a horizontal conveyor and inclined conveyor at constant speed and at a speed proportional to material flow.

Naturally, there is a difference between inclined and horizontal conveyor systems. With inclined conveyors, the lifting power can amount to about two-thirds of the total power required. This is not dependent on speed, but on conveying capacity, lifting height, and gravity alone. In this case, savings can only be achieved on the remaining one-third of the required power lost through friction and hysteresis. Nevertheless, when the conveyor is operating at 75 % rated capacity, approximately 5 % of the rated power can be saved by reducing the speed to 75 % rated speed. The situation is different, however, for horizontal conveyors, for which the required power depends only on friction and hysteresis. In this case, with 75 % rated capacity, up to 15 % rated power can be saved by reducing the operating speed to 75 %.

6. CONCLUSION

Variable speed AC drives are characterized by a large number of drive concepts, some of whose characteristics mutually supplement each other but which are in part conflicting. In the upper power range, various concepts employing thyristors nowadays dominate the field. In the future these will be superseded by equipment employing gate turn-off power semiconductors.

The significance of variable speed drives is signalized by the fact that intensive investigation and development continues in universities and in industry. This applies to all components of the drive system.

Further development of turn-off power semiconductors with improved switching characteristics and power ratings will make possible design of smaller, lighter and cheaper converters. It is already apparent that new technologies will extend converters depicted in Fig. 8. The aims of such development are:

- Higher power rating
- Higher pulse frequency
- Simpler design of control and protection circuits
- Improved efficiency.

New impulses also come from continuing development in the field of microelectronics. High-speed signal processing equipment, such as microprocessors, signal processors and semi- or full custom ICs allow implementation of advanced complex control structures and functions.

As a result, the technical characteristics of the system are improved and flexibility is achieved respecting matching of the drive characteristics to the application task integrated in automated manufacturing and process systems. The main points of development are:

- Realization of new control concepts
- Better system utilization and improvement of efficiency as a result of optimized control functions
- Self-adjusting systems with integral diagnostics.

New drive concepts are under development which are characterized by:

- Drives with converters reducing power system interference
  Employment of new materials for machines.

The aim of all these developments are:

- Improved system utilization
- Weight and volume reduction
- Improved overall efficiency.

In conclusion an example is given.

A completely new AC drive system has been developed for employment as a ship drive. The use of new hard magnetic materials based on cobalt samarium allows construction of permanent magnet synchronous machines , rated at 1.1 MW in this example. In contrast to conventional AC machines employing three-phase windings, the stator employs six phase windings (Figure 14).

These windings are fed with approximately square wave currents. In conjunction with the nearly square wave flux distribution of the permanent magnets it is possible to increase power utilization. Due to the number of windings the power rating of each inverter amounts only to 1/6.Therefore,GTO technologie with PWM inverter is possible. In contrast to two other ship drives of identical rating employing DC technology , the weight and dimensions (for the total drive system) could be reduced by 40 % and the losses at rated output could be reduced by 20 %. The reduction in power losses at partial loads is even better. This concept is further characterized by its redundant design, since the drive can still operate even when only one of the inverters is intact (with a corresponding reduction in output).

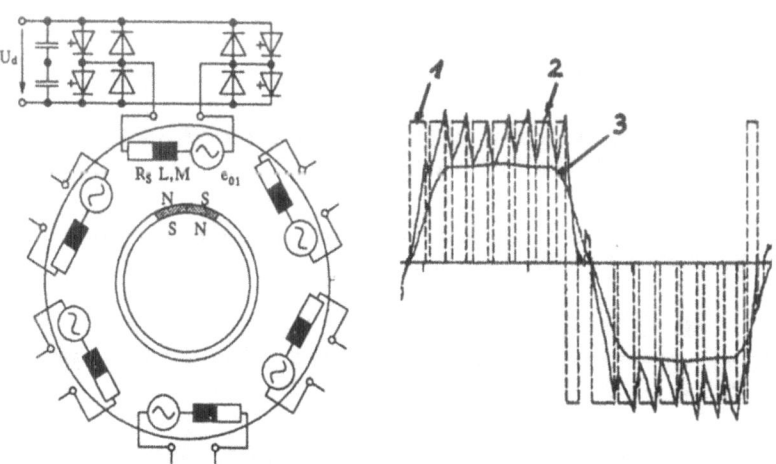

a   b

Figure 14.   New converter-fed permanent-field motor for employment as a ship
drive
a) circuit arrangement with PWM inverter

b) phase voltage 1, current 2 and no load e.m.f.3.

This prototype, which is at present being tested on board ship under practical
conditions, is the basis for a new drive concept for all applications, in which the gains
as regards weight, volume, efficiency and redundancy are of decisive importance and
which thus can justify the price increase by a factor of 2 as compared to conventional
drive concepts /9/.

REFERENCES

/1/ Schörner, J.: Speed controlled industrial drives. Proc. ICEM 86, p. 40.

/2/ Keiji, S., Hisakatsu, K., Ichiro, W.: State-of-the-art AC drive systems. Hitachi Reciew Vo. 36 (1987), No.1.

/3/ Von Musil, R.: Survey of large inverter fed motor designs for variable speed drives. Proc. ICEM 86, München, p. 32.

/4/ Herfurth, G., Schmidt, H., Wokusch, J.: Die Nutzung neuer elektrischer Antriebstechniken in der Stahlindustrie. "Stahl u. Eisen", 107 (1987), Heft 5, S. 193-198.

/5/ Yasuhiko, I., Tsutomu, Y.: Recent progress in power electronic devices. Hitachi Review Vol. 36 (1987), No. 1.

/6/ Leonhard, W.: Microcomputer control of high dynamic performance AC drives-a survey. Automatica, Vol. 22, No.1, p. 1-19, 1986.

/7/ Harprecht, W.: Lokomotive für Einphasenwechselstrom der Baureihe 120 mit Drehstromantriebstechnik. ZEV-Glas. Ann. 107 (1983) Nr. 8/9 Aug./Sept.

/8/ Kasparek, F.: Die elektrische Ausrüstung der neuen Wiener U-Bahn-Wagen. Eisenbahntechnik, 4/1985.

/9/ Fürsich, H., Thum, E., Hamann, J., Köllensperger, D.: New converter-fed permanent-field motor. Proc. ICEM 86, München, p. 50.

USE OF ELECTRONIC ADJUSTABLE SPEED DRIVES IN COMMERCIAL BUILDINGS

Anibal De Almeida

Joe Eto

Dep. Eng. Electrotecnica
Universidade de Coimbra
3000 Coimbra, Portugal

Lawrence Berkeley Laboratory
University of California,
Berkeley, CA 94720, USA

1. INTRODUCTION

Variable flow devices for formerly constant flow systems are popular energy saving measures that have both industrial and commercial building applications. The power to move a fluid grows roughly with the cube of the flow in centrifugal fans and pumps. A flow rate of 80% requires only half of the rated flow power. Commonly, flows are modulated by inlet vanes, discharge dampers, or other throttling techniques. Less well understood in commercial building applications are the additional electricity savings available through the use of adjustable-speed drives (ASD).

Industry has begun to employ ASD's, with great success [1,2]. Simple inspection of the idealized operating characteristics of ASD versus inlet vane control illustrates the potential for savings. Figure 1, for example, compares the part-load performance of inlet vane, discharge damper and ASD fan controls. The actual savings are defined by the length of operation at each part-load condition. For many industrial processes, these conditions are regularly monitored and evaluation is straightforward.

Evaluating the potential for ASD savings in commercial building applications is complicated by the diverse building and heating, ventilating, and air-conditioning (HVAC) system types in the sector, as well as climatic variations. Existing work tends to take the form of case-studies, from which it can be difficult to generalize [3], or, at the other end of the spectrum, cases in which completely hypothetical load conditions are assumed [4], and the reader is left to judge the degree to which these assumptions are realistic.

This paper evaluates the energy savings achievable with ASD's in commercial building fans and chillers versus inlet vanes. Parametric computer simulations of energy use for two prototypical commercial buildings located in five diverse U.S. climates is used.

In the next section, the commercial building prototypes used in the evaluation are described. This discussion is followed by brief reviews of the climates used, and the building energy simulation model. The final section describes the results. The results include comments on the energy and

A. T. De Almeida and A. H. Rosenfeld (eds.), Demand-Side Management and Electricity End-Use Efficiency, 321–333.
© *1988 by Kluwer Academic Publishers.*

322

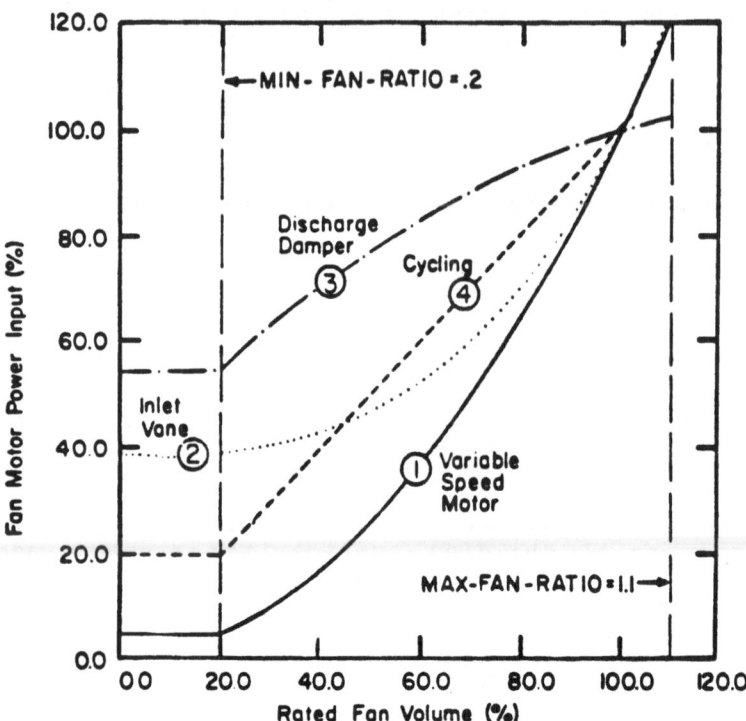

FIGURE 1. Part-load Performance for Several Variable Air Flow Techniques.
Source: DOE-2 Reference Manual [12].

peak demand savings from ASD's, as well as the economic
value of these savings relative to increased first cost.
Several sets of economic assumptions are used in an effort to
generalize the results obtained.

2. COMMERCIAL BUILDING PROTOTYPES

Two commercial building prototypes are used in our
evaluation. Table 1 summarizes major features of both
prototypes. The first prototype is a set of two single-zone,
retail stores in a strip store complex. The stores are
based on an actual structure, but have been modified to
comply with ASHRAE Standard 90-1975 [6]. The HVAC equipment
for the stores is a conventional variable air volume package
unit. The schedules are taken from a library of "typical"
schedules [7]. They specify 12 hours of operation six days per
week, and 10 hours of operation on Sundays and holidays.
The evaluation for this prototype consists of two
sets of simulations. The first modulates air flow with inlet
vanes; the second uses an ASD. Figure 1 compares the
performance of these two techniques for flow modulation. The
full-load efficiency of ASD's is higher, in spite of losses in
the ASD unit, since they avoid the high static pressure drop
associated with inlet vanes.
It is assumed an overall, full-load fan efficiency of 47%
for inlet vanes and 51% for ASD. An ASD for the cooling
system was not evaluated for the strip store because the
small size of these units rarely warrants the
relatively high first cost ($/horsepower) of ASD controls.
The second prototype is a medium office building. This
building, too, is based on an existing structure and has also
been modified to comply with ASHRAE Standard 90-1975. The HVAC
system for this prototype is a variable air volume dual-
duct system. The conversion of this kinds of systems from
constant volume to variable air volume has been a very
popular retrofit. The building is operated 10 hours per
day during the week, with limited occupancy on Saturdays,
and is closed on Sundays and holidays. Full-load, overall
fan and motor efficiency was specified to be 51% for inlet
vane control and 55% for ASD. The chiller for this prototype
is a hermetic centrifugal chiller and cooling tower. The full-
load coefficient of performance (COP) for the chiller alone
(not including electricity for operation of the cooling tower
and pumps) is set at 4.5 for a conventional chiller and 4.3 for
control with an ASD. For chillers, ASD control adds losses
that are not compensated for, since ASD control is used in
addition to (rather than in place of) inlet vane control.
Figure 2 compares the part-load performance of both a
conventional chiller and one controlled with ASD's. On this
graph, the loads have not been normalized (as in Figure 1)
and clearly illustrate the performance penalty for ASD's at
full-load.
The evaluation for the medium office building prototype
consists of three simulations. The first controls air volumes

TABLE 1. Summary of Commercial Building Prototypes

	Strip Store	Medium Office
Size	11,760 sq.ft.	48,600 sq.ft.
Shape	Two units (edge and adjacent unit), single story (18 ft),	3 floors, rectangular in cross section, approximately 16,000 sq.ft./floor.
Construction	Wood-frame construction.	Steel frame super-structure, exterior walls of 4" pre-cast concrete panels
Glazing	35% of wall area on southern and western exposures, no glass on northern and eastern exposures.	36% of wall area, evenly distributed.
Operation	10am - 10pm six days per week, reduced schedule of 10am - 8pm on Sundays and holidays.	8am - 6pm weekdays, with some evening work, 30% occupancy on Saturday, Closed Sundays and Holidays
Thermostat Settings	78°F Cooling 72°F Heating (night and weekend setback 62°F)	Identical to strip store
Internal Loads	1.9 W/sq.ft lighting 0.5 W/sq.ft equipment	2.5 W/sq.ft lighting 1.0 W/sq.ft equipment
HVAC	2 packaged rooftop single-zone, variable air volume, direct expansion units. Minimum fan ratio is 0.3.	Dual-duct variable air volume. Minimum fan ratio is 0.2.
Fan Efficiency	0.47 for inlet vane, 0.51 for ASD.	0.51 for inlet vane, 0.55 for ASD.
Economizer	62 F, dry bulb.	Identical to strip store.
Minimum Outside Air	5 cfm/person.	Identical to strip store.
Heating Plant	Electric resistance baseboard.	1 Gas-fired hot water generator (eff. 75%)
Cooling Plant	Reciprocating compressor with air cooled condenser (COP 2.4).	1 hermetic centrifugal chiller and cooling tower. Chiller COP is 4.5 for conventional and 4.3 for ASD control (does not include electricity used for cooling tower and pumping).

with inlet vanes and provides cooling with a conventional chiller. The second uses ASD's to control fan operations. The third combines ASD fan controls with ASD chiller control. This ordering replicates the likely stages for retrofit: first fans, then chillers.

The equipment was not oversized, in order to ensure that the estimates of the benefits from ASD's would be conservative. For both building prototypes, the sizing of fan and chiller is determined by a separate design calculation for each location. In each case, the installed size for a fan or chiller is exactly the peak or maximum expected load condition. Real-world sizing practice, by contrast, is based on an assumed over-sizing factor of, say, ten percent, for safety and may be further increased by the availability of equipment. For example, a 100-ton design load may become a 110-ton load for safety and then a 125-ton unit since this is the next incremental size chiller available. In these circumstances, equipment will rarely operate at full-load conditions, so the importance of efficient part-load operation is highlighted.

3. FIVE U.S. CLIMATES

The prototypes were simulated with weather data for five U.S. climates. The climates were chosen to represent a range of combinations for hot and cold, wet and dry US conditions. Briefly, El Paso (Texas) represents a hot and dry climate. Lake Charles (Louisiana) represents a hot and humid climate. Madison (Wisconsin) represents a climate with mild summers and very cold winters. Seattle (Washington) represents a mild climate with wet summers and cold winters. Washington DC represents an intermediate climate, with humid summers and mild winters. Table 2 summarizes heating and cooling degree-days to base 65 F (18 C) for each location.

WYEC weather data tapes were used to represent each climate in the simulations performed [8]. These tapes were developed specifically for building energy simulation analyses and include detailed meteorological data, such as solar insolation. The tapes were created from many years of historical weather measurements and have been designed to represent long-term averages for each site.

4. BUILDING ENERGY SIMULATION MODEL

The DOE-2 building energy analysis program (version DOE-2.1C) was used to study the additional energy savings from ASD's compared to conventional methods for flow modulation. The DOE-2 program was developed by the Lawrence Berkeley and Los Alamos National Laboratories for the Department of Energy to provide architects and engineers with a state-of-the-art tool for estimating building energy performance [9]. The DOE-2 program has been extensively validated [10].

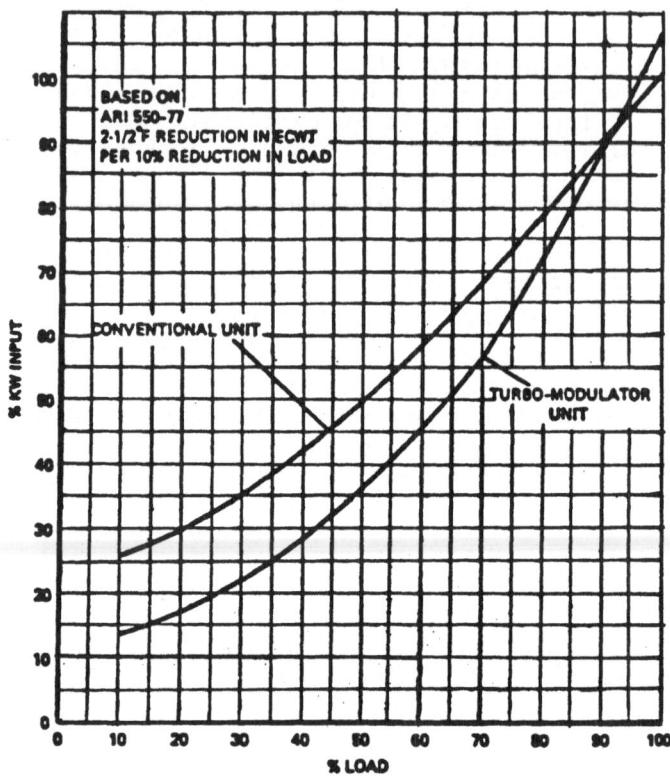

FIGURE 2. Part-load Performance Comparison for ASD vs. Conventional Chiller Control. Source: York International Corp. [13].

TABLE 2. Climate Parameters

Site	Heating Degree-Days	Cooling Degree-Days
El Paso TX	2,689	2,129
Lake Charles LA	1,535	2,696
Madison WI	7,710	447
Seattle WA	5,223	97
Washington DC	4,236	1,425

All degree-days calculated to base 65°F.

Three features make DOE-2 particularly applicable to the study of variable-flow devices for commercial buildings:

1. Heating and cooling loads are calculated on an hourly basis.
2. The structure and operation of a building can be entirely specified by user-inputs.
3. A special input-processor allows for the incorporation of manufacturer's data on equipment performance at part-loads.

5. ENERGY PERFORMANCE

The results of the simulations are presented in Tables 3 and 4 for the strip store and medium office building, respectively. For both prototypes, electricity consumption by fans is relatively stable across climates and often exceeds electricity consumption by the cooling system. For the base case strip store, fan electricity consumption represents 22% (Lake Charles LA) to 28% (Seattle WA) of annual electricity consumption. Cooling for the base case strip store represents 8% (Seattle WA) to 23% (Lake Charles LA) of annual electricity consumption. For the medium office building, the corresponding ranges for the base case are 14% (Lake Charles LA) to 16% (Madison WI, Seattle WA, and Washington DC) for fans, and 9% (Seattle WA) to 20% (Lake Charles LA) for cooling, where cooling includes cooling tower and pump operation in addition to the chiller.

For the strip store, the use of ASD's reduces fan electricity consumption by 17 to 27%. The largest savings occur in the warmer climates (El Paso TX and Lake Charles LA). These results indicate that these climates require longer hours of fan operation at low load conditions. Since decreased fan losses (higher overall fan efficiencies) lower heat gains to the air flows, small cooling energy savings are also realized. Chiller energy savings range from 2 to 6%. Differences in the change in total electricity consumption compared to the change in electricity consumption for cooling and fans combined often exceeds the change in total electricity consumption (see, for example, the results for Madison WI on Table 3). The difference is additional electric heat required due to the more efficient fans under ASD control. Less efficient fan controls contribute heat to the air, which must be replaced during the heating season when ASD controls are used.

For the medium office building, introduction of ASD fan controls saves relatively more electricity than the strip store, but exhibits less variation across climates, as internal gains tend to dominate. The range of fan savings is 26 to 30%. No clear pattern emerges linking the level of savings to a given climate. Recall that the range of variation in fan energy consumption in the base case was small. The associated chiller electricity savings are again small but noticeable, ranging from 3 to 5% of total

TABLE 3. Energy Results for Strip Store

Site	Base Case	ASD Fans (Δ)	ASD Fans (Δ %)
El Paso TX			
Electricity (MWh)	173.4	12.1	7
Fan (MWh)	42.3	11.2	27
Cooling (MWh)	31.1	1.0	3
Peak Demand (kW)	47.5	0.9	2
Lake Charles LA			
Electricity (MWh)	183.2	11.9	7
Fan (MWh)	40.8	10.9	27
Cooling (MWh)	42.5	1.0	2
Peak Demand (kW)	48.2	0.9	2
Madison WI			
Electricity (MWh)	168.8	7.5	4
Fan (MWh)	40.9	7.4	18
Cooling (MWh)	17.1	0.8	4
Peak Demand (kW)	73.5	0.0	0
Seattle WA			
Electricity (MWh)	157.5	8.1	5
Fan (MWh)	43.8	7.3	17
Cooling (MWh)	13.4	0.9	7
Peak Demand (kW)	50.5	0.0	0
Washington DC			
Electricity (MWh)	171.4	9.2	5
Fan (MWh)	44.0	8.6	20
Cooling (MWh)	26.3	0.7	3
Peak Demand (kW)	56.3	0.0	0

TABLE 4. Energy Results for Medium Office

Site	Base Case	ASD Fans (Δ)	ASD Fans (Δ %)	ASD Fans & Chiller (Δ)	ASD Fans & Chiller (Δ %)
El Paso TX					
Electricity (MWh)	741.2	34.0	5	43.0	6
Fan (MWh)	113.2	29.0	26	29.0	26
Cooling (MWh)	126.7	5.3	4	14.2	11
Peak Demand (kW)	295.5	4.3	1	1.2	0
Natural Gas (MBTU)	358.5	(14.3)	(4)	(14.3)	(4)
Lake Charles LA					
Electricity (MWh)	766.9	34.7	5	47.8	6
Fan (MWh)	110.9	29.7	27	29.7	27
Cooling (MWh)	155.7	5.2	3	18.3	12
Peak Demand(kW)	304.8	6.5	2	2.4	1
Natural Gas (MBTU)	255.4	(10.6)	(4)	(10.6)	(4)
Madison WI					
Electricity (MWh)	686.6	35.0	5	42.1	6
Fan (MWh)	107.7	32.6	30	32.6	30
Cooling (MWh)	74.8	2.6	3	9.7	13
Peak Demand (kW)	296.5	5.6	2	0.2	0
Natural Gas (MBTU)	729.1	(15.0)	(2)	(15.0)	(2)
Seattle WA					
Electricity (MWh)	671.6	33.0	5	41.0	6
Fan (MWh)	109.7	30.2	28	30.2	28
Cooling (MWh)	57.9	3.1	5	11.1	19
Peak Demand (kW)	286.8	5.5	2	0.0	0
Natural Gas (MBTU)	767.2	(20.8)	(3)	(20.8)	(3)
Washington DC					
Electricity (MWh)	723.5	33.7	5	44.4	6
Fan (MWh)	114.0	30.2	27	30.2	27
Cooling (MWh)	106.6	3.5	3	14.1	13
Peak Demand (kW)	296.5	5.6	2	0.2	0
Natural Gas (MBTU)	599.4	(5.2)	(1)	(5.2)	(1)

All savings are calculated relative to the base case.

cooling electricity consumption. The combination of ASD fan and chiller controls does not alter fan savings, but increases cooling electricity savings to 11 to 19%. In this case, the mildest cooling climate (Seattle WA) produces the largest percentage savings. Again, additional heat is required during the heating season to compensate for the more efficient ASD controls versus inlet vanes for fans. In this case, the additional heat is generated from natural gas.

The simulations for both the strip store and medium office indicate small reductions in peak demands for ASD's for fan control, because of higher full-load efficiencies relative to the base case. For the medium office, these savings are offset by the lower full-load efficiency of ASD controls for the chiller. Variable flow controls (inlet vane and ASD's) do not, as a rule, reduce peak electrical demands since the benefits stem from greater overall efficiencies at part-load conditions.

In closing, it must be reiterated the significance of the conservative equipment-sizing assumption. This conservative assumption tends to result in understated savings for ASD's. Even greater savings can be expected from an analysis that more closely followed conventional sizing practices.

6. ECONOMIC ANALYSIS

The electricity savings from adjustable speed drives are only valuable if these savings meet or exceed the additional cost of the drives. Our analysis uses payback time as a simple figure of merit for making comparisons. As a rule of thumb, paybacks of less than three years indicate a good investment opportunity, independent of such important considerations such as the time-value and after-tax cost of money.

Tables 5, 6 and 7 summarize the results of the payback analyses. Payback times have been calculated assuming a low and high electricity price ($0.07 and 0.10 per kWh) and, for the medium office, a constant natural gas price ($3.50/MBTU). Similarly, the cost of ASD's has been estimated in the form of ranges. Based on informal surveys of manufacturer's literature (summarized in Reference 11), a low cost of $150 per horsepower was assumed, and a high cost of $300 per horsepower. These costs are intended to reflect the full installed cost of an ASD, which include labor and, in the case of retrofit, the salvage value of a conventional motor starter. Of course, these are only estimates; actual costs and energy prices may vary considerably. Our intent is to present results for a likely range of these figures in order to identify the boundary of favorable economic conditions.

The results for the strip store indicate that ASD's are generally cost-effective for either electricity price with the low ASD cost (see Table 5). With the high ASD cost, the investment is marginal, though still somewhat attractive. Payback times are around five years with the high electricity price.

TABLE 5. Payback Analysis for Strip Store - ASD Fans

Equipment Cost ($/horsepower)	150		300	
Electricity Price ($/kWh)	0.07	0.10	0.07	0.10
El Paso TX	2.9	2.0	5.7	4.0
Lake Charles LA	2.8	2.0	5.6	3.9
Madison WI	3.8	2.7	7.6	5.3
Seattle WA	3.6	2.5	7.3	5.1
Washington DC	3.5	2.4	7.0	4.9

TABLE 6. Payback Analysis for Medium Office - ASD Fans

Equipment Cost ($/horsepower)	150		300	
Electricity Price ($/kWh)	0.07	0.10	0.07	0.10
El Paso TX	4.2	2.9	8.4	5.8
Lake Charles LA	4.0	2.8	8.1	5.6
Madison WI	3.9	2.7	7.8	5.4
Seattle WA	4.4	3.1	8.8	6.1
Washington DC	4.3	3.0	8.6	6.0

TABLE 7. Payback Analysis for Medium Office - ASD Fans and Chiller

Equipment Cost ($/horsepower)	150		300	
Electricity Price ($/kWh)	0.07	0.10	0.07	0.10
El Paso TX	8.3	5.8	16.6	11.6
Lake Charles LA	7.7	5.4	15.4	10.7
Madison WI	8.6	6.0	17.1	11.9
Seattle WA	8.7	6.0	17.4	12.1
Washington DC	8.7	6.0	17.3	12.1

Natural gas price is $3.5/MBTU.

The results for ASD fan control in the medium office building are also encouraging (see Table 6). Again, the investment is generally acceptable with the low ASD cost, but is not acceptable with the high ASD cost. ASD controls for both the fan and chiller in the medium office building are much less cost effective (see Table 7). None of the sets of economic conditions evaluated yield paybacks of less than five years for this combination.

ASD fan controls are more cost effective than ASD chiller controls because fans operate all year, while the majority of chiller operation is concentrated in the summer. In general, longer hours of operation would make the investments more attractive for both prototypes. We would expect, for example, that ASD chiller controls would be much more cost effective for applications which operate 24 hours per day. In this case, night-time operation under, usually, relatively low load conditions would enhance savings greatly.

7. CONCLUSIONS

A set of parametric computer simulations was completed to evaluate the cost-effectiveness of ASD controls for commercial building fans and chillers. The evaluations examined the additional electricity savings of ASD's compared to conventional flow modulation (inlet vanes) for two prototypical commercial buildings in five U.S. climates. ASD controls for fans were found to be generally cost effective for the low set of ASD costs and either set of electricity prices evaluated. ASD controls for chillers were not found to be cost-effective. Our sizing assumptions for the fans and chillers were strict compared to conventional practice and tend to understate the benefits of ASD controls.

ACKNOWLEDGMENT

The work described in this paper was funded by the Assistant Secretary for Conservation and Renewable Energy, Office of Building and Community Systems of the U.S. Department of Energy under Contract No. DE-AC03-76SF00098.

REFERENCES

1. Hickok, H. "Adjustable Speed - A Tool for Saving Energy Losses in Pumps, Fans, Blowers, and Compressors", IEEE Transactions on Industry Applications, Vol. IA-21, No. 1, January/February, 1985.
2. Rozner, J., Myers, P. and Robb, C.," The Application of Adjustable Frequency Controllers to Forced Draft Fans for Improved Reliability and Energy Savings", IEEE Transactions on Industry Applications, Vol. IA-21, No. 6, November/December, 1985.
3. Colburn, B. and Duthu, D.,"Fan/Pump Laws Applied to Energy

Conservation Design", Presented at the AEE 6th World Energy Engineering Congress, Atlanta, GA, November, 1983.

4. "AC Drives Fan Savings Up to 50 Percent", Plant Energy Management", July/August 1981.

5. Bogart, J., "Application of AC Inverters Drives to Textile Fans", IEEE 1983

6. The American Society of Heating, Refrigeration, and Air-Conditioning Engineers, Inc. "ASHRAE Standard 90-75: Energy Conservation for Buildings", 1975.

7. US Department of Energy." Standard Building Operating Conditions" DOE/CS-0118, November, 1979.

8. Crow, L., " Development of Hourly Data for Weather Year for Energy Calculations (WYEC), Including Solar Data, at 21 Stations Throughout the U.S., ASHRAE Transactions, 1981.

9. Curtis, R., Birdsall, B., Buhl, W., Erdem, E., Eto, J., Hirsch, J., Olson, K. and Winkelmann, F., " The DOE-2 Building Energy Use Analysis Program", Lawrence Berkeley Laboratory, LBL-18046, April, 1984.

10. Diamond, S., Cappiello, C. and Hunn, B.," DOE-2 Verification Project, Phase 1, Interim Report*U, Los Alamos National Laboratory, April, 1981.

11. Usibelli, A., Greenberg, S., Meal, M., Mitchell, A., Johnson, R., Sweitzer, G., Rubenstein, F. and Arasteh, D., "Commercial-Sector Conservation Technologies", Lawrence Berkeley Laboratory, LBL-18543, February, 1985.

12. Lawrence Berkeley Laboratory, "DOE-2 Reference Manual, Part 1, Version 2.1C", Lawrence Berkeley Laboratory, LBL-8706, Rev. 4, May, 1984.

13. personal communication from York International Corp.

PRESENT-DAY LIGHT SOURCES

J.A.J.M. VAN VLIET

Nederlandse Philips Bedr. B.V.
Lighting Division
Building EDW-4 Eindhoven

1. INTRODUCTION

Present-day light sources have become one of the greatest technological benefits of our century, creating optimum working and environmental conditions with artificial light. Without present-day light sources our modern society would be disrupted, especially after sunset. For without lighting nighttime trafic would grind to a halt, the role of commerce and industry would be severely curtailed, crime rates would go up, domestic life would be reduced to its former primitive level and leisure would no longer provide the pleasure it offers today.

People - confronted with lighting every hour of the day - persist in their completely unfounded belief that lighting swallows up large amounts of energy; lighting is sometimes thought to be the most notorious guzzler of energy under the sun. Infact, present-day lighting is only a moderate energy consumer and uses no more than 4 per cent of all world-wide primary energy (Figure 1).

Figure 1. Worldwide energy
consumption for lighting:
4% of total primary energy.

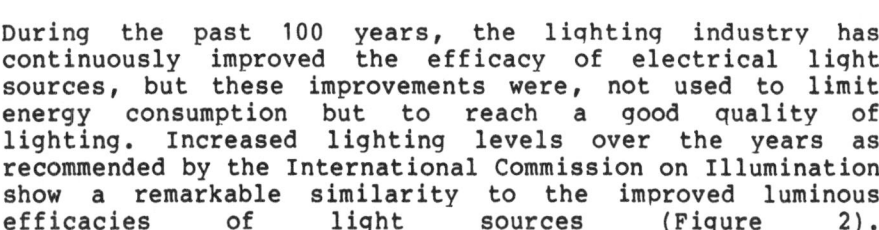

This does, however, not alter
the fact that in particular
applications, such as for
example in office buildings,
lighting can account for 40
up to 50 per cent of the energy costs.

During the past 100 years, the lighting industry has continuously improved the efficacy of electrical light sources, but these improvements were, not used to limit energy consumption but to reach a good quality of lighting. Increased lighting levels over the years as recommended by the International Commission on Illumination show a remarkable similarity to the improved luminous efficacies of light sources (Figure 2).

335

A. T. De Almeida and A. H. Rosenfeld (eds.), Demand-Side Management and Electricity End-Use Efficiency, 335–352.
© 1988 by Kluwer Academic Publishers.

Figure 2. Highest values for recommended lighting levels in lux for reading and writing tasks (broken line) and luminous efficacies of light sources during the years.

Fluor = fluorescent
Inc. = Incandescent
Hal. = Halogen
SL, PL = compact fluorescent
MH = metal-halide
HPS = high-pressure sodium
LPS = low-pressure sodium
HPMV = high-pressure mercury

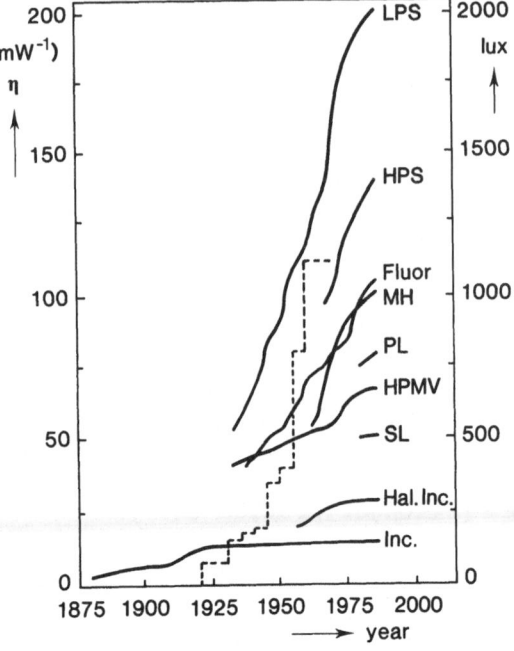

The improved luminous efficacies were used to increase the lighting levels and to improve visual performance, comfort and amenity[1]. However, during the past decade more attention has been paid to energy savings.

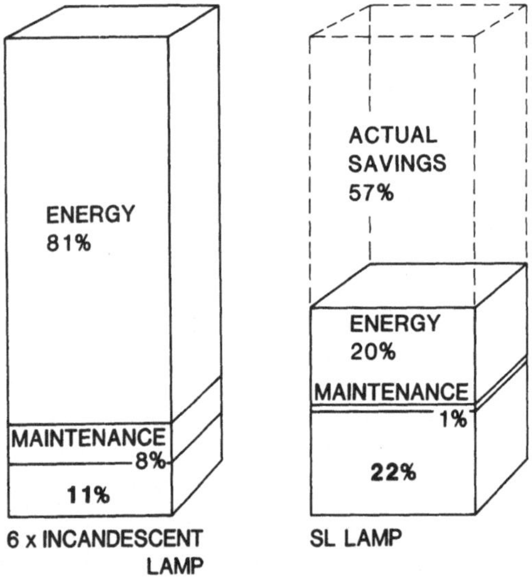

Figure 3. Bar A shows the relationship between energy, maintenance (including lamp replacement) for 6 incancescent lamps and Bar B shows that, despite a higher initial purchase price, a single SL will give the same light output over the same period of time, yet save at least 50% in total costs.

ENERGY 81%

MAINTENANCE 8%

11%

6 x INCANDESCENT LAMP

ACTUAL SAVINGS 57%

ENERGY 20%

MAINTENANCE 1%

22%

SL LAMP

If we look at the cost elements of an incandescent lighting system (Figure 3), we see that energy costs are nowadays about 81% of the total costs, whereas the lamp price is only about 11%, i.e. energy cost is more than 7 times lamp cost. Improvement of the luminous efficacy of light sourceshas therefore always been one of the driving forces for lighting industries to develop new or improved light sources even if these lamps are more complex and thereforemore expensive.

2. PRESENT-DAY LIGHT SOURCES AND THEIR HISTORY

The history of electric light sources is characterised by three mainstreams of development:
- the carbon-arc lamp
- the incandescent lamp
- the gas discharge lamp.

2.1 The carbon arc lamp

At the beginning of the last century Humphrey Davy experienced with the carbon arc. In 1813 he succesfully demonstrated the carbon arc for the Royal Society in London. A brilliant light is generated by the tips of the carbon electrodes, which are brought to incandescence by the power dissipated in the arc between them.

However, many technological difficulties stood in the way of the general introduction of this light source; therefore the carbon arc did not come into practical use until about forty years later. For general applications the heyday of the carbon arc came during the closing years of the last century, after which it was quickly superseded by the incandescent electric lamp. In particular applications, where a highly concentrated light source of extremely high intensity was required, such as theatre stage lighting or cinema projectors, the carbon-arc played a significant role until the introduction of the short-arc xenon discharge lamp in 1951.

2.2 The incandescent lamp

After numerous false starts, the first practical incandescent lamp came into being in 1879. Development continued rapidly until 1936. During these years efficacy gains of about a factor 10 could be realised mainly due to the use of better materials, a more sophisticated filament design and variation of the noble gas filling. No further significant improvements could be recorded until 1958 when Zubler and Mosby succeeded in making the first practical halogen incandescent lamp.

In normal incandescent lamps the tungsten evaporated from the filament settles on the coldest spot - the bulb wall - and causes lamp blackening. Higher fill pressures can reduce wall blackening. However, smaller bulbs have to be used to avoid explosion at the higher fill pressures, cancelling the gain by their reduced inner surface. Higher fillgas pressures can be used with succes in smaller bulbs in tungsten halogen lamps. In a tungsten halogen lamp, bromine in the form of an organic compound is usually added to the fill gas. This causes the tungsten vapour to combine with the halogen to form tungsten bromide, which stays in the gasphase instead of settling on the tube wall, provided the temperature of the latter is sufficiently high. When the tungsten bromide molecules arrive near the hot filament, decomposition follows. The result is that the tungsten is deposited on the filament and the free bromine is left behind in the fill gas. This process is called the "halogen regenerative cycle".

Because of the necessity of a high wall temperature these halogen lamps are smaller than the normal incandescent lamps and they are made of silica glass or pure fused silica (quartz) with its higher melting point. The smaller bulbs - as used for halogen lamps - allow higher pressures of the fillgas and higher efficacies can be reached because of that.

During the past years, finally, another considerable gain has been achieved by application of infrared-reflective coatings which reduced the biggest loss mechanism, i.e. infrared radiation. By using infrared reflective layers on the bulb of tubular incandescent lamps for floodlight (Figure 4) we can influence the power flow in such a way that the input power can be reduced by 40% while the lumen output remains the same.

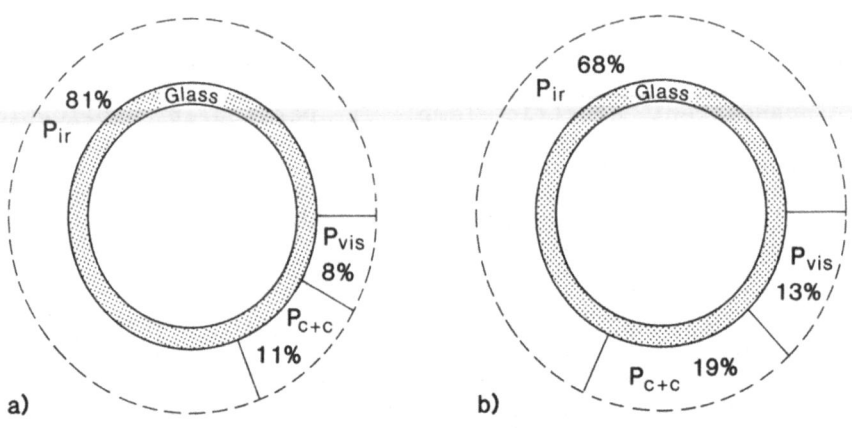

Figure 4. Influence of infrared reflecting filters on the power flow in incandescent lamps for floodlighting.
A. without infrared reflecting filter
B. with infrared reflecting filter.

The infrared losses are reduced from 81 to 68 per cent. Nowadays, the incandescent lamp family - including the halogen type - is extremely diverse and covers nearly all areas of lighting applications. However, the gas discharge lamp has brought the supreme reign of the incandescent lamp in many application areas to an end.

2.3 The gas discharge lamp

The systematic investigation of electric gas discharges started in the middle of the last century, but it was not until the end of the nineteenth century that gas discharge light could be used for lighting purposes. Development of discharge lamps has taken place along two different lines: light sources incorporating either the low pressure gas discharge or the high pressure gas discharge. Besides rarefied gases sodium and mercury were used as the radiating medium. Sodium and mercury discharges are similar in many aspects. Both discharges reveal two peaks in their curvesrepresenting the luminous efficacy as a function of the vapour pressure (Figure 5).

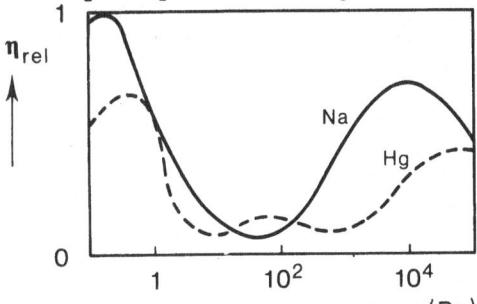

Figure 5. Comparison between sodium and mercury discharges showing the relative luminous efficacies η_{rel} of sodium and mercury discharge lamps as functions of the sodium or mercury vapour pressure (p). A fluorescent powder is used for the mercury discharge lamps to convert the ultraviolet radiation into light. Data from British Lighting Industries (ref: Light and Lighting 1966).

One peak occurs at a pressure lower than 1 Pa (the low pressure types) and one at a pressure exceeding 10 kPa (the high-pressure types). Mercury and sodium were obvious choices as fill gases for metal vapour discharge lamps. This had to do with the fact that of all metals they are the only ones with an appreciable vapour pressure at normal temperatures and that a fair portion of their radiation lies in the visible part or ultraviolet part of the spectrum. The ultraviolet radiation can be converted into light by suitable fluorescent powders.

2.3.1 Low-pressure discharge lamps
Low-pressure mercury discharge lamps[2]
The first practical low-pressure discharge lamps were constructed by Peter Cooper Hewitt in 1901 (Figure 6).

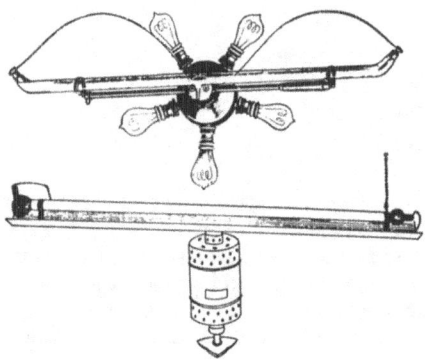

Figure 6.
Cooper-Hewitt low-pressure mercury lamps, fitted with a resistor ballast (top) or with carbon lamp is started by pulling the chain and so tilting it.

He employed a 1 m long tube with an electrode of iron or graphite at one end. A pool of mercury formed the other electrode. The lamp was started by tilting the tube, allowing the mercury to make initial contact between both electrodes. Because of the negative voltage characteristic a series resistor or ballast was necessary to limit the lamp current. Carbon-filament lamps were sometimes used as a ballast to improve the colour characteristics.Two important discoveries were largely responsible for the eventual succes of the low-pressure mercury discharge lamp. These were firstly the coating of the tube wall with fluorescent material to convert the strong ultra violet radiation of the mercury discharge into visible light. The coating thus serves to increase the efficacy and to improve the colour characteristics of the light. Secondly, the use of preheated, oxide-coated electrodes, which have a much higher electron-emission rate and therefore render ignition easier, reduces electrode losses and improves maintenance and efficacy.

Besides mercury, a buffer gas consisting of argon or a mixture of argon with neon or krypton is usually added to the fluorescent lamp. Its pressure is so chosen that the non-radiative losses are kept to a minimum.
The most important design parameters of these lamps are thus mercury pressure, buffergas pressure, radius and current. For an optimized discharge the electrode losses and non-radiative losses, i.e. wall and volume losses are about equal (Figure 7).

Figure 7.
Power flow in an
optimized 36W
fluorescent lamp.

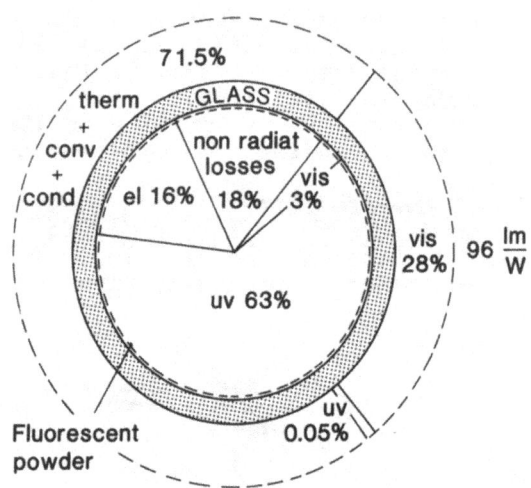

Together they amount to roughly one third of the power input. All the rest is ultraviolet and visible radiation, which means that the radiation efficiency is very high. Present-day low-pressure mercury discharges and their radiant efficiencies are essentially the same as fifty years ago. However, it has been known for a considerable time that high-frequency operation, which reduces electrode losses, could enhance radiation efficiency further.

Advantage of this was taken only recently, because now the large-scale application of high-frequency supplies for fluorescent lamps has become technically and economically feasible.

Significant progress has been made in the improvement of the phosphors. Early fluorescent lamps used phosphors based on zinc beryllium silicates. These compounds were later found to be a serious health hazard. In 1948 the safe halophosphate phosphors were first used in fluorescent lamps and they were further developed and perfected in the following decades.

The next major breakthrough came in the nineteen seventies. It was found that by limiting the phosphor emission to specific narrow bands it was possible to increase the luminous efficacy and improve colour rendering. This idea was turned into practice by the development of new narrow-band phosphors (Figure 8.)

Figure 8. Examples of spectral energy distributions of fluorescent lamps provided with
(a) halophosphates and
(b) narrow band phosphors
and their colour characteristics.

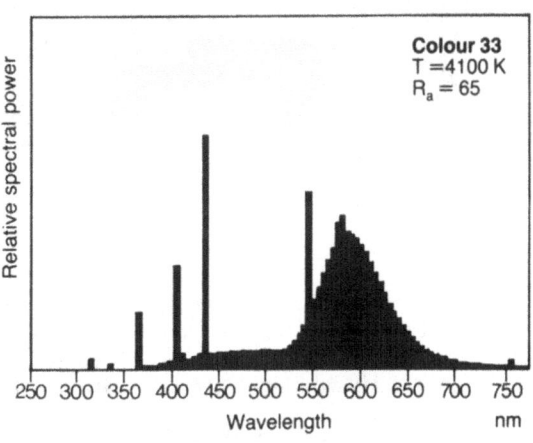

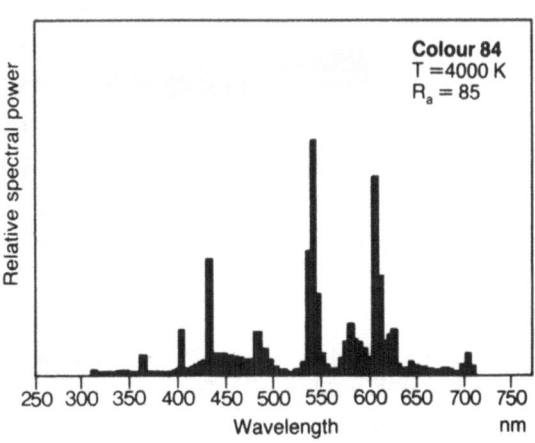

Application of these narrow-band phosphors lifted the luminous efficacy of fluorescent lamps up to the 100 lumens per Watt level.
Up to that time none of the existing discharge lamps could replace the less efficient but compact incandescent lamps with outputs below 2000 lm.
These new narrow-band phosphors, which are capable of withstanding higher ultraviolet radiation levels,

opened the way to miniaturization of the fluorescent lamps and filled up this gap of compact gas discharge lamps with low lumen output levels. Small-bore single-ended fluorescent lamps with discharge tubes folded into compact shapes were designed (Figure 9) to replace incandescent lamps.

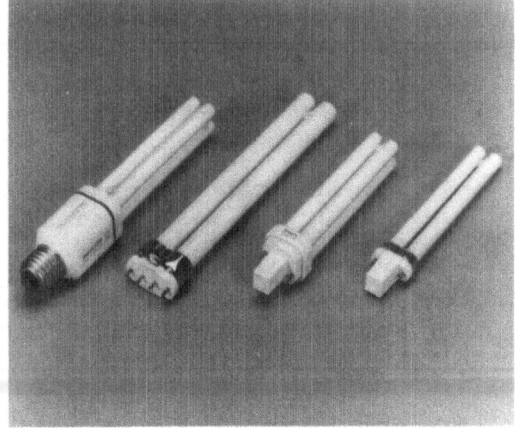

Figure 9. Examples of the new generation of compact fluorescent lamps

How attractive these new gas discharge lamps can be, is illustrated by a cost comparison of an incandescent and compact fluorescent lighting system (figure 3). Operational savings of about 60 per cent are possible.

Low-pressure sodium lamps[5]
Let us now switch over to the low-pressure sodium lamps (Figure 10).

Figure 10. Examples of low-pressure sodium-lamps.

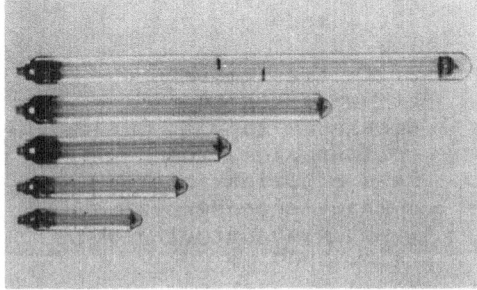

Characteristic differences exist between the low-pressure mercury and sodium discharges. Sodium is more reactive than

mercury. Furthermore, the temperature needed to obtain the same vapour pressure is higher for sodium than for mercury. So the requirements for the discharge tube material with respect to thermal and chemical stability are much more severe for sodium than for mercury. A borate layer on the inside of the discharge tube protects the soda-lime glass against sodium attack. To reach an adequate vapour pressure of sodium a discharge tube temperature of 530 K is necessary, and the corresponding losses through radiation and heat transfer are considerable at least without countermeasures. Therefore, the first commercial low-pressure sodium lamps - which appeared on the market in 1932 - only had a luminous efficacy of about 40 lm W despite the fact that the spectral luminous efficiency of the yellow D-lines of sodium for the generation of light is extremely high.

The constraint of heating the discharge tube to its optimal operating temperature limits the efficacy. The severity of this constraint can be decreased by heat insulation of the discharge vessel. Most of the improvements are associated with better insulation, ranging from a plain outer bulb to the sophisticated infrared-reflective layers of today (Figure 11).

Figure 11. Luminous efficacy of the low-pressure sodium lamp for different heat insulations with sinusoidal and square-wave current supply

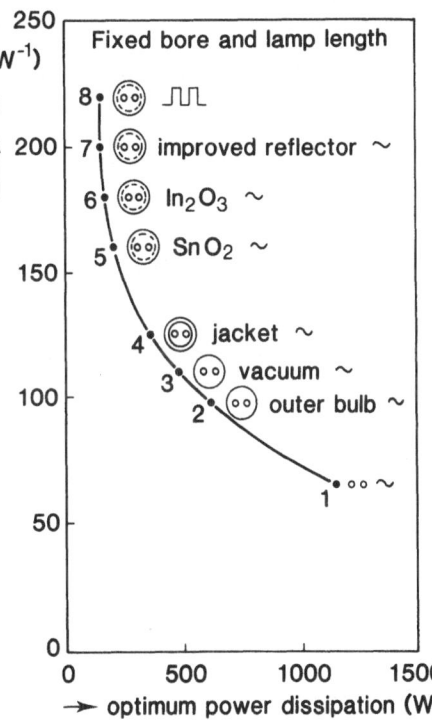

In this way the input power has been decreased by a factor eight to reach the optimum operating temperature. At the same time the efficacy increases more than four fold due to the decreased losses. In spite of this considerable efficacy increase the non-radiative discharge losses of for example a 180 SOX lamp are still higher (45%) than the discharge radiation (42%) (Figure 12).

344

Figure 12. Power
balance of the
180W low-pressure
sodium lamp.

Further improvements
by better heat insula-
tion are thus still
possible.

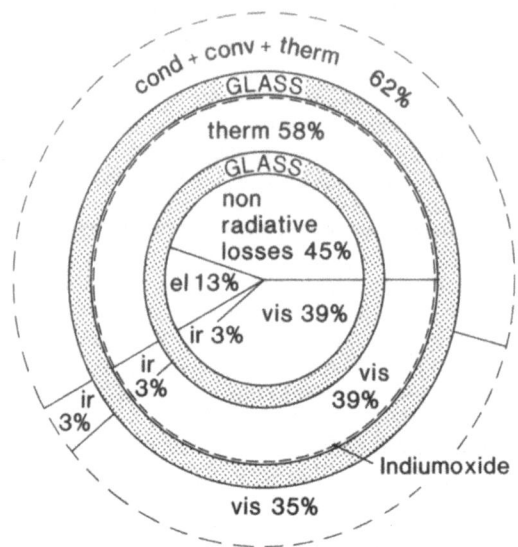

2.3.2 High pressure gas discharge lamps

Let us now step from the low-pressure region of a few
Pascals to the high-pressure gas discharge lamps, where
pressures above 10 kPa are used. In order to realize these
higher pressures tube temperatures up to 1500 K had to be
realised. Therefore small discharge tubes from other
materials than glass had to be manufactured.

The first high-pressure gas discharge lamp dates from 1906,
when Küch and Retschinsky carried out their first
experiments with high-pressure mercury discharges. As with
so many developments, technological problems with
manufacturing a reliable gas-tight seal between the lead-in
wires and the quartz discharge tube stood in the way of
mass production. It was only in the early thirties that
technological problems were solved. The first high-pressure
mercury discharge lamp came onto the market in 1934.

A new high-pressure gas discharge lamp saw the light in
1944: the xenon arc lamp. This - nearly pointlike - light
source appeared on the market in 1949. It excels in its
good colour rendering properties and high brightness.

An improvement of the high pressure mercury discharge lamp
can be realised by adding metal halides to the mercury
in the discharge tube. This development started in the
sixties. It represented a considerable improvement over the
traditional high-pressure mercury lamp in terms of both
increased efficacy and better colour rendering.

It took a long time before the application of the
high-pressure sodium discharge as a light source became

possible. Because of the very aggressive nature of sodium at high temperatures and pressures, the first experiments were conducted in a metal discharge tube with a sapphire window (Figure 13).

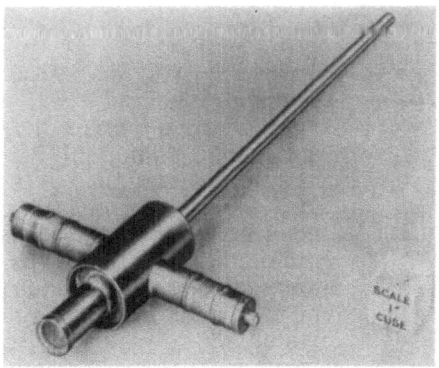

Figure 13. High-pressure sodium discharge vessel made in 1959 by Clarke and Moore. The arc ran between tungsten electrodes mounted on spark plug insulators in the side arms, the light being emitted through the circular sapphire window. The nickel tube opposite this window controlled the sodium vapour pressure.

After the development of a new translucent, gas-tight and sodium-resistant polycrystalline alumina the high-pressure sodium discharge lamp appeared on the market in 1965. At that time the difficulty of making a reliable electrical feedthrough on both ends of the discharge tube had been solved (Figure 14).

Figure 14. Typical high-pressure sodium lamps.

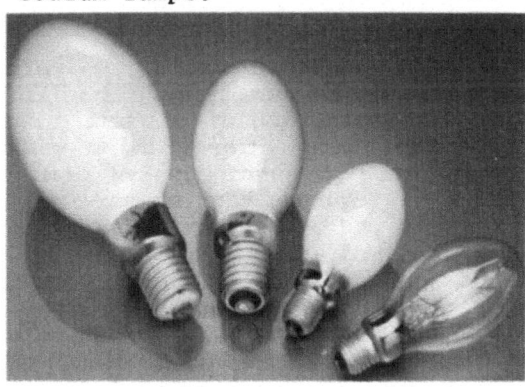
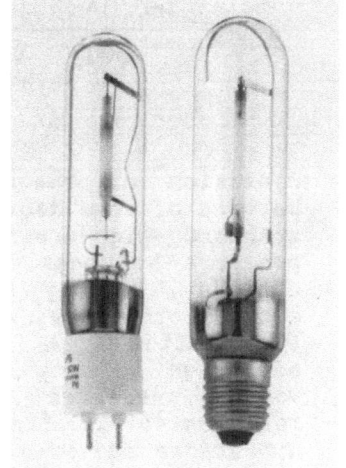

Let us look in more detail at two of the most important lamps within the family of high-pressure discharge lamps: the high-pressure sodium and the metal halide lamp.

High-pressure sodium lamps[4])

The discrimination of colours is not possible with the yellow colour of the monochromatic light of the sodium D-lines as radiated by the low-pressure sodium lamp. However, the pressure is increased in such a way in high-pressure sodium lamps that strong line broadening and self-reversal of lines occur (Figure 15).

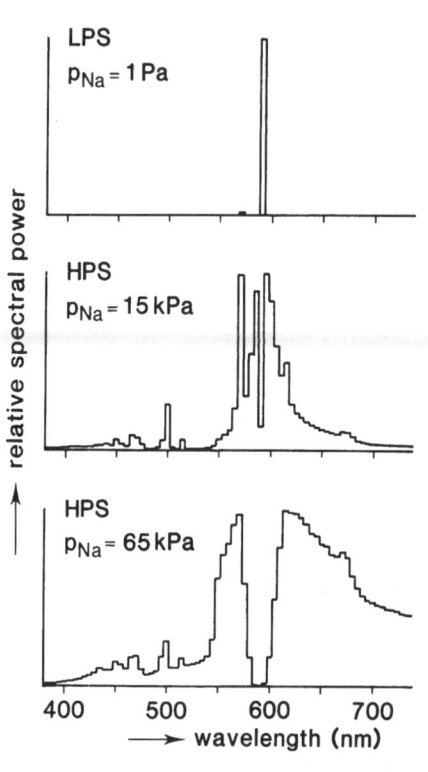

Figure 15. Visible spectra of (a) low-pressure sodium (LPS) lamp (b) standard high-pressure sodium (HPS) lamp and (c) "white" HPS-lamp with good colour rendering properties, showing the influence of the sodium vapour pressure (P_{Na}) on the spectrum. The spectra have been measured with integration intervals of 5 nm. The sodium D-lines at 589.0 and 589.6 nm show a strongh self reversal in the HPS-lamp.

These phenomena strongly influence the visible spectrum of the high-pressure sodium lamp. Because of the broadening of its spectrum the standard high-pressure sodium lamp has a golden white colour appearance and, in contrast to the monochromatic yellow radiation of the low-pressure sodium lamp, its light permits of some degree of discrimination of colours. In the standard high-pressure sodium lamps the sodium vapour pressure has been optimized with respect to the luminous efficacy. Mercury is added to improve the electrical and optical properties and xenon at a low-pressure functions as the starting gas. These standard lamps can be used to advantage in many outdoor applications. At higher sodium vapour pressures nearly all colours are sufficiently represented in the spectrum to approximate the colour appearance and colour quality of an incandescent lamp. This means that this lamp will become of importance for indoor applications. The relatively small size of the discharge tube enables us to construct spotlights with even narrower beams than those realized by the incandescent reflector lamps, and with a significantly lower energy consumption.

An improvement of the light quality of high-pressure sodium lamps by the increase of the sodium vapour pressure means that the luminous efficacy decreases due to a less good matching of the spectrum to the eye sensivity curve. In general, a maximum colour rendering index of 85 (incandescent lamps having a colour rendering index of 100) is achieved against a loss of about 40% in luminous efficacy. However, this lamp is still 4 times as efficient as the incandescent lamp.

The power flow in a 400W standard HPS-lamp can be split up into three constituent balances, one inside the PCA-arc vessel, one outside the vessel but within the outerbulb, and finally the balance which the user experiences outside the outer bulb (Figure 16).

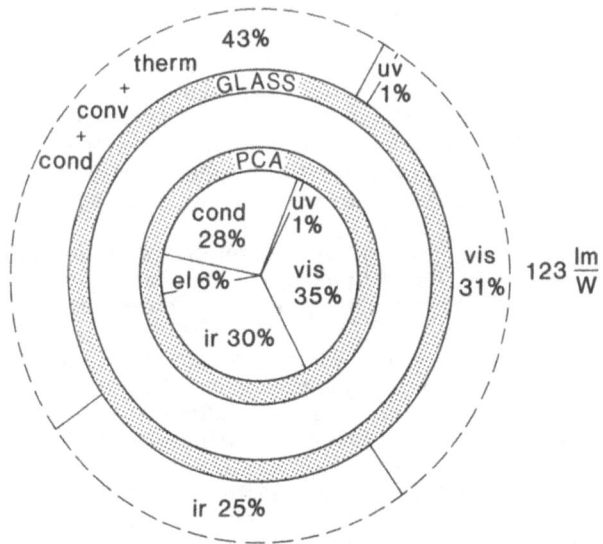

Figure 16. The power flow in a standard 400W HPS-lamp. Inside the PCA discharge tube the input power is converted into ultraviolet (UV), visible (VIS), and infrared (IR) radiation, electrode power (el) and conduction power(cond). The greater part of the (near) ultraviolet, visible and infrared radiation passes through the glass outer bulb (P_{uv}, P_{vis}, P_{ir}), the rest of the power leaving the outer bulb as thermal radiation (P_{th}) and by way of conduction and convection(P_{c+c}). The efficiency with which the visible radiation generates light is given by the product of P_{vis} (visible radiant power) and V_s (luminous efficacy of visible radiation) resulting in 123 lm W^{-1}.

Let us first look at the balance inside the PCA-tube. As always we have the electrode losses of about 6%, which are the lowest of all lamps types discussed here. Then we have losses through heat conduction and re-absorption of radiation in the cool outer mantle of the discharge. These losses amount to about 28% of input power. A further loss mechanism in high-pressure lamps is infrared radiation, which is neglible in the power balance of low-pressure sodium lamps. Here infrared radiation amounts to about 30% of lamp input power. The visible part amounts to about 35% and ultraviolet radiation can be practically disregarded. This power flow inside the tube is then altered by absorption and reflection effects of the PCA-tube and the outer lamp bulb. Finally, we find that the heat of the bulb is transferred to the surroundings by thermal radiation (30%) and by conduction and convection. The visible radiation output amounts to somewhat less than one third of the lamp power. Multiplying this by the efficiency of visible radiation V_s we end up with a weighted visible radiation output of about 18% representing a luminous efficacy of about 120 lm/W. This high efficacy was reached by not only optimising the sodium pressure but also other parameters, e.g. the mercury vapour pressure and the dimensions of the discharge tube. I will not go into the details of this optimisation process but try to discuss one further big improvement which could be achieved by reducing the conduction losses and modifying the spectral distribution.

A considerable amount of a gas with a lower considerable than sodium, e.g. Xe, can be added to the discharge to reduce the heat conductivity of the plasma. Xenon pressure has to be increased to about 10 times the sodium pressure. The visible efficiency increases from 31% to 36% and the conduction losses decrease accordingly. On the whole we see that HPS-lamps with high Xe-pressures are very efficient lamps which are especially well suited for outdoor lighting, e.g. road lighting.

The widespread introduction of these lamps to replace the less efficient high-pressure mercury lamps is hindered by the fact that retrofitting in existing mercury installations is difficult.

High-pressure mercury lamps have a starting gas that ignites more easily and an internal starting electrode. Therefore they can ignite without a high-voltage peak, simply on the mains voltage. High-pressure sodium lamps with Xe as a starting gas, on the other hand, need an ignition peak of some kilovolts generated by a starter device.

To replace mercury lamps, therefore, special lamps with a Penning mixture are used, because this starting gas is capable of igniting at the line voltage without high-voltage peaks. On the other hand, the Penning mixture has a much higher heat conductivity than xenon and therefore the efficacy of these lamps is considerably lower. The visible radiation is only about 25% and conduction losses amount to about 50% for these special lamps, considerably higher than for standard lamps with xenon as starting gas even at low pressures.

This example serves to show that energy efficient design can be hindered by the circumstances of the application, in this case retrofitting mercury lamps. A possibility to overcome this application constraint imposed on lamp design is offered by electronics. A small electronic ignitor, built into an adaptor between lamp and screw-in base can alow the efficacy gains of high Xe-pressure to materialize also in retrofitting of mercury lamps. Thus HPS-lamps with an ignitor in the lampbase, are able to replace mercury lamps with a considerable increase in lumen output and reduction of lamp power.

Metal halide lamps[5]

By covering the inner wall of the outer bulb with a fluorescent powder. The spectral distribution remains far removed from that of an incandescent lamp, unless the ultraviolet radiation is converted into red radiation. With the addition of metal halides the mercury lines in the spectrum are weakened. The strong resonance lines of e.g. sodium, thallium and indium in the orange, green and blue part of the spectrum can then predominate. The metal halide lamp with sodium, thallium and indium jodide surpasses the mercury lamp both in luminous efficacy and colour rendering. By using different additives we can tailor the spectrum of metal halides lamps according to the requirements of a certain application. The Olympic Games in Montreal would not have been brought onto your television screen in such brilliant colours if other types than metal halide lamps had been used.

The application of scandium tri-iodide or other rare-earth alkali metal halides can fill the gaps in the mercury spectrum better than the metal halides because of their multi-line spectral distribution. The scandium lamps are preferably used in the United States, whereas the rare-earth metal halides, such as the dyprosium, holmium and thulium halides are preferred in Europe.

Another possibility to approximate the continuous spectrum of an incandescent lamp or Planckian radiator is the use of the quasi-continuous radiaton of molecules which do not dissociate at the high temperatures in the axis of the discharge tube, such as the tin halides. The quasi-continuum of the tin monohalide extends over almost the entire visible part of the spectrum.

However, all high-pressure discharge lamps designed to replace incandescent lamps show too little radiation in the red. The molecular radiation of some monohalides of rare-earth alkali metals - such as dyprosium mono-iodide - can radiate effectively in the red part of the spectrum in special conditions, thus improving colour rendering.

The photometric properties and power balances are very different for various metal halide lamps. (Table 1).

Table 1. Power balances of different metal halide lamps in percentages of input power

	400W Na/Tl/ln iodide	400W Sn- halide	400W Dy iodide	400W Na/Sc- iodide	400W Na/Sc- iodide
Electrode losses	9	10	9	9.5	13
Non-rad. losses	38.5	37	27	35.5	33
UV-radiation	4	3	6	11.5	5
IR-radiation	24.5	27	26	9.5	15
Visible radiation	24	23	32	34	34
η (Lm/W)	87	60	81	97.5	85
R_a	67	91	85	65	65
T_c (K)	4500	4100	5250	4860	3200

Lamps with high colour rendering indices have only a modest luminous efficacy, but lamps with a moderate colour rendering index can reach a luminous efficacy of nearly 100 lm/W.

The development of metal halide lamps (Figure 17) for television, studio, flood and industrial lighting applications is an increasing succes. Attemps to design lamps with a warm colour, reasonably good colour rendering, high luminous efficacy and long lamp life for indoor lighting continue all the time.

Figure 17. Examples of metal-halide lamps.

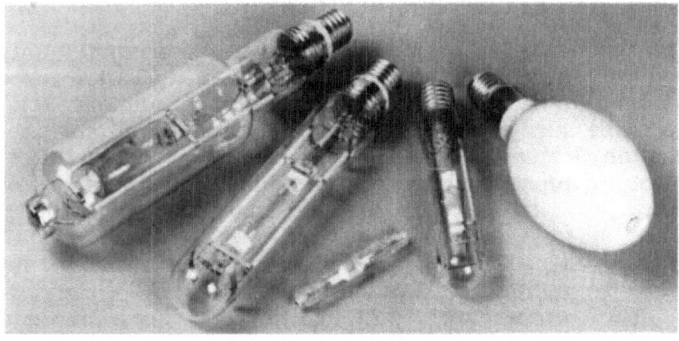

The chemical processes in metal halide lamps are complex and numerous. The consequences of these reactions are sometimes fatal and lead to short lamp lives, especially in high loaded metal halide lamps with chemically highly

reactive salts. The flexibility of the optical properties also have negative consequences. Colour differences can lead to problems in indoor lighting. If the composition or coldest spot temperature varies due to production spread, disappearance of a component during lamp life or a change in burning position. But present-day developments look very promising.

3. SUMMARY AND OUTLOOK

Continuous improvement of the luminous efficacy of existing lamps and the development of new light sources have prevented lighting becoming the most notorious energy guzzler under the sun.
Although the annual light production per capita has been strongly increased between 1972 and 1984, the energy consumption for lighting has shown only a slight increase due to this continuous improvement (Table 2.)

Table 2. Annual energy consumption for lighting per capita (Excl. China and USSR) in 1972 and 1984 in kWh.

	1972	1984
Incandescent lamps	93.4	89
Reflector lamps	2	3
Fluorescent lamps	125	138
High pressure mercury lamps	37	40
High pressure sodium lamps	1.25	10
Low pressure sodium lamps	2.5	2.5
Total energy consumption	261	283

Annual light production per capita* in 1972 and 1984 in K-lumen x hour

	1972	1984
Incandescent lamps	1258	1200
Fluorescent lamps	8462	10920
High pressure mercury lamps	1715	1864
High pressure sodium lamps	125	1000
Low pressure sodium lamps	325	358
Total light production	11885	15342

What will future developments bring ? Is further improvement of the luminous efficacy of lamps possible ? Small improvements at the heart of the light source - the discharge - are possible. (Table 3).

Table 3. Possibilities for future energy efficient design.

	Fluorescent	Low pressure sodium	High pressure sodium	Metal halide
Discharge optimization				*
Electronic ballasts	*	*	*	*
Heat insulation		*	*	*
Colour filters			*	*
Materials	*		*	*

There is the possible optimization of the discharge of metal halide lamps, or improved heat insulation of low- and high-pressure sodium lamps and metal halide lamps. One other possibility is the use of thin infrared reflecting layers on the inner or outer wall of the outer bulb. They can selectively reflect thermal radiation back toward the discharge. They are already being used in low-pressure sodium and incandescent lamps. Filters can be used to tailor the spectral distribution and improve colour rendering or adapt colour temperature. Discharge tube materials which can better withstand high operating temperature will lead to new improvements. Will the future also offer new phosphors efficiently converting one ultraviolet photon not into one but into two visible photons ?
This would mean another revolution in the fluorescent lamp field. One thing is certain, the way to electronification of light sources lies wide open. Electronification offers new possibilities, such as the enhanced luminous efficacy through high-frequency operation of low-pressure discharges, a reduction of the ballast losses and the addition to the lighting unit of several control functions.

I wish to thank many of my colleagues for their support in making this paper possible.

REFERENCES

1. Boer, J.B. de, Developments in illuminating engineering in the 20th century, Lighting Research and Technology, vol. 14, no. 4, pp. 207 - 217 (1982)

2. Jack, A.G. and Vrehen, Q.H.F., Progress in fluorescent lamps, Philips Techn. Rev. 42, no 10/11/12, pp. 342-351, (Sept. 1986).

3. Denneman, J.W., Low-pressure Sodium Discharge Lamps. I.E.E. Proc. Pt. A., vol. 128, pp. 397-414 (1981).

4. Groot J.J. de and Vliet J.A.J.M. van, The High-Pressure Sodium Lamp, Philips Technical Library/Kluwer Technische Boeken B.V., Deventer (1986).

5. Dobrusskin, A, Review of Metal Halide Lamps. Fourth Int. Symp. on the Science and Technoloy of Light Sources, Karlsruhe F.R.Germany (1986).

FROM LIGHT SOURCE TO LIGHTING UNIT

J.A.J.M. VAN VLIET

Nederlandse Philips Bedr. B.V.
Lighting Division
Building EDW-4 Eindhoven

1. INTRODUCTION

The achievement of Thomas Alva Edison was not only the manufacturing of the first useful electric lamp. Above all, his success was due to his vision regarding the distribution of light. Lighting was only viable if electricity was generated at a central point and from there distributed to the public in the same way as gas had been supplied years earlier. For that purpose Edison designed - in spite of all the criticism of his ideas on the distribution of light - an incandescent lamp of high resistance. These high resistance lamps - connected in parallel and connected to a high voltage from a dynamo - could be switched off independently. In this way Edison tried to optimize the total lighting unit consisting of supply voltage, switch and light source.

The lighting unit incorporating a ballast - the current-limiting device for proper operation of gas discharge lamps with their negative voltage current characteristic - as known at the beginning of this century has not undergone substantial changes, with the exception of the light source. The supply voltage remained the same. The passive copper/iron ballasts were only reduced in weight (Figure 1) and volume by improved insulation and winding techniques without important changes in the losses. Only the efficacy of the light source has been improved many times by new developments.

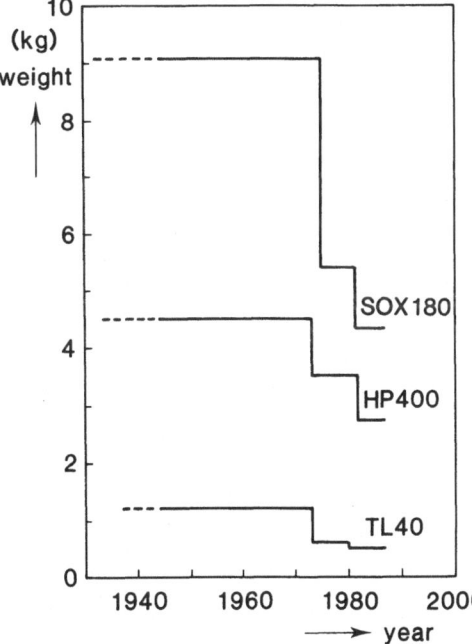

Figure 1. Reduction in weight of various copper-iron ballasts over the years.

A. T. De Almeida and A. H. Rosenfeld (eds.), Demand-Side Management and Electricity End-Use Efficiency, 353–366.
© 1988 by Kluwer Academic Publishers.

From the seventies the lighting industry, driven partly by the energy crises, partly by the availability of new modern electronic components, again focused its attention on lighting unit as an entity, in the same way as Edison had done a century earlier.

New trends in the development of lighting units are:

- miniaturization of light sources, both to make the replacement of incandescent lamps by discharge lamps possible and to improve the optical control of light for extreme directional lighting.
- use of electronics in the control circuitry, which is attractive because of the possible energy savings in both the light source and control gear, and the integration of other functions, like light regulation, in the same control circuitry.
- integration of the various components of the lighting unit as a logical continuation of its miniaturization and the use of electronics and the integration of the lighting units in a total lighting system.

2. MINIATURIZATION

What was quite normal in the development of incandescent lamps, namely their miniaturization to suit a wide range of applications, does not apply to gas discharge lamps. The miniaturization of these lamps did not get under way until after the energy crises in the seventies. In 1972, incandescent lamps accounted for roughly 35 per cent of all electricity used world wide for lighting (Table 1).

	1972	1984
Incandescent lamps	93.4	89
Reflector lamps	2	3
Fluorescent lamps	125	138
High-pressure mercury lamps	37	40
High-pressure sodium lamps	1.3	10
Low-pressure sodium lamps	2.5	2.5
Total energy consumption	261	283

Table 1. Annual energy consumption for lighting per capita (excl. China and USSR) in 1972 and 1984 in kWh.

Replacing of all incandescent lamps by gas discharge lamps would reduce energy for lighting enormously. Therefore, after the energy crises in the seventies and the invention of new narrow band phosphors, withstanding higher ultraviolet radiation levels, a new trend in gas discharge lamps became visible: the miniaturization of fluorescent lamps incorporating the copper-iron or electronic ballast in the lamp cap. How attractive these new compact fluorescent lamps are, is illustrated by the following relative cost comparison between a standard 75 W incandescent and an 18 W compact fluorescent lamp having comparable luminous fluxes (Table 2).

	Incandescent	Compact fluorescent
Lamp power (W)	75	18
Life (hrs)	1000	5000
Number of lamps	5	1
maintenance and		
Initial costs	19%	23%
Energy costs	81%	20%
Total costs	100%	43%

Table 2. Cost comparison between a standard 75 W incandescent and an 18 W compact fluorescent lamp both having 900 lm luminous flux. Costs are related to the total costs of the incandescent lighting unit.

For general lighting purposes where light sources are required to have a luminous flux of less than 6 klm, only compact fluorescent (PL, SL) lamps are the only present-day solution. They can replace incandescent (Inc., Hal. Inc.) lamps having luminous fluxes below 2 klm (Figure 2).

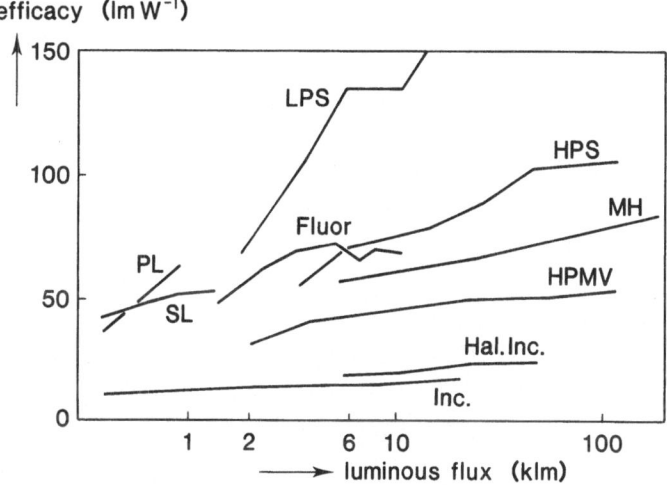

Figure 2. Efficacy of various light sources (including ballast losses) as a function of their luminous flux (Fluor = fluorescent lamps).

Between 2 and 6 klm the low-pressure sodium (LPS), high-pressure sodium (HPS) or mercury (HPMV) discharge lamps are no real substitute for incandescent lamps, because of their inferior colour-rendering properties. Up-grading of the luminous flux of compact fluorescent lamps to values above 2 klm meets thermal difficulties. Therefore, real future alternatives for incandescent lamps with a luminous flux between 2 and 6 klm are metal-halide (MH) and white high-pressure sodium lamps.

In spite of the trend toward making compact high-pressure and low-pressure discharge lamps available for interior lighting, one important distinction between the two types of discharge lamps remains: the small physical dimensions of the high-pressure gas discharge lamp facilitates the precise optical control of the luminous flux by a luminaire with a well defined reflector. Optical control of the luminous flux of the low-pressure discharge lamps - having much larger burners - into light beams is very difficult, bearing in mind the very large fixtures which would be necessary. Low-pressure discharge lamps are therefore excluded from use in those applications where beamed light is required, such as for example shop lighting.

The recently developed 50 W white high-pressure sodium lamp is a good replacement for halogen incandescent lamps were extreme directional lighting is involved. It is to be expected that metal halide lamps, which can be manufactured with very small discharge tubes and a luminous efficacy that is nearly independent of the luminous flux, will be employed in this application field. The compact fluorescent lamps are better suited for general lighting purposes.

3. ELECTRONIC CONTROL

Up to the seventies discharge lamps were - except for some special applications - operated at a supply frequency of 50 or 60 Hz although it was known that important gains in system efficiency could be realised by using higher supply frequencies. The energy crises made the change-over from conventional copper/iron ballasts to electronic ballasts easier. Because of the increased energy prices, the pay-back time of the more expensive electronic control gear was reduced and thus brought within acceptable limits so that electronic gear became economically viable.

3.1 Full electronic control gear

Electronic supply improves the system properties of the lighting unit in several ways. The main advantages are:

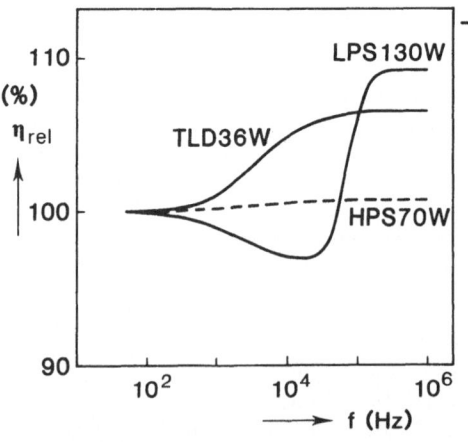

- The efficacy of the lighting unit is enhanced. At high frequency, the luminous efficacy of low-pressure discharge lamps increases due to reduced electrode losses (fluorescent and low-pressure sodium lamps) and due to an increase in the efficiency of the discharge (low-pressure sodium lamp).

Figure 3. The relative change in luminous efficacy of low and high-pressure gas discharge lamps brought about by an increase in the operating frequency.

For the low-pressure gas discharge lamps with lower lamp voltages, the increase in luminous efficacy is even greaterbecause the electrode losses account for a larger part of the power input. For high-pressure discharge lamps no substantial gain in the luminous efficacy is found. The use of electronic control in conjunction with high-pressure discharge lamps at high frequencies is hampered by the absence of any gain in luminous efficacy. Furthermore, the high-frequency operation of high-pressure discharge lamps can lead to arc distortions at frequencies correlated to the resonance frequencies of standing acoustic pressure waves (Figure 4).

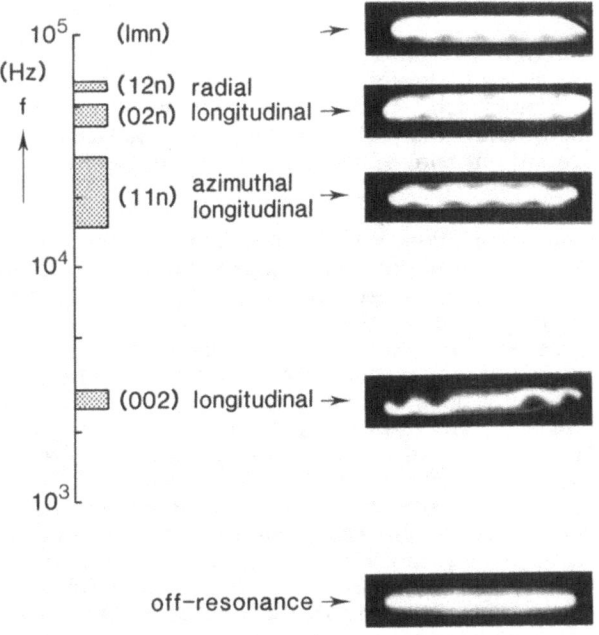

Figure 4.Photographs of the discharge paths in a 250W HPS lamp at various frequencies of the current supply. The letters (l.m.n.) indicate the mode of resonance occuring inthe batched frequency region.

Therefore, full electronic control of discharge lamps is nowadays limited to the low-pressure discharge lamps. The power dissipation in high-frequency ballasts is lower than that in conventional circuits for all discharge lamps regardless of types. At higher frequencies the coil impedance needed for stabilisation can be reached at a much lower inductance.
The improvement in the efficiency of a practical unit made possible by the use of electronics to control fluorescent lamps is illustrated by comparing the power dissipation in

high-frequency and conventional lighting units (Table 3).

Lamp type	'TL' D 26 mm	'TL' D 26 mm	'TL' 38 mm
Control gear	HF Electronic	Conventional	Conventional
Nominal lamp power	50 W	58 W	65 W
Nominal mains power (warm)	111,5 W	144 W	158 W

Table 3. Comparison of high-frequency and conventional lighting units provided with two fluorescent lamps.

An additional advantage of the electronic ballast is that it can reduce the total power needed to light an air-conditioned area. Since the total power for lighting is converted into heat, the use of electronic ballasts will reduce the load placed on the air-conditioning system.

- Light regulation
Practical experience over the years has shown that manual lighting control by switches, as introduced by Edison, does not always result in an efficient use of artificial light. People are not educated to switch off light as daylight levels increase or when leaving the room.
The introduction of electronic ballasts, however, opens the way to eliminating many of the problems associated with the management of energy in lighting installations using automatic controls. Light regulation is no longer left to the initiative of the individual user. The level of artificial lighting can be adjusted to complement the amount of daylight entering via the window. The presence of occupants in an office can be detected, and switching off or dimming the light shortly after the occupants have left the room, is now possible. The lighting control can also be integrated (with the help of a computer) into the management of the total environment of a building.

- Elimination of hum and disturbing light fluctuations. The sensitivity of the human eye and ear depend on the frequency of light and sound. In the case of electronic ballasting, the operating frequency is always chosen far above the sensitivity limit so that hum and disturbing light fluctuations are avoided. Also fluorescent lamps can be started without the flicker phenomena occuring with conventional lighting units provided with glow-switch starters. In electronic control gear, starting is performed after preheating of the electrodes. The warm start guarantees starting without flicker and prolongs the switching life of the lamp, thereby minimizing blackening of the lamp ends and improving maintenance of the light output during the life of the lamp.

High-frequency operation of light sources also modulates
the infrared radiation. A correctly chosen supply frequency
is therefore important to prevent interference with
infrared receivers for remote control, such as are used in
television sets and video recorders. The frequency must
be above the audible limit but not so high as to
interfere with infrared receivers. The advantage of a low
supply frequency is that electromagnetic radiation effects,
which decrease with a fall in frequency, can be kept below
relevant international standards. However, there is no
general agreement about the choice of the operating
frequency.
Electromagnetic interference via the mains wiring can be
eliminated by a low-pass filter. Other parts of the
electronic ballast (Figure 5) are the rectifier, needed to
supply the correct d.c. voltage to the transistors and to
the electrolytic capacitor.

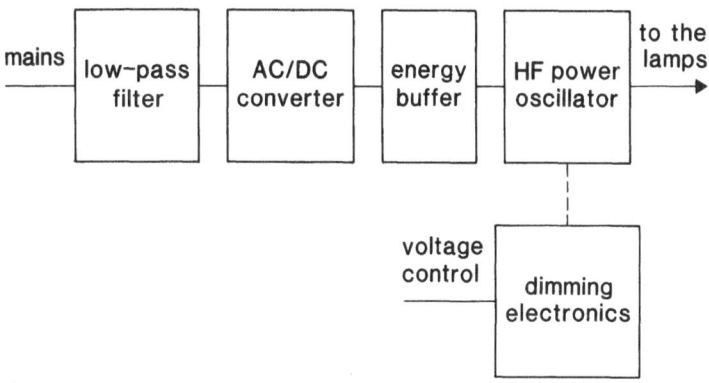

Figure 5. Block diagram of a high-frequency ballast.

This buffer capacitor provides energy to the light source
during those moments that the rectified a.c. voltage is
unable to do so. This absence of current zero periods also
serves to suppress any stroboscopic effect that would
otherwise be present.

3.2 Hybrid circuits

The price of a full electronic solution to delivering all
the control functions for the operation of discharge lamps
is in several cases prohibitive. However, using electronics
in combination with the conventional copper-iron control
gear in a hybrid circuit for selected functions or to
supply additional functions can be profitable. Examples of
partial electronic solutions are the hybrid circuits for
the low-pressure sodium lamps and the high-pressure white
sodium lamps.

Low-pressure sodium lamps

During the operation of low-pressure sodium lamps on sinusoidal currents the ground state of the sodium atoms is depleted for a large part of the period near current maximum. Operation on square-wave currents diminishes depletion of the ground state, allowing the luminous efficacy to be increased by up to 10 per cent. A partial electronic solution is found by separating the functions of the conventional auto-leakage transformer. The ignition and reignition is performed electronically; the current stabilisation is performed by a linear inductance and a saturating inductance in combination with a capacitor introducing third harmonics in the current waveform. In this way the ideal square-wave current waveform is approximated. The electronic starter provides the necessary high-frequency pulses for initial ignition and for reignition during every reversal of the current. Not only is the hybrid control gear much lighter and smaller in volume, the power dissipation in this lighting unit for the same luminous flux is greatly reduced (Table 4).

Lamp type	SOX-E 36 W	SOX-E 36 W	SOX-55 W
Ballast type	Electronic	Hybrid	Auto-leakage
Nominal lamp power	35 W	35 W	56 W
Nominal mains power (warm)	40 W	46 W	76 W

Table 4. Comparison of high-frequency and conventional lighting units provided with low-pressure sodium lamps.

High-pressure white sodium lamp

Recently, new types of high-pressure sodium lamps with maximised colour rendering properties were introduced. The life of these white high-pressure sodium lamps are mainly determined by the fall-off in the quality of the light emitted. Production tolerances on these lamps and their ballasts, and deviations of the mains voltage from the nominal value would critically influence their performance were no countermeasures taken. This is due to the fact that these high-pressure sodium lamps are operated at saturated sodium and mercury vapour pressures. Small changes in the coldest spot temperature are strongly correlated with fluctuations in the vapour pressures, and thus with the quality of the light emitted. This problem is solved by employing hybrid control gear in which the lamp voltage and current are electronically controlled (figure 6).

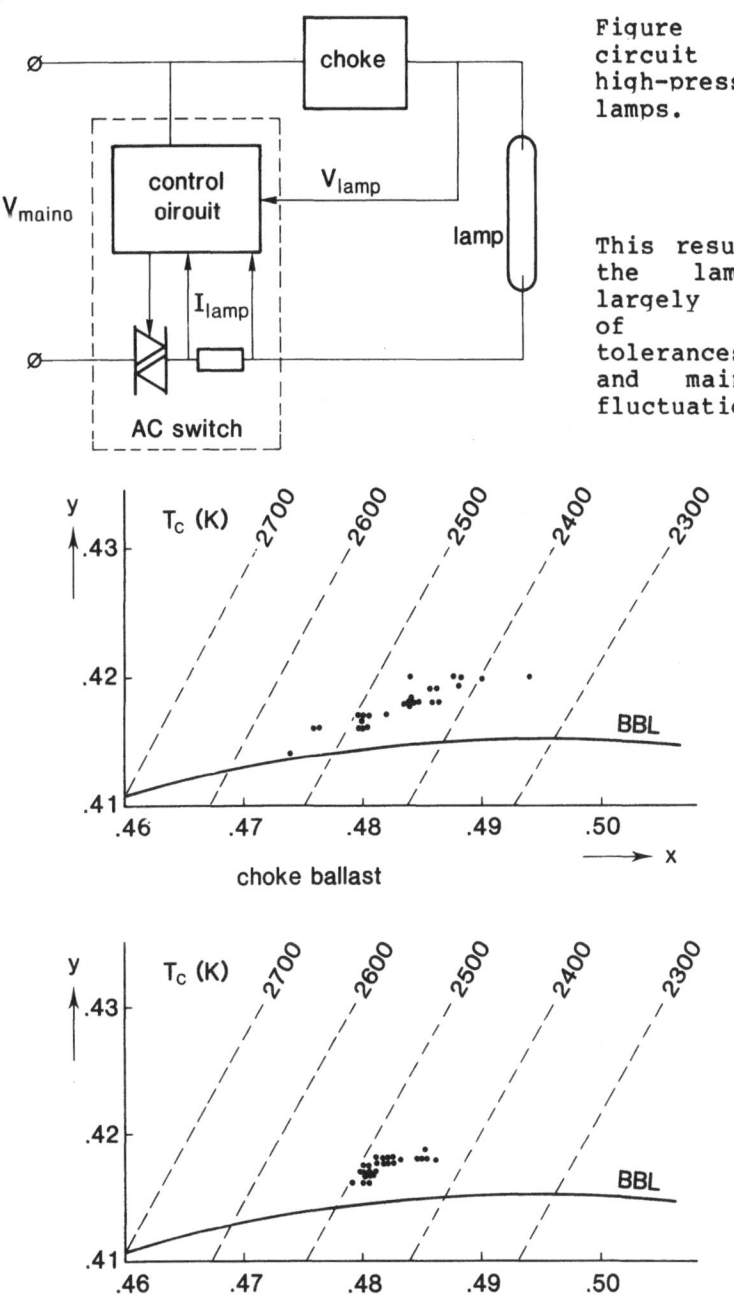

Figure 6. Hybrid circuit for white high-pressure sodium lamps.

This results in that the lamp becomes largely independent of production tolerances (figure 7) and mains voltage fluctuations.

choke ballast

ballast with control circuit

Figure 7. The spread in colour coordinates of a series of 50W white high-pressure sodium lamps if stabilisation is performed by a) a choke ballast, b) hybrid ballast.
The hybrid ballast largely diminishes the spread in colour coordinates.

4. INTEGRATION

Miniaturization and electronic control of the light source facilitate the physical and functional integration of the various components of the lighting unit. Until the seventies the development of the lighting unit was more characterised by optimization of the various separate components of the lighting unit. The interaction between the various components, such as the light source, starter, ballast switch and luminaire were small.

This situation has changed during the last decade. Not only are existing components in the lighting unit integrated, but the lighting unit itself has become part of the lighting system.

4.1 Integration of lighting unit components

The opinion is growing that the efficient use of energy for lighting is not just a matter of getting the highest luminous efficacy from a light source so as to minimize the installed power for lighting. It also involves employing as the wise use of the luminous flux by employing an efficient luminaire. The luminaire forms an integral part of the lighting unit. An efficient luminaire brings the luminous flux of the light source to the right place to give the correct illuminance without glare. It is outside the scope of this paper to go into more detail on this subject.

Building the various components of the lighting unit into one housing makes the physical, electrical and mechanical interactions between the various components much stronger. Thermal problems, in particular, can arise such as occur in controlling the mercury vapour pressure in compact fluorescent lamps and avoiding the premature failure of electronic components.

Control of the mercury vapour pressure

For compact fluorescent lamps (figure 8) where the starter, control gear and light source are integrated to form one lighting unit, it is extremely difficult to realise the optimum mercury vapour pressure needed for the efficient generation of radiation if no special measures are taken. In such compact fluorescent lamps use is made of an amalgam such as In-Bi-Hg. The optimum mercury vapour pressure above the amalgam is now reached at a far higher temperature of $90°C$ instead of $40°C$ above pure mercury. In addition, the temperature range within which the lamp operates close to its optimum mercury vapour pressure, is significantly increased. When the lamp is switched on at room temperature the presence of an amalgam results in a lower light output than would have been the case with only mercury.

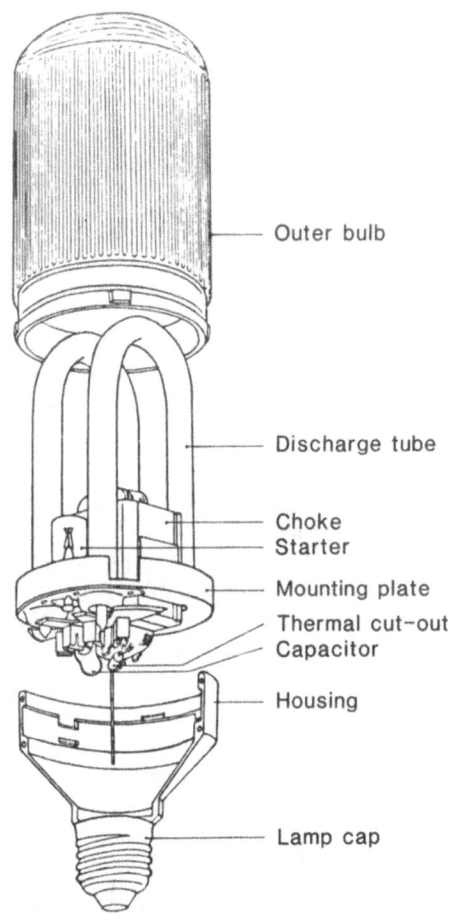

— Outer bulb

— Discharge tube

— Choke
— Starter

— Mounting plate

— Thermal cut-out
— Capacitor

— Housing

— Lamp cap

Figure 8. An exploded view of a compact fluorescent lamp. The small diameter tube is folded and integrated into one unit.

Thus, an auxiliary amalgam system is often necessary to generate a sufficiently high mercury vapour pressure immediately after ignition. Only with a well controlled mercury vapour pressure is optimum generation of radiation in the discharge possible.

Life time and failure rate of control gear

In general the failure rate of control gear as a function of operating time is characterised by three periods:
- an initial failure rate due to manufacturing defects; these defects are easily recognized during preoperation;
- a failure rate during actual use; the quality of the active and passive components of copper/iron ballast is good enough to hold the failure rate low during this this period, even if the operating temperature is increased.
- an increasing failure rate at the end of life time, which value is strongly temperature dependent.

The expected median life time of copper/iron ballasts provided with polyesterimide insulated wire is shown in figure 9 by an Arrhenius plot. The same type of plots describe the temperature dependence of median life time of passive and active components in electronic control gear.

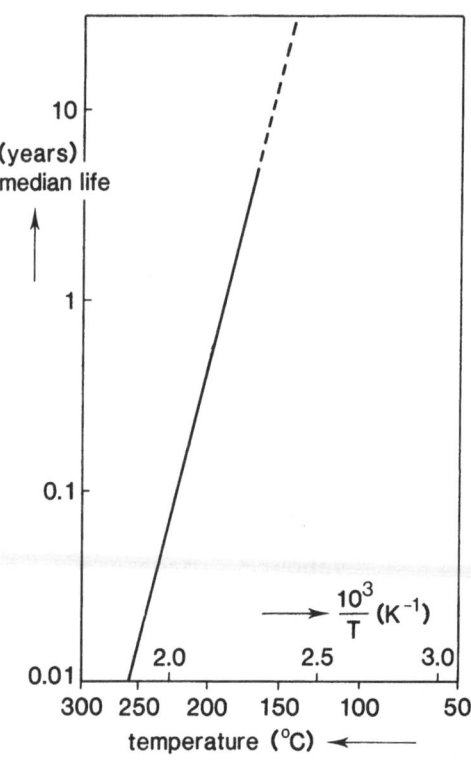

Figure 9. Relationship between median life (50% failure rate) and temperature of coils with wound windings using polyesterimide insulated wire.

The most critical component in the electronic control gear is the electrolytic capacitor, its medium loss by evaoration being strongly temperature dependent. It will be clear that physical integration of light source and control gear calls for a design that minimizes the thermal stress on the components of the lighting unit to extend its life time.

4.2 Integration of lighting unit in lighting system

Up to now we have only spoken about the energy efficiency of the lighting unit by:
- improvement of the luminous efficacy of the light source, for example by h.f. operation;
- replacement of less efficient light sources by new ones: incandescent by fluorescent lamps;
- reduction of the control gear losses by h.f. operation;
- enhancement of the lighting efficiency of the luminaire.

However, energy efficient lighting also includes the efficient use of the installed power at the right time. For energy is the product of power and time. This can be achieved by integrating lighting units in a total lighting system. An example of a lighting system is shown in the block diagram of figure 10. The luminous flux of groups of lighting units can then be controlled:
- centrally and automatically by the control unit, which serves the automatic functions such as switching at preset times,
- by sensors that deliver an input to the control unit. The luminous flux delivered by the lighting units varied in response to the level of day-light available, as monitored by the sensors.
- manually by the use of infrared transmitters and receivers. Manual control can overrule automatic control so that the user can create an optimum lighting level to work by.

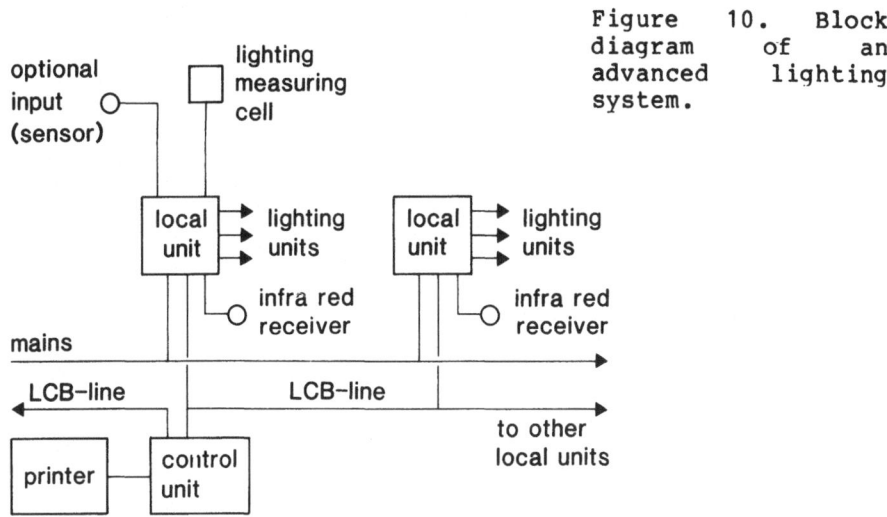

Figure 10. Block diagram of an advanced lighting system.

Energy savings brought about by adaptating the artificial lighting to the available daylight, changing visual tasks conditions and occupancy patterns are very dependent on the local situation. General rules cannot therefore be given.

The benefits of such lighting systems are not only the reduction in lighting costs of the lighting installation. It also affords more comfort and flexibility to the user, and simplifies the design of the installation as a whole. One example is the flexibility with respect to functional changes in the use of the internal space of buildings, because switching cables become superfluous if infrared remote control is used.

5. SUMMARY AND OUTLOOK

Driven by the energy crises in the seventies and the development of modern electronics, the lighting unit has changed enormously during the last two decennia. The basic ideas underlying these changes were not only to increase the efficiency of light source and control gear, but also to improve the efficient use of the lighting unit by bringing the luminous flux to the right place at the right time. Miniaturization and the use of electronics in the light source were essential for the realisation of these ideas.

The development of electronic control gear will continue. It is expected that the development of new high-power electronics will also make it possible to operate high-pressure discharge lamps on currents widely different in frequency and waveform to those presently used.

In the office of the future, and later on in our homes, the conventional switch introduced by Edison could be completely removed from the lighting unit and replaced by automatic daylight-controlled switching or infrared remote control.

The new developments will be characterised by a more integral development of lighting units, which are easily incorporated into a multi-functional system.

I wish to thank many of my colleagues for their support in making this paper possible.

THE DEVELOPMENT OF NEW EFFICIENT ELECTRICITY END-USE EQUIPMENTS AND PROCESSES IN THE FRENCH INDUSTRY

Denis FOUQUET and Maurice ORFEUIL

SEPAC, Direction Générale d'Électricité de France

1. THE FRENCH ENERGY CONTEXT

Since the 1973-74 oil shock, the french energy policy has been rather consistent throughout the years and clearly stated.

This policy includes two major components :
– a substantial energy conservation program,
– an intense effort to diversify energy sources and reduce the dependence on imported oil (in 1973, 3/4 of the energy consumed in France was imported and oil accounted for 85% of it).

For electricity, this energy policy has meant :
– on the production side :
. the conversion of existing thermal power stations from oil to coal, whenever possible,
. the initiation of an important nuclear plants program (40 000 MW commissioned in the last 10 years).
– on the demand side :
A new strategy which could be summarized by these two guidelines :
. less energy per use,
. more uses of electricity.

In more quantitative terms, it means that, thanks to the success of EDF nuclear program and to its plants using hydraulic power and coal, total electricity sales should be in France in the range 350-370 TWh by 1990 against 171 TWh in 1973 or, more significantly, that they should exceed by about 50 TWh (30 TWh of which in industry) what is considered as the «natural trend», i.e. the trend which would have probably been observed in the absence of a marketing policy aimed at electricity penetration.

Two other major objectives of this electrification program should also be stressed :
– make french industry more competitive and profitable through the use of highly efficient electrotechnologies ;
– strenghten the position of manufacturers of electricity using equipments.

To achieve these goals, a comprehensive marketing and R & D program has been implemented by EDF. This paper will mainly concentrate on the development of efficient Electricity End-Use equipments and Processes in industry.

2. AN APPROACH WHICH SUITS THE MULTIPLICITY OF TECHNOLOGIES AND APPLICATIONS

Electricity is versatile, even protean. It has a remarkable property which strongly differentiates it, for its use, from fossil fuels. It can be mobilized in extremely varied forms – induction, resistance, conduction, infrared, arc, plasma, laser, microwave, motive power, electrolysis... – and these can be harmoniously combined.

It now offers a vast range of technologies which can satisfy the energy requirements of companies, particularly the thermal ones. But every rose has its thorn and this multiplicity of possibilities, a substantial advantage over conventional fuels, also implies, to cope with the acceleration of technological progress, considerable efforts toward innovation to make industrial structures evolve, to modernize existing technologies, to develop new ones and to disseminate them in the various sectors of industry.

Under these conditions, to incite innovation in electricity applications, it is necessary to combine two complementary approaches :
– The improvement of existing electric equipments and the development of new ones, particularly intended for «unit operations» (concentration, melting, drying...) which are elementary links in production chains ; this action means developing equipments based upon the major electric technologies – electromagnetic radiations, resistances, plasma, heat pumps... – and to adapt them to the processing of various products.
– The analysis of energy systems, and their optimization ; the purpose of this action is essentially to study how new processes may be designed thanks to inserting suitable electric technologies into various

A. T. De Almeida and A. H. Rosenfeld (eds.), Demand-Side Management and Electricity End-Use Efficiency, 367–381.
© 1988 by Kluwer Academic Publishers.

production phases, analyzing the interactions among the various unit operations, and optimizing these new production systems from an economical point of view. This work concerns, for example, the analysis of the consequences of introducing electric technologies into industries whose thermal processes are based upon the use of steam or the technical and economical implications of inserting plasmas in steel production lines.

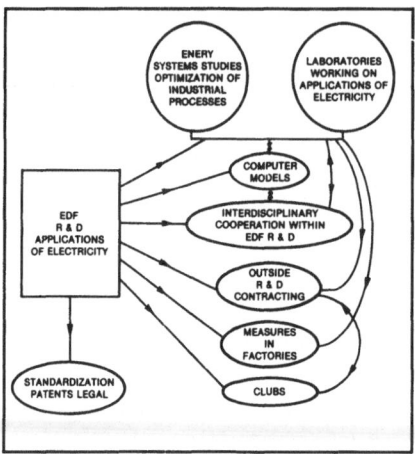

FIGURE 1. EDF approach to R & D for industrial electrotechnologies.

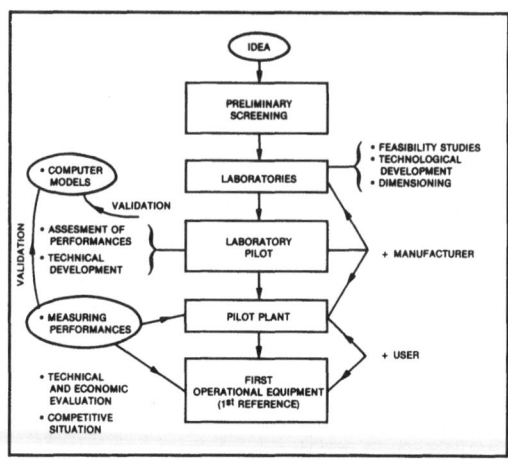

FIGURE 2. From an idea to an operational technique.

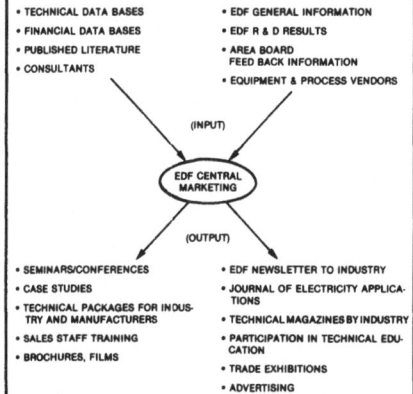

FIGURE 3. Marketing programs to promote the use of efficient electrotechnologies in industry.

Figures 1 and 2 summarize this EDF approach to the development of new technologies in the field of Electricity Applications. Needless to say, there are very strong links between these R & D efforts and marketing programs designed to promote the use of highly efficient electro-technologies as shown on figure 3.

Among all these actions to develop new processes, two keys points should be stressed : .

– Efficient interdisciplinary cooperation : heat science and electrical engineering are essential in applications of electricity. It quickly became clear however that progress would be impossible without the contribution of other fast developing disciplines such as electronics, material science, automation, numerical analysis, computer science... and their integration into electricity using equipments.

So, for instance, replacing steam with electricity for motive power in industries such as chemistry or refining are only at their very beginning because it is only recently that high power, variable speed electric motors have become available which can technically and economically compete with steam turbines. It is the progress which has occurred both in motor design and in the performance and reliability of power electronics (particularly on thyristorized static frequency converters) and control electronics which have made it possible to give industry motors with variable speed, high power and high speed. Since 1984, a number of motors from 10 to 15 MW, turning at speeds from 3000 to 6000 rpm, with variation fields of approximately 1000 rpm, have been placed into service, particularly for driving compressors in the petrochemical industry. Motors with higher powers, one at 33 MW and 3780 rpm, which should be the most powerful in the world, have recently been designed and put into service.

– Very close collaboration between the partners concerned : the success of this innovation work will be the fruit of intense collaboration between all the partners concerned – equipment manufacturers, engineering firms, users, Électricité de France, Public Authorities, financial institutions and researchers of all kinds –, of a good overall organization of this work and of a good amount of imagination.

The recent development of a new induction boiler, baptized the thermo-inductive steam generator, is a typical example of the possibilities opened by this collaboration between various partners. The thermo-inductive steam generator, is designed using the same architecture as for a transformer. It consists of a conventional magnetic circuit, generally three phase, on whose cores there is a primary winding made up of wires with dry insulation, directly powered at mains voltage, plus a secondary winding consisting of stainless steel tube coils. This secondary is short circuited and the actual heating body. So it is a heating by indirect induction because the fluid contained in the tubes is heated by conduction and convection by thermal energy exchange between the walls of the tubes which are themselves heated by the inductive currents. The secondary circuit is formed by a number of elementary tubular circuits connected hydraulically and coupled electrically by thyristors, which enables a very fine regulation and easy control.

FIGURE 4. 3000 kW thermo-inductive generator for the production of superheated or pressurized hot water, steam and other fluids (Compagnie Générale de Chauffe - Alsthom Doc.).

FIGURE 5. Cracked gas compressor driven by a 14.2 MW variable speed motor (4200 to 5100 rpm) replacing a steam turbine (Jeumont-Schneider Doc.).

The advantage of this operating principle is that it unites in a single modular, very compact and easily transportable apparatus, the power supply transformer and the boiler itself. The manufacturing technique for this equipment is very similar to that used for transformers.

The development of this original and efficient boiler is mainly due to the close cooperation between a transformer manufacturer and a firm specialized in the operation of thermal installations with the support of EDF, each of which contributed its experience and know-how to design a totally new product, easily adaptable to the requirements of various markets (production of hot water and steam in industrial processes, reheating of liquid and, probably later, gaseous fluids which may be thermal or reactional, heating of residential premises, and space heating in the tertiary or industrial sectors...).

This cooperation has also been very effective at all stages of the marketing process — market research, segmentation, promotional strategy, sales development...

3. DEVELOPING EFFICIENT EQUIPMENTS AND PROCESSES : A TRIPLE STRATEGY

Three main routes can be taken to develop these new electrotechnologies.

3.1. Modernizing existing techniques

Improving the technical and energy performances of existing equipments or processes often makes them more competitive with modest research and development investment.

This modernization can take very diverse paths, but will often consist of modifying existing equipment by the integration of components with higher performances or lower costs, developing their automation by having recourse to systems controlled by microprocessors or programmable automatons, and improving their quality and reliability thanks, for example, to using computer codes which guarantee a fine representation of the physical phenomena and which provide a better mastery of their sizing or their operation in dynamic operating mode.

Many examples can be mentioned in applications of induction heating thanks to the development of magnetothermal computer codes, to the emergence of new static frequency converters based on more potent electronic components and assemblies or to the design of new coils strongly increasing the efficiency of this type of heating. The improvement of electric boilers, of high power hot gas generators or of resistance furnaces also falls into this category.

3.2. Transposing electrical techniques existing in one sector into other sectors

Electrical techniques are often widespread in certain sectors, but totally unknown in other branches of activity, either for historical reasons (the industry considered developed around a process composed of interdependent individual operations which at the time had a certain coherence) or, more often for various economic reasons.

The evolution of the absolute level and of the relative magnitude of electric energy and fuels prices together with the progress achieved by techniques which did not seem attractive enough at an earlier time can however lead to a reexamination of previous situations.

Mechanical steam compression for concentration for example first developed in the dairy industry at the end of the 70 s. Conditions in this industry were favorable : substantial quantities of materials to be concentrated, specific consumption of fossil fuels rather high when using conventional concentrators, intensive research on this technique, peak activity often taking place during EDF summer low rate period, hygiene and quality restrictions...

Due to their energy and cost savings ability, major research projects have been undertaken to transpose these techniques to other industries and to bring technological improvements to the equipments.

Many other examples of inter-sector technology transfers can be given such as the adaptation of infrared radiation to the treatment of new materials, extending from active carbon to composites and including the most diverse plastic coatings or the treatment of papers, textiles and paints and the transfer of induction heating techniques, a now conventional process in metal transformation industries, to the heating of tanks, reactors and concentrators in the chemical or food industries.

3.3. The emergence of very innovative techniques

Modernizing electric techniques and disseminating them throughout industry while adapting them to the needs of each firm should not lead to sterilize the search for deeper innovations. These can be limited to individual unit operations or involve complete production processes whose various phases are considerably modified.

For example, in the first family can be placed the development of high temperature heat pumps with a condensation temperature from 120 to 130°C instead of 55 to 60°C for the conventional heat pumps and cooling equipments from which they are derived, the design of «transverse flux» induction coils which give excellent efficiency when heating sheetmetal with little resistivity (aluminium, copper) while heating the same materials by a solenoid coil is impossible or provides mediocre efficiency, and the industrialization of microwave or high frequency dielectric heating techniques.

In the second family are techniques which might bring production processes into deep question such as the plasmas or the modern separation processes.

These different points will be better illustrated by examining a few examples of recent developments.

4. DEVELOPMENT OF NEW ELECTRIC END-USE EQUIPMENTS AND PROCESSES : A FEW SELECTED EXAMPLES OF PROMISING ROUTES

4.1. The development of dual energy systems

The «Dual Energy» concept can be in many instances an appropriate answer for better demand side management since it is a logical consequence of the increase in the differentiation of electricity prices throughout the year which themselves reflect the differences between the marginal costs of production and distribution at different time periods. Figure 6 shows, for example, the structure of proportional prices for an optional tariff which is proposed to large customers.

FIGURE 6. High voltage tariff with 8 time periods	WINTER December January, February	MID-SEASON March, November	SUMMER April, May, June September, October	July, August
From 9 to 11 a.-m. and 6 to 8 p.-m., monday to friday	Peak hours PH			Off-peak hours JA
	Normal hours NHW	Normal hours NHM	Normal hours NHS	
From 1 to 7 a.-m. monday to friday + saturday + sunday + public holidays	Off-peak hours OHW	Off-peak hours OHM	Off-peak hours OHS	

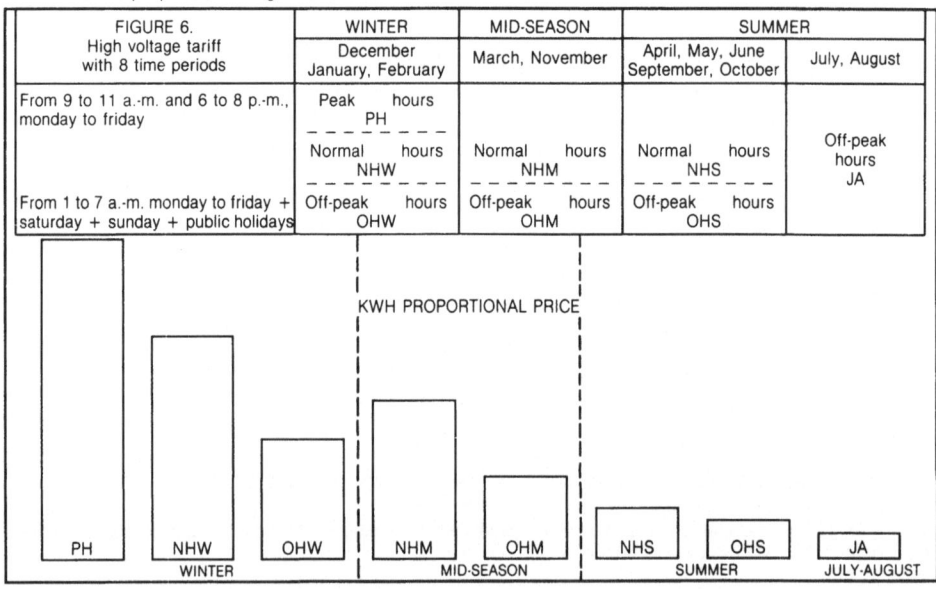

It can therefore be profitable for an industrial energy consumer to invest in a new equipment which will use electricity when its price is advantageous while the existing equipment will burn a storable fuel when the price comparisons show it is favorable.

For the last five years, EDF has promoted this «Dual Energy» concept which has been an instant success since more than one third of the electricity consumption increase comes from dual energy systems ; they rely on different technologies, mainly :
. Electric boilers : electrodes, resistances, induction...
. Hot air generators for atomizing towers, drum dryers...
. Plasma : generally coal electricity dual energy...
. Heat pumps with simultaneous or alternative dual energy.

A broad range of hot gas generators has, for example, been developed. They use different types of heating element — metal tubes, metal sheathed metallic resistances, specially cut metal sheets, sheated graphite elements... the most powerful presently operating generator is rated at 25 MW and used for drying alfalfa and sugar beet pulps. Most of these equipments have been commissioned in the food and chemicals industries. It should be pointed out that, beyond energy savings aspects, other benefits came from the use of electricity, especially increased production capacity and productivity thanks to air temperatures which are higher than those obtained by the steam boilers previously existing in the plants.

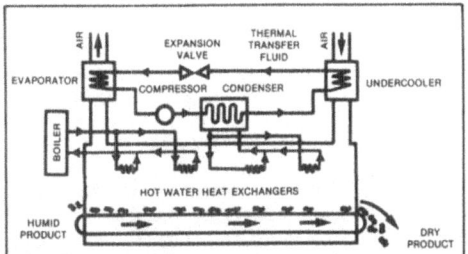

FIGURE 7. Air-water heat pump and simultaneous dual energy for drying.

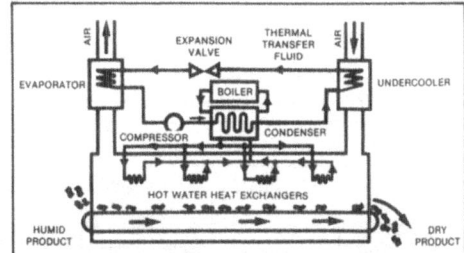

FIGURE 8. Air-water heat pump and alternate dual energy for drying.

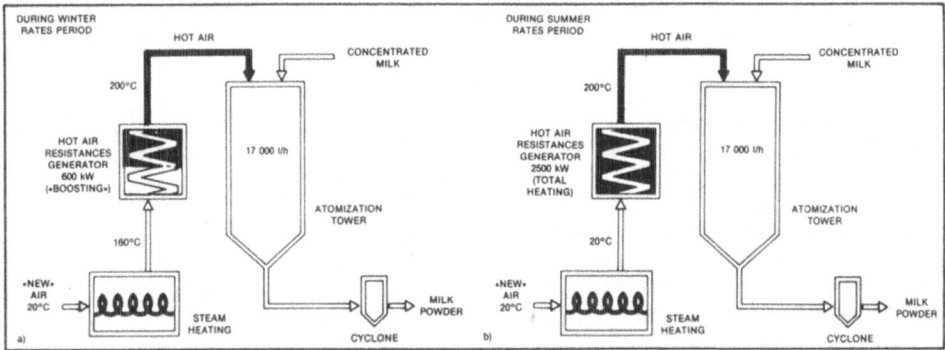

FIGURE 9. Atomization tower for milk drying with an electricity-steam dual energy system.
a) during the winter rate period, only 25% of the resistances are connected to the mains and the electric generator is used to increase the air temperature, which is limited by the available steam temperature, and therefore the production.
b) during the summer rate period, all the resistances are connected and no steam is used for drying air reheating.

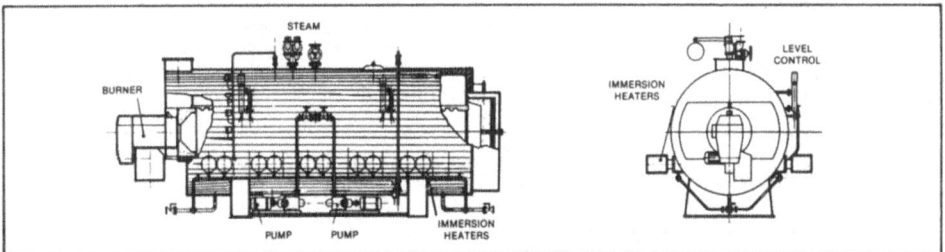

FIGURE 10. Electric and oil fired steam boiler.

372

4.2. The optimization of the liquid materials concentration and drying processes

4.2.1. A few ways to reduce concentration-drying costs using electricity

The high cost of dehydration by thermal means is linked to the large quantity of energy which this operation requires. Its reduction therefore involves the development of high performance processes for saving energy and optimizing the series of operations leading to a dry material.

The concentration of a material by mechanical means (filtration, pressing, centrifuging) consumes much less energy than its concentration by thermal means. Furthermore, processes of concentration by thermal means such as mechanical steam compression or multiple effects which enable the recovery of all or part of the latent heat in the evaporation of materials generally absorb less energy than drying which in most cases is related to a single effect process without recovery of latent heat.

To reduce the costs of dehydrating liquid materials, one path therefore seems promising, whenever it can be used ; first of all push the mechanical processes for concentration as far as they can go both technically and economically, optimize the exploitation of high performance thermal concentration techniques, develop final drying processes which are adapted to a high initial concentration of the materials to be dried. Furthermore, superconcentration can give birth to new highly concentrated products which will be marketed in the place of ordinarily dried products.

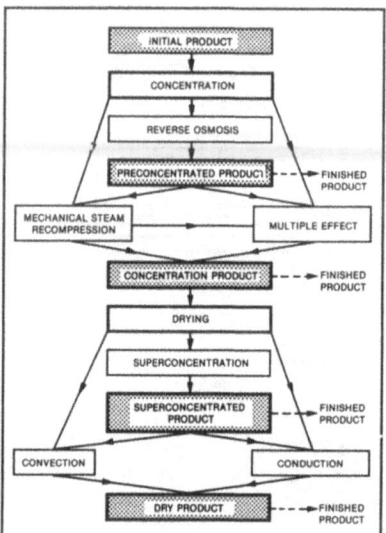

To reach this objective, one must attempt to optimally insert into this concentration-drying system techniques like :
- separation on membranes such as reverse osmosis, ultrafiltration or microfiltration which only use mechanical energy,
- the pairing of mechanical compression of vapor and evaporators,
- thin film centrifugated superconcentrators,
- drying cylinders using resistance or induction heating, or steam generated by a mechanical compression energy recovery system,
- flash dryers and dual energy systems.

Furthermore, all these techniques could play an important role in the emergence of new activities such as biotechnologies or, in the case of membranes, in operations where they are now practically absent as the separation of gases constituents.

More attention will be given to mechanical steam compression since this technique has been rather successfully implemented in french industry during the last ten years and more developments are still to come.

FIGURE 11. General diagram of an optimized concentration and drying process.

4.2.2. Mechanical steam compression (MSC) in industrial processes

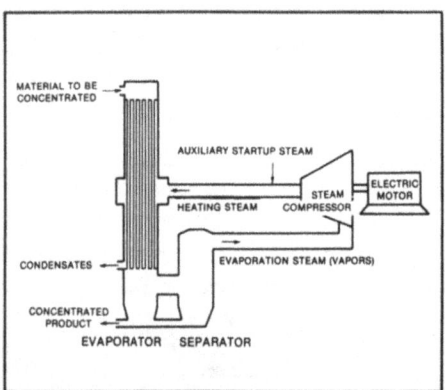

This technique uses as heating steam for an industrial process the evaporation vapors produced by that process during its operation. The vapors are collected and compressed by a steam compressor in order to increase their temperature and allow an efficient heat exchange. The mechanical energy absorbed by the compressor is only a small part of the heat energy exchanged during the condensing of the heating steam. This technique is analogous to the heat pump one with the particularity of using the solvent from the process itself (for example the water steam) and not an auxiliary thermal transfer fluid.

FIGURE 12. Single effect evaporator with mechanical steam compression.

4.2.2.1. Mechanical compression of steam for concentration, crystallization and distillation processes

Mechanical compression of steam has achieved a high rate of growth for concentration by evaporation of liquid materials and, at a lesser extend, for crystallization and distillation. This success can mainly be attributed to its very high energy efficiency compared to more traditional multiple effect concentrators as shown on figure 13, but also to the intense development work which has been done to optimize the equipments and to adapt them to the treatment of various products and to important marketing efforts.

EQUIPMENT TYPE	ENERGY CONSUMPTION PER TON OF WATER EVAPORATED	
	STEAM kg/t	ELECTRICITY kWh/t
Single effect	1050	10
Double effect	550	10
Triple effect	350	< 10
Single effect with thermocompression	550	< 10
Double effect with thermocompression	350	< 10
Triple effect with thermocompression	280	< 10
Single effect with MSC	10 to 20	15 to 30
Double effect with MSC	10 to 20	11 to 25
Triple effect with MSC	10 to 20	11 to 25

FIGURE 13. Specific consumption of energy for concentrators with multiple effect evaporators and mechanical steam compression.

For concentrators, temperature differential are generally small, less than 10°C, and the coefficients of performance are very high, in the order of 20 or more. It should also be pointed out that the electricity consumption of multiple effect concentrators cannot be neglected in comparison with those of MSC processes.

The principle of crystallization by evaporation is the same as the one for concentration, but its special features is that highly concentrated solutions are involved and the boiling temperature lag can reach 20°C or more.

In these crystallization processes, in order to limit the heat exchangers surface, temperature differences of about 25 to 50°C between the heating steam and the product are used. The coefficient of performance is therefore lower than for MSC concentration, but still exceeds 10.

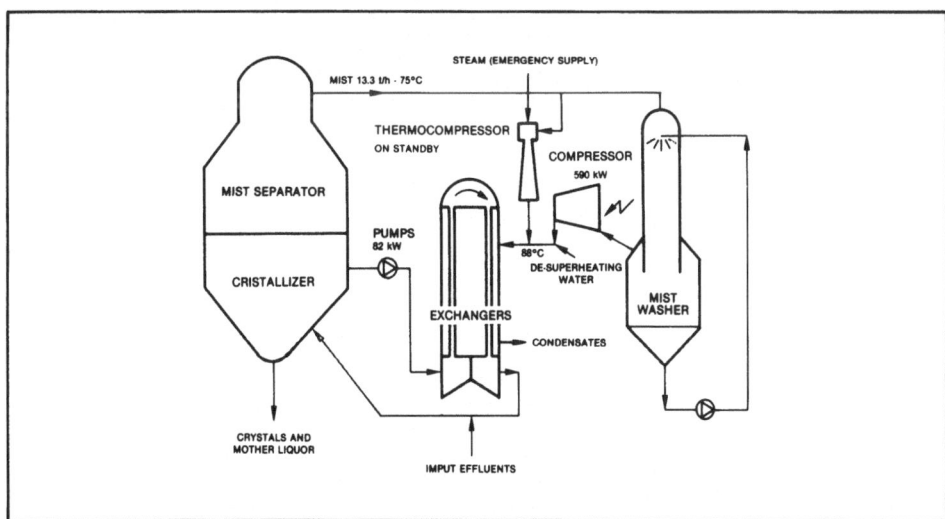

FIGURE 14. Crystallizer with mechanical steam compression.

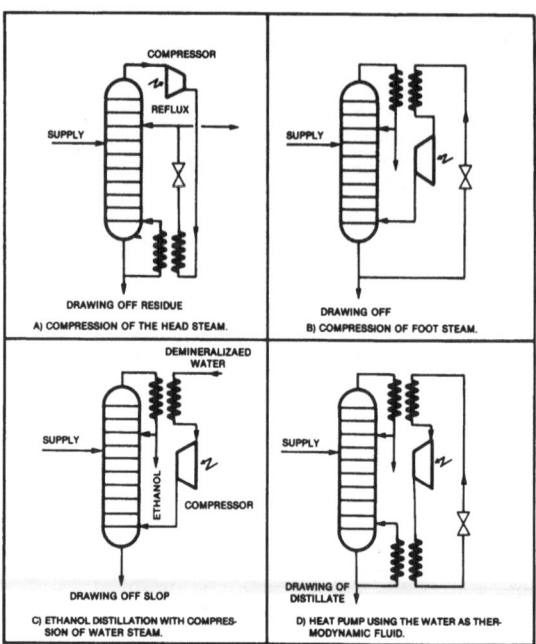

FIGURE 15. Distillation with mechanical steam compression.

For distillation, MSC is mainly used on columns for distilling or rectifying ethanol (beet alcohol) and on certain columns in the chemical industries where the temperature differences between the foot and the head of the column are not very great. The adaptation of MSC to existing distilling columns generally requires to change the exchange surfaces of the condensors at the head and sometimes those of the boil-off units at the foot of the columns, as well as the operating pressures and temperatures, which entails considerable investments. However, the energy savings is frequently very high and leads to rather low payback times.

Various configurations are possible. Compression of vapors at the foot or at the head, compression of water steam produced by exchange with the head vapors. The last two configurations are to be considered when thermodynamical, physical or chemical reasons prevent the direct compression of head vapors.

Figure 16 gives, to illustrate the economical aspects of MSC processes, a few data about the first french paper mill which has been equiped with a MSC system to concentrate black liquors.

FIGURE 16. Concentration by evaporation of black liquors.	Earlier system		Present system	
	Triple effect and ejection-compression	Triple effect finisher	MSC	Triple effect finisher
Evaporated water quantity (t/h)	20	21	27	14.5
Steam consumption (t/h)	17.5		7	
Electricity consumption (kWh/h)	120		850	
Annual operating cost saving (kF) (8000 h/year, 0.25 F/kWh) • 115 F/t of steam • 70 F/t of steam			8200 4400	

When this investment has been decided in 1983, it was the first MSC for black liquors concentration and the gross pay back time was about 20 months. With present oil and electricity prices, the payback time would be about 3-4 years, but proper financing of the project, for example energy prices variation risk sharing with EDF, would substantially increase its profitability.

By the end of 1986, there was about 90 MW of MSC processes in operation in France for concentration, crystallization, distillation and miscellaneous operations and they contribute to save at least 300 000 t of imported oil per year.

MSC processes are mainly used in the following industries :
- Dairy industry : concentration of whey and milk products.
- Sugar industry : concentration, boiling and crystallization of solutions, distillation of alcohol.
- Beverage industry : boiling of wort in breweries, distillation of alcohol, crystallization of mineral water.
- Canning industry : concentration of tomato products.
- Pulp and paper industry : concentration of black and spent liquors.
- Wood industry : concentration of effluents.
- Mining and minerals : concentration of waste and crystallization of effluents.
- Chemical industry : distillation and concentration of solutions, concentration of effluents.
- Biotechnologies : concentration of solutions.

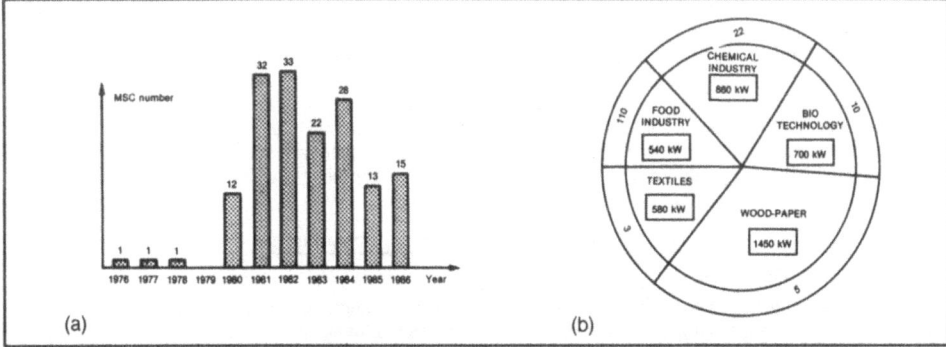

(a)   (b)

FIGURE 17. (a) Yearly flow of new mechanical steam compression equipments. (b) Repartition of Mechanical Steam Compression Equipment per industry and average power (☐ average power).

4.2.2.2. Mechanical steam compression for drying

For liquids and the previously mentioned operations, the use of mechanical steam compression is rather well mastered today. For drying, the development of techniques based on mechanical steam compression is quite recent.

Traditionally, industrial drying has been a major energy consumer and has required high temperature differences. The installation of MSC has therefore been made difficult by the high compression rate required. In spite of this, the introduction of double screw or multi-stage centrifugal compressors seems to be able to bring a remedy to this problem. Drying whith superheated steam can illustrate these new developments of MSC.

• Drying with superheated water steam

The originality of this process is the use of the superheated water steam as the drying fluid, which makes it possible to place a steam compressor in the circuit. In fact, this means a technological transfer to drying of the principles applied to the concentration of liquids by evaporation.

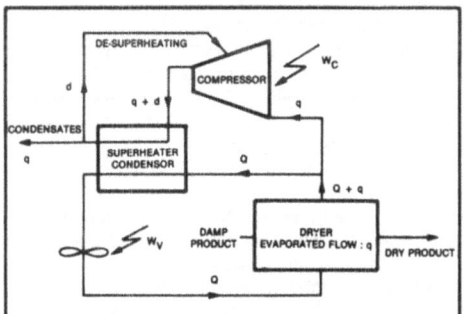

FIGURE 18. General diagram of drying in superheated water steam.

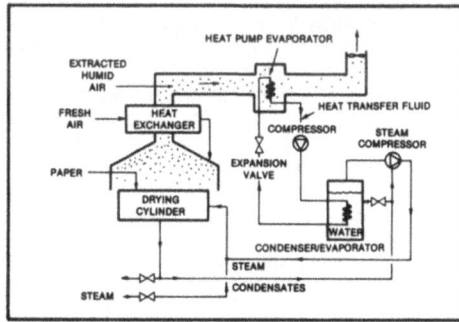

FIGURE 19. Drying cylinder with heat pump and mechanical steam compression.

The water steam circulates on closed loop in the installation. From the superheated steam, flow Q is introduced into the dryer. Upon contact with the damp product to be dried, the steam de-superheats, giving up a part of its «sensible heat», and thus vaporizing the water contained in the body to be dried. At the outlet from the dryer, a quantity of saturated steam equal to the quantity of water evaporated from the product, flow q, is branched off to the compressor.

Once it is compressed, this steam is routed to a condenser-superheater in which, as it condenses, it releases the quantity of heat necessary to resuperheat the saturated steam coming back from the dryer. A fan circulates the steam in the installation.

An industrial equipment has been built for drying beet pulps and alfalfa. The evaporated flow is closed to 20 t/h for a consumption of 180-200 kWh/t of evaporated water. Many other applications can be imagined : wood, ore, chemical, food industry or household wastes, textiles...

● Other developments in MSC drying

Other developments are now in progress to adapt MSC to drying processes. For example, it could be profitable to associate a mechanical steam compression and an indirect heat pump to recover energy on drying cylinders as shown on figure 19.

The heat pump evaporator is used to recover the energy contained in the humid air extracted above the drying cylinder and its condenser produces low pressure and temperature steam. The compressor of the MSC system rises the temperature level of the steam in order it can be used to heat the drying cylinder. The heat pump and the MSC are both working with fairly small temperature differences and the coefficients of performance and the overall efficiency should be rather high. Developments have now been undertaken and feasibility studies have shown that the gross payback, for example in the paper or the food industries, should be in the range 2-4 years.

4.3. Electric resistance furnaces : from the high performance furnace to the «intelligent» one

4.3.1. A quickly changing conventional technique

Resistance furnaces are the best known and most widespread electrothermal equipment. They nevertheless continue to undergo many improvements, particularly by the integration in these furnaces of the most recent contributions from material science, electronics and monitoring-control.

The use of technical ceramics in the form of insulating fibrous refractories, materials originally developed for the needs of the aerospace industry, have, for example, made it possible to reduce the thermal inertia of the furnaces used in industries such as metal transformation or ceramic products production, by a factor which varies, according to the case, from 2 to 10, reducing by comparable proportions the processing cycle time in discontinuous furnaces, dividing specific energy consumption by 2 or 3 and substantially lowering the cost of electricity consumed thanks to a greater use of energy during off-peak hours.

The most significant progress doubtless comes however from the increase in the power density of furnaces and the improvement of their control systems.

4.3.2. Developing resistance furnaces with higher power density

In a thermal process, high power density is often synonymous with high productivity. Moreover, high temperatures contribute to obtaining high power densities and enlarge the range of applications of resistance furnaces.

In a resistance furnace, once the temperature exceeds 600 to 700°C, the energy transfer from the heating elements to the charge takes place essentially by radiation and is essentially proportional to the difference of the fourth powers of the absolute temperatures of the resistors and of the charge as well as to various characteristics of the resistors and of the charge (surface, form factor,...).

So to increase the furnaces power densities, two main paths have been explored :

− High temperature non-metallic resistors : when heating elements which withstand very high temperatures (1750°C for disilicide of molybdenum, 1600°C for silicon carbide, 1750°C for lanthanum chromite, 2800°C for graphite in inert atmosphere) are used in medium temperature furnaces (900°C) or relatively high temperature furnaces (up to 1300°C), it is possible to install very high power densities, 4 to 10 times greater than what is possible with the usual metal resistors, while guaranteeing a high service life for the elements since they operate at a temperature which is considerably less than the maximum use temperature.

Works conducted in laboratory, then in industrial conditions, in collaboration with furnace manufacturers, heating element makers and industrial users are beginning to bear fruit since a number of furnaces equipped with molybdenum disilicide, silicon carbide or lanthanum chromite resistors have recently come into production for applications such as reheating before forging, heat treatments and the melting of metals or the firing of ceramics. The power densities are often from 30 to 60 kW/m² of furnace wall while conventional furnaces cannot exceed 10 to 20 kW/m² or are even excluded from certain applications which remain the exclusive preserve of fuel furnaces.

Other studies are continuing, particularly to develop original graphite based resistors and to diversify the use of all high temperature resistors, promoting the transfer of these technologies to new applications and new sectors.

− High density power metallic resistors : non-metallic resistors are relatively expensive. So research teams have also tried to increase the power density of metallic resistor furnaces although their maximum working temperature, from 1100°C to 1350°C for the usual alloys, is far below that for non-metallic elements. These projects have made it possible to multiply by 2 to 3 this power density which now reaches 30 to 70 kW/m² of furnace wall for charge temperatures varying from 900 to 1150°C while before, it was difficult to exceed 20 kW/m².

FIGURE 20. Aluminium casting ladles heating using welded tubular resistors has divided by 10 the energy consumption when compared with gas firing.

FIGURE 21. Mobile hearth furnace, high power density type (42 kW/m²), with machine welded tubular resistors and fibrous ceramic refractories for the heat treatment of metal parts.

These first results were obtained by mobilizing the new resources offered by electronics, modern regulation, fast response temperature probes and resistors which are better adapted to thermal radiation, such as machine welded tubular resistors. Regulators on cascade on the resistors and the furnace charge together with low inertia probles actually make it possible **to continually modulate** the power using a thyristorized static power supply and this maximizes, **at each instant,** the energy from the heating elements while remaining compatible with their safety.

It is however possible to go further in modernizing resistance furnaces and to add to the productivity increase a better operating flexibility and product quality.

4.3.3. The needs for better controls

The quality of manufactured products often depends upon strict respect of thermal cycles which they must undergo and it is the role of the furnace programming and regulation system to provide this. Beginning at a given thermal state of the charge and of the furnace, this system must therefore pilot the power supplied to the resistors in order to bring the temperature of the charge to a given point without going beyond it, nor oscillating, while respecting a temperature evolution curve, and at the same time protecting these resistors against any overheating.

For furnaces with modest performance, the «go-no go» type regulations actuating electromechanical contactors have long been sufficient and still provide many services. But, faced with new requirements for furnace performance, particularly in terms of power density, much finer regulations of the Proportional-Integral-Differential controller type (PID) acting on thyristorized static contactors capable of very accurately and continually modulating the electric power fed to the heating elements are now seen to be indispensable both for increasing temperature accuracy and the intensity of thermal transfers and for ensuring the safety of the resistors. Thanks to various characteristics such as their universality, their sturdiness, their accuracy, their limited cost and their possibility of association in cascade for the simultaneous control of the charge and resistors temperatures, these systems are therefore very well suited to control resistance furnaces.

Nevertheless, these systems may have limitations, such as :

— Difficulty in setting and adjusting : the PID controller actions do not directly correspond to physical parameters which can be intuitively grasped. Their settings and adjustments are therefore often arduous and require a good knowledge of the consequences implied by modifying an action. In the presence of a thermal transfer function which is reacting slowly, obtaining the desired setting may turn out to be laborious and this adjustment difficulty increases in case of regulations in cascade.

— The absence of accounting for the non-linear nature of thermal transfers in a furnace ; the setting of a PID controller will be correct around a given operating point, but will no longer be correct in another temperature range because of this absence of linearity of the thermal phenomena. For the same reasons, this setting will be adapted to a certain type of charge, but not necessarily to a charge which differs too much due to its dimensional and thermal characteristics. This is a problem since existing controllers do not provide the possibility of easily varying these actions and these phenomena are accentuated when the specific power of the furnace goes up, particularly above 30 kW per m² of wall.

— Available PID controllers are always monovariable. There are regulators which can regulate several temperatures, therefore several zones, but these are considered as totally independant. The accounting for interactions between zones is never considered, an approximation which is not acceptable in certain furnaces.

378

Consequently, it would be extremely attractive to have regulations which could, by themselves, without human intervention, adapt to varied situations in furnace operation, that is, to a large range of setpoint temperatures and to charges which vary considerably, to take into account interactions between the various zones of a furnace, to quickly track and find the setpoints imposed, without oscillations, or going beyond them, and to provide very good temperature accuracy. With modern automation methods, these demands need no longer be considered as utopian.

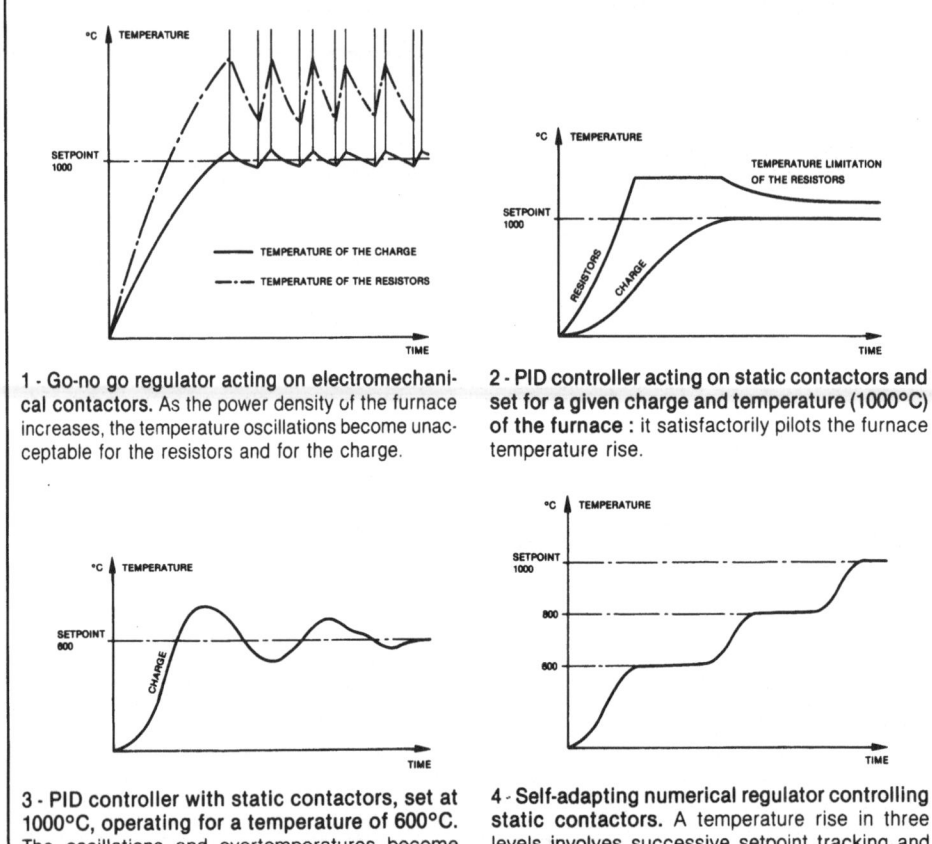

1 - **Go-no go regulator acting on electromechanical contactors.** As the power density of the furnace increases, the temperature oscillations become unacceptable for the resistors and for the charge.

2 - **PID controller acting on static contactors and set for a given charge and temperature (1000°C) of the furnace** : it satisfactorily pilots the furnace temperature rise.

3 - **PID controller with static contactors, set at 1000°C, operating for a temperature of 600°C.** The oscillations and overtemperatures become excessive with high density power furnaces or furnace charges which are different from those taken into account during the settings.

4 - **Self-adapting numerical regulator controlling static contactors.** A temperature rise in three levels involves successive setpoint tracking and encountering without going beyond, nor oscillating, although the dynamics of the furnace may totally change from one level to another.

FIGURE 22. How modern automation and electronics can give the resistance furnace higher performance and make it more «intelligent».

4.3.4. Towards the «intelligent» furnace

To better understand these new regulation methods without going into their theoretical elements, two successive stages in their development should be presented.

• Numeric control with prior identification

Before going into the operation phase, an «identification protocol» needs to be applied to the furnace and this will make it possible to determine its static and dynamic behavior, that is, set up a mathematical representation still called a transfer function. This phase can be entirely automated and handled by the numeric system of monitoring-control.

The regulator can then come into operation because, thanks to the mathematical model of the furnace which is stored in memory, it is capable of predicting its behavior and therefore, thanks to algorithms designed for that purpose, generating the control necessary for obtaining the desired temperature response.

In comparison to PID controller, this method saves the user or manufacturer all the sometimes arduous work of adjusting the parameters of the regulator because it is the identification algorithm which takes care of this and nothing opposes taking interactions between zones into account.

Nevertheless, the furnace is identified with a type of charge at a given temperature. So the model is ill adapted when the charge or the temperature change and it is necessary to go further to respond to the requirements mentioned earlier.

• Numeric control with online identification

This time, the regulator and the identifier operate permanently. During furnace operation, the identifier regularly estimates its transfer function and its structural variations. This automatic and constant updating of the mathematical representation of the furnace characterizes the self-adapting of the regulation system. The regulator itself then generates the control to obtain the desired temperature.

Although this control system is founded on extremely complex mathematical developments, it should be noted that it is totally transparent for the user. No manual intervention is really necessary during the operation of the regulator except at the beginning to initialize the setpoints and launch the system.

The tests conducted on furnaces of the EDF laboratory at Les Renardières have shown that these regulation concepts were well adapted to controlling resistance furnaces. The performances of these systems were then checked on an automobile industry carbonitriding furnace, and their superiority over conventional regulators was demonstrated during these tests.

The first numerical self-tuning temperature regulator to enter normal industrial operation has been implemented in 1987 on a micro-computer controlling a 1600 kW - 60 m³ low thermal mass moving earth resistance furnace for sanitary ceramics firing. First results show that the difference between the theoritical temperature rise curve (the cooling is not controlled for the moment) and the actual one is almost imperceptible. The temperature homogeneity in the furnace is moreover exceptional since between the six controlled areas, the temperature difference is less than 2°C and there are no oscillations or overtemperatures at the end of the heating-up.

This development represents a major step towards the «intelligent» furnace and should make it possible not only to increase the productivity of resistance furnaces, but also to substantially improve the quality of the products.

4.4. The maturing of arc plasma technology

4.4.1. Plasma and industrial processes

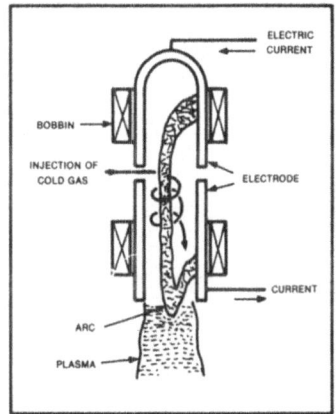

FIGURE 23. Diagram of a plasma torch.

Thermal plasmas, produced by electric arc, are high temperature gases consisting of a mixture of molecules, atoms, electrons and ions, usually in excited states. The gas temperature is raised by transferring energy from the electric arc maintained between the electrodes of the generator to the gas. The advantages of plasma heating are due to characteristics such as :
- high energy concentration : 2 to 5 kWh/Nm³ for industrial use,
- high temperature : 2000 K to 20 000 K,
- low thermal inertia : plasma generators with powers up to a few MW can be started within a few seconds, stopped instantaneously (less than a second) and finely regulated,
- flexibility in the choice of the plasma gas...

The use of Arc Plasma heating have been researched for many years, but this technology is just beginning to mature into a more fully understood industrial process. Recent studies have shown that increased implementation of plasma could be one of the most attractive possibilities to extend the electrification of manufacturing industries. Moreover, the flexibility of combining electricity, in the form of plasma, and other fuels, notably coal, still adds to the interest of this technique.

In the short term, the most promising areas lie within the iron, steel and ferro-alloys industries for upgrading conventional systems, such as blast furnaces, cupolas and even arc furnaces, or replacing some of them by new processes. Thus, a few plasma systems have recently reached the stage of the pilot plant or of commercial demonstration. For example, the feasibility of plasma fired cupolas and their ability to melt machine chips at high productivity levels has been demonstrated. The use of plasma torches to supply hot reducing gas by injecting coal into the hot air stream has been tested succesfully on a blast furnace in industrial operation, thus showing that electricity and low-quality coal can replace a high percentage of the coke. Pilot plants have also been built for direct reduction of iron ore, ironmaking, ferrochrome production, and recovering zinc from steelmaking dust. A few new routes for iron and steel making can be imagined including such other miscellaneous utilizations as induration of pellets, heating of soaking pits and rolling-mill furnaces, temperature adjustments during ladle refining and laddle preheating.

In the long term, plasma processing could provide new opportunities in the chemical industries for the processing of coal and hydrocarbons, the synthesis of special materials such as titanium dioxide, silica and refractory powders or the destruction of the wastes generated by many industries. This will however require an important research and development effort to overcome the technical problems. Further penetration of this technology will also remain largely dependent on the economic situation and on the price of electricity as compared to that of other energies.

EDF has been very active these last years to help plasma to leave laboratories for the «shop floor» and closely collaborated with manufacturers, research laboratories, government bodies and potential industrial users to develop this technique. Among the different actions undertaken should be mentioned :

− the creation of a plasma laboratory at EDF research Centre ;
− the development with a manufacturer of a highly reliable torch in the several MW range ;
− the creation of a transportable plasma system rated at 2 MW to demonstrate in industrial plants the possibilities of plasma ;
− the development of mathematical models for understanding the phenomena and dimensioning the equipments ;
− the support of university laboratories working on plasma ;
− the technical and economical optimization of the insertion of plasma in the processes of different industries (steel, ferro-alloys, foundries, cement, waste destruction...).

The first industrial successes have been met in the ferro-alloys and steel industries.

4.4.2. Plasma and the blast furnace

High energy concentrations and heat levels have always been a major concern of steel makers whether for making cast iron in blast furnaces, or for remelting the steel in arc furnaces.

Since 1979, the French Steel Industry Research Institute (I.R.S.I.D.) and EDF have been looking into the possibilities of plasma to substitute electricity for coke for making pig iron in blast furnaces. Various approaches were examined :

− substituting plasma for the injection of fuels at the nozzles (fuel oil, gas...),
− superheating the furnace blast,
− recirculating and superheating blast furnace gas,
− blast superheating together with a massive injection of coal.

This examination indicated that from the cost point of view, the superheating of the blast together with injections of coal was by far the most advantageous solution under today's economic conditions, although the other approaches presented no technical problem for their implementation upon initial examination. However, the simple superheating of the furnace blast by 100°C, makes it possible to economize 10 kg of coke for a plasma electricity consumption of 60 kWh per ton of pig iron, but remains very limited (100-200°C according to the blast furnace), except for the production of ferro-manganese for which a superheating of 600°C can be technically and economically attractive.

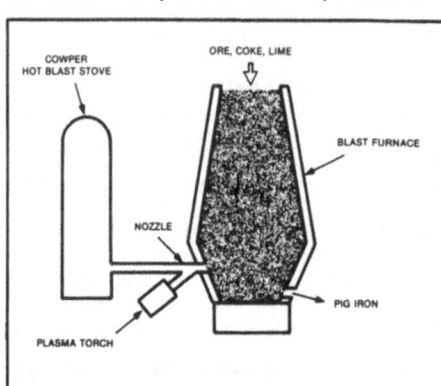

FIGURE 24. BLAST FURNACE EQUIPPED WITH A PLASMA TORCH.

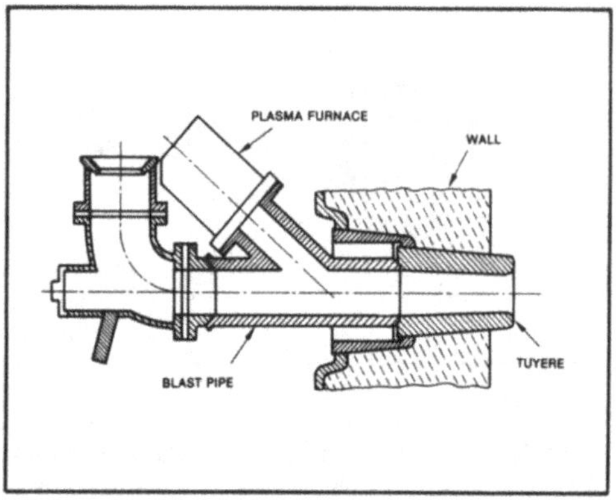

FIGURE 25. PLASMA FURNACE-BLAST PIPE-TUYERE UNIT.

To implement the combined injection of plasma and coal in blast furnace tuyeres, it was however necessary to answer several questions :

− how will a multitorch plasma system behave on a blast furnace ?
− will the torch - blast furnace (blast pipe) interface, as well as the tuyere withstand the very high termal stresses (plasma at 4000 K) ?
− will the combustion (gasification) of the coal be significantly improved by the superheating of the blast ?

− will the blast furnace reach a stable operating point ?

Satisfactory answers were brought to these questions thanks to laboratory work and the use of mathematical models. Only full scale experimentation could however have brought definitive answers.

A ferro-alloy company, producing ferro-manganese from ore and coke in blast furnaces, then decided to go further and ordered three plasma torches rated at 2 MW for blast superheating. The operation of these torches confirmed their reliability and six more torches were implemented two years later.

The most significant results of this first coupling of a plasma system to a blast furnace are :

− a good operation of the blast furnace equipped with plasma torches,
− a reduction of coke consumption quite close to what was forecast,
− a good stability of the interface blast pipe after several thousand hours of operation,
− a plasma torch electrode service life of more than 1000 hours (1200 hours for the upstream electrode, 2500 hours for the downstream electrode),
− a quick adaptation of the personnel to the new technology.

Following these results, a steel company has decided to invest in a plasma system to assist massive coal injection at the tuyeres. This system should make it possible to reach 200 g of coal per Nm³ of hot blast while the present ratio without plasma is about 130 to 140 g/Nm³.

According to the simulation results, the thermal balance for a pig iron production of 1600 tons per day should be :

− a coke savings of 60 kg/t of pig iron,
− an increase in the coal injection of 50 kg/t of pig iron,
− an extra electricity consumption of 140 kWh/t of pig iron.

To prepare this final operation, complementary experiments have been conducted at EDF plasma laboratory and have shown the feasibility of such massive injection of coal.

The plasma equipment, as for the previous company, should be operated only 7 months per year (on EDF summer rate) and is expected to start operation in summer 1987.

5. CONCLUSION

The previous examples demonstrate that the possibilities of new electrotechnologies are considerable and can greatly contribute to a better management of the electricity demand. These techniques not only open the way to a better mastery of the use of energy, but also, to rapid modernization of industry by responding to certain requirements for tomorrow competitiveness such as improved product quality, production automation and the rational use of human resources.

C.L. LOPEZ-CACICEDO AND G.W. BRUNDRETT

THE ELECTRICITY COUNCIL RESEARCH CENTRE, CAPENHURST, CHESTER, CH1 6ES.

1. INTRODUCTION

Electricity is an alternative energy source for drying processes which use the traditional techniques of conduction, convection and radiative heat transfer. However, it can be used in a variety of other ways, such as high frequency, electro-magnetic volumetric heat transfer which cannot be matched by other energy sources. It also has a very effective role in heat recovery and in heat pumping processes.

This paper concentrates on some of the less well known drying methods, particularly those which have been developed recently or are active research topics at the ECRC Laboratories and are finding favour with manufacturers and plant managers.

2. DEWATERING

Non-thermal water removal is much cheaper than thermal drying. Pressure and vacuum filtration are well established but only the two new techniques will be considered.

2.1 Air knives

The use of this well known principle, where a thin jet or blade of air at high velocity displaces water or solvent from the wet material has increased significantly over the last few years, mainly because of its low cost [1].

The principle is shown in Figure 1. Air velocities of between 30 and 60 m/sec (100-200 ft/sec) are used with pressures of 10 to 70 kN/m² (1½ to 10 psi).

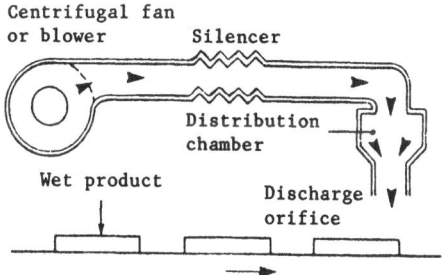

FIGURE 1. Air Knife System

The most important engineering design element is the orifice or slot which should have a discharge coefficient as close to unity as possible. For any particular process, a great number of variables must be considered: orifice width, pressure, number and spacing of knives, angle of incidence,

A. T. De Almeida and A. H. Rosenfeld (eds.), Demand-Side Management and Electricity End-Use Efficiency, 383–390.

product shape and line speed etc., and therefore the whole system design is very dependent on practical experience.

There are well over 200 installations in the UK, most of them since 1980, drying products as diverse as green bricks, food cans, metal components and strip, plastics and washed vegetables before packing.

2.2 Electrokinetic dewatering

The combination of two phenomena, electrophoresis or particle movement and electroosmosis or water pumping by an electric field, can be used to dewater emulsion or suspensions of products like PVC, clay, sludges, etc.

The principle for a continuous process is outlined in Figure 2. Here particles of about 0.1-10 μm which are in suspension - and generally negatively charged, migrate towards the anode by the imposition of a DC electric field, forming a solid cake, whilst water is attracted to the cathode and by means of a physical barrier which stops migration of the water-shrouded solids, an electroosmotic pressure can be generated and used for water removal.

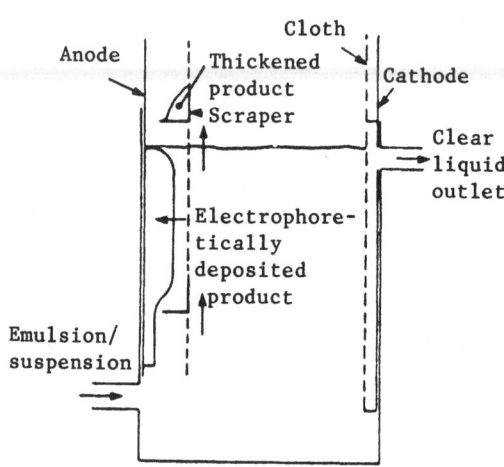

FIGURE 2.   Dewatering Cell Schematic

A typical application is PVC dewatering. The feed suspension (called emulsion) is about 30 to 40% solids, the cake out of the cell is 80 to 85% solids and the water less than 0.1% [2]. The process requires close pH and temperature control as those parameters affect the product, and the rates of particles' movement and electroosmotic water flow rate.

An important characteristic of this process is that the cake product retains the properties of the solids in suspension, since no external force, e.g. pressure, is exerted on the cake.

Cell throughputs can be very large, typically a 1 m^3 cell with 10 m^2 electrode area can dewater 1.5 tonnes/h of PVC with an energy consumption of between 100 and 200 kWh/tonne depending on initial concentration and pH. There is a recent installation of this technology in the USA.

3. HEAT RECOVERY

Heat pumps using conventional refrigerants can be used to reduce the energy costs in drying processes by recovering and upgrading the waste heat [3].

Malt drying is a popular example because it is an energy intensive process and is carried out over the whole year. The batch operation is in two parts. A steady drying period of around 12 hours at 60-65°C is followed by caramelisation with temperatures rising up to 100°C. The actual temperatures are carefully chosen to meet the particular requirements of the kind of malt desired. Large electrical heat pumps have been used to recover the heat from the outgoing moist air and preheat the incoming cold air. The first system had a payback period of just over 2 years.

Multi-stage heat pumps, particularly when combined with an air/air heat exchanger, offer an improved coefficient of performance, controllability without efficiency penalties, optimum use of different refrigerants and at surprisingly little extra capital cost above that of a single stage heat pump.

An example of a 375 kW electric multi-stage malt dryer is illustrated in Fig. 3. The performance predictions are outlined in Table 1. The first two stages are electrically driven compressors using refrigerant R22, while for the higher temperature the latter two use R12.

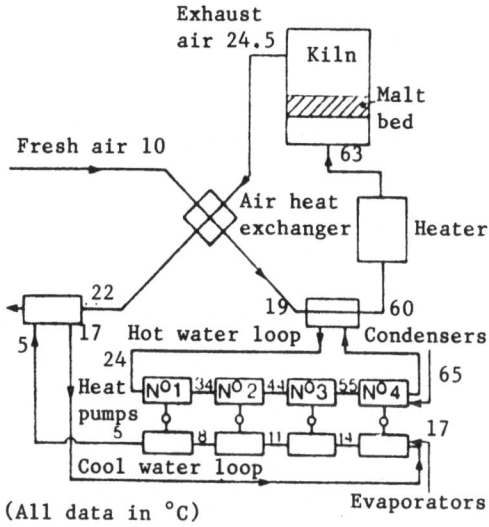

FIGURE 3.   Malt Dryer/Heat Pump System

TABLE 1. Performance of multi-stage heat pump

Heat Pump Stage		1	2	3	4
Temperature difference	(°C)	35	42	50	57
Coefficient of performance		5.3	4.6	4.0	3.6
Compressor Power	(kW)	75	88	101	111
Total Heat Output	(kW)	1600			
Total Power Input	(kW)	375			
Average COP		4.3			

4. DEHUMIDIFICATION

The use of heat pumps in dehumidification is well established and tens of thousands are used in commerce and industry. They tend to be used at room temperatures and are typically 15-50 kW in size [3].

The effectiveness of a dehumidifier is best defined as the ratio of the amount of water evaporated to the energy used to achieve it. This is termed the Specific Moisture Extraction Ratio SMER.

$$\text{SMER} = \frac{\text{amount of water evaporated}}{\text{energy used}} \quad \text{kg/kWh}$$

The theoretical maximum value for conventional drying is the latent heat of evaporation at $100°C$, i.e. 1.55 kg/kWh. In practice the values are typically 0.5-1 kg/kWh. The heat pump dehumidifier can achieve very much higher values of SMER because the evaporator operates below the dew point of the humid air, thereby condensing water on it and lowering the relative humidity of the drying chamber. This latent heat is then recycled, together with the compressor heat to the condenser. The cycle is illustrated in Fig. 4. There are many examples of such effective drying for textiles, ceramics, food confectionery products and commercial applications such as swimming pools. The industrial processes achieve Specific Moisture Extraction Ratios of 3 in practice.

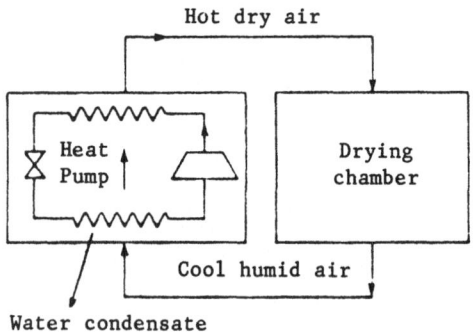

FIGURE 4. Heat Pump Dehumidification Drying

The commercial application to swimming pools is a particularly good one because the design objective is to limit the maximum relative humidity to 60%. Traditionally the air supply was designed to provide sufficient warmed outside air to achieve this control and energy consumptions values were typically 11,000 kWh/m² of pool per year. The combination of a heat pump dehumidifier, controlled ventilation and ventilation heat recovery can lower the energy consumption to well below 3000 kWh/m² of pool per year [4].

The upper temperature limits of refrigeration technology in the late 1970s, restricted industrial applications to 60-65°C. Work at ECRC concentrated on identifying suitable heat pump fluids for higher temperature operation, together with suitable oils which operated with minimum modification to existing equipment. The refrigerant R114 had suitable thermal stability, absence of toxicity, non-flammability and thermodynamic acceptability. The use of this with appropriate lubrication

enabled the dehumidification temperature range to be raised to 120°C.
The main application to date has been batch timber drying in well insulated
kilns. Work at ECRC, using comparable batches of timber, measured the
energy consumption for timber drying in both a modern conventional kiln
and in the well insulated dehumidifier dryer at 80°C. Hardwoods showed
an energy saving of 40%, while for softwoods it was 65% [5].

Further work at higher operating temperatures has shown that a number
of fluids are available up to 200°C but that at these temperatures the
water vapour itself is a suitable refrigerant and this leads to the concept
of steam recompression drying [6].

5. STEAM RECOMPRESSION

The most efficient steam recompression system for a continuous contact
dryer is illustrated in Fig. 5. Air is excluded from the dryer and the
water vapour is generated at near atmospheric pressure within the dryer.
This water vapour is cleaned and then compressed up to the supply steam
pressure and then fed to the dryer heating surfaces.

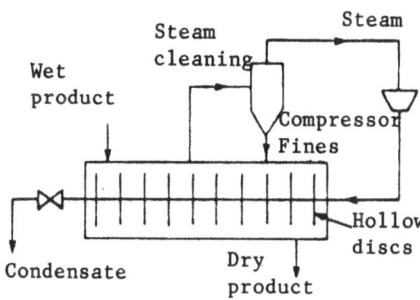

FIGURE 5. Open Cycle Contact Dryer with Heat Pump

The pilot plant at ECRC has a heated jacket with rotating paddles inside
the machine to ensure good heat transfer between wall and product. Both
solids feed and discharge must be airtight since air intake and steam
loss would affect performance adversely. The field measurements from
this pilot plant have been used to draw up a feasibility study for a
full scale plant with a drying rate of 4000 kg/h. The SMER was 5 kg/kWh
for the steam recompression plant compared to 1 kg/kWh for the conventional
equipment and the payback period was 1.4 years.

There are several variations to the basic cycle to cater for special
needs. Additional heat exchangers can be provided if the initial vapour
is corrosive. Vacuum operation can be used to limit the maximum
temperature for those products which are heat sensitive. Many convective
dryers operating with air can be modified to operate satisfactorily with
superheated steam. Such a cycle is illustrated in Fig. 6. The drying
times are very similar to those on air although the drying curves are
different. The SMER varies between 3.6 to 6 kg/kWh for inlet temperatures
of 150°C and 130°C respectively [7]. An industrial demonstration is
in progress for board drying.

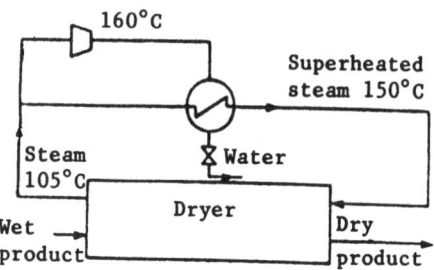

FIGURE 6.   Convective Drying with Open Cycle Heat Pump

6.   ROTARY INDUCTION DRYING

Following a survey of the food and chemical industries, a simple continuous dryer was identified as having a a reasonable market potential. Since electrical transformer principles and technology offers one of the lowest capital costs, we set out to develop a dryer based on this principle [8].   The result is a Rotary Induction dryer which is based on axial current induction principles operating as a contact rather than as a convective dryer.   Figure 7 shows schematically the principle.

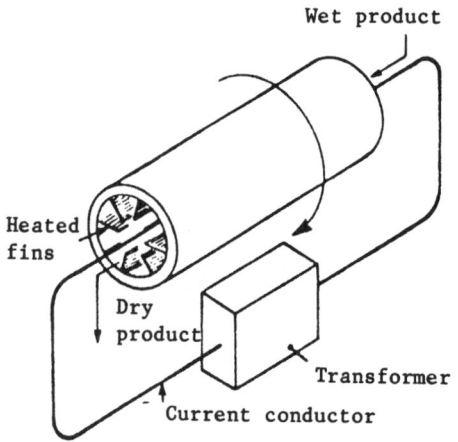

FIGURE 7.   Contact Rotary Induction Dryer

A low voltage high current conductor through the centre of the dryer induces a current on the stainless steel fins inside the dryer heating them uniformly.   The contact mode of operation is achieved by fin design and rotation of the dryer, generally the opposite of a conventional convective dryer, thus ensuring that the particulate product is gently transported, agitated and contacted with the hot fins.   Furthermore, a high degree volume utilisation can be achieved.

The electrical efficiency, i.e. the heat generated in the dryer as a fraction of the power input, is around 90% with power factors of 0.92 or thereabouts.   Control of fin temperature can be easily achieved by controlling the current flow in the conductor.

Product throughput can be adjusted by the angle of the drum, the rotational speed and the feed rate.   Solids transportation experiments show that backmixing is very small.

Sectionalised dryers can be built with different heating zones along its length to match the drying needs of the product. Variations in local power input are achieved by selecting the thickness of the fin to give the required induced current. Vacuum operation is readily achieved.

Typical sizes are between 1 to 5m in length and in diameter up to 0.6m. Power inputs are between 20 and 2000 kW. Capital costs are equivalent to those of a fluidised bed dryer and well below band dryer costs. Running costs are within ∓20% of the conventional drying systems with electricity costs at 3.1p/kWh and gas at 1.2p/kWh.

7. RADIO FREQUENCY DRYING

Radio frequency drying distinguishes itself by heating the water selectively and in preference to most materials. It is a well established technique for materials where flexibility and product quality are particularly important [9,10]. Applications include the final drying and moisture profiling stage in manufacturing paper and also in the drying and baking processes in the biscuit industry [11].

A combined electric convective and radio frequency dryer is used for high quality fabric or board drying. The hot air nozzles dry the surface while the small amount of radio frequency accelerates the water diffusion from the body of the product to the surface. The concept is called ARFA (Air Radio Frequency Assisted) and the hot air nozzles are also the radio frequency electrodes, Fig. 8 [12,13]. The presence of moisture on the product surface allows higher air temperatures to be used without affecting quality. The net result is either a considerable increase in production or a much smaller dryer. An illustrative dryer uses 450 kW of hot air in combination with 50 kW of radio frequency heating to evaporate 300kg of water per hour.

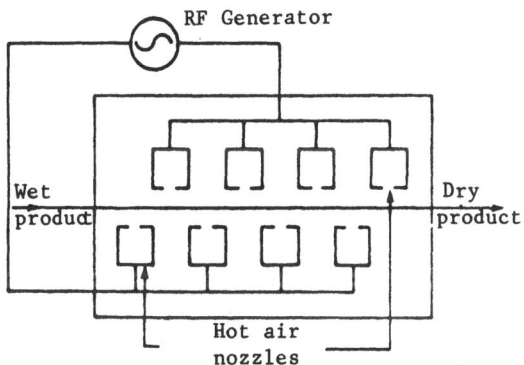

FIGURE 8. RF/Convective (ARFA) Drying Principle

8. CONCLUSIONS

Electricity can provide all of the conventional drying techniques of radiation convection and conduction heat transfer. In addition it has the ability to selectively and preferentially heat the water itself as in radio frequency heating. It also has the most convenient, clean and controlled way of inductively heating drying cylinders. It also has the ability to provide mechanical power which can be used for compressing air in air knife dewatering or for compressing refrigerant fluids as in heat recovery or dehumidification processes. New developments in electrophoresis and electroosmosis are cutting the costs of dewatering.

The next stage is to look at each drying process carefully and match the most suitable range of electrical techniques.

9. ACKNOWLEDGEMENTS

This paper summarises the work of the following Research Officers: R.H. Green, R. Benstead, A.V. Heaton, J.G. Sunderland, S.M.L. Hamblyn, J.T. Griffiths, P.L. Jones and F.W. Sharman.

10. REFERENCES

1. Ganden, C. Electrical Review, 1 October, 1982.
2. Sunderland, J.G. 3rd International Drying Symposium, Birmingham U.K., Vol. 1, 417-436, 1982.
3. Brundrett, G.W. Handbook of dehumidification technology. Butterworths, London, 1987.
4. Braham, G.D. The energy factor. 51st Annual Conference of the Institute of Baths and Recreational Management, Septemper, 1981.
5. Anon. Timber drying. Timber Trades Journal, p.24. 25 July, 1981.
6. Bertinat, M.P. International Journal of Refrigeration, 9, (1), 43-50, January, 1986.
7. Heaton, A.V. and Benstead, R. Proceedings 2nd International Symposium on Large Scale Applications of Heat Pumps. York, England, 25-27 September, 1984. Published by BHRA Cranfield, Bedford, U.K.
8. World wide Patents applied.
9. Jones, P.L. J. Society of Dryers and Colours, 98, 248, 1982.
10. Anon. R.F. systems for textile processing. Int. Dyer & Textile Printer, July 1985.
11. Jones, P.L. Drying Technology, 4 (2), 1986.
12. U.K. Patent 212 3537 World wide Patent applied.
13. Swift, G. and Jones, P.L. Paper Technology and Industry, 24 (6), 224, October, 1983.

ELECTRICITY SAVINGS IN REFRIGERATION PROCESSES

PREBEN BUHL PEDERSEN,

Physics Laboratory III, Technical University of Denmark
DK-2800 Lyngby, Denmark.

1. INTRODUCTION

The annual electricity consumption in Denmark is about 25,0 TWh (1985) or a little less than 5,0 MWh per capita. Of this about 12,5% is used in refrigeration processes. It is convenient to split-up the refrigeration processes into domestic, commercial, and industrial processes. Today more than half (57%) of the electricity used for refrigeration is used in the households. Approximately 12% is used in commercial cooling plants and approximately 31% is used in industry, mainly the food-industry.

In Denmark, the efforts in saving energy have been concentrated on heating energy. Recently, however, political pressure has convinced the Ministry of Energy to start a general electricity saving investigation. Before that only smaller efforts have been directed towards electricity conservation. However, at Physics Laboratory III, the Technical University of Denmark, we have been working on the subject for some years (3,4). In 1983, research made it clear that approximately 200% increase in electricity end use efficiency could be achieved if the political will was there. Furthermore, we designed and built a prototype of a low energy refrigerator which uses today 44% of the electricity consumption of the most efficient model on the market, or about 25% of the average model in use today (2). Also an American type combined refrigerator-freezer with corresponding savings has been developed at Physics Laboratory III (5).

These projects encouraged us to go further in building energy efficient prototypes in other areas of refrigeration. The first step was a pilot research into the possibilities of saving energy in refrigeration. This work was sponsored by the Danish Ministry of Energy. It has just been completed and includes all three of the above-mentioned areas of refrigeration (1). The pilot project includes no practical experiments but is a summary of the theoretical savings potential found by studying the literature and doing simple calculations.

This paper outlines the content of the pilot project.

2. BASIC RELATIONS

Usually the efficiency of a refrigeration process is expressed by the coefficient of performance (COP), which is defined as the cooling load (Q_0) divided by the electricity used (P):

$$COP = \frac{Q_0}{P} \qquad (1)$$

A. T. De Almeida and A. H. Rosenfeld (eds.), Demand-Side Management and Electricity End-Use Efficiency, 391–397.
© *1988 by Kluwer Academic Publishers.*

In thermodynamic terms, the COP can also be written as the carnot-efficiency (η_c) multiplied with the theoretical maximum value of COP, the carnot power factor (ε_c):

$$COP = \eta_c \cdot \varepsilon_c \qquad (2)$$

The electricity consumption can therefore be expressed as:

$$P = \frac{\dot{Q}_O}{\eta_c \cdot \varepsilon_c} \qquad (3)$$

where ε_c is defined as:

$$\varepsilon_c = \frac{T_O}{T_c - T_O} ,$$

T_O: evaporation temperature (Kelvin)

T_c: condensing temperature (Kelvin)

Equation (3) describes the three main factors which are essential to electricity consumption, and shows us that the electricity consumption can be reduced by:

- reducing the cooling load
- increasing carnot-efficiency, which means improving the machinery, e.g., electric motors, compressors, fans, pumps etc.
- increasing the carnot power factor, which is basically dependent on the actual evaporator and condenser temperatures.

3. HOUSEHOLD REFRIGERATORS AND FREEZERS

In Denmark we have today approximately 3,7 million refrigerators and freezers for 2,3 million households, which are estimated to use 1850 GWh/yr. This is about 7% of the total Danish electricity consumption. Of these 3,7 million refrigerator appliances, 44% are refrigerators, 19% combined refrigerator-/freezers and 37% freezers (1984).

3.1 Reduction of cooling load

As mentioned above, the cooling load is essential for electrical consumption. The cooling load is determined mainly

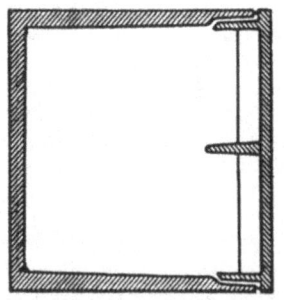 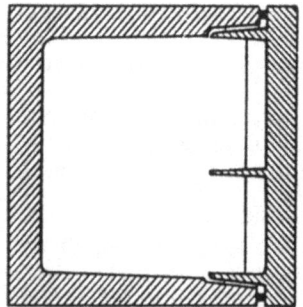

FIGURE 1. Cross-section of a refrigerator showing insulation thickness. Left: average today, right: increased insulation i.e. reduced load, as in (2).

by three conditions: 1) heat transmission through cabinet walls, 2) heat transmission through door seals and 3) heat transmission through cold bridges as lead-in's and the door-frame. By increasing insulation thickness to a little more than double, the heat transmission is reduced by approx. 40%. The heat transmission through door seals and cold bridges (making up 10-15% of cooling load) can be reduced by bigger "shoulders" on the door and maybe double door seals. It is, therefore, realistic to save about 45% of the electricity consumption through reduced load alone. The cabinets, though, will have to be around 20 cm (~ 8 inches) taller to maintain the net volume if width and depth is to be maintained.

3.2 More efficient machinery

The main factors determining the carnot-efficiency are motor efficiency, compressor efficiency, throttling device, and control of the system.

Carnot-efficiencies for household refrigerators and freezers are as low as 0,1-0,2 while for large refrigeration plants it can be around 0,6. A substantial potential is therefore in sight.

Electric AC-induction motors (60-500 W) have efficiency at approximately only 70%, but surveys (6) have shown that efficiency can be increased to around 85% by using:
- more copper in windings
- more and better iron in stator and rotor
- run capacitor
- more efficient bearings

In reduced load refrigerators and freezers, even the smallest hermetic compressors on the market are on-duty only 10-20% of the time. This leads to a low efficiency. Efficiency can, however, be increased by semi-direct intake because of lesser overheating of suction gas, and as mentioned before, by using better bearings. Laboratory test has shown a 15% reduction in electricity consumption when using both run capacity and semi-direct intake (2). Many other types of small hermetic compressors are being developed (rotary-, swing-, and vibration-compressors), which might improve efficiency, because of smaller compressors and thereby better use of heat exchangers.

The on- and off-duty has to be carefully controlled because of losses in off-periods and declining carnot power factor when running electricity consumption reaches a minimum at a certain running time.

3.3 Increasing carnot power factor

As shown in section 2, the COP can be increased by lowering condensing temperature and raising evaporator temperature. This means increasing heat transfer coefficients and/or prolonging the effective time for heat transfer in the heat exchangers.

Larger heat exchangers will diminish the necessary temperature differences and thereby increase the coefficient of performance. Introducing larger heat capacities in the heat exchangers will prolong the effective time for heat transfer, which also reduces the necessary temperature difference and hence increases COP even more. By these means, up to 35% savings can be obtained (1).

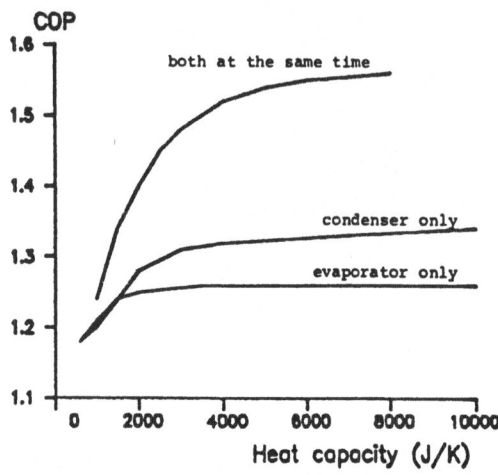

FIGURE 2. Relation between increased heat capacities in heat exchangers and COP.

3.4 Total savings in household refrigeration

The total electricity saving potential through the above mentioned means is summarized to be 75% for refrigerators and 63% for freezers. Weighted according to the actual distribution of types, the total potential is approximately 70% or about 1300 GWh/year in Denmark.

4. COMMERCIAL REFRIGERATION PLANTS

The total electricity consumption in Denmark for commercial refrigeration is estimated to be between 300-500 GWh/year including all commercial plants, ranging from large central supermarket plants to small ice-cream boxes in ice-cream stalls. Saving potentials for specific plants are between 45%-75%, depending on the size of the plant and the type. Seen in relation to the total stock of the respective types, the total saving potential will be 50-60%, corresponding to approximately 200 GWh/year.

Quality of the goods (food) has a clear relation to the temperature conditions. Alternating temperatures affect the quality and temperature level is essential to tenability, since low temperatures as a rule prolong tenability. Better packing of the goods can remedy the temperature problems. In addition, the use of night covers and closed (glazed) cabinets has a positive effect on both quality and electricity consumption.

The design of the display cabinet is of great importance to electricity consumption which can vary by a factor three in cabinets of same volume, depending on the type of the cabinet (7). Regarding open display cabinets, the convection and radiation losses are the greatest loss giving factors. The closed display cabinets are the least electricity consuming and open cabinets most consuming. Convection and radiation losses in open display cabinets can be counteracted by using night covers which reduce electricity consumption by approximately 35% (8).

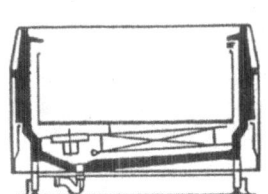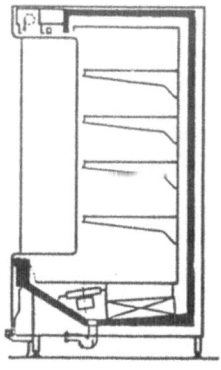

FIGURE 3. Commercial display cabinets. Left & middle: usual types, right: glazed cabinet (electricity consumption reducing).

In addition, radiation losses can be reduced in opening hours by using radiation shields (regarding isle cabinets like fig.3 left).

An appropriate choice of compressors with adapted valves and ports can reduce electricity consumption by 5% (9). In shops with a large supply of frozen foods, variants of two-step compression plants can reduce electricity consumption by up to 25% (10). Moreover, with high efficiency motors, it is possible to reduce electricity consumption further by 10-20% (11).

Looking at the condensers, it is possible to achieve reasonable savings in decentralized plants by combining all the condensers and running fans according to need. In this way, it should be possible to save 60% of the fan energy (9).

The above mentioned means are quite obvious, if for example, a supermarket is changing from a decentralized to a central plant, which normally leads to a 15% reduction in electricity consumption (12). If the condensers, moreover, are placed outdoors, the lower condensing temperature will lead to another 15% reduction, and if the condensed liquid is subcooled, for example by cold water, another 5-10% reduction could be in sight (9). In the same way, better utilizing of evaporators and maybe mixed refrigerants might give savings.

Development of electronically improved expansion valves with a wider area of operation, together with better regulation of capacity, are expected to give 10-25% savings (13).

Electricity use for defrost and frame heating can probably be reduced by two-thirds if these are used only when really needed (14). It should also be investigated whether defrosting can be done by reversed run of fans combined with new design of air intake. The circulation of cooling air in the display cabinet can be reduced with benefits by reduced speed drives or intermittent operation in periods of lower needs for cooling (night and weekend). Hereby, approximately 15% of the electricity consumption can be saved (8).

5. INDUSTRIAL REFRIGERATION PLANTS

The electricity consumption for refrigeration in industry is estimated to be around 1000 GWh/year. It is assumed that approximately 30% of this can be saved. Savings are first of all achieved through more attention to electricity consumption of each specific plant (15).

Most of the improvements suggested in the preceding section are, of course, also usable in the industry, especially in connection with decentralized and relatively small industrial plants. Furthermore, there are some more specific improvements to be mentioned in this section.

In cold stores, the cooling load can be reduced by 30% at least through increased insulation, reduced air infiltration and more efficient lighting (16). Some cooling services can be achieved without using refrigeration machinery, just by using cold water or air.

A large part of the year, power need is reducible by lowering condensing temperature as much as possible and using the existing condensers. This requires little or no changes in the plant but it is still only sparsely utilized (17).

Moreover, components should be chosen to make it possible to run the plant efficiently even at part load. This implies among other things that central plants are chosen, which allow condenser areas to be used optimally. The use of computer based supervision and steering will (in central plants) make it possible to reduce electricity consumption considerably by assuring that the above-mentioned conditions are kept at an optimum.

There are great variations in the design of industrial refrigeration plants, and, therefore, the savings potential will vary much from plant to plant. Since, therefore, in this study we have not been able to investigate industrial refrigeration technology in detail our overall estimate for savings potential of 30% in industry is probably on the conservative side.

REFERENCES

1. Pedersen PB: Forprojekt vedrørende elbesparelser ved køleanlæg (Pilot project concerning electricity saving in refrigeration plants, part 1,2 & 3) (in Danish). Ministry of Energy's Research Programme, journal no. 151/85-49. Published by Physics Lab. III, DTH, March 1987.
2. Guldbrandsen T et.al.: Development of energy efficient appliances, part one: Refrigerators. EEC contracts no. EE-A-3-025-DK(G) and EE-A-3-068-DK(SD), Sept. 1985.
3. Nørgård JS: Improved efficiency in domestic electricity use. **Energy Policy**, Vol.7, No.1, March 1979.
4. Nørgård JS, Holck J, Mehlsen K: Langsigtede tekniske muligheder for elbesparelser (Long term technical potential for electricity savings) (in Danish). Physics Laboratory III, Technical University of Denmark, 1983.
5. Pedersen PH, Schaer-Jacobsen J, Norgard JS: Reducing electricity consumption in American type combined refrigerator-/freezer. Physics Laboratory III, Technical University of Denmark, 1986.
6. Aalbregtse R, Schroeder GH: A new approach to improved efficiency compressors for household refrigerators and freezers. Proceedings of the 1978 Purdue Compressor Technology

Conference, 1978.

7. Schwartz P:Jahreskostvergleich unterschiedlicher Tiefkuhlmo-
bel und Kalteanlagensysteme. **Ki Klima-Kalte-Heizung** 12/1985.

8. Guldager F: Nyt prisbilligt system til køle- og frysemøbler
sparer energi året rundt i hidtil ukendt målestok (New low
cost system for refrigeration- and freezing display cases
saves large amounts of energy all year round) (in Danish).
Knudsen Køling A/S, **Scandinavian Refrigeration** 2/1982.

9. Haaf S: Entwicklungstendensen bei Kalteanlagen fur Super-
markte. **Die KALTE und klimateknik** 4/1985.

10.Quast U, Kruse H: Der energieverbrauch von Schaltungsvarian-
ten fur verbundanlagen in Supermarkten. **Ki Klima-Kalte-
Heizung,** 10/1985.

11.Seppings A(ed): Saving on energy in refrigeration. Institut
International du Froid, Paris 1980.

12.Overgård B, Christensen F, Knudsen Køling A/S. Personal
communication, March 26, 1986.

13.Tuesen SE, Winther J: Improved Control of Liquid Injection
into Display Case Evaporators. Danfoss, ASHRAE 1985.

14.Muller ED: Moglichkeiten der Energieeinsparung durch Einsatz
von Bedarfabtaureglern in gewerblichen Umluft-Kuhl-anlagen.
Ki Klima-Kalte-Ingenieur 4/1977.

15.Perry EJ, Gluckman R: The potential for energy saving in
refrigeration. **Refrigeration and Air Conditioning,** January
1981.

16.Union of Commerial Coldstores in Denmark: Kølefrysehuse,
energibesparelser, rapport, (Coldstores, energy savings,
report) (in Danish), August 1975.

17.Lorentzen G: Energiøkonomisering i kjøleteknikken (Energy
economizing in refrigeration) (in Norwegian). **Scandinavian
Refrigeration** 4/1981.

Heat Pumps – Basic Principles

Karl Holzapfel
Director
IEA Heat Pump Center

With the increasing concern about sufficient and reliable energy supplies in industrialized countries initially caused by the oil shortages in the early seventies, the heat
pump technology and its possible application for efficient low temperature heat
supply has received much attention. The basic physical principle of this technology is
completely different from the processes traditionally used to produce heat by simply
utilizing heat from the burning of solid, liquid, or gaseous fuels in boilers or furnaces.
In heat pumps a thermodynamic cycle is used that basically transports heat from a
low temperature heat source to a high temperature heat sink. According to the first
and second law of thermodynamics this heat transport from a low temperature level
to a high temperature level is not possible without additional energy input driving the
process. This drive energy input, however, is much smaller than the heat energy delivered at the higher temperature level. This is the desired effect making the heat
pump completely different from conventional combustion devices which always have
a heat output that is lower than the energy supplied by the fuel.

For a better understanding of the potentials of this technology some basic thermodynamic considerations are necessary. A heat pump is essentially a heat engine operating in reverse between two temperature levels (fig. 1).

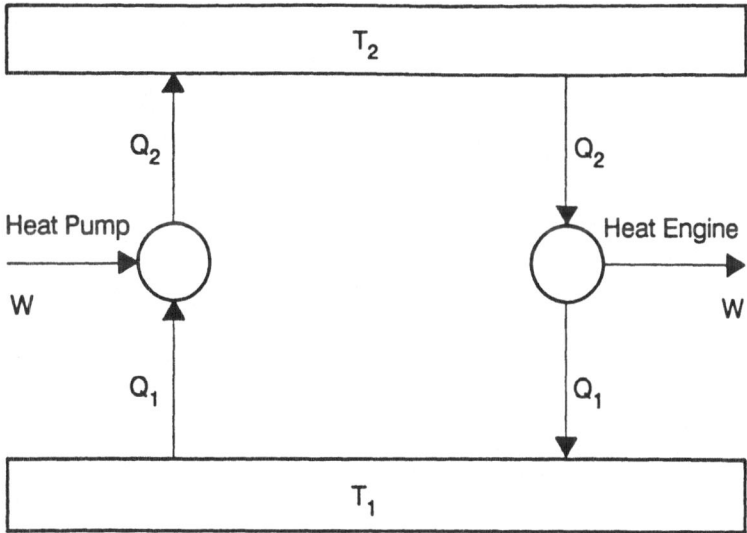

Fig. 1: Heat pump and heat engine operating between the temperature levels T_2 and T_1.

A. T. De Almeida and A. H. Rosenfeld (eds.), Demand-Side Management and Electricity End-Use Efficiency, 399–406.
© *1988 by Kluwer Academic Publishers.*

A heat engine produces work W by extracting heat Q_2 from the high temperature T_2 and delivering heat Q_1 to the low temperature T_1, whereas a heat pump delivers heat Q_2 at the high temperature T_2 by extracting heat Q_1 from the low temperature T_1 and requiring work input W. A refrigerator operates in exactly the same way as a heat pump with the only exception that the desired effect being not the heat delivered at T_2 but the heat Q_1 extracted at the low temperature T_1. The first law of thermodynamics gives the relation between heat and work involved in these processes by:

$$Q_2 = Q_1 + W$$

The second law of thermodynamics states that the work output (W) continuously produced by the heat engine can never be greater than the work input (W) required by a heat pump operating between the same temperature levels. This results in the following relationship between the temperatures and the heat transferred:

$$Q_1 / T_1 = Q_2 / T_2$$

The efficiency of these processes is defined as the ratio of the useful heat or work output to the necessary input. In the case of a heat pump the efficiency is called the Coefficient of Performance of COP and is defined as:

$$COP_H = Q_2 / W$$

The theoretically maximum possible COP is given by a Carnot Process operating between temperatures T_1 and T_2. This COP_{HC} depends only on these temperatures as:

$$COP_{HC} = T_2 / (T_2 - T_1)$$

Although it is not possible in reality to construct a heat pump with real working fluids operating with such a COP (completely reversible cycle), this COP is often used for comparative calculations because it is a very convenient way of quickly calculating a theoretical maximum COP for a given application only requiring knowledge of the temperature levels.

In practice a mechanical vapor compression heat pump more closely approximates a reverse Rankine Cycle. The COP_{HR} of the ideal heat pump Rankine Cycle can be calculated for a specific refrigerant with the equation:

$$COP_{HR} = (h_3 - h_4) / (h_3 - h_2)$$

where h is the enthalpy of the refrigerant per unit mass at the different state points shown in fig. 2. The process steps in this ideal cycle are as follows:

At state point 1 the refrigerant enters the evaporator of the heat pump with low pressure and low temperature T_1 in the two phase region. By extracting heat from the low temperature heat source the refrigerant is isothermically and isobarically evaporated to state point 2. The saturated vapor is then isentropically compressed by the compressor to state point 3 which is in the superheated vapor region. The desuperheating occurring in the condenser is isobaric up to state point 3. Further condensation of the refrigerant is isothermal to state point 4, saturated liquid. The high pressure is then isenthalpically reduced in a throttling device (expansion valve) and the refrigerant enters the evaporator again at state point 1, completing the cycle.

In real heat pumps a number of deviations from the ideal cycle will further decrease the COP. The single most significant deviation is introduced by the compressor.

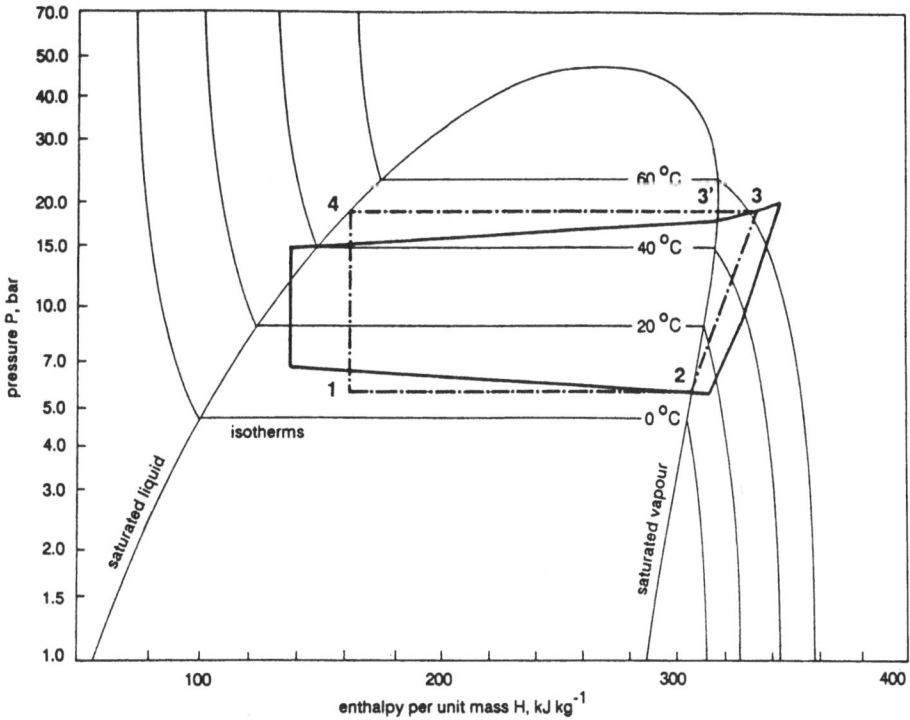

Fig. 2: Rankine heat pump cycle (dotted lines) in a pressure enthalpy diagram for refrigerant R 22. The solid lines show deviations of the real heat pump from the ideal cycle (exaggerated).

Condensation and evaporation temperatures need to be higher than the heat sink temperature and lower than the heat source temperature to allow heat transfer in limited area heat exchangers. Pressure drops in the refrigerant lines cause further losses. In order to prevent liquid refrigerant from entering the compressor the refrigerant is superheated. This can be done by using an additional heat exchanger which at the same time subcools the liquid leaving the condenser. The superheating must be kept to a minimum, however, because the increased specific volume of the refrigerant decreases the compressor capacity and increases the required work input and compression end temperature.

In fig. 3 the COP of a Carnot cycle, COP_{HC}, (independent of the refrigerant used), and of an ideal Rankine cycle, COP_{HR} (for R 22), are shown for different temperature lifts $T_2 - T_1$ (x-axis), for condensation temperatures (T_2) of 65 °C (338 K) and 35 °C (308 K). This diagram shows that for the considered condensation temperature range the COP_{HR} hardly changes with the condensation temperature, however, it is very sensitive to alterations of the temperature lift between evaporation temperature T_1 and condensation temperature T_2. Due to the deviations of the real process from the ideal cycle, the real COP is still lower than the COP_{HR}. The effectiveness of a heat pump can be defined as the ratio of the real COP_H to the ideal COP_{HR} for a given working

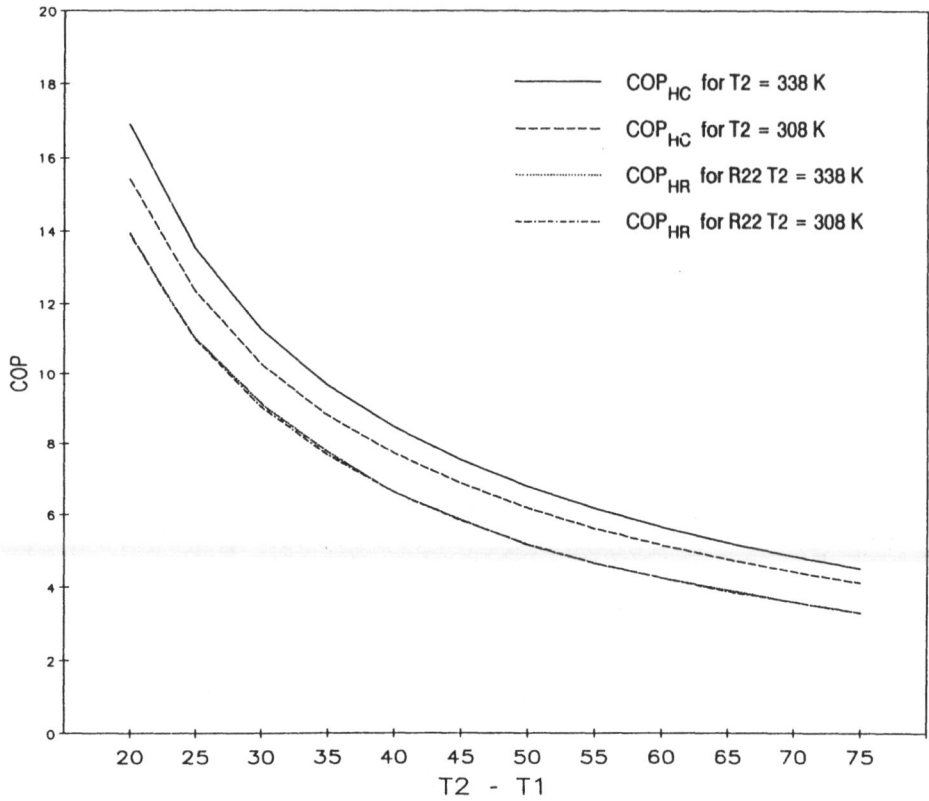

Fig. 3: Heat Pump COP for a Carnot cycle and for an ideal Rankine heat pump cycle for condensation temperatures of 65 °C and 35 °C as a function of the temperature lift $T_2 - T_1$.

fluid. This effectiveness is low for low temperature lifts, approximately 0.5 for $T_2 - T_1$ = 20 K and high for high temperature lifts, approximately 0.85 for $T_2 - T_1$ = 60 K.

There are mainly two different types of heat pumps being used today, the vapor compression heat pump with a mechanical compressor requiring mechanical drive energy, and the sorption heat pump using instead of a mechanical compressor a thermodynamic cycle requiring thermal drive energy. The drive energy for compression type heat pumps can be delivered by an electric motor supplied with electricity or by an engine supplied with liquid or gaseous fuels. If these different heat pump types are compared, for example on the basis of their COP, the quality of the supplied energy has to be taken into consideration. A COP of 4 for an electric heat pump looks much better than a COP of 1.6 for an internal combustion engine driven heat pump. However, if one takes into consideration the losses occurring in the electricity generation process at the power plant and the distribution losses, this COP of 4 of the electric heat pump comes down to a COP of about 1.4 calculated on the basis of heat pump heat output and necessary fuel input in the power plant. This figure is now lower than the engine driven heat pump COP which was assumed to be 1.6.

Basically this consideration is justified, however, it is also necessary to include the differences in the quality of fuel, used in large power plants to the quality of light oil or natural gas used in the engine driven heat pump plant. Furthermore waste heat from power generation is often used for district heating systems and should therefore be included in the comparison, since the internal waste heat utilization of the engine driven heat pump also contributes significantly to its former assumed COP of 1.6. The ability to directly use the waste heat from the conversion of fuel energy into mechanical energy in small engines is actually the main reason for their comparatively good COP even though the efficiency of the engine itself is lower than the efficiency of electricity generation in a power plant.

A direct comparison however is possible between electric resistance heaters with an efficiency of at most 1 and electric compression heat pumps having efficiencies (COP) of 2, 3, 4 or even 5. In a similar way the efficiency of an oil or gas boiler, about 0.85, can be compared to the efficiency (COP) of an absorption heat pump or an engine driven heat pump which can be more than twice as high.

A mechanical vapor compression heat pump consists of six basic parts as shown schematically in fig. 4. The heart of the heat pump is the compressor with its drive motor. The counterpart of the compressor is the throttling device, usually an expansion valve or a capillary tube. The two heat exchangers (evaporator and condenser), connect the heat pump to the heat streams from the heat source and the heat sink respectively. Finally a working fluid or refrigerant is used to transport the heat in the closed heat pump cycle. Additional equipment such as monitoring, control, and safety devices; additional internal heat exchangers; accumulators; oil separators; and refrigerant dryers is necessary to build a real heat pump.

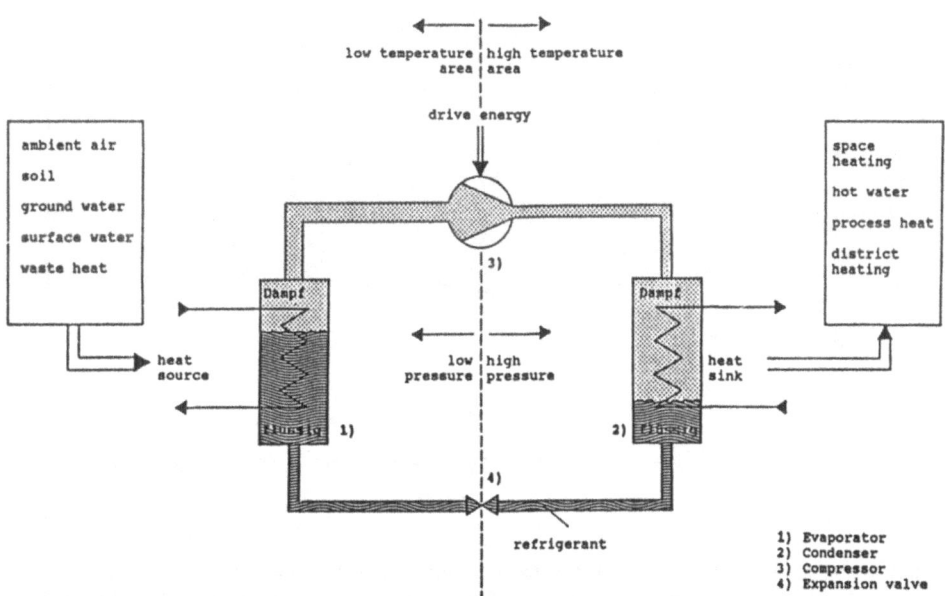

Fig. 4: Schematic of mechanical vapor compression heat pumps.

Initially heat pumps were equipped with compressors designed for refrigeration machines. With the increasing heat pump market and with the necessity to improve the efficiency of heat pumps, compressors have been developed specifically for heat pump applications. Compressors are being built as:

- "open" compressors, where motor and compressor are in separate casings with a coupling device connecting the shafts;

- semi hermetic compressors where the compressor and motor casings are coupled together and the refrigerant is used to cool the motor;

- hermetic compressors where the compressor and motor are contained in one welded casing.

A variety of different kinds of compressors exist and can be used for heat pumps depending on the application. For small heat pumps applied in the domestic sector mainly three types of compressors are in use:

- Reciprocating piston compressors
- Rolling piston compressors
- Scroll compressors

The reciprocating piston compressor has a long tradition in refrigeration application whereas the two rotary type compressors have been developed in the past decade, particularly in Japan. The introduction of rotary type compressors contributes significantly to the overall improvement of heat pump COPs. This is shown for Japan as an example in fig. 5.

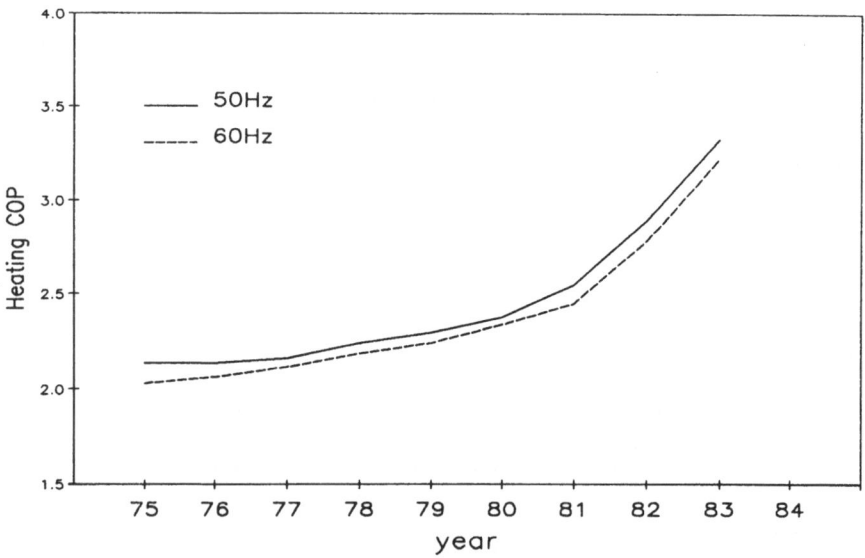

Fig. 5: Development of heating COP for heat pump room air conditioners in Japan. (source: National Position Paper Japan, IEA Heat Pump Center Workshop on Electric Heat Pumps for Retrofit in Small Residential Buildings, Graz, 1985).

The throttling device expands the liquid refrigerant from a high pressure level in the condenser to a low pressure level in the evaporator. This expansion results in a strong temperature decrease to a temperature below that of heat source. Very simple capillary tubes have been used in particular for heat pumps having a small heating capacity and a limited variation in the operating temperature conditions. A thermostatically controlled expansion valve (fig. 6) is used if an automatic control for variations of the operating conditions is required. In recent years the electronically controlled expansion valve has been developed to allow better adjustment of the refrigerant flow to variations in the external operating conditions.

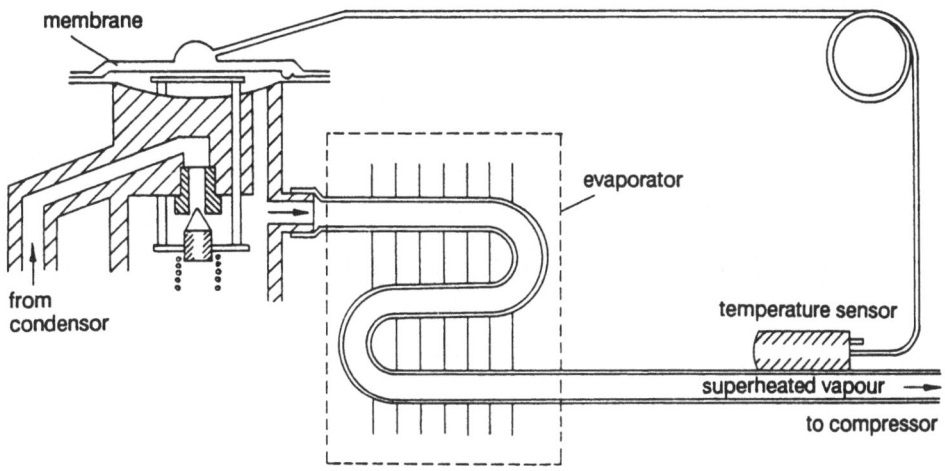

Fig. 6: Schematic of thermostatic expansion valves.

As an example for thermally activated heat pumps a schematic of an absorption heat pump is shown in fig. 7. This heat pump is very similar to the mechanical vapor compression heat pump in that it also has a condenser, expansion valve, and evaporator. However, in this case the mechanical compressor closing the heat pump cycle is replaced by another thermodynamic cycle. This part is frequently called a thermal compressor. The drive energy for the absorption heat pump is supplied in the form of high temperature heat to the generator. In general a small amount of electric power is also needed to drive the solution pump.

The thermal compressor also requires a working fluid pair, consisting of the refrigerant and a solvent, instead of a single working fluid. The solvent is used in the absorber to absorb the refrigerant vapors leaving the evaporator. This is an exothermic process resulting in heat release at a useful temperature level to the heat sink. The rich solution leaving the absorber is pumped by the solution pump to the high pressure level of the generator. By supplying high temperature heat to the generator refrigerant and solvent are separated again. The hot refrigerant gas is then led to the condenser and continues through the circuit in the same manner as in a mechanical vapor compression heat pump. The poor solution leaving the condenser is expanded to the low pressure level of the absorber. A real absorption heat pump has additional heat exchangers, for example ones between rich and poor solution, and additional control and safety devices.

The COP of absorption heat pumps is calculated as the ratio of the useful heat extracted from absorber and condenser to the high temperature heat supplied to the generator. COPs of presently available absorption heat pumps for domestic applications vary around 1.4. For comparative calculations the electric power required to drive the solution pump has to be included as well. This energy amounts to approximately 7 - 8 % of the heat output in heat pumps utilizing the refrigerant solvent working pair of ammonia and water.

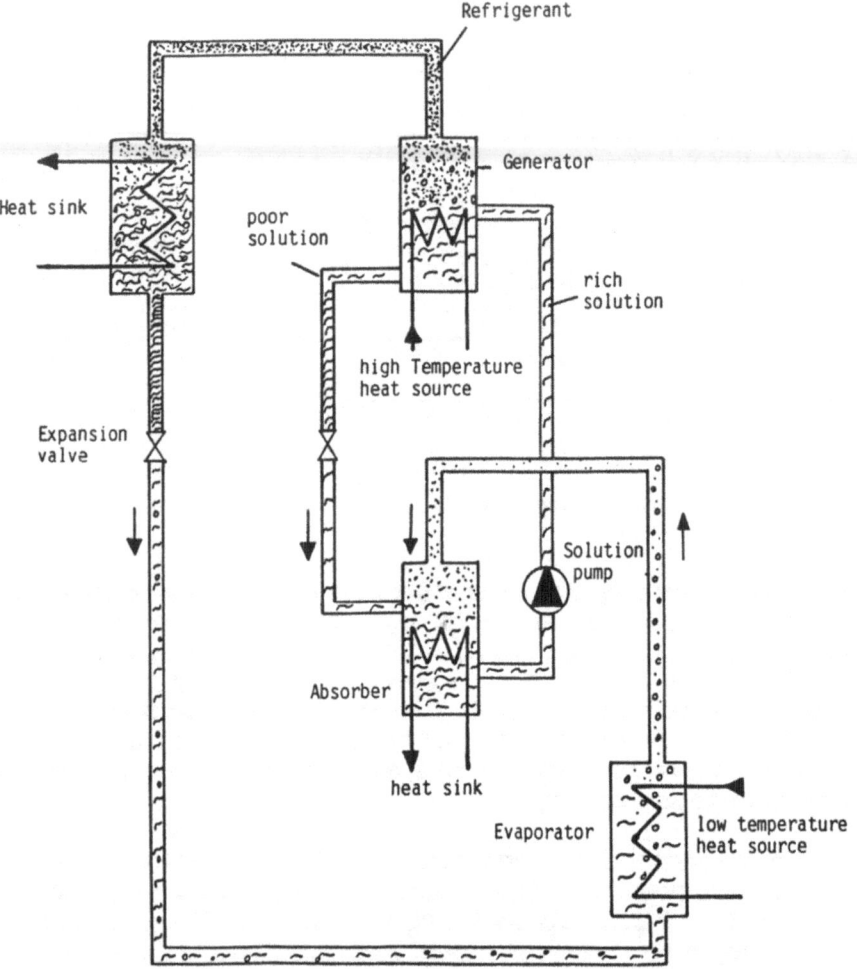

Fig. 7: Schematic of a single stage absorption heat pump.

Application of Heat Pumps

Karl Holzapfel
Director IEA Heat Pump Center

Heat pump technology is already a rather old technology. In 1852 Lord Kelvin proposed that heat pumps be used for space heating. Nevertheless, this application of heat pumps is still in an early stage of development since only recently has there been a need to use heat pumps in this manner. Only the oil shortage of the 1970's and the explosive increase in energy prices which followed provided the impetus to seriously attempt introduction of this technology into the large market for low temperature heat supply.

Since its invention in the 19th century, the traditional application of heat pump technology has been in the commercial cooling and refrigeration market. Here this technology did not have any serious competition. As standards of living increased after WWII a new market for this technology developed in the area of building air conditioning. This occurred mainly in the highly industrialized countries, such as Japan and the United States, which have regions with moderate to hot and humid climates. According to common terminology these units are called air conditioners rather than heat pumps since the heating effect (heat from the condenser) is not used. It was, however, a small step for manufacturers to produce the heat pump air conditioner, which could be used for cooling in summer and for heating in winter. For this reason the first major heat pump markets for space heating applications began to develop in the 70's in Japan and the USA in regions where, in addition to the cooling demand, a heating demand existed. (Fig. 1). In Japan preference was given to small single room air to air heat pumps while in the USA preference was given to centralized systems. These latter systems are also of the air to air type, however they deliver warm and cool air to the various rooms in a house from a central heat pump via a duct system.

In other countries, in particular those in Europe, hydronic heating systems are in use in the majority of residential buildings. These centralized systems are supplied with heat from a boiler fired with oil, gas, or solid fuels. In general no air conditioning equipment for cooling or dehumidification is in use in residential buildings. Considerable efforts have been made to develop heat pumps suitable for these applications. However, compared to the before mentioned applications in the US and Japan there are two major draw backs.

1.) The heat pump has to provide a higher temperature lift, due to low outdoor temperatures and high heating water temperatures.

2.) The investment for the heat pump can only be amortized while supplying heat, since its cooling capability is not utilized.

A. T. De Almeida and A. H. Rosenfeld (eds.), Demand-Side Management and Electricity End-Use Efficiency, 407–413.
© *1988 by Kluwer Academic Publishers.*

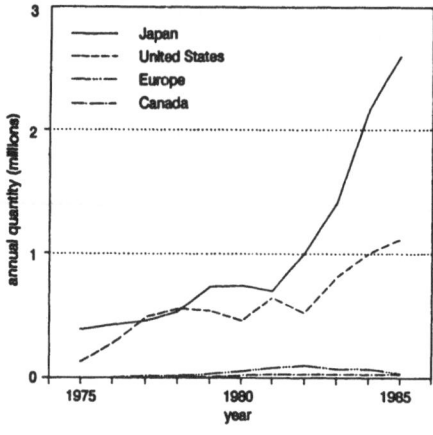

Fig. 1: Heat pumps for space heating

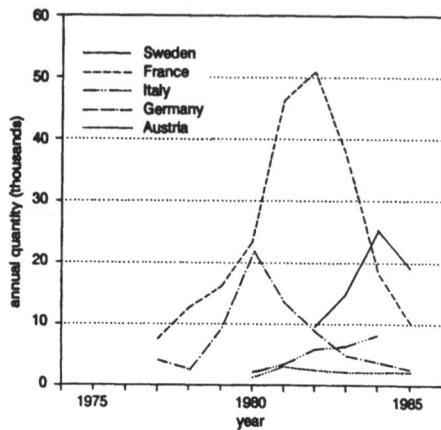

Fig. 2: Heat pumps for space heating in Europe

The second point means that heat pumps have severe competition from conventional boilers since they are just as capable of providing the necessary heat for an agree-able room environment as heat pumps are. The attractiveness of a heat pump heating system therefore depends heavily upon its potential energy cost saving which should also result in an overall heating cost saving.

Quite a number of differing approaches and concepts were developed in order to respond to the requirements of the heating only markets in Europe.

In the Skandinavian countries and in particular in Sweden preference, was given to large electric driven heat pump systems with heat pumps supplying heat into a district heating network and backed up by conventional boilers. While in Germany, Austria, and France preference was given to smaller electric heat pumps for individual hou-ses. Fig. 2 shows the development of electric heat pump sales in various European countries.

Besides these small electric heat pumps emphasis was placed upon the develop-ment of heat pumps utilizing fossil fuels directly instead of electricity as the driving energy. As a result internal combustion engine driven heat pumps were developed and installed mainly in Germany and total at present approximately 1000 units. These heat pumps have been successful only in the several hundred kW heating capacity range. Attempts to offer such heat pumps, as well as absorption heat pumps, in the small capacity range for individual homes have so far not been successful. The reason for this is that the rather high maintenance requirements and high first cost of engine driven heat pumps in the under 50 kW capacity range for space heating makes them uneconomical. Absorption heat pumps on the other hand have so far not been able to achieve a high enough efficiency in order to amortize their higher investment compared to conventional boilers in space heating applications.

Besides heat pumps for space heating a variety of other application areas have emerged. These include industrial process heat supply applications at temperatures below 100 °C. Although very important and also expandable in the near future, the number of heat pumps installed in this sector is very small compared to the installation numbers for space heating and cooling purposes (Fig. 3).

In the residential sector heat pumps for water heating have become important in some regions. This is in particular true for Germany were in the past years more tap water heat pumps have been sold than heat pumps for space heating. These heat pumps in general have a heating capacity below 3 kW and are often sold as compact units combined with a hot water tank of 200 to 500l capacity. Besides a considerable market in Germany heat pumps for water heating are also important to a certain extent in Austria, in Italy, and in the USA (see Fig. 4).

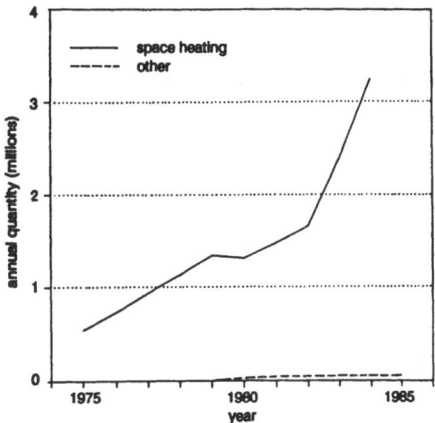 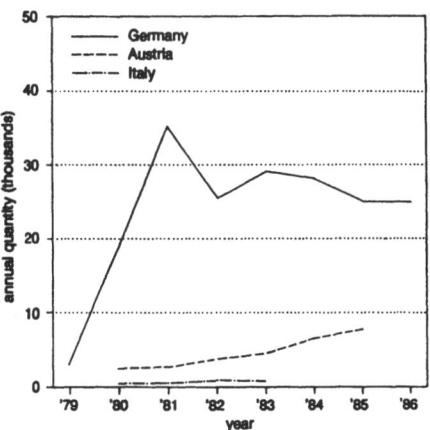

Fig. 3: International heat pump use   **Fig. 4**: Heat pumps for water heating

As mentioned above, heat pumps in the heating only markets have to meet rather high temperature lift requirements. For this reason, a variety of systems have been investigated and developed that are able to operate under these conditions.

The use of ambient air as a heat source for heat pumps is very convenient because of its general availability. However, the lowest temperature of this heat source occurs when the heating demand is highest. This creates a problem not only because the COP is low under such operating conditions, but also because the heating capacity increases rapidly as the outdoor temperatures increase and heating water temperatures decrease. This is particularly bad if the heat pump is sized for the design heat load of the building (at -12 °C). In this case the heating capacity of the heat pump will be more than 10 times as high as the heat load of the house at an outdoor air temperature of +10 °C. The design heating capacity is used for only a few hours per year and this has a rather negative impact on the economics of such systems. Different approaches are being used to overcome this problem. The simplest being, from the users point of view, using an electric resistance heater as a backup for low outdoor temperatures and sizing the heat pump for a balance point at higher temperatures (for example -3 °C). However, electric resistance backup heaters are in general not permitted by the electric utilities in central Europe. Therefore, air source heat pumps are in general combined with conventional boilers. In this arrangement the heat pump operates only in the higher outdoor temperature range. Its balance point being at about 50% of the buildings design heat load. The boiler is used to supply the necessary heat at lower outdoor temperatures. Such systems represent approximately 50% of space heating heat pump sales in Europe. The high heating water temperatures (more than 60 °C) required by older buildings at low outdoor temperatures can easily be provided by the boiler.

Heat sources other than ambient air are used in the remaining 50% of heat pump space heating installations. Ground water, for example, is very suitable because of its high and rather constant temperature level throughout the year. As a rule of thumb the groundwater temperature at a depth of 10 m can be assumed to be equal to the annual average ambient air temperature at the same location. A typical installation scheme that is being used in Germany is shown in fig 5. There are normally two wells required, one to extract the water from the ground, which is then pumped through the heat pump evaporator and cooled by a few K, and a second where the water is reinjected into the ground.

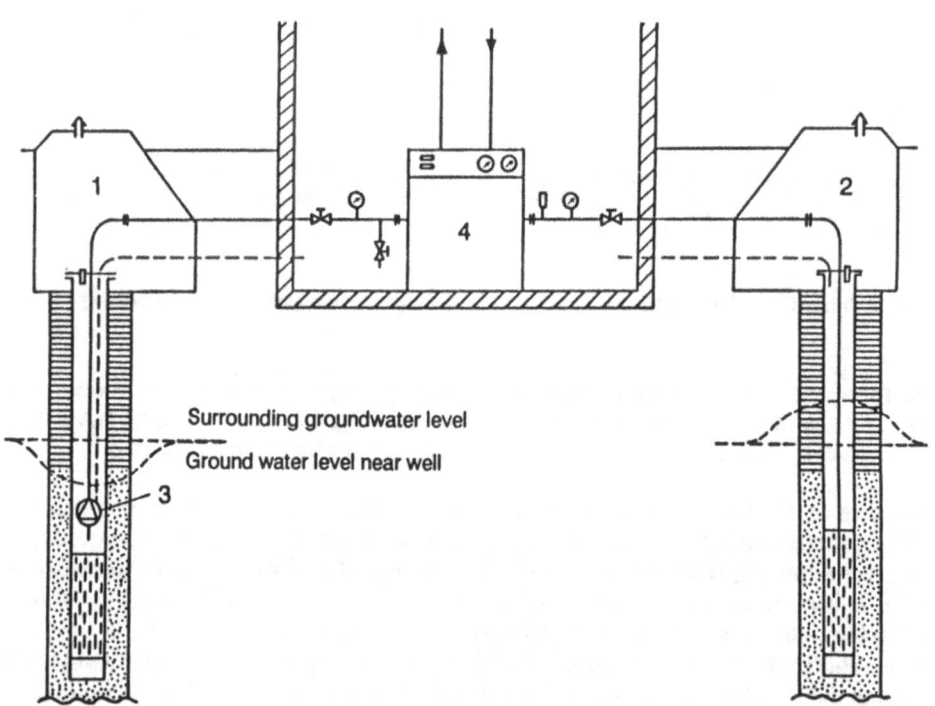

Fig. 5: Groundwater source heat pump: supply well (1); reinjection well (2); submersible pump (3); water-to-water heat pump (4)

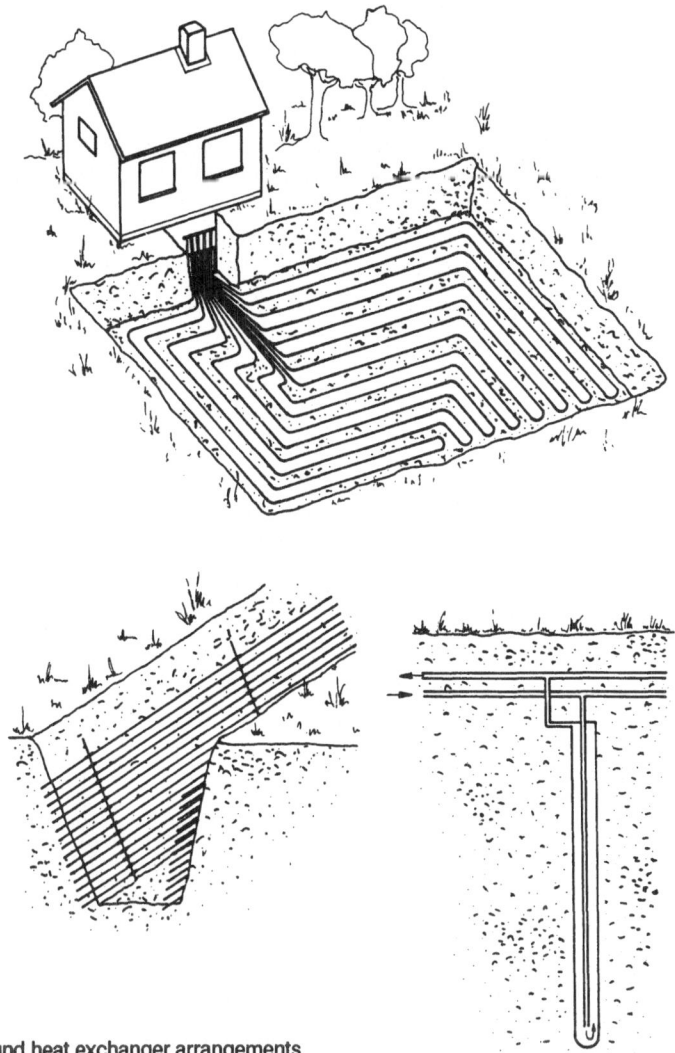

Fig. 6: Ground heat exchanger arrangements

The ground itself is also being used as a heat source if no ground water can be extracted or if it is not desired to extract water. In this case heat exchangers are embedded in the ground in a shallow (1 to 1.5 m below the surface) horizontal arrangement or in vertically drilled holes. Heat is extracted from the soil by either the refrigerant directly cycling through the ground heat exchangers or by an intermediate closed brine circuit. Some of the possible arrangements are shown in fig. 6.

In order to provide a better understanding of the situation of monovalent heat pump heating systems typically used in Central Europe, a ground source heat pump for a new detached house in Germany will be discussed. For this case, the design heat load of the building is 10.5 kW at -10 °C. The heated floor area is 160 m^2 and the size of the lot is about 500 m^2. For heat distribution in the building a floor heating system is used consisting of plastic tubes embedded in the concrete slab. The flow temperature required at design temperature (-10 °C) is 35 °C. This rather low value can only be achieved with good thermal insulation under the concrete slab, dense packing of the heat exchanger tubes in the slab, and sufficient thermal insulation of the walls.

The applied heat pump has a heating capacity of 11.9 kW with a COP of 3.8 at a heat source temperature of 0 °C, and a flow temperature of 35 °C. The ground heat exchanger is installed horizontally and consists of eight loops of polyethylene tube (outer diameter 25 mm) with a length of 100 m each. The tubes are connected to the manifold at the outside of the basement wall. Fig. 7 shows the flow scheme of the heating system.

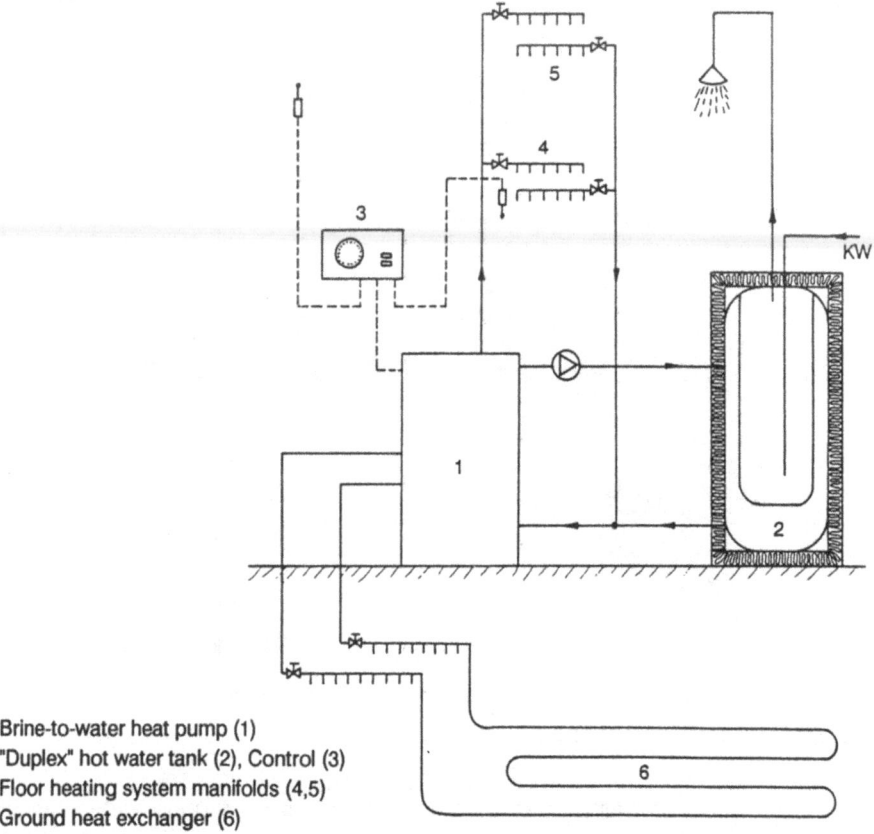

Brine-to-water heat pump (1)
"Duplex" hot water tank (2), Control (3)
Floor heating system manifolds (4,5)
Ground heat exchanger (6)

Fig 7: Flow scheme of a ground source heat pump heating system including tap water heating.

In addition to the space heating load, the heat pump also covers the hot water load. For this purpose a "duplex tank" is installed with a 330 l hot water inner tank and a 170 l heating water outer tank. The heating water transfers the heat from the heat pump condenser to the fresh water in the inner tank.

According to the German technical rules the annual heat demand of the building can be calculated by multiplying the design heat load with the full load hours. For the given location in northern Germany about 1650 full load hours can be assumed. This results in a net heat demand of about 17 000 kWh per year. This heat demand is distributed over the outdoor temperature as shown in fig 8.

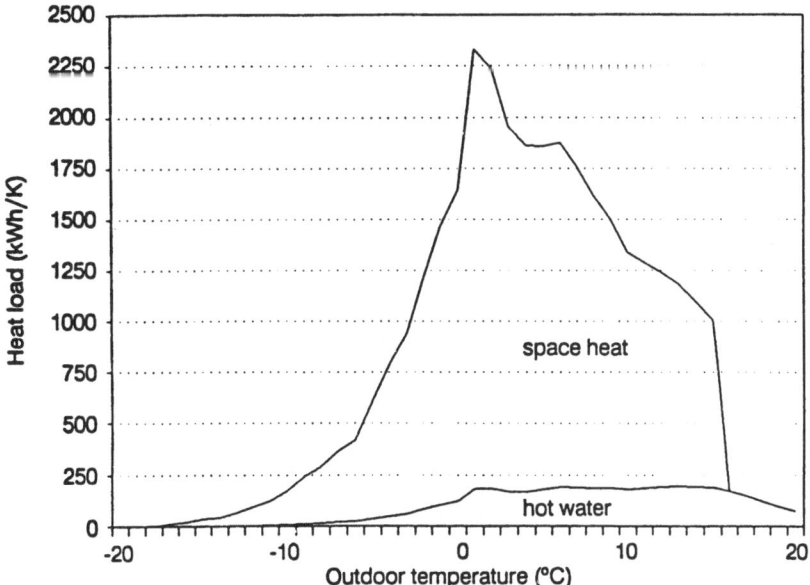

Fig. 8: Annual net heat load for space heating and tap water heating

The following table shows the energy demands of the heat pump and an alternative oil boiler for supplying a heat demand of 21 000 kWh/a. If both alternatives are compared using average prices for electricity (high/low tariff) and fuel oil, the heat pumps energy costs are 400 DM/a less than the energy costs for operating the oil boiler.

	heat demand kWh/a	SPF	energy demand kWh/a	electr. price DM/kWh/a	energy cost DM/
Space Heat	17 000	4.0	4 250		
Hot Water	4 000	3.3	1 215		
Heat pump	21 000	3.8	5 465	0.15	820
Oil boiler	21 000	0.85	24 700	0.05	1 235

The first cost for this heat pump system, excluding the floor heating system, is about 25 000 DM. This is 5 000 to 10 000 DM more than the cost of a modern oil boiler system including oil storage. These figures show that without subsidies such heat pump systems can hardly be economic with the present energy price situation. The second factor influencing this rather bad result is the first cost of the system. Theoretically, a considerable reduction of the first cost should be possible in particular if component prices in Europe are compared to those in Japan or in the US. However the very low sales figures and the special difficulties in integrating heat pumps into hydronic systems do not make such a first cost reduction very likely.

RESEARCH AND DEVELOPMENT IN THE AREA OF HEAT PUMP TECHNOLOGY

Karl Holzapfel
Director IEA Heat Pump Center

1. General:

Heat pumps are used today in a wide range of applications. Except for Japan and partly the United States of America where heat pumps have actually penetrated the residential air conditioning market to a great extent, heat pumps only have a very small share of their potential markets. This is in particular true for space heating of small residential buildings in heating only areas and for all kinds of industrial process heating applications.

This small market share is due to a number of reasons. The most important being that users are not convinced that a heat pump system will offer them the best cost benefit ratio from among a variety of potential alternatives. Thus the most important goal for current research and development activities is to

improve the cost benefit ratio of heat pump systems.

This important goal can be achieved through a number of different measures. These are grouped as follows into three subtasks:

(1) Reduce first cost by:
- decreasing the production cost
- simplifying the system design
- simplifying and improving the control system
- increasing the knowledge of installers and planning engineers
- developing improved concepts for heat source systems other than ambient air

(2) Increase efficiency by:
- improving compressors and heat exchangers
- reducing the need for defrosting (air source heat pumps)
- developing improved concepts for capacity modulation
- developing new working media (refrigerants)

(3) Increase reliability by:
- developing improved system control concepts
- developing components for longer service life
- developing better system integration concepts
- developing guidelines for preventive maintenance

A. T. De Almeida and A. H. Rosenfeld (eds.), Demand-Side Management and Electricity End-Use Efficiency, 415–421.
© 1988 by Kluwer Academic Publishers.

Besides research and development activities to achieve this top priority goal further additional activities are aimed towards achieving the following goals:

Increase the positive environmental effects of heat pump technology as compared to conventional heating systems by:

- developing improved concepts for cleaning of exhaust gases from heat pump drives using internal or external combustion
- investigating possibilities for replacing working fluids which have a negative impact on the environment (atmosphere, ground water)
- developing highly efficient heat pump systems in order to reduce primary energy consumption, which in turn reduces the production of energy conversion by-products (pollutants)
- decreasing operating noise levels of heat pumps

Widen the potential application range of heat pumps by:

- developing working fluids and components for high temperature applications
- developing working fluids and components for higher temperature lifts

2. Increasing the Cost Benefit Ratio of Heat Pump Systems

This overall goal is the most important of current research objectives. Its aim is to remove the most severe obstacle to further heat pump penetration in the huge market for space heating in regions with moderate to cold climates. This market is characterized by the general need for heating only in residential buildings as opposed to moderate to hot (and humid) regions that require cooling as the only or dominant space conditioning requirement.

The reason that heat pumps have had difficulty penetrating the heating only residential market is that this market has been covered for a long time by conventional techniques of producing heat through the burning of fossil or biomass fuels. In addition these conventional heating technologies have been developed to the point that they are very reliable and require little attention form the home owner.

The only reason that home owners began to look for another technology was the rapid increase in heating costs caused by the oil crises in the seventies. Since this is an economic reason a replacement system can only be successful in the long run if it actually offers an economic advantage to the respective decision maker. There are certainly additional reasons for seeking an alternative heating system such as general concerns about energy conservation or environmental protection. However it seems doubtful that these last two reasons, without the additional economic advantage, can provide a solid basis for a strong and continuous move away from well established combustion heating systems and into the new heat pump technology.

The following examples of ongoing development activities in the heat pump field will demonstrate the large improvement potential.

2.1 Reduction of First Cost

As a rule of thumb the first costs of installed heat pump systems for residential space heating applications are composed of 50% for the heat pump and 50% for the installation, excluding the cost of the hydronic heat distribution system.

Therefore, reducing the installation cost is just as important as reducing the heat pump cost. Both parts depend heavily on the market volume. However there are additional cost reductions possible which are independent of the market volume.

As opposed to conventional boilers and furnaces the price of a heat pump depends strongly on its size or capacity. In addition its heating capacity depends strongly on the temperature difference between heat source and heat sink. The heat load of a building depends on the outdoor temperature and is at its maximum for only a few hours during the year. For the major part of the heating period a heating capacity somewhat less than 50% of the design heat load is required. The consequence of this is that the part of the total investment that is required to provide the peak heating capacity will be paid back very slowly because of the small number of peak capacity utilization hours. In order to achieve a high profitability, this part of the investment must be as small as possible.

One alternative is to use inexpensive seasonal storage devices that are charged with energy from the environment when temperatures are high and are discharged when the ambient temperature is low. This type of heat storage flattens the heat pumps load characteristic which allows the size of the heat pump to be reduced. In addition its COP is increased because of the lower temperature lift required (higher heat source temperatures). However, The main first cost reduction effect results from avoiding the need for a conventional backup boiler and its auxiliary equipment such as fuel storage tank and chimney.

Such storage devices can be applied on the hot side of the heat pump, storing heat at the temperature level required for space heating, or on the cold side of the heat pump storing heat at a lower temperature level. In the latter case the low temperature heat is later transformed by the heat pump to the required temperature level when needed. For both concepts research activities are directed towards developing storage materials and devices where energy is stored as latent heat from phase changes or chemical reactions.

Another possibility providing similar results is the development of improved concepts for utilizing heat sources other than ambient air. One such heat source that is widely available is the ground. It is a natural seasonal storage for solar energy and has similar advantages as the before mentioned artificial seasonal storage devices. As a cold storage it can provide latent heat from the phase change of water to ice which allows the year round heat source temperature to stay at or above 0 °C.

Although the ground has been used as a heat source for more than 20 years, it never really achieved a market breakthrough. The largest hindrance to this is that the installation of the heat exchanger in the ground requires considerable on site work. Some of the other problems and special requirements are as follows. For horizontally embedded heat exchanger tubes large ground areas are required that typically are either not available or are used in a way that prohibits the installation of such heat exchangers. Vertically installed heat exchangers have been rather expensive and additionally no sound design and sizing criteria are available. These systems are further complicated by the need for an intermediate closed loop heat transfer circuit that transports heat from the ground heat exchanger to the evaporator of the heat pump.

Current development is aimed at replacing this intermediate brine loop by the refrigerant circuit directly (direct expansion systems) and at developing cheaper installation concepts for vertical systems. Also the development of well based dimensioning criteria and rules, which take into account the differences in soil and ground characteristics, and climatic conditions, will allow a further cost reduction by helping avoid unnecessary oversizing.

Despite its disadvantages ambient air as a heat source for cold regions is still attracting a lot of research and development interest. Most activities aim at increasing the heat pump heating capacity at low outdoor temperatures without increasing the first cost of the heat pump too much. In the past a commonly used method was to integrate an electric resistance heating element in the heat pump in order to cover a peak demand. This concept is a very inexpensive and simple method from the point of view of first cost. However, with high electricity prices the operating costs are rather high and in addition this system is very often not accepted by electric utilities because of its negative impact on the utilities load characteristic.

Other concepts include the use of refrigerant mixtures with the possibility of varying the composition according to the operating conditions. In this way heating capacity at low ambient temperatures can be increased for a very low additional investment. A loss in efficiency is possible, but the COP is still higher than if the electric resistance heater was used to increase capacity.

Another approach is to equip electric heat pumps with a frequency inverter by which the compressor speed and subsequently the heating capacity can be varied over a wide range. The main advantage of this variable speed operation is that a compressor can be operated, if designed appropriately, at a capacity that is two to three times its rated nominal capacity. The necessary changes in the electric motor and in the compressor itself require only a low additional investment. If this technique is applied in heating only heat pumps, considerable cost reductions seem to be possible if a single speed heat pump is compared to a variable speed heat pump on the basis of their respective maximum heating capacities at lowest operating temperatures.

This technology is being widely used in Japan today for heating and cooling heat pumps in moderate to hot climatic regions. These regions represent almost 80% of the Japanese air conditioning market. Inverter driven heat pumps have already achieved a 50% share of all heat pumps sold in this market and this amounts to more than 1 million inverter driven units sold in 1986. The indicated cost reduction effect for heat pumps to be developed for cold climates (heating only operation) can be deduced from the prices of these mass produced components. An analysis undertaken by the Heat Pump Center at the Analysis Center in Graz, Austria shows that this positive effect can actually be achieved for European heating-only applications.

3. Increasing the Efficiency of Heat Pumps

The investigation of inverter driven heat pumps also fits in with other activities aimed at increasing heat pump efficiency. Although the COP of a heat pump will be reduced by adding a frequency inverter between the electric supply and the motor, the seasonal performance factor (SPF) increases. This is because with the adjustable hoating oapaoity, looooo oauood by on-off operation and by operation with condensing temperatures higher than necessary and evaporation temperatures lower than necessary will be avoided. Since these improvements might also be accompanied by a reduction in first cost it becomes obvious that this concept deserves more attention.

Some of the preliminary findings of a computerized simulation and analysis of this concept applied to heat pumps for space heating with hydronic systems in small residential buildings in Austria are as follows:
Large SPF improvements can be obtained if the heat pump is sized for monovalent operation and has to operate with a wide heating capacity range. Presently available compressors that are designed for variable speed operation can only be used in applications with a low temperature heat distribution system and with heat source temperatures that are not too low. For the higher temperature lifts required in applications having high temperature distribution systems and low temperature heat sources, such as ambient air in monovalent operation, suitable compressors for variable speed operation need to be developed.

The following diagrams show the preliminary results of simulations by H. Halozan comparing ground water source heat pumps without an inverter (fig. 1 and fig. 2) to those with an inverter (fig. 3 and fig. 4). The heat pumps, in this case, are applied to a detached house in Austria which has a low temperature heat distribution system. As shown in figure 1 the capacity of the heat pump without an inverter is large at higher outdoor temperatures. This extra capacity, since it is not required by the house, leads

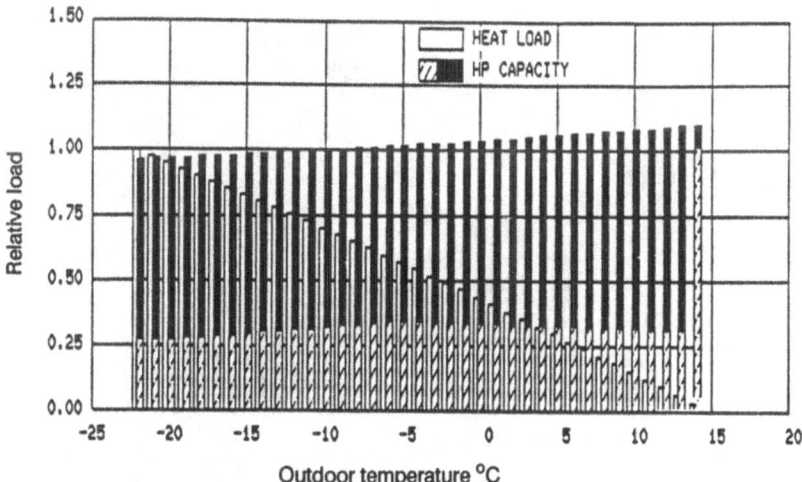

Fig. 1: Heating capacity of a ground water source heat pump without inverter and heat load of a detached house in Graz, Austria in dependence of outdoor temperature.

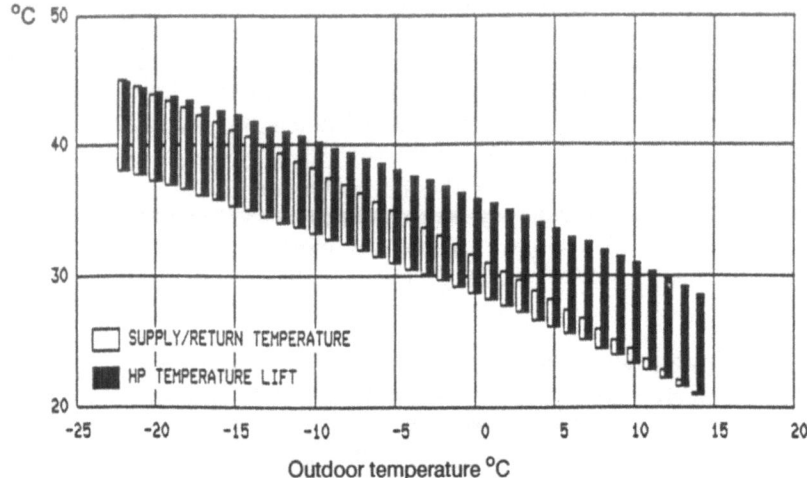

Fig. 2: Flow and return temperatures of a hydronic floor heating system in a detached house in Graz, Austria, (same house as in fig. 1) and heat pump supply temperature (without inverter) in dependence of outdoor temperatures.

to considerable losses. The seasonal performance factor (SPF) in this case was calculated to be 3.1. If the same house is supplied by a ground water heat pump having the same maximum heating capacity, however equipped with a frequency inverter that allows for a compressor speed operating range of 1:5, the losses can be reduced significantly. Figure 3 shows the resulting heat pump heating capacity characteristic and figure 4 the corresponding heating water temperatures. In this case the calculated SPF is 4.1 which represents an improvement of more than 30% compared to the system without an inverter. As an intermediate case an inverter with a speed operating range of 1:2.5 was simulated resulting in a SPF of 3.6.

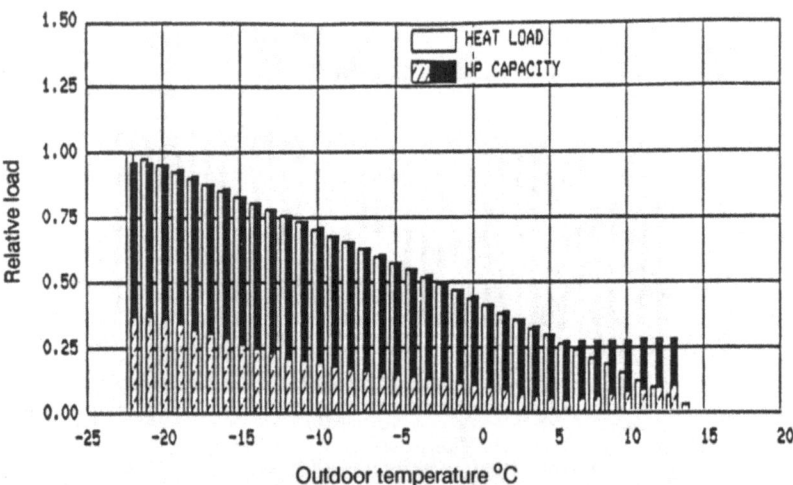

Fig. 3: Heating capacity of a ground water source heat pump with inverter with compressor speed range 1:5 and heat load in the same building as in fig. 1.

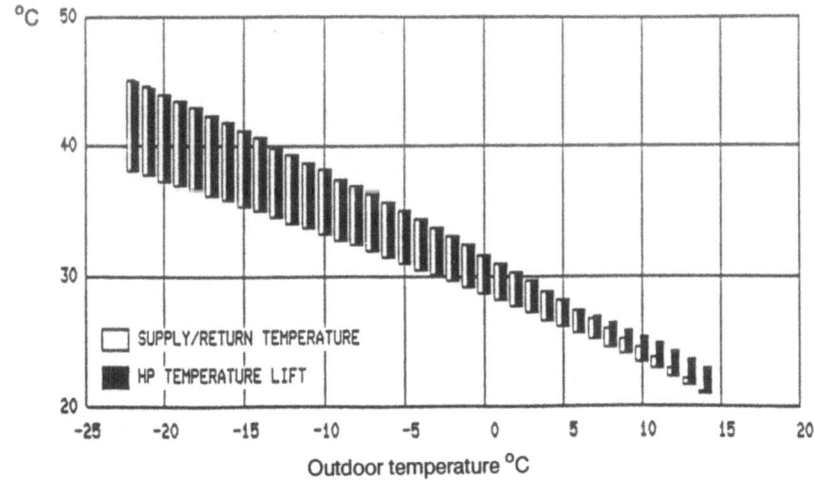

Fig. 4: Supply temperatures of a heat pump with inverter (1:5) in the same floor heating system as in fig. 2.

4. Increasing the Reliability of Heat Pump Heating Systems

Developments and investigations with the objective of increasing the reliability of heat pumps are of similar importance as the two former ones (2 and 3) in increasing the cost benefit ratio of heat pump systems. Because of the special characteristics that clearly distinguish heat pumps from conventional heat generating devices, special attention has to be given to the integration concept and overall heating system control. For example, insufficient heating water mass flow through the condenser will rapidly and inevitably reduce the SPF and may even result in a heat pump failure. This problem may arise if the heat pump and heat distribution system are mismatched. One the approach used to avoid such problems is to decouple the heat pump and the heat distribution system through the use of a hot water storage buffer, hydraulic decoupling hot water volume, or an appropriate flow scheme. These concepts however add to the first costs and complicate on site installation. A cheaper alternative may again be the use of heat pumps with continuous capacity modulation to compensate for the mismatch between heat pump capacity and heating requirements.

As indicated before, most of the present work in the area of heat pumps is aimed at developing components and systems which will make this technology competitive in the low temperature market. Theoretically the heat pump is the answer to the requirement of maintaining the standard and quality of life and simultaneously saving energy. The heat pump technology provides a tool for intelligent end use management. It is however still necessary to put considerable effort into further development of presently available concepts and components in order to actually make this technology attractive and competitive.

DESIGN OF EFFICIENT HVAC SYSTEMS

J.W. JONES

Department of Mechanical Engineering
The University of Texas at Austin
Austin, Texas, U.S.A.

INTRODUCTION

The acronym HVAC is a shorthand designation for heating, ventilating and cooling often used in describing air conditioning systems. The HVAC designation is used as the common notion is that air conditioning is just cooling. However in the broadest sense air conditioning encompasses all those heating, cooling, humidity control, ventilation and air filtration, and air distribution processes necessary to maintain a healthy, comfortable, and productive environment in buildings. An efficient air conditioning, or HVAC, system will also incorporate the design characteristics and the control functions necessary to achieve these goals at a minimum cost and with the minimum expenditure of energy.

The primary objective in building design is to provide a healthy and productive environment for human occupancy with a certain level of amenity. The definition of what is required to meet these objectives, and what the building owner is willing to pay to achieve them, has varied over time. Both occupant perceptions of the level of comfort and amenity required and the advance of technology have significantly changed HVAC system design over the past 50 years. The advent of the energy crisis in 1973 also brought a change, an increased awareness of the energy use and energy cost of air conditioning. Although energy concerns have receded in recent years the cost of energy and problems associated with peak demand do remain items of some concern. The intent of this discussion is to examine the function of HVAC systems and to consider how systems may be designed to meet a buildings objectives; comfort, productivity and amenity, in a cost effective and energy efficient way.

THE HVAC SYSTEM

Thermal comfort requires the control of temperature, humidity and air movement. These parameters must be held within a range of variation which is generally much less than is common in the external environment. A building's envelope is intended to provide a means of moderating climate variability. However in most buildings an air conditioning system is also required to meet current levels of occupant expectation. Heating or cooling loads are generated by differences between indoor and outdoor temperature and humidity levels, solar gain, and the heat release from lights, equipment and occupants within the conditioned space. In the situation where heating is required there is more heat lost through the building envelope than is provided by the solar and internal gain. Cooling is required when there is more heat from solar and internal gain than can be dissipated to the environment through the building envelope. In either case energy is required to add or extract heat and to provide air movement within the space. Figure 1 indicates the processes involved in accomplishing this. The varying size of the circles used to schematically represent these processes indicates a variation in the energy use or energy transport associated with each of these processes.

In Figure 1a the three circles represent; first the energy required to maintain thermal comfort, second the energy transported for this purpose plus the energy required to transport the heating energy (fans and/or pumps) and third the energy input to the primary heating equipment. The operation of an HVAC system in the heating mode is relatively straight forward, the system simply provides and transports the amount of energy needed to offset the net heat loss from the building and satisfy ventilation requirements. In this

A. T. De Almeida and A. H. Rosenfeld (eds.), Demand-Side Management and Electricity End-Use Efficiency, 423–434.
© 1988 by Kluwer Academic Publishers.

case the energy input from the fans and pumps is a benefit both in terms of transport and temperature rise. The only losses are those associated with any inefficiency in the energy conversion process in the heating equipment. In some circumstances additional energy may be required to maintain an acceptable minimum level of humidity in the space.

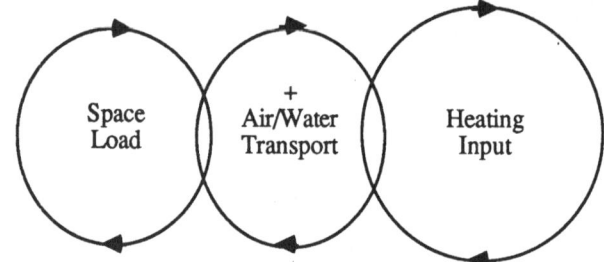

Figure 1a Energy Flow in an HVAC System - Heating Mode

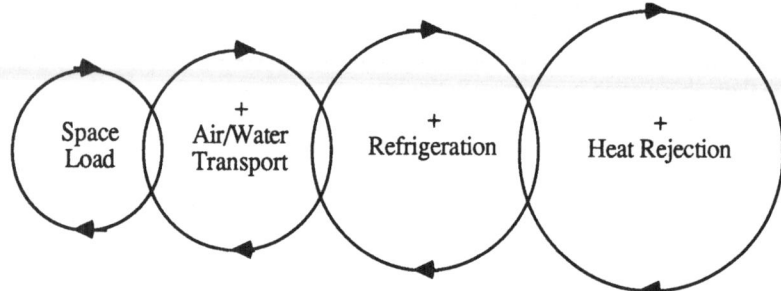

Figure 1b Energy Flow in an HVAC System - Cooling Mode

The cooling operation of a HVAC system is somewhat more complex. Figure 1b illustrates the four energy flows involved in the cooling process. The first circle represents the energy which must be extracted from the space, and the second the energy extracted plus that required for transport. In the cooling case the third circle represents the extracted energy, the transport energy, plus the energy required to operate the refrigeration system compressor. The forth and largest circle represents the total energy that must be dissipated from the cooling system to the environment. This energy flow is somewhat larger than that of the third element due to additional transport energy (again for fans and pumps).

HVAC System Characteristics

The ideal air conditioning system would provide just enough heat to offset the net heat loss from a space or extract just enough heat from the space to offset the net heat gain to the space. It would also assure appropriate levels of air quality, air movement and humidity within the space with the minimum expenditure of additional energy. Some simple single space systems can approach the ideal in terms of the heat addition or extraction but they often fail in terms of maintaining air quality, air movement and humidity control. In a building of any size these single space systems are distributed (multiple unit) systems and also suffer disadvantage in terms of control and maintenance. Larger central systems come closer to meeting the air quality, air distribution, control and maintenance criteria but often do not do as well in terms of minimizing energy use for heat addition or extraction. In the end HVAC system design is a matter of balancing all the criteria to achieve a satisfactory solution for comfort, operation, cost and efficiency. Achieving this balance is in the end the real challenge in HVAC system design.

One of the primary challenges in this design effort is the selection and control of the air distribution system. Figure 2 provides a schematic representation of the most basic HVAC system configuration that might be found in a commercial building. There are three air flows of concern in all systems: the outside air, which must provide enough fresh air to meet the ventilation requirements; the return air, which is that portion of the air flow through the conditioned spaces that is returned to be

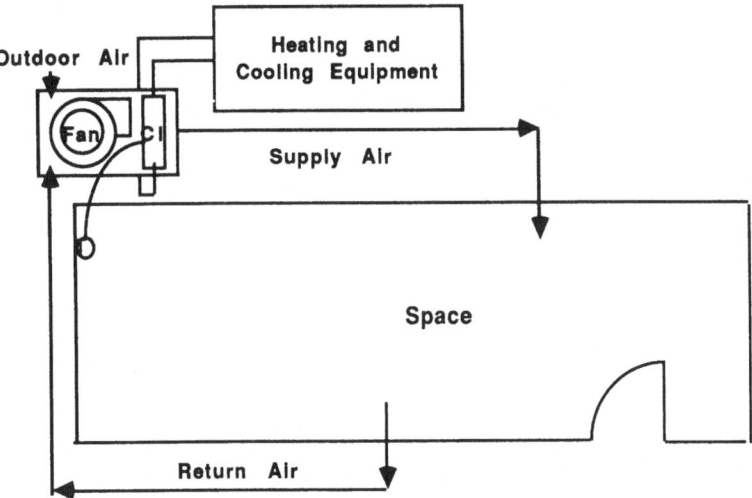

Figure 2   Basic Single Space HVAC System

mixed with outside air, recooled and recirculated to the spaces; and the exhaust air, which vents approximately the same amount of air as is brought in from the outside. The balance of the outdoor air and return air flow depends upon ventilation requirement, which sets the minimum outdoor air flow, and the temperature and humidity content of the outdoor air. When the outdoor air has a lower energy content than the return air it is more economical to cool it than to recool the return air. Systems that monitor outdoor air conditions and vary the amount of outside air to minimize energy use in the cooling system are referred to as economizer systems. The effectiveness of an enconomizer is quite dependant on climatic conditions. This will be discussed further later in this presentation.

HVAC System Control

One of the primary challenges in designing an efficient HVAC system is in the selection and control of the air distribution system. The system shown in Figure 2 uses a single duct to supply a single space, or a number of relatively open spaces. In such an arrangement whatever heating and cooling is necessary is done centrally and the conditioned air is distributed directly to the space, or spaces, under the control of a single thermostat. The control usually cycles the fan and refrigerant compressor or heating unit on and off. The system may also have a continuous fan and cycle the compressor or heating unit. It may operate in either a constant volume or variable volume mode although the vast majority are constant volume.

Single duct systems are however not limited to single control point applications. The many of the thermal distribution schemes found in commercial buildings are expansions or modifications of the basic configuration shown in Figure 2. These modifications have been developed to serve multiple spaces with varying loads,comfort, and ventilation requirements more effectively. The primary distinctions between the systems commonly used are whether there is more than one supply air distribution path, whether the fan is constant volume or variable volume, and the manner in which heating is provided and controlled. Figures 3 and 4 provide illustrations of two common single duct multiple space system configurations. Figure 3 shows a constant volume, variable temperature reheat system and Figure 4 a variable air volume (VAV) system. There are a considerable number of variations on each of these systems in use. Only the generic characteristics of these systems are described here.

The single duct constant volume reheat system shown in Figure 3 provides central cooling and either terminal reheating at the air distribution box or some form of in-space heating such as electric baseboard heaters. The terminal reheat systems are quite common as they provide a great deal of flexibility and good humidity control. The supply air is first cooled and dehumidified in the central system and then tempered (reheated) to meet the load in the individual spaces. These systems are unfortunately energy wasteful as they simultaneously cool and heat air to meet the load in a space. There are several control options that can be utilized to minimize the energy use which should be considered.

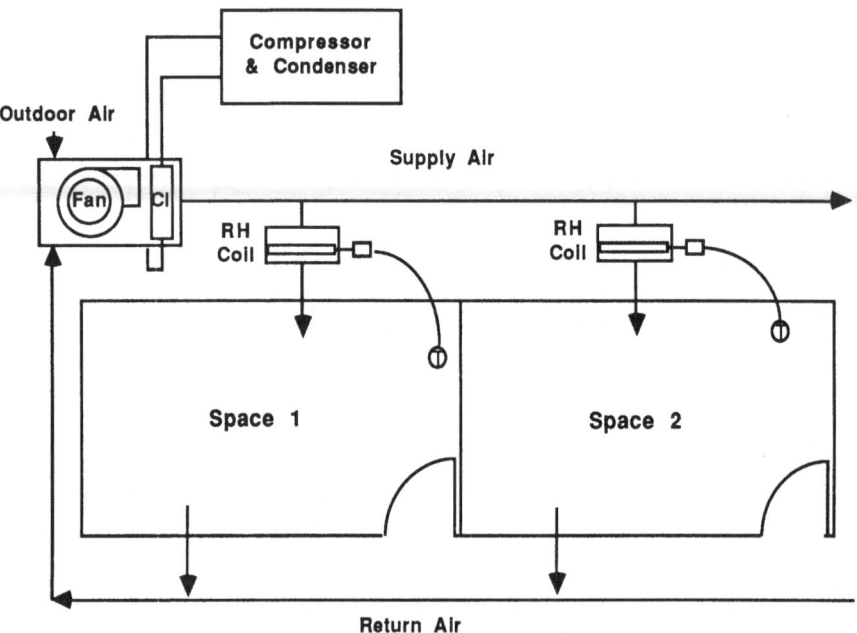

Figure 3 Multiple Space Single Duct Constant Volume System

Figure 4 illustrates a single path, variable volume system. The VAV box shown for the interior zone simply regulates the volume flow of air at a constant temperature. As the loads in internal zones are usually quite uniform this arrangement is usually satisfactory. If however some heating is required the VAV box can be equipped with a reheat coil which operates only at the lower end of the air flow range. The VAV box illustrated for the exterior space where heating is regularly needed indicate another approach for providing heating with a minimum of energy use. In this case under heating conditions the volume of cooled air from the central system first drops to the minimum required for ventilation. At this point air from the space, or from the plenum above the space, is drawn in by the small local fan in the VAV box and heated as necessary to offset the heating load. The small fan uses little energy and the coil responds directly to the heating needs of the individual space thus reducing overall energy use.

The differentiation between constant volume and variable volume operation can best be illustrated by considering the basic relationship in energy transport:

$$Q = c \cdot V \cdot \Delta T$$

where   Q   = the energy transport rate   c   = the product of the density and
            V   = the volume flow rate   the specific heat
           ΔT = the temperature difference

It is clear from this relationship that the energy flow rate can be varied by varying the volume flow rate or by varying the temperature difference. There two primary advantages in varying the volume flow rate. First the fan power requirement is proportional to the cube of the volume flow rate. Thus a 50% reduction in volume flow rate provides an 87% reduction in fan power. Second varying the flow rate minimizes simultaneous heating or cooling or energy use for reheat. Unfortunately the VAV air distribution boxes which control the air flow into the conditioned space are more complex than those used in constant volume systems. Thus additional first cost and maintenance costs may offset some of the energy cost savings in VAV systems.

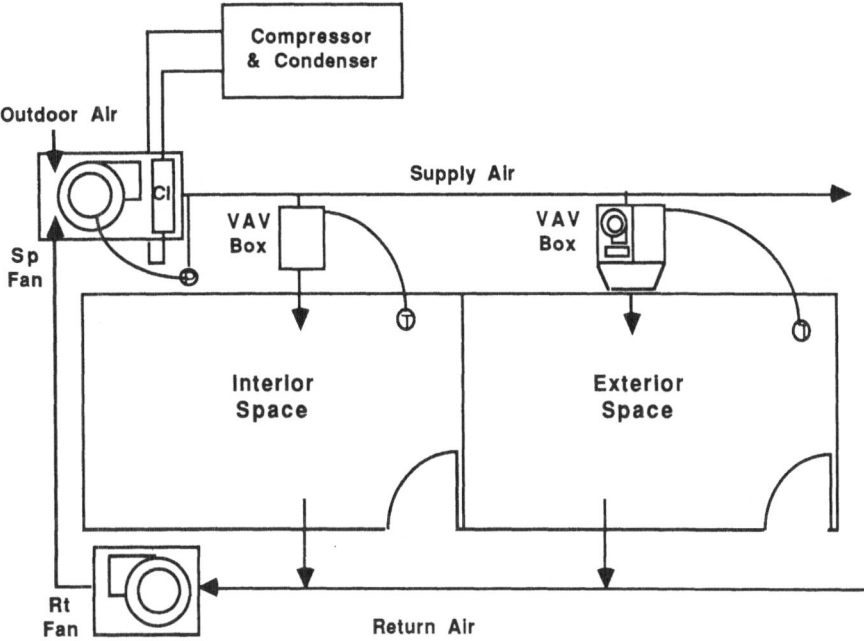

Figure 4   Single Duct Variable Air Volume System

The primary occupant objection to the use of VAV systems is the variable nature of the air flow into the space. If the air flow drops too low the space becomes uncomfortable. The type VAV box illustrated in Figure 4 provides better air circulation within the space than with the straight VAV systems and it rates well in terms of occupant comfort. It is also more economical to reheat space or plenum air than to reheat supply air, even at the minimum volume flow rate. Unfortunately VAV boxes of this sort represent a larger first cost and maintenance costs.

Figure 5 shows a dual duct (dual supply air path) system. Dual path systems, provide for both central heating and cooling. The streams of warm air and cool air are mixed in terminal boxes at the space to provide the proper air temperature for conditioning the space. The mixing is accomplished with a simple mechanical damper providing relatively maintenance free control. The dual path systems are very flexible and are able to accommodate significant variations in space use and load. They are generally highly rated in terms of thermal control and air movement for occupant comfort. Unfortunately since they again simultaneously heat and cool air they can be inefficient in the use of energy. Dual path systems are obviously more expensive to install than single path systems because of the increased cost of duct work and labor for installation.

Dual path systems can operate in the constant volume or variable volume mode. Most existing dual duct systems are constant volume but many are being converted to variable air volume to reduce energy use. Some designers argue that there are alternative reset control schemes that can provide comparable savings without the complexities of VAV.

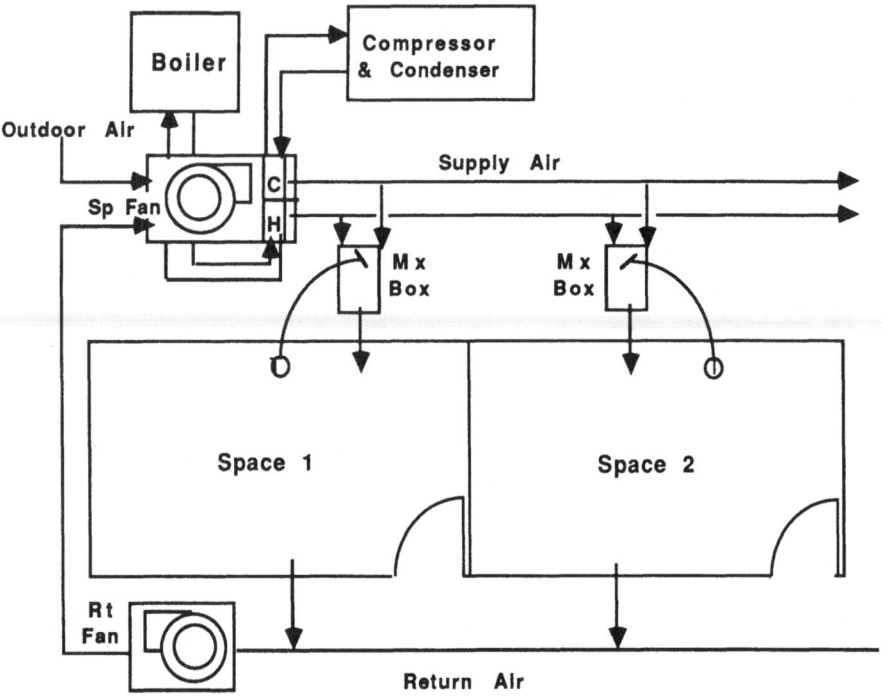

Figure 5 Dual Duct Air Distribution System

Design Criteria

The HVAC systems of Figures 2 through 5 illustrate only a few generic configurations to point out a number of basic design options. A designer could use several of these systems, or combinations of the system concepts represented by these figures in a single building. There are an almost infinite combination of system components and controls. Despite this wide variety of HVAC systems there are a few basic rules of good design that can always be applied. These rules are simple and represent what might be called the 'common sense' approach to design. Briefly stated they are:

- Avoid simultaneous expenditure of energy for heating and cooling in the same space.
  Use separate systems for spaces expected to operate on widely differing operating schedules.
  Use separate systems for spaces which have widely differing load patterns (interior versus perimeter spaces, spaces with large internal gains etc.).
- Provide control for operating in an occupied mode and an unoccupied mode (control for scheduling or sequencing fans, compressors, ect.). These controls should provide for a gradual change in space setpoints to avoid large startup loads (load

peaks). In the unoccupied mode ventilation and exhaust systems should be off if at all possible.
- Utilize heating or cooling recovery and economizer systems wherever possible
- Provide for reset of control setpoints to minimize energy use. Use anticipatory control wherever possible.
- Minimize energy expenditure for energy transport (minimize flow quantities). Energy should be transported by the most efficient means possible. In order of priority

    Electric Wire or Fuel Pipe
    Two Phase Fluid
    Single Phase Fluid
    Air

    Maintenance and operation considerations are often the deciding factors in deciding what means of energy transport are possible.

Again these rules have to be applied within the context of productivity, cost and amenity criteria.

In addition, flexibility of operation and adaptability to changes in the function of the space in a building are major concerns. The lifetime of a building may range from 50 to 100, or in some cases even hundreds of years. The useful lifetime of the HVAC equipment in the building is much shorter, perhaps on the order of 15 to 20 years. The lifetime of the fluid transport systems (air and water) are somewhere in between. The air and water distribution system usually can be replaced or modified, but at considerable expense.

At the other extreme the use and occupancy 'lifetime' for functions within the building may be on the order of a few years, and in some cases perhaps even a matter months. The design of an efficient and effective HVAC system must weigh the likelihood for change in a particular building and provide some reasonable measure of flexibility to accommodate potential change. This need for flexibility and adaptability is a major issue in HVAC system design. Unfortunately it is an issue that is not handled well in many instances.

HVAC SYSTEM ENERGY USE CHARACTERISTICS

HVAC System Component Performance

HVAC system component performance can be best illustrated through the use of a specific example. Figure 6 provides such an example. The space loads and energy flow shown are for a cooling system in a building located in the Southwestern U.S. The first column indicates the space cooling load (sensible and latent). This is the load attributable to

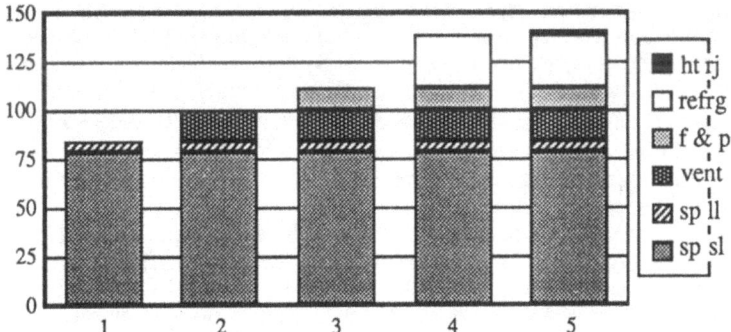

Figure 6 Components of Cooling System Load

the requirement for thermal comfort in the space. The second column adds to this the outside air loads (sensible and latent) showing the increase due to the requirement for ventilation. The third column adds the power requirements of fans and pumps. This increment represents both an additional load on the refrigeration system and a direct power requirement for transport. This total of the space and ventilation loads plus the energy input to the fluid stream by the fans and pumps represents the load on the refrigeration system. The forth column adds to this the refrigeration system power input and indicates the load on the heat rejection system (condenser). The final column adds the power required to operate the heat rejection system. The total then represents the heat load on the environment resulting from the space cooling loads. These data are shown without units have been normalized to give a value of 100 for the space plus ventilation load to simplify the example which follows.

Tao[1] proposed a useful procedure for examining the efficiency of an air conditioning system and each of its components in a 1984 article. He developed his procedure as an extension of the air transport factor concept introduced in ASHRAE Standard 90-75[2]. The following definitions are required to follow through the procedure:

ATF	= air transport factor	$= Q_s / P_f$	Q_s = sensible heat removed by air
WTF	= water transport factor	$= Q_t / P_p$	(inc. infiltration. but exc.ventilation)
RTF	= refrig. transpt. factor	$= Q_t / P_r$	Q_t = the total system cooling load
HRF	= heat rejection factor	$= Q_t / P_{hr}$	(inc. vent. but excl. fans and pumps)
STR	= sensible heat gain to total		P_f = fan power input
	system cooling load	$= Q_s / Q_t$	P_p = pump power input
CSP	= cooling performance	$= Q_t / P_t$	P_r = refrigeration comprsr power
			P_{hr} = condensr fan and pump power
			$P_t = P_f + P_p + P_r + P_{hr}$
			= total system power

The individual components of the system can be rated and compared to other possible component selections in terms of the ATF, WTF, RTF, HRF and their overall combination rated in terms of the cooling performance factor CSP. The one drawback of this procedure is that it does not examine the consequence of the selection of some of the distribution system control elements.

Transport Factor Analysis Tao's procedure is best described with an example. The data in Figure 1 is displayed in Table 1. Table 2 provides data based on seasonal data. Again the data has been normalized and is displayed without units. In these illustrations System A is a central system consisting of a variable air volume air distribution system, a single centrifugal chiller and associated pumps, and a single induced draft counterflow cooling tower. System B consists of a constant volume, variable temperature air distribution system with two air handlers, two water cooled reciprocating chillers and two induced draft, crossflow cooling towers. Each of the components of System B have half the capacity of the comparable components of System A. System C consists of a constant volume air distribution system with multiple air handling units and four air cooled direct expansion condensing units. Data for component efficiencies were taken from Tao[1] and Section 10 of ASHRAE/IES Standard 90.1P[3]. There are many combinations of components that could have been selected. The three systems described were chosen simply to illustrate the process.

The system analysis results indicate that there are significant differences between component performance for the three systems. The major differences occur for the fans and refrigeration equipment. The VAV fans have a lower AFT at peak load primarily due to the additional pressure loss associated with their inlet vane control. The annual VAV system AFT (AFTS), on the other hand is considerably higher as a result of the power savings associated with the lower average flow volume (assumed to be approximately 75% of the peak volume rate) for the VAV system on an annual basis. The seasonal comparisons of the refrigeration systems are based on an integrated part load performance value (IPLV) procedure developed for the ASHRAE/IES Standard 90.1P. The part load performance criteria were developed to represent, at least in an approximate way, the seasonal performance characteristics of the equipment. The IPLV values indicate for example that the seasonal efficiency of a centrifugal chiller is higher than its peak load efficiency while the seasonal efficiency of a reciprocating machine with suction valve unloading is less than its peak efficiency. Again the combined effect of the equipment selection is reflected in the annual CSP (CSPS).

Table 1 Peak Cooling Loads and System Comparison for Three Systems

Peak Cooling Loads	A	B	C		A	B	C
space sensible load	78	78	78				
space latent load	7	7	7				
ventilation load	15	15	15				
total cooling load	100	100	100		A	B	C
fan energy	9.5	8	8	AFT	8.3	9.8	9.8
pump energy	2	1.5	0	WFT	50	67	
refrigeration energy	30	38	43	RFT	3.3	2.6	2.3
heat rejection energy	3	2	0	HRT	33.3	50.0	
total power	44.5	49.5	51	CSP	2.2	2.0	2.0

Table 2 Annual Cooling Loads and System Comparison for Three Systems

Annual Cooling Load	A	B	C		A	B	C
space sensible load	83	83	83				
space latent load	7.5	7.5	7.5				
ventilation load	9.5	9.5	9.5				
total cooling load	100	100	100		A	B	C
fan energy	4.2	8.5	8.5	AFTS	19.7	9.8	9.8
pump energy	1.5	1.5		WFTS	67	67	
refrigeration energy	28.5	42	49	RFTS	3.5	2.4	2.0
heat rejection energy	2	1.3	0	HRTS	50.0	76.9	
total power	36.2	53.3	57.5	CSPS	2.8	1.9	1.7

The magnitude of the various component terms shown in both Tables 1 and 2 also show the relative energy intensiveness of each of the HVAC system components. The pumps and heat rejection components have high factor values and represent relatively small energy requirements. The fans and the compressors with low factors represent relatively large energy use. This of course focuses attention on those components that use the most energy.

The CSP and CSPS values indicate that there is a greater variation in terms of seasonal performance than there is at peak load for the systems selected for the comparison. Of course there are ways to modify the performance values for each of the components by selecting different equipment or control options. However the point of the example is to illustrate a procedure is useful in examining the energy consequences of choosing one or another component in the design of HVAC systems.

System Control Impact

The control strategy used for temperature setpoint and the selection and control of an air transport system has a clear impact on energy use. These options need to be carefully examined during design. The implications of a few common control and system options will be briefly considered here to illustrate possible approaches to the analysis of system an control characteristics. This will be accomplished by examining the cooling coil loads for a five zone floor of an office building in a Southwestern U.S. city. Again the data has been normalized to simplify presentation.

Temperature Setpoint Control The space temperature setpoint options relate to scheduling and the selection of a setpoint, or setpoints if a dead band thermostat is used.

The cooling coil discharge temperature is another important control setpoint where the options are a fixed cooling coil discharge temperature versus a coil discharge temperature reset on zone cooling demand or outside air temperature.

Figure 7 shows the cooling coil loads for a summer day with and without a unoccupied period temperature control schedule.

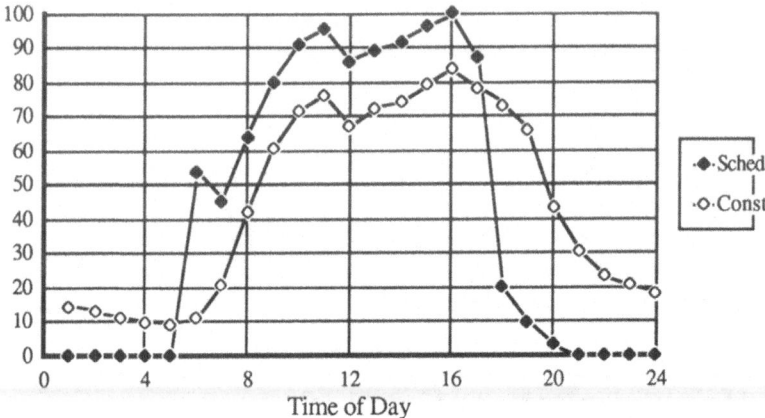

Figure 7 Summer Cooling Coil Loads With Constant and Scheduled
Space Temperature Setpoints

The impact of resetting the cooling coil discharge temperature can be illustrated by considering hourly loads for a summer day. Figure 8 shows cooling coil load data for the five zone floor of the office building with a constant volume terminal reheat system and

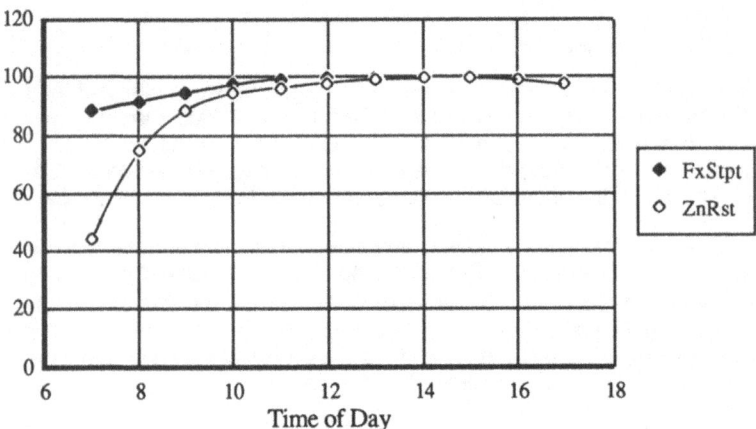

Figure 8 Constant Volume Terminal Reheat System Control Options - Peak Day

a scheduled space temperature. The CAV/TR is not an energy efficient system option but it provides good air movement and humidity control and seems a reasonable starting point for several comparisons. The data are for a summer day in which the peak cooling load occurs. The first option shown in the figure is the CAV/TR with a fixed temperature maintained at the cooling coil discharge. The second option illustrated is the same CAV/TR system with the cooling coil discharge temperature reset to the maximum air temperature that will still meet the cooling load in the warmest of the five zones. These results show some benefit of the temperature reset in the early morning hours. The reset control scheme is generally of greater benefit than is shown by this single peak summer day comparison.

Transport System Options There are a number of air transport system options that might be examined. Among these the use of a constant volume versus variable volume transport system and/or the use of an economizer are often considered. Figure 9 illustrates the impact of an outside air dry bulb temperature controlled economizer on the cooling coil loads of the CAV/TR system on a sunny fall day. The fall day was selected for this comparison as a dry bulb controlled economizer will generally have no impact on the cooling coil loads under peak summer load conditions. These results show significant savings during the morning hours when the outside air temperature is low enough for the economizer to operate (below about 14 C).

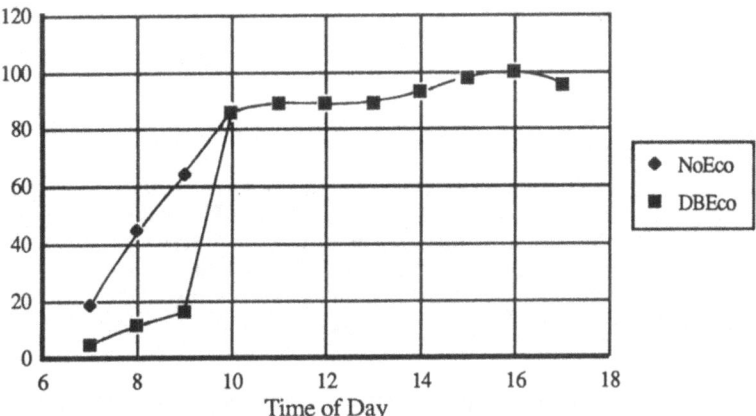

Figure 9 Fall Day Cooling Coil Loads for a CAV/TR System With and Without a
Outdoor Air Dry Bulb Temperature Controlled Economizer

The economizer savings that accrue over an entire cooling season are generally significant except in hot and humid climates. Savings for commercial buildings are usually greatest in cold climates.

The final figure, Figure 10, shows a comparison of the cooling coil loads for CAV/TR system with those of a VAV for the same five zone floor and the same fall day time period. Both the CAV/TR and VAV systems are operating with dry bulb economizers. These results again show additional savings with the VAV system, particularly during the mid part of the day. The results shown in Figure 5 tends to underestimate the benefits of a VAV system somewhat. The five zones input to this analysis consisted of four perimeter zones and a large interior core zone (60% of the total floor area). The core zone operates with very nearly constant internal load and therefore at essentially full air volume. VAV is of little benefit in such a zone.

SUMMARY

The results of one approache to evaluating air conditioning system performance has been presented. It represents an initial effort to draw together a procedure that will aid building designers in the selection of energy efficient and cost effective HVAC systems and equipment. Although this result does provide some interesting insights, there is clearly work yet to do.

The large hour by hour computer based building simulation and analysis codes provide more specific peak and annual energy use numbers. However the shear mass of

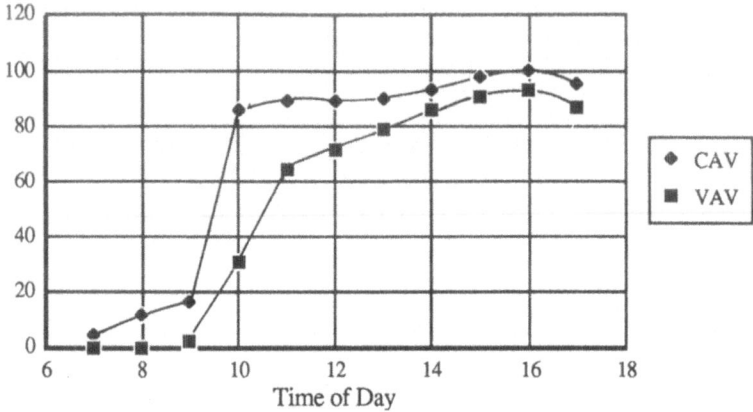

Figure 10 Comparison of CAV/TR and VAV System Cooling Coil Loads - Fall Day

data output from these programs, and the difficulty of amassing the necessary input detail, often obscures rather than clarifies the processes involved in arriving at the end numbers. The real need at this point in time is for a tool, or a series of tools, that provide understanding and a general insight in the early stages of the design process. It is hoped that further exploration of the kinds of analysis presented here might lead to the development of such tools.

REFERENCES
1 Tao, Willian F., "CSP Cooling System Performance" Heating Piping and Air Conditioning Reinhold Publishing, May 1984
2 Energy Conservation in New Building Design ASHRAE Standard 90-75, American Society of Heating Ventilating and Air Conditioning Engineers, August 1975
3 Energy Conservation in New Building Design ASHRAE / IES Standard 90.1P, American Society of Heating Ventilating and Air Conditioning Engineers and the Illuminating Engineering Society of North America, Draft, June 1987

TECHNOLOGY DEVELOPMENTS IN HOME APPLIANCES

G.P. ABBATE

Zeltron-Istituto Zanussi per la Ricerca S.p.A.

ABSTRACT

The consumption of electrical energy, attributable to household appliances, is approximately half of the total domestic consumption. The domestic consumption is one quarter of the total consumptions of a nation such as Germany.

Consequently, creating more efficient household appliances means a greater decrease in consumptions. At times, this also signifies less pollution and minor raw material waste.

Two parallel processes, evolution and innovation, have improved product performance.

Innovations such as the new washing method, the JET-SYSTEM in particular, enable notable savings.

In the future, the introduction of electronic technologies, with the development of the Home-System and of the Integrated Kitchen, will enable us to obtain savings levels which no single electrical household appliance may reach.

1. INTRODUCTION

According to Howard Geller of the American Council for an Energy-Efficient Economy, approximately 50% of the total domestic energy consumption in the United States is determined by electrical household appliances, and particularly refrigerators, air-conditioners and water-heaters. This percentage was confirmed to me recently, also by authoritative Japanese researchers referring to their domestic market.

I personally feel that in Europe, the situation is better above all because of the minor diffusion of air-conditioners, but in any case the consumption of electrical household appliances is greater than one third of the total consumption at home. According to the ZVEI statistics, in Germany during 1983, approximately 26% of the total national electrical energy consumption was used for domestic purposes.

Therefore, the realization of more efficient products means less consumption.

In practice, two cases exist:
a) products such as the refrigerator, where the consumer may save only on electrical energy;
b) products such as the washing-machine, where greater efficiency permits

A. T. De Almeida and A. H. Rosenfeld (eds.), Demand-Side Management and Electricity End-Use Efficiency, 435–448.

energy savings, water and detergent as well as reduces pollution.

Before reviewing some evolution examples of different products, I would like to exploit this second case in all of its implications.

Let's analyse the washing cycle. In order to obtain a certain quantity of detergent, we must begin with certain raw materials and process them chemically, using energy. In order to have the exact quantity of water, we must start from a natural deposit, for example a lake, purify and distribute the water into the acqueduct and (this requires energy), then we must bring locally the pipe-lines to the desired pressure until we reach the water-tap. Again, this uses energy. Now, we load the washing-machine with dirty laundry and the detergent, and turn on the water-tap: washing begins. We must load the water, bring it to the desired temperature, wash the clothing, unload the water, re-load clean water for rinsing, and this work goes on until clean laundry is obtained. Obviously, the dirty water which is unloaded, in some way, returns to the sea.

Now, let's consider a more efficient washing-machine. The first result is:

- less time
- less water
- less detergent
- less electrical energy for washing

As a consequence, we also have:

- less pollution
- less energy consumed in the beginning for water
- less energy consumed to produce detergents
- minor consumption of raw materials

Furthermore, a more efficient washing-machine usually washes better and therefore, we assume that the laundry will be less damaged and this equals further energy savings.

As noted in this example given, the global advantage drawn from more efficient electrical household appliances is far greater with respect to the electrical energy savings at home.

2. EVOLUTION AND INNOVATION

Two processes exist which improve a product's performance: evolution and innovation.

Designers always try to better the product as well as the production process: this brings about hundreds of small technical changes which cause the product to evolve towards improved quality levels. This is why each new product has a greater efficiency level than the preceeding one, even though the two models look alike to the person who is outside of this technical field.

The second process is instead characterized by the introduction of a new component, or a new technology, or a new way of imagining the whole product.

In that case innovation is visible to everyone and achieves a real performance jump, opening-up a new product era.

Further on in this paper, I will state some examples of both these processes, which normally take place in parallel, for all products.

3. REFRIGERATOR AND DISHWASHER EVOLUTION

Freezers and refrigerators have not changed much from their origins, at least in appearance. In reality, a modern freezer consumes approximately 55% of energy which was necessary for its predecessor in 1972.

Comparing some Zanussi statistics with those of the American Council for an Energy-Efficient Economy, for a top-mount, automatic-defrost refrigerator-freezer, we can see the changes in consumption and the differences between large-sized American products and small-sized European ones.

This evolution is principally related to:
a) better isolating materials
b) more efficient compressors
c) a more rational design
d) new temperature control techniques

In the last years we started talking about innovations which could lead to a jump in history of this product: rotating compressors, new isolating materials, electronic controls. However, up until now, the technical and economical conditions necessary to introduce this innovation in mass production have not been carried out.

Vice-versa, the no-frost technique with forced, cold-air ventilation has caused greater electrical energy needs, which is the price we have to pay in order to have qualitative type benefits.

In any case, the product, both in the traditional direct-cooling version and the no-frost version , continues its slow but progressive evolution.

The same is valid for the dishwasher. Introduced on the market in the 60's, it is the most recent among the great electrical white goods. After the first ten years of adjustment,it started its descending curve of consumption, going from 70 liters to approximately 25 liters of water and from more than 2,5 kWh to less than 2 kWh of electricity. A further reduction of consumption is foreseen thanks to the introduction of a greater reduction in dimensional standards.

It is interesting to note how much cheaper it is to wash by machine than by hand.

The major problem to resolve is the heating of hot water; the energetic balance of a dishwasher is the following:

> 41% for water heating
> 21% for dish heating
> 18% for start-up heating
> 16% for various dispersion
> 4% for drying

More than 60% of energy is used for heating water and dishes: therefore, either we find new heating techniques or we must find a new way of washing at a lower temperature.

4. THE INNOVATION (NEW METHOD): THE JET-SYSTEM

In order to understand the evolution of laundry washing in Europe, at least three factors must be kept in mind.

The first is the usual progressive perfectioning of the machines.

The second is the ever changing habits in washing.

In 1960 only 20% of laundry loaded was made up of mixed loads; these days it is standard to load white and coloured laundry of any type.

Furthermore, wash at 60 °C is always in ascent while in Italy, for example, wash at 90 °C has changed from 75% to 45%.

The third factor is ecology.

From 1980, many countries have introduced laws to limit the content of phosphorus in detergents and in parallel all manufacturers have begun studying the creation of more efficient machines in order to consume less detergent.

The result of this evolutionary product is that we have moved from 3,5 - 4 kWh, in the past, currently approximately 3 kWh using a traditional product.

But in the middle of the 80's, Zanussi introduced a new method of washing: the JET-SYSTEM.

No more traditional bath washing, but:
- bathless washing
- linen soak by continuous spouting of lye through water recirculation
- short spin speed phases help the water-detergent solution to pass continuously through the fabric
- lye heating obtained outside of the tub
- temperature control outside of the tub
- water loading controlled automatically and proportionally to laundry load

This truly methodical innovation modifies the above mentioned figures and causes a jump to less than 2 kWh. If we want to measure the overall effects of this innovation, the energy, water and detergent savings in percentages for each different load, go from 35, 20 and 30 percent for 4,5 kilos to 60, 35 and 70 percent for 1 kilo.

The most important thing to note is that the base technology of a JET-SYSTEM machine is not different from that of the traditional one.

5. THE INNOVATION (NEW TECHNOLOGIES):MICROWAVE OVENS

In the U.S.A. and Japan, microwave ovens are already a standard in the kitchens. In Europe, and especially in the U.K., they are in rapid expansion.

They represent a double innovation but do not lessen the importance of the traditional cooking appliances. The first innovation in technology was the introduction of a magnetron in the domestic ambient. The second concerns the way of conceiving the cooking process.

Up until now, a housekeeper selected only the type of cooking desired:

boiled, fried, roasted and so on. These days, one can select between more appropriate sources of heating: gas heating, electric heating,or microwave. In the future there will be the possibility of selecting a magnetic type of cooking (by induction system).

Everyone knows that microwave ovens have limits, but if integrated with other means, they can represent a notable advantage in terms of time and energy without losing quality and flavour in foods.

Comparing time and energy in the preparation of different foods, we can see how this technological innovation permits us, in certain cases, savings of 300%: a real jump in quality.

6. THE INNOVATION (NEW ARCHITECTURE): THE INTEGRATED KITCHEN

Up until now, we spoke about evolution or innovation of single products, but a new perspective exists which opens doors to a new horizon: the evolution of the whole system.

Obviously, the modification of a system brings about the innovation of the single products involved.

For example, Zanussi is developing the Integrated Kitchen for Eureka's Interactive Home System project. The system is composed of all white goods connected by means of a digital bus. The system's controller is connected to this bus and acts as a supervisor; it is also connected to the general Home-bus in order to exchange information with the other sub-systems.

Furthermore, the products used are furnished with a microprocessor control, which alone improves the single product efficiency.

But the system's controller can do more, it can manage the different load factors in an intelligent and integrated way and furthermore, for some services, it can switch on the power when electrical energy costs less: e.g., during the night, for those countries which have the possibility of paying less for energy in established hours.

Zanussi is developing a demonstrative prototype of the potentiality of this approach, but the fervor of activity which we encounter in this sector around the world is a guarantee that these studies will soon mature.

This last example of innovation is therefore different from the preceeding ones because it cannot be explained with the usual product evolution nor with their technological innovations.

The only means of understanding this whole process is to jump to a system level and to think in terms of temporal-space integration of the different functions carried out.

7. CONCLUSIONS

The future offers two parallel roads to take. On the one hand, the evolution and innovation of the single products must continue, improving their performances in a global way and, as a consequence, increasing their efficiency.

On the other hand, the introduction of electronics and information systems could determine a new system condition which enables the realization of greater energy savings. Electronic technologies are not yet able to develop these new systems, but within Eureka's IHS project, the major European industries, including Electrolux and Zanussi, are developing all that which is necessary for this new product generation.

REFERENCES

1. A. Burello: 2 World Conference on Detergents, Oct. 8, 1986, Montreaux
2. Proceedings of the 38th Annual International Appliance Technical Conference, 12-14 May, 1987, Columbus (Ohio)
3. Progetto Eled: Zeltron, Internal Reports
4. Progetto IHS (EUREKA), Internal Reports
5. Sommer: 32 Referatetagung Waschereiforschung, April 1986, Krefeld

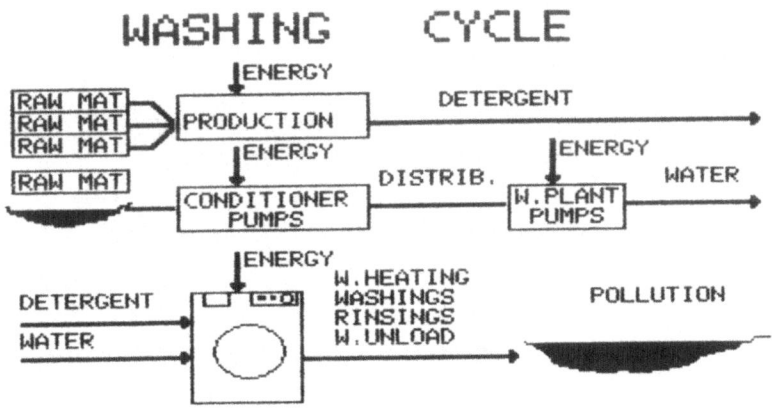

FIGURE 1

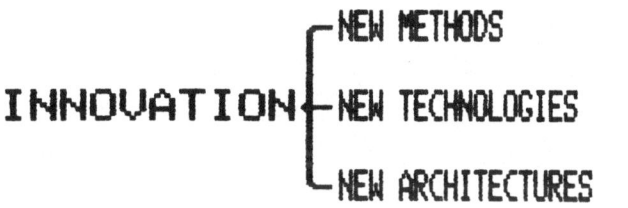

FIGURE 2

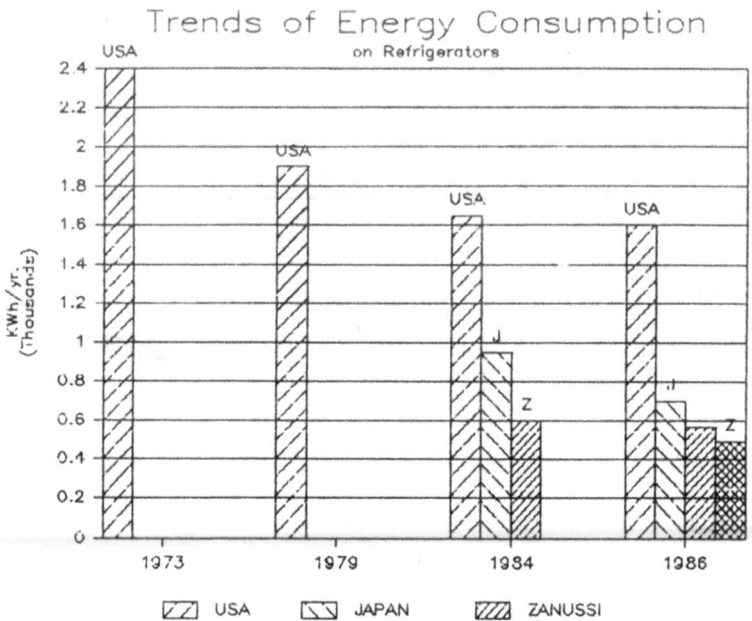

FIGURE 3

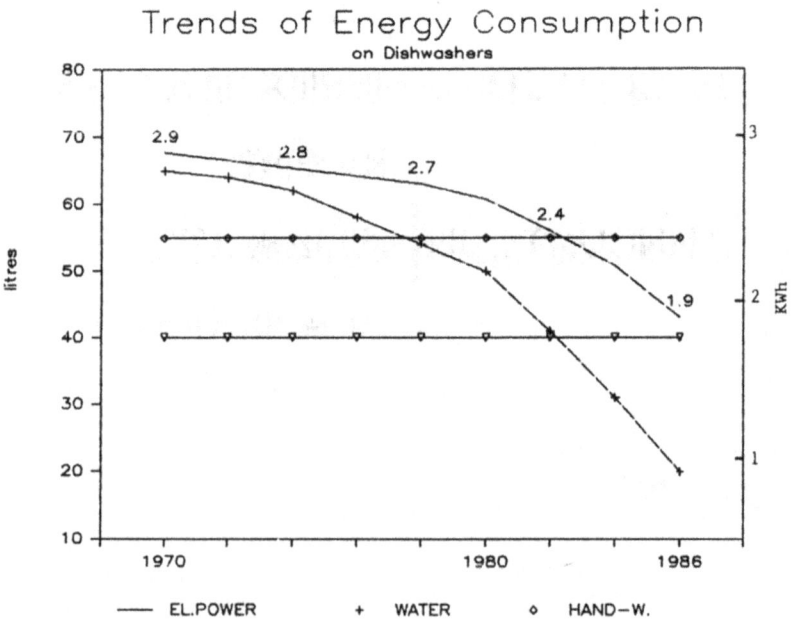

FIGURE 4

ENERGY BALANCE IN D.W.

41%	WATER HEATING
21%	DISH HEATING
18%	MOTION
16%	LOSSES
4%	DISH DRYING
100%	

FIGURE 5

Mixed Loads Washed in W.M.
% with Mixed Loads
100
90
80
70
60
50
40
30
20
10
1960
1970
1980
1986

FIGURE 6

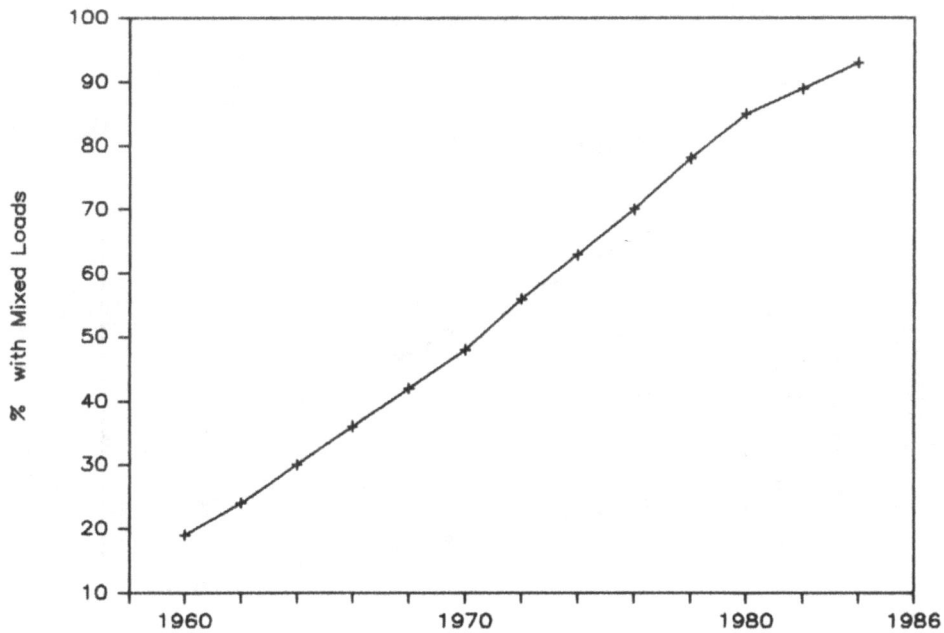

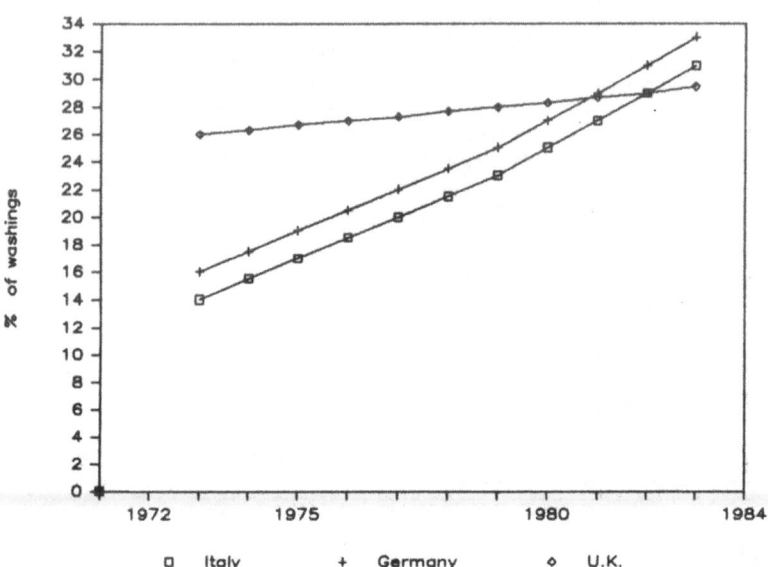

FIGURE 7

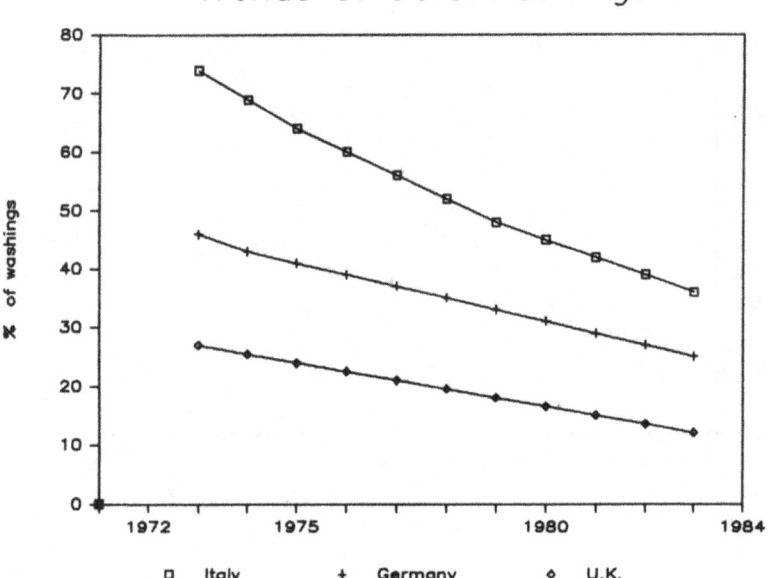

FIGURE 8

ECOLOGICAL PROBLEM

COUNTRY	LAW LIMITS (PHOSPHORUS)	LAW OPERATING FROM :
ITALY	2.5 % 1.0 %	1.7.1986 1.4.1988
SWITZERLAND	0.5 %	1.7.1985
NORWAY	3.0 %	1.1.1986
GERMANY	5.0 %	1980
HOLLAND	0 %	1987
FRANCE , U.K. : NO REGULATION OR LAW		

FIGURE 9

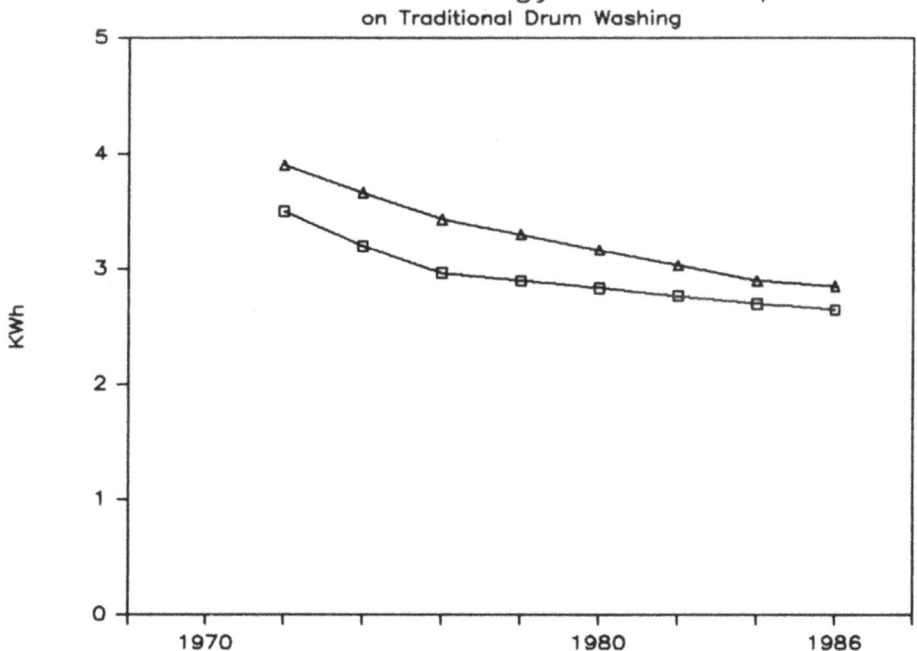

Trends of Energy Consumption
on Traditional Drum Washing
KWh
0
1
2
3
4
5
1970
1980
1986

FIGURE 10

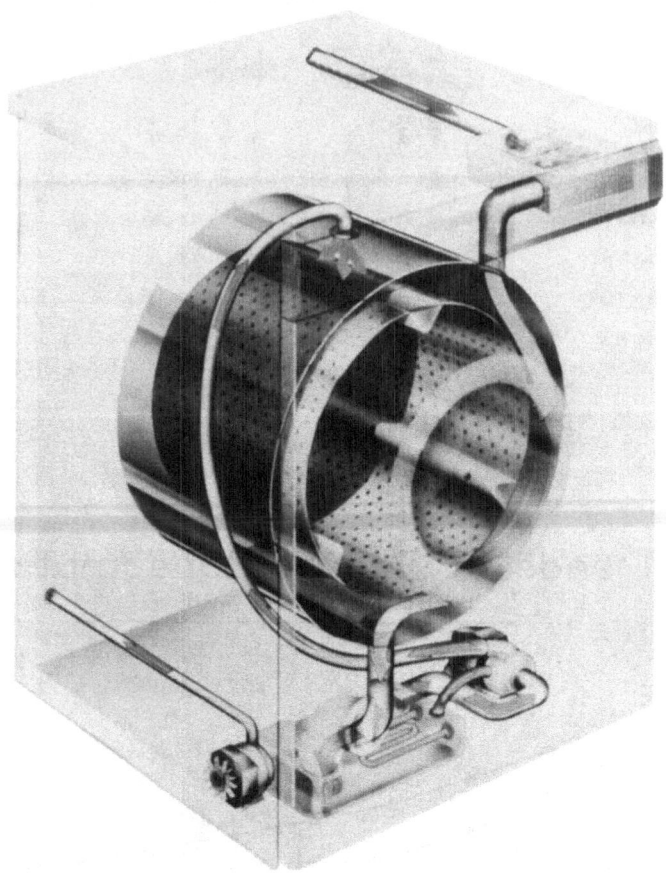

FIGURE 11 JET-System washing machine

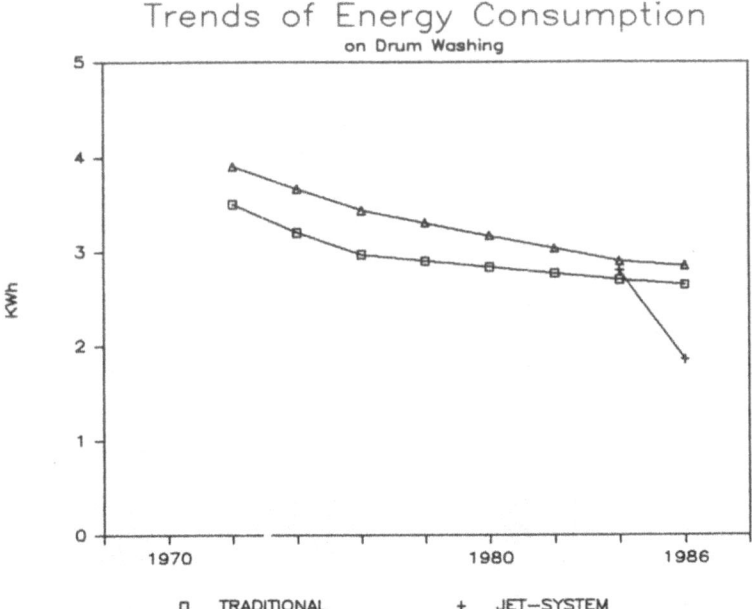

FIGURE 12

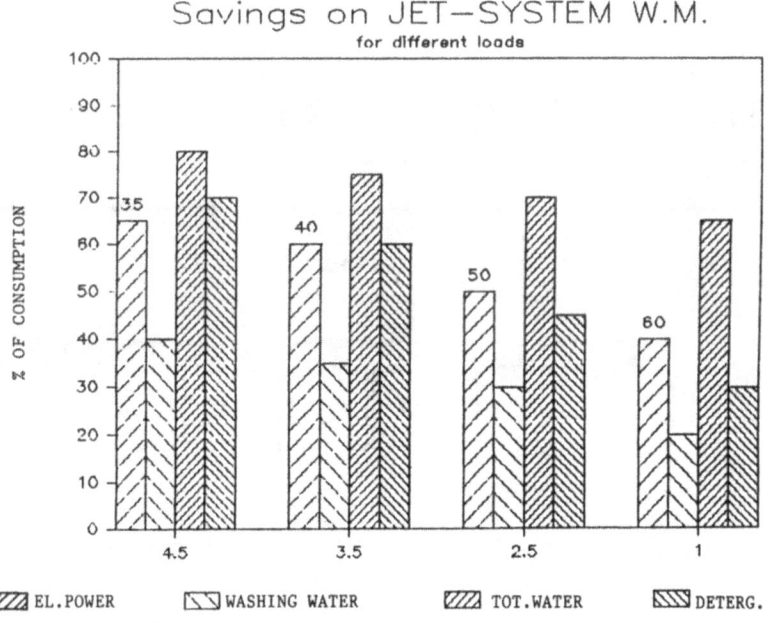

FIGURE 13

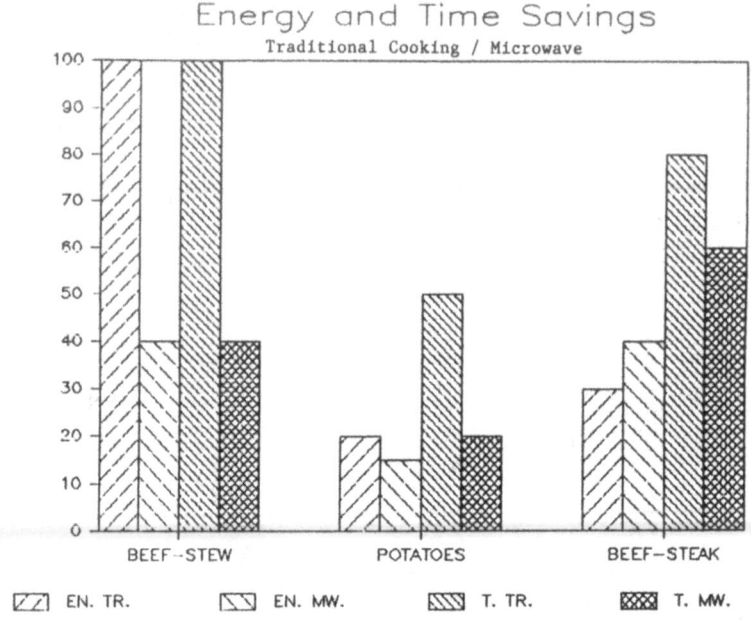

FIGURE 14

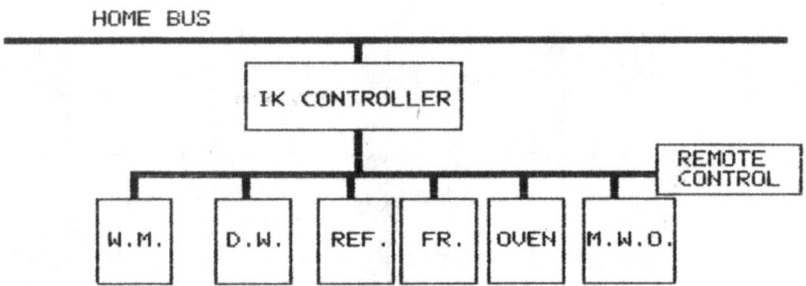

FIGURE 15

LOW ELECTRICITY HOUSEHOLD FOR THE FUTURE

JORGEN S. NORGARD and PREBEN BUHL PEDERSEN

Physics Laboratory III, Technical University of Denmark
DK-2800 Lyngby, Denmark.

1. INTRODUCTION
This paper is based on a contribution to a conference in 1980 (1), edited and updated to the situation today.

1.1 General background
One may easily argue that, from a global point of view, the rich nations should cut down their material standard of living, and thereby reduce their heavy consumption of the world's resources. Alternatively, the rich nations may reduce their consumption while maintaining their contemporary material standard of living by using these resources more cautiously and efficiently. In this paper, we describe a private household in which the most precious form of energy, electricity, is used efficiently to achieve a standard of living, which materially can be described as being around the present average in Denmark. This is assumed to be a standard that is acceptable to most people in the rich parts of the world. It may also serve as a reasonably achievable goal for many of the developing countries. With only a few modifications, the low-electricity stock of appliances described below could serve as a framework for a household in any part of the world.
All the technical improvements suggested here are based on well known, simple, and reliable technologies. The fact that significant improvements in efficiency can be realized, reflects a general neglect in that area by consumers as well as appliance manufacturers in the past.

1.2 Determinants of electricity consumption
Electricity consumption in private households is determined by the following three factors:

1) The <u>Stock</u> of electric equipment, such as number of washing machines, TV-sets, lamps, refrigerators, vacuum cleaners, etc.
2) The <u>Utilization</u> of the stock, such as number of washes per month on each program, hours of watching TV per day, level of lighting, etc.
3) The <u>Efficiency</u> of the stock of appliances, expressing the amount of service and comfort obtained per kWh of electricity consumed. For example, how much light we get per Watt of electric power, or - expressed reciprocally - how much electricity is used for each hot wash, how much power does the TV use, etc.

A. T. De Almeida and A. H. Rosenfeld (eds.), Demand-Side Management and Electricity End-Use Efficiency, 449–457.
© 1988 by Kluwer Academic Publishers.

In the following three sections we shall briefly describe the assumptions leading to values of Stock, Utilization and Efficiency for the low-electricity household.

2. STOCK OF ELECTRIC APPLIANCES

One of the important elements of the modern material standard of living is the stock of electrical appliances. The low-electricity household described here lives in a one-family house. The size of the household is chosen to be 2.5 persons (2 adults and one small child) which is slightly higher than the present Danish average of around 2.2 persons.

The appliances included in the household are listed in Fig.1.

In the low-electricity household proposed here most, of the cooking is done with gas (bottled butane gas, biogas, piped natural gas, or piped synthetic gas). However, the oven is heated by electricity. The kitchen is furthermore equipped with a 200 litres automatic defrosting refrigerator, a 150 litres chest freezer, and some small appliances like a toaster, a mixer, a coffee machine etc. The household also has a washing machine. A colour TV, a hi-fi system and a radio are also part of the stock of appliances. Besides the small appliances in the kitchen, the household also possesses a number of other appliances for which the electricity consumption is of minor importance. For example an electric shaver, a vacuum cleaner, a sewing machine, and some electric hand tools.

The number of lamps in the household is not specified in this paper as a factor in determining the electricity consumption for lighting. The level of illumination is used directly as a determinant of electricity consumption for lighting.

The house is equipped with a hot water radiator heating system like most houses in Denmark. If the heat source (oil furnace, gas furnace, storage tank, district heating, etc.) is placed at a low level in a basement room, no pump is necessary to circulate the water in the system. This was the circulation principle for most early central heating systems in Denmark. A small pump is, however, included in the low-electricity household.

3. UTILIZATION OF ELECTRICAL APPLIANCES

We will assume a slightly lower utilization of the appliances (0-10%) compared to what has been the case in the last 5 years in Denmark. This decline in utilization could be considered a decline in the standard of living, but for example, as time is also saved by using the washing machine less often, this does not obviously mean a lower standard. It is rather a matter of getting into a few other habits of organizing the household, promoted for instance through higher prices for electricity.

The oven is used for cooking in the same way and as often as we estimate for today, namely 3 times a week.

The washing machine is started only 15 times per month on the average in the low-electricity household, compared with typical 18 times today (3). This is achieved by always filling up the machine before starting, so the assumed amount of laundry per month is unchanged. Also important for the energy consumption is the elimination of hot wash (around 90°C), which is used today for about one third of the laundry. Instead, we assume in this study a longer washing time and a certain soaking. This means that no higher washing temperatures than the 55°C of the hot water supply of the house are used. This washing pattern or even the use of cold water only is widespread today in the USA, for instance.

Equipment for entertaining, TV, hi-fi, radio etc., is assumed here utilized as often as today, which is estimated typical ly to be the same as in earlier studies, namely 3 hours per day for TV, 2 hours per day for radio and 10 hours per week for the hi-fi system (2).

Today, the circulation pump in the heating system is normally running all time through a heating season of 225 days. However, as the low-electricity house is assumed to be very well insulated also, the heating season is shorter. Furthermore, the system is designed for self-circulation such that the pump will be needed only during the coldest two months of the year. The furnace itself is on only half as much as today due to better insulation of the houses.

Small appliances are used as much as today.

Utilization of light is best expressed as the amount of light in lumen·hours used per year. It is the most uncertain part of electricity consumption in a household with estimates differing by a factor of two. For Denmark, a high estimate is around $4 \cdot 10^6$ lm·h per capita· year (3). For the household considered here, this would amount to 10^7 lm·h (around 750 kWh per year when incandescent lamps are used). However, we assume a 10% reduction in the level of lighting down to $9 \cdot 10^6$ lm·h, achieved through minor changes in habits. This level corresponds to three 50 W incandescent lamps for each person, turned on an average of 5 hours per day.

4. EFFICIENCY OF ELECTRIC APPLIANCES

The potential for substantial improvements in the electric efficiences of domestic appliances has been pointed out (2), (3). The following text describes briefly how efficient the appliances could be designed, mainly by using well known technology such as better insulation. Figure 1 shows the main results for the low electricity appliances, and compares them to the most efficient models on the market in 1986 and to the average models in use today.

4.1 Electric oven

A small oven (about 20 litres in volume) is assumed to be sufficient for 90% of the oven cooking. If this small oven is designed to be inserted in a normal large (50 litres) oven, it can be well insulated and still leave the large oven available for a few occasions. The small oven can also be constructed with a much lower heat capacity than the large one, which is important when considering electricity consumption. 0.30 kWh is consumed per average use of the small oven compared to 1.00 kWh for a typical one in use today (7). With 4 uses per week, the annual electricity consumption in the oven will be 60 kWh compared to a current average value of 200 kWh. One of the best ovens on the market consumes 130 kWh per year.

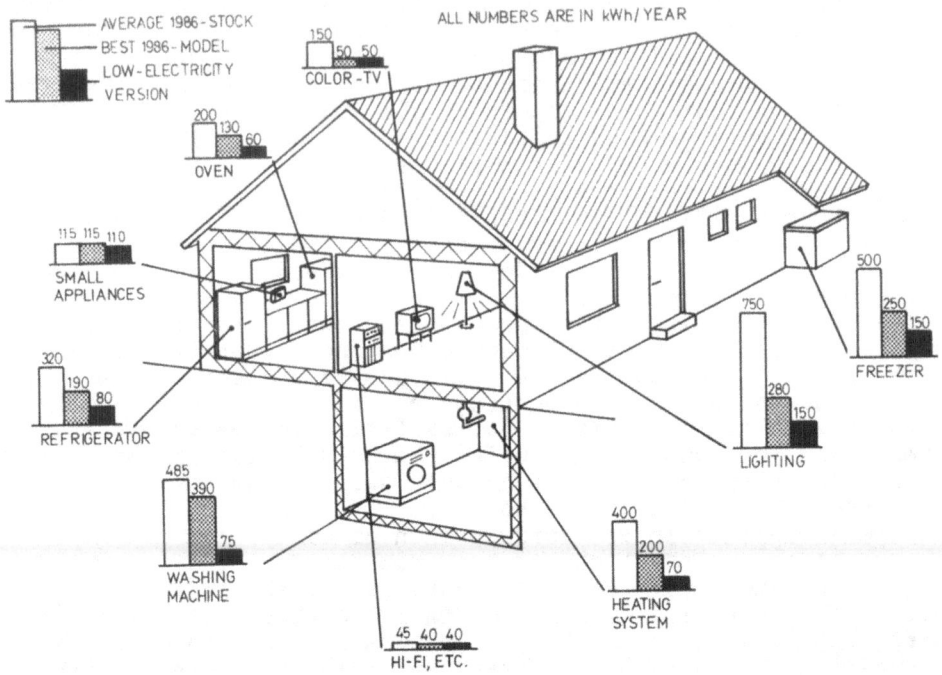

FIGURE 1. Electricity consumption pattern in the low electricity household compared to two cases: with most efficient 1986-models and with average 1986-stock.

4.2 Refrigerators

Refrigerators sold in 1986 were significantly improved compared to models of 1975 and earlier, when 550 kWh/yr was normal for a 200 litres refrigerator. Some of the best 1986-models use only 190 kWh and average for refrigerators in use today is 320 kWh (9).

A 200 litres low-energy refrigerator with an annual electricity consumption of only 80 kWh has been developed and tested (6,9,10). The COP is 2.0 or around twice as high as earlier models. This is achieved by increasing the surface areas as well as the heat capacities of the condenser and the evaporator. Also, the compressor used is improved by using semi-direct intake and run capacitor.

The heat load is reduced to half, from 320 kWh/year to 160 kWh/year by increasing the thickness of the polyurethane foam insulation from 3-3.5 cm to about 6.5 cm. This increase in insulation thickness leads to a 20 cm higher cabinet than the normal 200 litres refrigerator, since we maintained standard width of 60 cm.

4.3 Freezer

A low electricity 150 litres chest freezer is suggested. Better insulation here is also a basic means for reaching a high efficiency. The freezer is assumed to be equipped with 15 cm of polyurethane insulation compared to 6-9 cm today. The heavy insulation requires extra space which is found to be acceptable for a freezer, since it is assumed to be placed in a secondary, non-heated room such as a basement or an outhouse. The freezer will, with its interior temperature of $-18^{\circ}C$, benefit in two ways from being placed in an environment with an average assumed temperature of only $10^{\circ}C$, rather than the usual $20^{\circ}C$. First the heat load will be down by about 25%, and secondly the condenser is cooled better than at normal room temperature. The better cooled condenser together with larger surface and higher heat capacity of the condenser and the evaporator makes it possible to reach a COP of more than 1.5.

The result is a 150 litres low-electricity freezer using only 150 kWh/year. This is compared in Fig.1 to a situation where the household has an average 150 litres freezer which is estimated to consume 500 kWh/year and to a situation using the best 1986-model using 250 kWh/year.

An extra benefit of the low electricity freezer is that a temperature rise from $-18^{\circ}C$ to $-12^{\circ}C$ in case of a cut-off in electricity will take 3-4 days, if the freezer is full.

4.4 Washing machine

Typically a washing machine in Denmark today consumes 485 kWh of electricity per year (4). About 85% of this, however, is used for heating water, and only 80 kWh/year is needed for the motor and the automatic control (2).

With the utilization pattern described for the low-electricity household, the $55^{\circ}C$ water from the hot water system will be sufficiently warm for all washes, which means that the electric heating element in the washing machine may be eliminated. In order to maintain a temperature above $50^{\circ}C$ during the washing, it might be necessary to insulate the tub. The longer washing times suggested here do not mean longer running time for the motor. The efficiency of the motor can be improved by an estimated 15% and total annual electricity consumption is down to 75 kWh only, compared to an average of 485 kWh for the present stock, and around 390 kWh for the most electricity efficient models on the Danish market today, using present washing pattern. More on low electricity washing is in (8).

4.5 TV, hi-fi etc.

The energy efficiency of the group of electronic entertainment appliances, such as TV, radio and hi-fi, has improved significantly over the last decade through the introduction of semiconductors.

A typical average electricity consumption for the colour TVs in use in 1970 was 400 W. This average today is estimated to be down to 150 W. With the assumed 3 hours of TV watching a day, the 150 W amounts to around 150 kWh/year. The most efficient TV on the market in 1986 consumes only around 50 W, which means an annual consumption of <u>50 kWh/year</u>. This is the value used here, also as the future option, since we have not looked for further saving possibilities.

Hi-fi, radio and similar equipment is estimated to use 45 kWh per year today, and <u>40 kWh</u> in the future household.

4.6 <u>Small appliances</u>

An estimated average of 15% increase in efficiency is assumed for the very mixed group of small appliances. This reduces the estimated electricity consumption from 115 to <u>100 kWh per year</u> (2),(3).

4.7 <u>Heating system</u>

The 65 W pumps typically used for circulating water in domestic heating system in the sixties and seventies have been shown to be much too powerful compared to what is needed. A 13W pump would be sufficient, especially in the case presented here, in which natural convection will contribute to the circulation. Today 25 W pumps are available for that purpose.

With a running time of only 2 months a year, the annual electricity consumption for the 13 W pump would be 20 kWh only. The ventilator and oil pump of todays furnace typically consumes 120 kWh per year (3). Minor changes in design could improve the efficiency of ventilator and oil pump by 20%. As the furnace will be on only about half as much as today, the furnace in the low electricity house will use only 50 kWh per year. The total heating system thus accounts for <u>70 kWh</u> of electricity per year, compared to 400 kWh typical today, and 200 kWh, if the best equipment on the market is used (11).

Finally, we assume the furnace to be of the double-chamber type now available, in which the use of oil or gas can be supplemented with use of solid fuels like waste paper, wood, etc. Use of solid fuels requires no electricity, and a small stock of wood, coal, coke etc. can prevent interruption in heating in case of an electricity cut-off.

4.8 <u>Lighting</u>

Electricity in private households was originally promoted as a means for generating light. Today, the same incandescent lamp technique dominates domestic lighting as it did when it was invented a hundred years ago. Efficiency, expressed in lumens per watt, has increased since then from around 3 to now typically 13 lm/Watt. But the efforts to develop new high efficient lamps providing up to 130 lm/Watt have mainly been devoted to large scale illumination purposes such as streets, high-ways, stores, factories, and sports stadiums.

In the seventies, however, various attempts were made to develop small scale high efficiency lamps (1). The technique, which has so far been most successful, is based on small fluorescent light tubes folded to almost normal bulb size. Many variants of such compact fluorescent lamps have been marketed, some of which fit into a normal screw-in socket. The efficiency of these lamps is typically 50 lumen/W (5). The colour quality is approaching that of incandescent lamps, while some start delay and flickering might still be found annoying for some domestic use. These problems are, however, also being overcome

with high frequency electronic ballast, which improves also
the efficiency beyond the 50 lumen/W. Such ballasts are on the
market already but the present price around 35 US dollars seems
prohibitive for domestic uses. For the future low-electricity
household we assume that 90% of the domestic light is provided
by such efficient lamps providing 60 lumens per Watt. The
remaining 10% of the light comes from some of the best incandescent lamps - maybe half of the lamps in the house, but those
which are used the least.

By also exploiting some of the many options for more efficient lamp shades, the annual electricity consumption in the
near future can be 150 kWh as shown in Fig.1.

5. ELECTRICITY CONSUMPTION

Fig.1 shows the annual electricity consumptions for the
appliances presented in this paper and compares them with
corresponding values of appliances in use today and the best on
the market today. Fig.2 summarizes the results.

The total electricity consumption of the low-electricity
household is 785 kWh per year. A household, similarly equipped
with the most efficient new models of 1986, would consume today
1650 kWh or more than twice as much. Equipped with the
average models in use today, the consumption would be around
3000 kWh per year or 3.8 times that of the low-electricity
household. The Danish average for all households in 1985 was
around 3900 kWh/year per household, including electric heating,
hot water and electric cooking in some households. Electric
consumption in the low-electricity future household presented
here would be around 20% only of what is typical for a
Danish average household today.

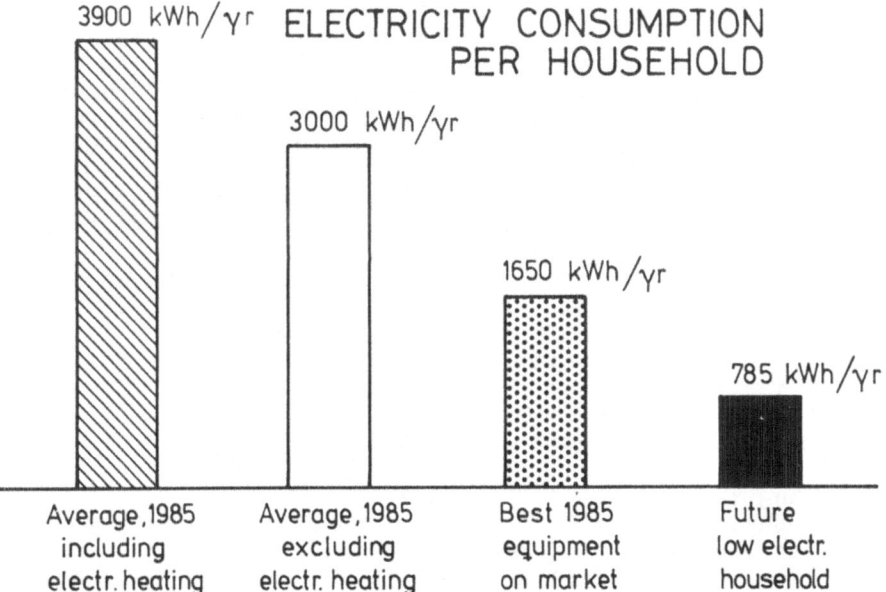

FIGURE 2. Annual electricity consumption of a Danish household with different efficiencies of appliances.

456

6. CONCLUSION

In the future, an average household could live in a one-family home, maintaining present Danish high material standard of living with an electricity consumption of only 1/5 of what an average household uses today.

The savings can be achieved through the introduction of more efficient technology combined with minor changes in habits. Generally, the technological knowledge necessary for the improvements is available today.

A closer look at the electricity consumption pattern reveals that a 3 day cut-off in electricity supply would have no serious consequences for the low-electricity future household presented here. The freezer would stay cold and the heating system could continue to work. Large investments in electricity supply systems could be saved, if a higher risk of cut-off than today became acceptable.

Several low-energy houses have been established all over the world, but usually with emphasis on a low heat consumption, and often with little attention to the electricity consumption. Compared to those low-heat experimental houses, the establishment of a low-electricity house as outlined here is an inexpensive experiment and demonstration. It could have a profound impact on the dominating sceptical attitudes towards significant electricity savings.

REFERENCES

1. Norgard, J.S.: "Low-Electricity Future Household" p. 159 in IEE Conference Publication No. 186, Effective Use of Electricity in Buildings, London 1980.
2. Norgard, J.S.: "Improved efficiency in domestic electricity use", Energy Policy, March 1979.
3. Norgard, J.S.: "Husholdninger og Energi" (in Danish), Polyteknisk Forlag, Lyngby, 1979.
4. Møller, J.: "Prognose for boligsektorens elforbrug og elbelastning" (in Danish). Danske Elværkers Forenings Udredningsafdeling, Lundtoftevej 100, DK-2800 Lyngby, Denmark 1986.
5. Mehlsen, K.: "Bedre lys og luft i servicesektoren" (in Danish). Physics Laboratory III, Technical University of Denmark, DK-2800 Lyngby, Denmark, 1986.
6. Comm. of the European Communities: "Development of energy efficient electrical household appliances, part one: Refrigerators". Report EUR 10449 EN, 1986.
7. Comm. of the European Communities: "Development of energy efficient electrical household appliances, part two: Cooking". Report EUR 10449 EN, 1986.
8. Comm. of the European Communities: "Development of energy efficient electrical household appliances, part three: Washing mashines". Report EUR 10449 EN, 1986.
9. Pedersen, P.B.: "Forprojekt vedrørende elbesparelser ved køleanlæg" (pilot projekt concerning electricity savings in refrigeration plants) (in Danish). Ministry of Energy, Research Program, j. nr. 151/85-49. Physics Laboratory III, Technical University of Denmark, March 1987.

10. Guldbrandsen, T., and. J.S. Norgard,: "Achieving substantially reduced energy consumption in European type refrigerator". From the 37th Annual International Appliance Technical Conference, Purdue University, May, 1986.
11. Styregruppen for Forsyningskataloget: "Forsyningskatalog '85", vol. 34-35, Danish Energy Agency, 1985.

AFFORDABILITY AND PLEASURE: THE LESSONS FROM LOW ENERGY HOUSING

G.W. BRUNDRETT

ELECTRICITY COUNCIL RESEARCH CENTRE, CAPENHURST, CHESTER, CH1 6ES, U.K.

1. INTRODUCTION

The choice of a new home requires a decision on a very complex mixture of influences which include location, price, site layout, the house design itself and finally to a relatively minor role, the heating equipment.

1.1 The three historical development stages are:

1.1.1 <u>Shelter</u>. The first stage is to minimise the discomfort of the weather by providing shelter from the rain and wind and providing warmth by means of a fire.

1.1.2 <u>Warmth</u>. The second stage is to provide warmth on a room basis through some form of central heating. This is the stage for most of the new homes in Britain.

1.1.3 <u>Pleasure</u>. This stage provides warmth and pleasant air conditions. These development stages are also modified by financial factors such as land price and interest and mortgage rates. These factors have a strong influence on housing density and on the physical size of the house. In general the house sizes were growing until the 1970s when the typical size reduced from $100m^2$ to $75m^2$.

Let us look at those factors which influence the space heating energy.

2. SOCIAL FACTORS

2.1 Clothing

The trend has been for people to wear lighter clothing. Surveys show that people compromise between economy and comfort when they pay for the energy. This means that a comfortable home living room temperature is likely to be 2-4°C cooler than the temperature which the same people report to be comfortable at work. This has resulted in a rise in average living room temperature of 1°C every ten years (fig. 1).

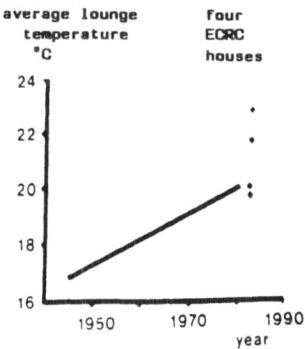

FIGURE 1. Average living room temperatures are rising.

A. T. De Almeida and A. H. Rosenfeld (eds.), Demand-Side Management and Electricity End-Use Efficiency, 459–466.
© *1988 by Kluwer Academic Publishers.*

2.2 Behaviour

Two behavioural factors have emerged from British surveys. The first is that the key room for thermal comfort is the living room. Television watching occupies most of the evening recreational time and this is viewed in the living room. The other rooms tend to float in temperature. The second factor is the window opening habits of the householders. Surveys over the last 30 years have shown that in Britain most houses have a window open during the heating season (fig. 2). The energy penalty of this action is not normally recognized.

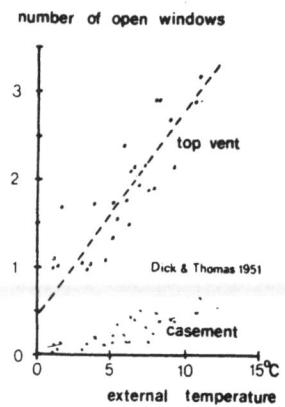

FIGURE 2.   Relationship between number of open windows and outdoor temperature.

2.3 Expenditure

One consistent characteristic of family expenditure patterns over the last 30 years has been the proportion of expenditure spent on energy in the home. This averages 5% of all household expenditure but it is proportion- ally more for the lower income groups and less for the higher income groups. The amount is £8/week for people with the lowest 20% income and £12/week for people with the highest 20% of income. The transition in the 1960s from the coal fire to central heating in Britain brought about a sharp change in comfort and convenience and a change from coal to oil but the expenditure was the same.

3. PHYSICAL FACTORS AFFECTED BY THERMAL INSULATION
3.1 House temperatures

Thermal insulation of the house shell results in two changes in temperature. The first is that the unheated bedrooms become warmer. The average house temperature therefore increases with reducing heat loss (fig. 3). The second is that the areas outside the occupied and insulated zone of the house run cooler. In particular the loft will be very cold and water services should be particularly well insulated with a heat leak from the occupied zone or even better brought into the warm occupied zone. The outer bricks will also run colder and will be more liable for spalling damage as they cycle through the freezing conditions.

3.2 Growing importance of ventilation

Until recently the heat loss from a house was 75% through the fabric and 25% through ventilation losses. The ventilation losses were never estimated by calculation because of the vagaries of infiltration and an arbitrary amount of 1-1½ air changes/hour were added. Improvements in fabric insulation highlight the growing importance of the ventilation term (fig. 4).

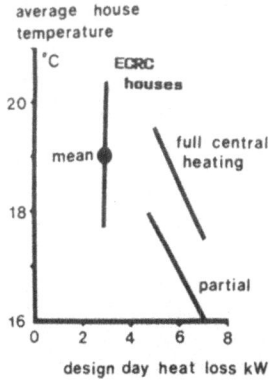

FIGURE 3. The influence of insulation on average house temperatures.

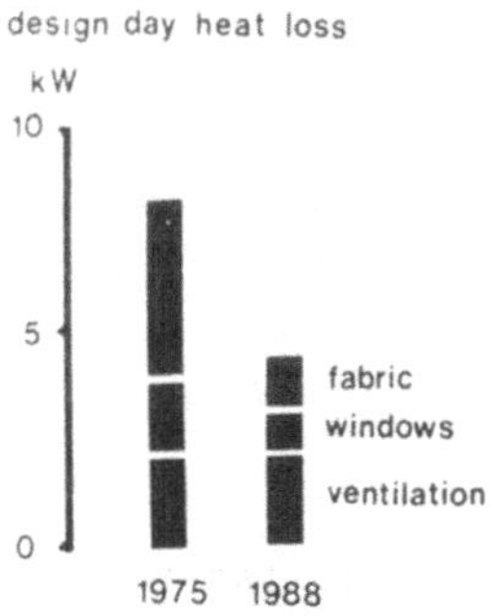

FIGURE 4. The changing balance
of heat loss.

3.3 Heating equipment

The traditional British approach to space heating has been to size the equipment to meet the design day condition of -1°C outdoor temperature and add a 25% extra margin to deal with intermittent operation. This is inappropriate in well insulated buildings. First, the energy savings by operating the building intermittently on a daily basis are insignificant. A well insulated building tends to behave more as a heavy building and is slow to cool. This effect is illustrated in fig. 5. The second is that if from time to time the room will be heated intermittently, for example, in a bedroom, then the heaters have to be sized for response not steady state heating. The third effect of insulation is that it enables the free heat of the house to provide the bulk of the space heating energy and the heating equipment tops up the final amount. This means that the heating has to be controllable. This also has implications in predicting the energy bill because the heating system is influenced by both the weather and the life style of the occupants. Variability of weather means that while the energy consumption is expected to be small, it will be very variable from year to year. This variability will become bigger in relative terms for the better insulated houses. The effect is illustrated in fig. 6. The base temperature for British degree day weather is normally 15.5°C and the value will fall for increasing levels of thermal insulation.

4. FIELD TRIALS

The Electricity Council decided to explore the customer response to a well insulated, all-electric house which was designed, constructed and weatherstripped to give minimal air infiltration. Planned ventilation with heat recovery was provided to give control of the air quality and direction through the house. All this work was done in cooperation with two national builders and the houses were built, sold and, with the owners' permission, monitored (fig. 7).

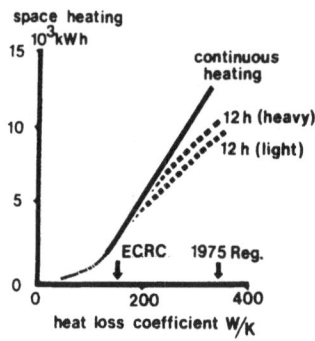

FIGURE 5. Energy savings for
intermittent operation.

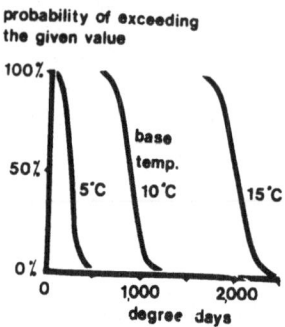

FIGURE 6. The year to year
variability of degree days.

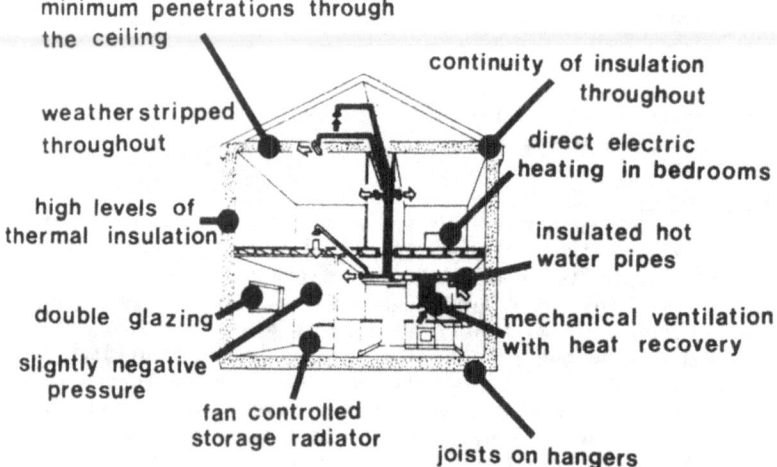

FIGURE 7. Components of the low energy house.

The thermal design was based on the most practicable masonry practice
with a widened filled wall cavity and fibrous insulation in the roof space
and under the floor wherever the floor was suspended. The thermal
transmittance was 0.35 $W/m^2°C$ for all the opaque surfaces of walls and
floor, and 0.25 $W/m^2°C$ for the upstairs ceiling. Double glazing was used
throughout. The airtightness was checked by pressurisation and had to be
less than 7 air changes/hour at an over-pressure of 50 Pascals.

The ventilation system comprised two fans, one supplying fresh outdoor
air to the living room and bedrooms, while the other extracted slightly
more stale air from the bathroom, toilet and kitchen. A cross-flow heat
exchanger linked the two air flows so that energy could be recovered from
the stale outgoing air and used to preheat the cold incoming air. The
temperature efficiency of the heat exchanger was 60% but because the
outgoing expelled air was humid, the actual energy efficiency was only 40%.

The householder could choose from three fan speeds which gave 0.3, 0.5 and 0.75 air changes per hour for the house. There was a special damper on the kitchen cooker hood to rearrange the air flow to give proportionally more air extraction from the cooker hood during cooking.

The windows were conventional in design but fitted with electrical contacts which recorded whenever the windows were open. Another electrical circuit monitored when either the front or rear door was open. A storm porch was recommended to provide an air lock to the house. The houses were all-electric on the Economy 7 tariff which provided off-peak, low cost electricity for the 7 hours between midnight and 7.0 a.m. (2p/kWh) while electricity used for the rest of the day was 5.5p/kWh.

5. USER BEHAVIOUR

The four low energy houses were marketed in the conventional way, through estate agents and the site offices. The price was approximately £30,000 each and these houses were priced at £750 above the similar conventional houses on the estate.

The energy saving features were not a major influencing factor to prospective purchasers. One of the most important features appeared to be site location and once a site looked attractive then the prospective purchasers sought a house with the layout and features which they could afford. Some purchasers bought their house with some misgivings about the electric heating system. All the householders had purchased at least one house previously and all were accustomed to paying normal fuel bills.

The first year of occupancy for all the houses was recorded in detail in terms of moisture generation rate, ventilation, temperatures and energy consumption for different purposes. The householders all had some initial apprehension but relaxed when the running costs became known and they then deliberately operated their houses to be unusually warm. The seasonal average lounge temperatures for the four houses are plotted in fig. 1 compared with the national trend line. Average house temperatures are linked to the design day heat losses and our four houses are in line with predictions based on field surveys. This is illustrated in fig. 3. Families were sensitive to small temperature falls of 1-2°C during the day in the hall.

The householders generally used the ventilation system on the lowest fan speed with short excursions to the higher speeds when cooking. This reflects the time of day when food is prepared and shows the regular use of the boost speed for short periods each evening. All the householders found it unnecessary to open the windows for most of the heating season. The total time for which windows were open for 30 weeks of the winter was less than 1 hour. Towards the end of April, as the heating season was ending, there were a few days when the windows were open for an hour or so. The two doors to each house were open for a total of 20 minutes each day. This time was remarkably similar for all four houses, despite wide differences in family size and age of the children. The average ventilation rate was 0.7 ac/h. This is illustrated in fig. 8.

The householders' appreciation of the air quality inside their houses was an unqualified success but often in unexpected ways. They recognized that the heat recovery component meant that the ventilation system could be used without fear of high energy costs. It provided an indoor air quality better than they had experienced before. It enabled moisture to be controlled and the benefits of this were more widespread than initially envisaged. They were aware of the ability of the system to remove odours, particularly in some cases as cigarette smokers themselves. Individual

total ventilation rate
a.c.h.

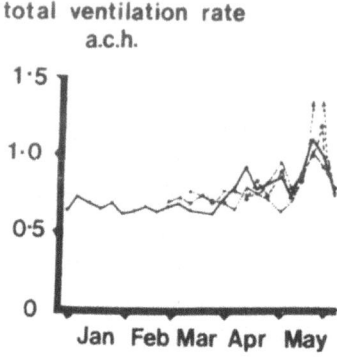

FIGURE 8.   Total ventilation rate for each of four houses over six months.

householders used their ventilation controls imaginatively.   Boost speed
with cooker extract restricted meant that the bathroom mirrors cleared
immediately after a bath or shower and could then be used for grooming.
Damp towels dried quickly too.   Numerous minor benefits included the fresh
sensation when entering the living room in the morning after entertaining
friends who had been smoking on the previous evening.   They also included
having dry cupboards where the sugar and salt would always flow freely,
where crisp breakfast cereals stayed crisp for several days after opening
the packet, and enjoying a clear, uninterrupted view out of all the windows
throughout the year.   The cooker hood extract also meant that the kitchen
decorations stayed cleaner for much longer because the rate of grease
deposition was much lower than normal.

The combination of double glazing and mechanical ventilation protected
the household from external noises and this was particularly welcomed by
one family who worked unsocial hours.   However, the exclusion of outdoor
noise did make the indoor noise much more noticeable, particularly at
night.   One householder was very sensitive to the noise which came from the
thermostat which controlled the bedroom heater.   The ventilation system
itself was inaudible at its lowest speed which was used at night.

Minor chores were readily accepted.   The cooker hood grease filter had
to be cleaned monthly and this could also be done in the dishwasher.   Towel
fluff collected in the bathroom extract and had to be removed occasionally.
Airborne outdoor dust too fine to be stopped by the air filter, caused
slight pattern staining around the ceiling inlet grilles but was so slight
that the owners did not think additional painting between normal
redecoration periods would be needed.

All the families liked their houses.   The householders were revisited
after two years and they had retained their favourable views of their
houses.   Their overall energy consumption has stayed relatively constant
for the first 4 years of occupation.   The energy analysis for each house
is summarized in Table 1.

TABLE 1. Data relevant to heating over a 32 weeks heating season

		22 AG 32 weeks	22 AG av. daily	31 AG 32 weeks	31 AG av. daily	8GC 32 weeks	8GC av. daily	9GC 32 weeks	9GC av. daily
Average temperatures									
Lounge	°C		20.0		22.8		21.7		19.7
Whole house	°C		17.7		20.4		19.3		18.5
Outside	°C		6.1		5.9		6.4		6.4
Difference	K		11.6		14.5		12.9		12.1
House heat losses									
Transmission heat loss coeff.	W/K		137		137		140		140
Transmission heat loss	kWh	8,512	38.0	10,752	48.0	9,715	43.4	9,148	40.8
Ventilation loss	kWh	1,857	8.3	2,363	10.6	2,668	11.9	2,327	10.4
Total		10,367	46.3	13,125	58.6	12,383	55.3	11,476	51.2
Other energies ('free' heat)	kWh								
Water heating	kWh	1,537	6.9	4,014	17.9	1,537	6.9	1,364	6.1
Fan energy	kWh	477	2.1	277	1.2	423	1.9	347	1.6
Cooker	kWh	498	2.2	603	2.7	796	3.6	493	2.2
Other electricity	kWh	1,558	6.9	2,920	13.0	2,534	11.3	1,662	7.4
Metabolic heat	kWh	1,523	6.8	2,195	9.8	1,611	7.2	1,142	5.1
Net solar heating	kWh	2,231	10.0	2,092	9.3	1,836	8.2	1,798	8.0
Total free heat	kWh	7,824	34.9	12,101	54.0	8,737	39.0	6,806	30.4
Space heating	kWh	5,365	24.0	6,428	28.7	5,280	23.6	5,680	25.4
Useful free heat	kWh	5,002	22.3	6,697	29.9	7,103	31.7	5,796	25.9
Lost free heat	kWh	2,822	12.6	5,404	24.1	1,634	7.3	1,010	4.5
% useful free heat			61.6		55.3		81.3		85.2
Family size		2 adults 2 children		2 adults 4 children		4 adults -		2 adults 1 child	

The householders held two important concepts. The first was affordability. The total cost for energy in the home must be affordable, i.e. not more than 5% of an average family's expenditure. The second concept was value. Once it was recognized that the total energy cost was well below conventional values then the householders relaxed and deliberately chose to take some benefit in the form of higher comfort standards. This higher thermal standard, when taken with previously unknown standards of air quality, produced a very high level of satisfaction.

6. CONCLUSIONS

(1) The customer satisfaction with the low energy houses is very high. This satisfaction is more related to a freedom to choose high comfort standards at low energy cost without worry, rather than trying to save the maximum energy. Each family chose to have a warm house.

(2) The mechanical ventilation system was an unqualified success. Not only did it provide an attractive indoor climate with low energy penalty but it removed the desire to open the windows. Unexpected but highly

popular benefits included an unmisted bathroom mirror, free running sugar and salt, and breakfast cereals which remained crisp when opened.

(3) Free heat from the occupants, sunshine, hot water and electrical appliances and light within the house provided a large amount of the space heating requirement. However, there was a wide difference in the degree of usefulness of these sources of energy. Energy estimates for low energy dwellings are more likely to be accurate if based on total energy rather than space heating energy alone.

(4) The energy saving benefit of daily, intermittent operation of the heating system reduces in the well insulated house because the rooms cool much more slowly. Steady living room temperatures were popular but the heating system in the bedrooms has to be sized for response rather than steady state losses, to enable it to cater for times when the bedrooms are used for social or work activities.

(5) The householders were well protected from external noise but this made internally generated noises more noticeable. More attention will be needed for silent household services and equipment.

(6) Our knowledge of people's preferred comfort temperature and our recognition that for many years now we have spent 5% of our expenditure on fuel and power for the house, should form the basis for an energy labelling scheme for houses which will work well and be popular to live in.

VI. Standards and Policies

THE IMPACT OF BUILDING ENERGY STANDARDS ON ENERGY USE AND DEMAND

J.W. JONES

Department of Mechanical Engineering
University of Texas at Austin
Austin, Texas, U.S.A.

INTRODUCTION

Although energy requirements, and the associated energy costs, have always been of some concern to building designers energy criteria are a relatively new addition to building standards. It took the formation of OPEC and the oil embargo in late 1973 to focus attention on energy use use in buildings and to initiate efforts to develop standards.

The first attempts at developing building energy standards in the United States began at the National Bureau of Standards in May 1973. The oil embargo in October of that year created a sense of urgency that brought two professional societies, the American Society of Heating, Refrigerating and Air Conditioning Engineers (ASHRAE) and the Illuminating Engineering Society of North America (IES) into the effort. ASHRAE, with the cooperation of IES, initiated the Standard 90 project, "Energy Conservation in New Building Design", early in 1974 and issued a standard in August of 1975. This standard was adopted by many code agencies in the U.S. over the next year. The implementation of this standard changed building design practice. Some have viewed the standard as a beneficial while others have viewed it only as a nuisance, just another regulation to comply with. Critics of the standards ask do energy standards make a difference? If so, is it enough difference to justify the effort involved in demonstrating compliance?

EVALUATION OF STANDARDS

Figure 1 shows the pattern of energy use in the residential and commercial markets in the U.S. from 1960 through 1986. In this market the majority of the energy use is for heating, cooling, lighting and hot water, all items effected by the standard. This figure illustrates a marked change in the rate at which energy use has increased since 1975. However the part energy standards have played in this change is difficult to assess. Although these observed trends in energy use are of interest, the examination of these data will not answer the question of the specific impact of standards. Much of the reduction in the rate of growth in energy use which began in 1974 has more to do with the rapid rise in the price of energy and a general sense of concern for the future than it has to do with the implementation of building energy standards. However it is clear that the implementation of the standards has effected energy conservation in buildings.

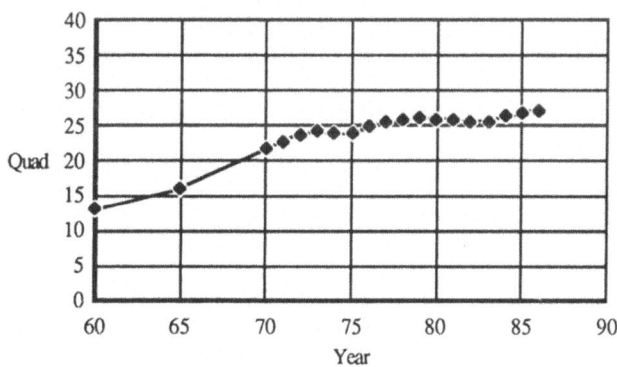

Figure 1 U.S. Residential and Commercial Energy Use

A. T. De Almeida and A. H. Rosenfeld (eds.), Demand-Side Management and Electricity End-Use Efficiency, 469–475.
© *1988 by Kluwer Academic Publishers.*

The only means of defining the possible impact of building energy standards is to begin with individual buildings. Two approaches have been used to do this. The first is to measure the energy use of a building over time. A number of efforts have been made to obtain measured data and then to relate the measured energy use to various building features. Unfortunately the variability of operating schedules and climate factors make interpreting measured data very difficult. The second approach is to simulate the performance of the building using one of the detailed computer analyses tools now available. Parametric studies using this approach have been quite helpful in estimating the impact of specific design strategies. These results do not provide the 'real' answers in the sense that measured data would, but they provide much more detail and are thus more useful in assessing the impact of individual design changes.

Two recent efforts, using the simulation approach, provide some insight into the impact that standards might have on building energy use and demand are related to ASHRAE's efforts to update Standard 90. The first referred to as ASHRAE Special Project 41 "Development of Recommendations to Upgrade ASHRAE Standard 90A-1980 'Energy Conservation in New Building Design'" began in 1980 and was completed in 1983. The second was a continuation of this evaluation effort at the Pacific Northwest Laboratory (PNL) operated for the U.S. Department of Energy by Battelle. DOE also sponsored, and Battelle managed, the ASHRAE SP41 work.

The objective of the ASHRAE SP41 work was to develop specific recommendation for the revision of Standard 90. A supporting objective was however to measure, both in energy and economic terms, the effectiveness of the standards considered. The effectiveness of the standards was tested by simulating the annual energy use of 10 reference buildings at five locations throughout the U.S. Plans and specifications for 10 real buildings constructed in the mid 70's were used as the basis of these simulations. These plans were selected from a sample of approximately 100 buildings obtained from architectural and engineering firms around the U.S. The buildings were selected to provide a variety of configurations and use types. Their characteristics were considered in terms of providing tests of the criteria in the standards and in being a reasonably representative sample of the then current design practices. Buildings that were unusual were purposely not included in the sample. Once the building plans were obtained building components were reconfigured to comply with each of the standards tested. The building form and function were not changed but the building envelope, HVAC and lighting systems and equipment were revised as needed to comply with the applicable standard criteria.

Simulation Procedures

Of the ten buildings used in the modeling and analysis process three were office type buildings, ranging from a small (2250 ft^2) suburban office to a 38 story urban office tower. Two retail stores, a small store in a strip shopping center and an anchor store in a shopping mall, were also considered. The other five buildings included a school, a warehouse, a multi- use assembly building, a convention type hotel and a high-rise apartment building. Each of the buildings was simulated with two to four different HVAC systems in five locations to assure a variation in climatic conditions. The alternative HVAC systems were simulated to provide a means of testing the system and control criteria of the standards.

Each of the building / system combinations were simulated using the DOE 2 program developed at the Lawrence Berkeley Laboratory (LBL) operated for the U.S. Department of Energy by the University of California. This computer simulation estimates energy use for heating, cooling, fans, auxiliaries, domestic hot water, lighting, vertical transportation, and miscellaneous equipment. The individual component estimates allow consideration of the relative magnitude of each use and the impact of the changes in each of the standards criteria on each component of energy use. Detailed construction, operation, and energy cost estimates were also made for each building / system combination. Finally, life cycle cost was estimated to provide a measure of the cost effectiveness of proposed changes in the energy standard criteria. The locations (climates) considered were El Paso and Houston Texas, Washington D.C., Milwaukee Wisconsin, and Seattle Washington.

Energy Use Patterns

Typical energy use distributions are shown for a medium size office building (3 story, approximately 49,000 ft^2) and the high-rise apartment building (approximately 490,000 ft^2) in two of the five locations in Figures 2 and 3. These energy use estimates are based on compliance with the 1980 version of the ASHRAE / IES Standard 90. These results provide an indication of the potential opportunities for conserving energy. Lighting, cooling, heating and fans are the major targets for energy conservation and load management. The 'other' category includes vertical transportation and miscellaneous equipment which also represents a significant energy use but is not covered by provisions of the standard.

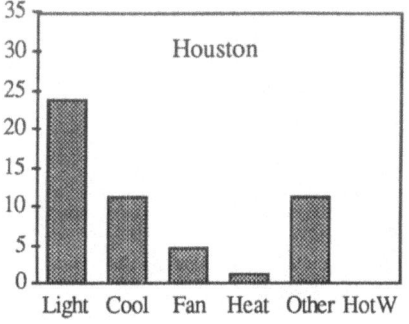

Figure 2 Office Building Energy Use

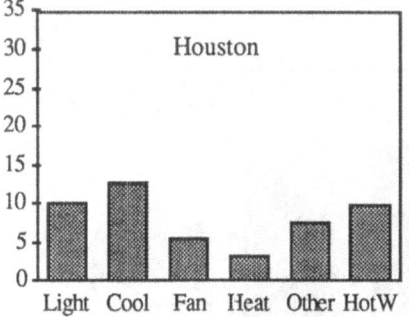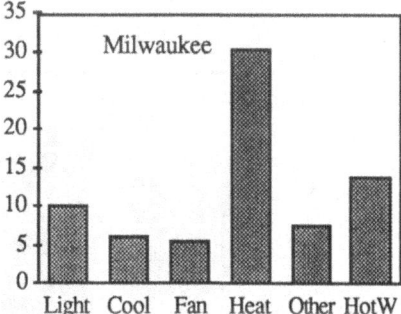

Figure 3 Apartment Building Energy Use

Impact of Standards

It is difficult to set a bench mark for design practice prior to 1975. However it seems reasonable to assume that design after 1975 was consistent with ASHRAE/IES 90-75 and examen the change that would be effected by future revisions to the standard. A revision to Standard 90 has been developed from the ASHRAE SP41 recommendations and the four year development and review of the ASHRAE /IES Standard 90 Project Committee. A simulation of the same buildings brought into compliance with the proposed criteria of the revised standard, again using the DOE 2 program, was carried out. The simulation results indicated that the revised standard would effect energy savings ranging from 7% to 29%, for the ten buildings considered depending on location and building type. In all but the apartment and warehouse buildings about half of the projected savings would come from modification of the lighting systems. A significant portion of the remaining savings would come from changes in the building envelope, principally related to controlling solar gain. The overall potential savings for five of the buildings in Houston and in Milwaukee are shown in Figure 4.

Although improvements in the efficiency of the HVAC equipment were shown to have an affect, it was smaller than expected. The selection of one HVAC system versus another however had a significant affect. As an example, the impact of HVAC system selection for the medium sized office building is illustrated in Figure 5. These results show the overall effects of the standard on annual energy use. However they still do not provide a convenient means of examining the impact of the individual criteria within the standard.

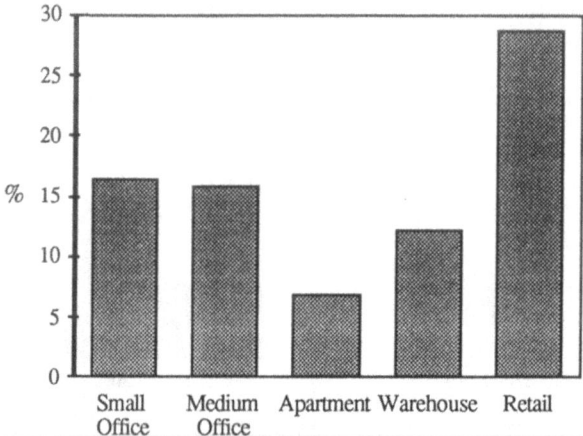

Figure 4 Percentage Energy Savings With Revised Standard

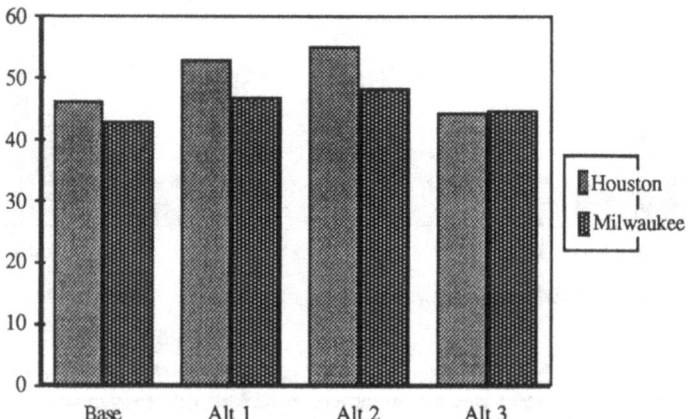

Figure 5 Effect of HVAC System Selection on Energy Use in an Office Building

EVALUATION OF STANDARDS CRITERIA

During the SP41 work a procedure for assessing the impact of envelope characteristics on heating and cooling loads was developed at LBL. This procedure was based on the analysis of 1100 DOE 2 simulations in which the thermal transmission, solar gain and internal load parameters of a prototype building were varied. The parametric analysis showed that annual heating and cooling loads and the peak cooing load varied in a consistent way as building and systems characteristics were changed. On this basis correlation equations were developed which expressed the cooling and heating loads attributable to thermal transmission, solar gain and internal gain as functions of envelope and internal load parameters. These correlations were developed by means of regression analysis using the DOE 2 simulation results as input.

As an example, the form of the equations developed at LBL for the annual load was:

Space Load = f(Transmission Loads+Solar Gain+Internal Loads)

where
Transmission Loads = $C_t(UA_t)$
Solar Gain = $C_s(ScA_g)$
Internal Loads = $C_i(I_pA_f)$

and

A_t is the area of the exterior wall of the space
Sc   is the shading coefficient of glass (and blinds or drapes included) for orientation i
A_g is the area of the glazed fenestration in space with orientation i
Ip   is the lighting + equipment power density in the space
A_f is the floor area of the space
C_t, C_s, C_i are the regression coefficients that link transmission, solar and internal heat gain to climate characteristics

Annual heating and cooling load and peak cooling demand correlations were developed using this form

Envelope Criteria

The correlation equations are quite useful in illustrating the relative effect of various building characteristics on loads. Table 1 provides some interesting comparisons for the annual cooling loads in a south facing space in two U.S. cities. It is clear that solar gain and internal loads are the principle areas of concern, even in what is considered a heating climate (Milwaukee Wisconsin). The regression coefficients C_t, C_s, and C_i were initially defined as constants for a given location in the LBL work

Table 1   Annual Cooling Load Components - South Facing Space

	Component	Base Parameter	Component Load		
			Base	Base x 1.5	Delta
Houston	Transmission	U = 1.5	49	74	25
	Solar	Sc x WWR = .15	230	346	116
	Internal	Ip = 15	119	178	59
	Total		398	598	200
Milwaukee	Transmission	U = 1.0	-12	-17	-5
	Solar	Sc x WWR = .15	123	184	61
	Internal	Ip = 15	58	87	29
	Total		169	254	85

However following the completion of simulation and regression analyses for several additional locations these coefficients were redefined as functions of basic climate variables and selected building parameters. Unfortunately the equations defining these variable coefficients have become rather complex. Work is continuing in an attempt to both simplify the form of the equations and to broaden their application. Table 2 shows similar results for the peak cooling correlations.

Table 2   Peak Cooling Load Components - South Facing Space

	Component	Base Parameter	Component Load		
			Base	Base x 1.5	Delta
Houston	Transmission	U = 1.5	120	181	61
	Solar	Sc x WWR = .15	120	180	60
	Internal	Ip = 15	39	59	20
	Total		279	420	141
Milwaukee	Transmission	U = 1.0	45	68	23
	Solar	Sc x WWR = .15	148	221	73
	Internal	Ip = 15	49	73	24
	Total		242	362	120

Again the simple format allows comparison of the impact of changes in the various components of the load. The peak cooling loads for a south facing space in Houston are lower than those for Milwaukee because of the latitude effect. The correlations for peak heating have not been included. The simulations indicated that the peak heating loads were primarily a function of operating schedule. The heating peaks were invariably morning pick-up loads in all but the high-rise apartment building.

HVAC Criteria

The annual energy use is effected by the HVAC systems selected. This is not surprising. However it does raise an interesting problem in the development of a standard. For a variety of reasons, many of them compelling economic reasons, there is no one best HVAC system for all buildings. A system that is well suited for a large multi-story office is usually not suitable for a small retail store. A general standard can not therefor be written to define the best system for the wide variety of buildings and operation schedules encountered. The ASHRAE / IES standard addresses this problem by specifying minimum criteria for a number of systems without specifying the use of one system or another. This leaves the designer with some flexibility and the responsibility of making a suitable choice.

The ASHRAE / IES Standard 90 do not yet provide an overall system criteria similar to those developed for the envelope although several approaches have been considered. Preliminary research carried out at the Center for Energy Studies (CES) at the University of Texas at Austin however indicates that it will be possible to characterize systems and equipment interactions in a straightforward way. Simulation results have indicated that for a particular system type the energy use is a linear function of the space load over a wide range of building envelope and internal loads. Figure 6 illustrates

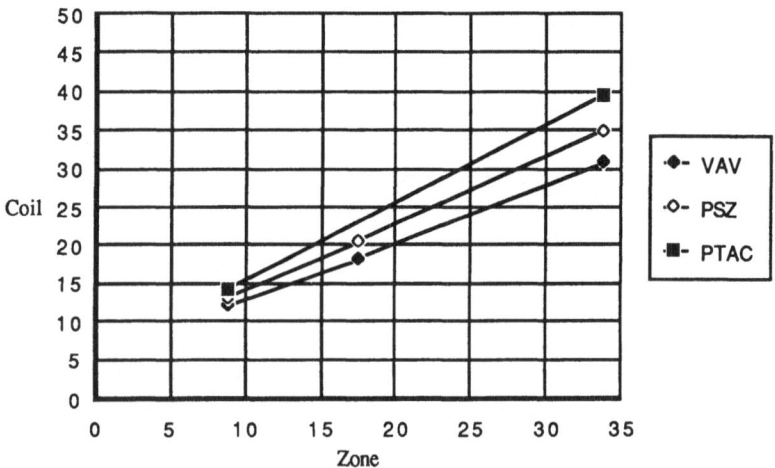

this relationship for two dissimilar systems. The slopes and intercepts of the lines that describe energy use as a function of load differ but the relationship is linear for both systems. An important point to note here is that any reduction of space load is translated into a reduction in energy use for the systems examined. This relationship may be expressed as follows:

$$\text{Cooling Energy Use} = C + (\text{Space Load}) \cdot (F_{system}) \cdot (F_{equipment})$$

Where

C	is a constant for a given climate and system selection
Space Load	is the function of (Transmission + Solar Gain + Internal Gain) in each space as indicated previously.
F_{system}	is a factor which varies with system type and climate
$F_{equipment}$	is an annual energy use efficiency for the equipment selected

If the F_{system} and the $F_{equipment}$ can be expressed as functions of specific system or equipment characteristics, the impact of individual criteria may be estimated in a simple and systematic way. The CES work has yet to be completed but the preliminary results show promise.

SUMMARY

The work described above leads to two conclusions. First energy standards do have an impact on energy use and on peak demand. In general those strategies that reduce annual cooling energy use also reduce peak cooling demand. Although peak heating demand is largely dependant on operating schedule it can also be said in general that most of those strategies that reduce annual heating energy use reduce peak heating demand. The possible exceptions are in lighting and the control of solar gain. These cases require further analysis, but appropriate selection of control strategies will reduce both use and peak demand.

The second conclusion is that energy standards will be most effective when they are formulated in such a way that the impact of their criteria can be evaluated. This approach will provide appropriate criteria and adequate guidance to assist designers in meeting those criteria. Some progress has been made in developing standards that will achieve this objective but further research and development will be required to complete the task.

ENERGY TARGETS: AN INNOVATIVE APPROACH TO BUILDING ENERGY STANDARDS

J.W. JONES

Department of Mechanical Engineering
The University of Texas at Austin
Austin, Texas, U.S.A.

INTRODUCTION

A number of building energy standards have been developed over the past 15 years. These standards have taken a variety of forms - ranging from statements of specific prescriptive criteria for building elements to general statements of expected annual building energy performance. Each approach has had its supporters and each has had its detractors. None of the standards has yet managed to satisfy all elements of the building design community. In addition the success of a building design has always been defined first and foremost by economic and aesthetic considerations wholly unrelated to energy. Therefore there is continuing resistance to any energy standard which will significantly increase the time and cost associated with the design of buildings. Considering the varying needs of the individuals involved in the design process perhaps no one approach will ever be satisfactory. However there is a persistent feeling in the design community, and in society in general, that energy is an important long-term concern and efforts to develop a more effective approach to setting building standards should continue. The objective of these continuing efforts should be to develop an energy standard that contributes to the design process in a positive way, not a simply a standard which imposes additions requirements at the conclusion of the design.

Most of the building energy standards that have been developed have to date been prescriptive in nature. The critics of this approach feel the prescriptive standards limit innovation and slow the development of new technologies. They maintain that performance standards, on the other hand, encourage both innovation and the development of new technologies. Unfortunately attempts to develop effective performance criteria have not been entirely successful. Many in the building community have expressed reservations, for example, about the practicality of both the Building Energy Performance Standards (BEPS) and Section 10 of ASHRAE Standard 90 -75 developed in the United States. In an effort to accommodate both sets of concerns the U.S. Department of Energy is sponsoring a project which is examining a new approach to defining building energy criteria.

A NEW APPROACH

The objective of this effort has been to define a more flexible, yet simpler, approach to establishing building energy performance guidelines which would encourage the design of energy-efficient buildings without creating the need for a complex enforcement mechanism. The guidelines would be presented in terms of design targets, or indexes of expected annual building energy performance. The targets will be set at levels that architects and engineers can achieve through careful integration of building systems and components in a cost-effective way and without interfering with other design goals.

This project, referred to as Whole Building Energy Design Targets for New Commercial Buildings, is being carried out by a team of design professional assembled by the American Society of Heating, Refrigerating and Air-Conditioning Engineers (ASHRAE), the American Institute of Architects (AIA) and the Illuminating Engineering Society of North America (IES) and is being managed by Battelle's Pacific Northwest Laboratory (PNL). To date only the first phase of this work, the definition of a new performance based approach and some proof-of-concept testing of this new approach, has been completed.

This Targets project is focused on developing a procedure for setting custom targets, applicable to a specific building and accommodating all the design constraints imposed on that particular building. The proposed Targets procedures will not specify performance criteria for individual components of the envelope, lighting or HVAC systems; nor will they specify methods, materials, or processes for

A. T. De Almeida and A. H. Rosenfeld (eds.), Demand-Side Management and Electricity End-Use Efficiency, 477–484.
© 1988 by Kluwer Academic Publishers.

achieving the target energy performance levels. Instead, all aspects of energy use within buildings are treated in an integrated, interactive way allowing the flexibility necessary for innovative responses in the design of energy efficient, and cost effective, new commercial buildings.

The procedure which is being proposed is built on the hypothesis that the energy use of a building can be defined as the sum of the energy required to meet the thermal comfort, ventilation and illumination needs of the specific sets of human activities, or functions, that occur within the various parts of a building. This approach further presumes that the comfort, illumination and ventilation needs of a particular activity can be defined in terms of a basic set of space configuration and function characteristics, independent of the needs of the other functions occurring in the same building. Thus the basic unit of analysis in this approach is an office space, a circulation space, or a retail space et cetera, rather than an office building, a store and so forth. There are two principle advantages to this approach. The first is that there are fewer separate functions encountered in commercial buildings than there are probable combinations of these functions in a given building type. The second follows from the first in that the energy needed to provide appropriate levels of amenity, thermal comfort and illumination for individual functions are better defined than are those of a whole buildings which might house a wide variety of functions. In this procedure the basic unit of analysis is an office space, circulation space, a retail space rather than an office building, a retail store and so forth. The target for a particular building project is determined by aggregating the targets for individual functions within the building on an area-weighted basis.

This approach allows the building designer to generate custom targets appropriate to the constraints and requirements of the specific building being designed. These custom targets will reflect building internal load levels, hours and patterns of occupancy, site and other design constraints rather than being based on 'typical' conditions that may have no relevance for the particular building. Another important feature of the proposed approach is that energy cost and building cost factors can be incorporated into a basic methodology for establishing targets. Then the Targets can then represent a balance between energy conservation and cost and economic considerations.

THE PROPOSED PROCEDURE

The proposed Target setting procedure is illustrated in Figure 1. The first element is an input block in which the constrained space function characteristics are specified (function, location, space orientation, amenity factors and energy costs and economic limitations). The second element, the major portion of the procedure, is the analysis block. It consists of four subelements; function characteristics, energy analysis, energy cost and building cost analysis, characteristic modification. The final elements of the procedure are the output and review blocks. The output block is simply a reporting function but the review block provides a mechanism for additional feedback. Each of the elements of the target setting procedure will be described briefly.

Input

The input block provides a means of specifying those space function characteristics that are required for the analysis and are constrained by non-energy design considerations and can not be modified within the analysis.

Analysis

Space Function Characteristics The Space Function Characteristics subelement reviews the input and then selects from a library the remaining energy-related physical characteristics required to complete the energy and cost analyses. The library is structured to provide values which are appropriate for the specified function and input constraints and are cost effective. The relevant characteristics of the space enclosure and the HVAC and illumination systems are defined in terms of the comfort and illumination needs of the intended function of the space. This is in generally straightforward as the temperature and lighting levels appropriate for a particular function are relatively well defined. The relationship between function and the amount of fenestration is not as well defined. However there seems to be some consensus as to an acceptable range for many of the functions that commonly occur in commercial buildings. Similarly there is a general consensus as to the types of HVAC and illumination systems which are acceptable within economic and amenity constraints for a particular space function and overall building configuration (building size, number of stories etc.).

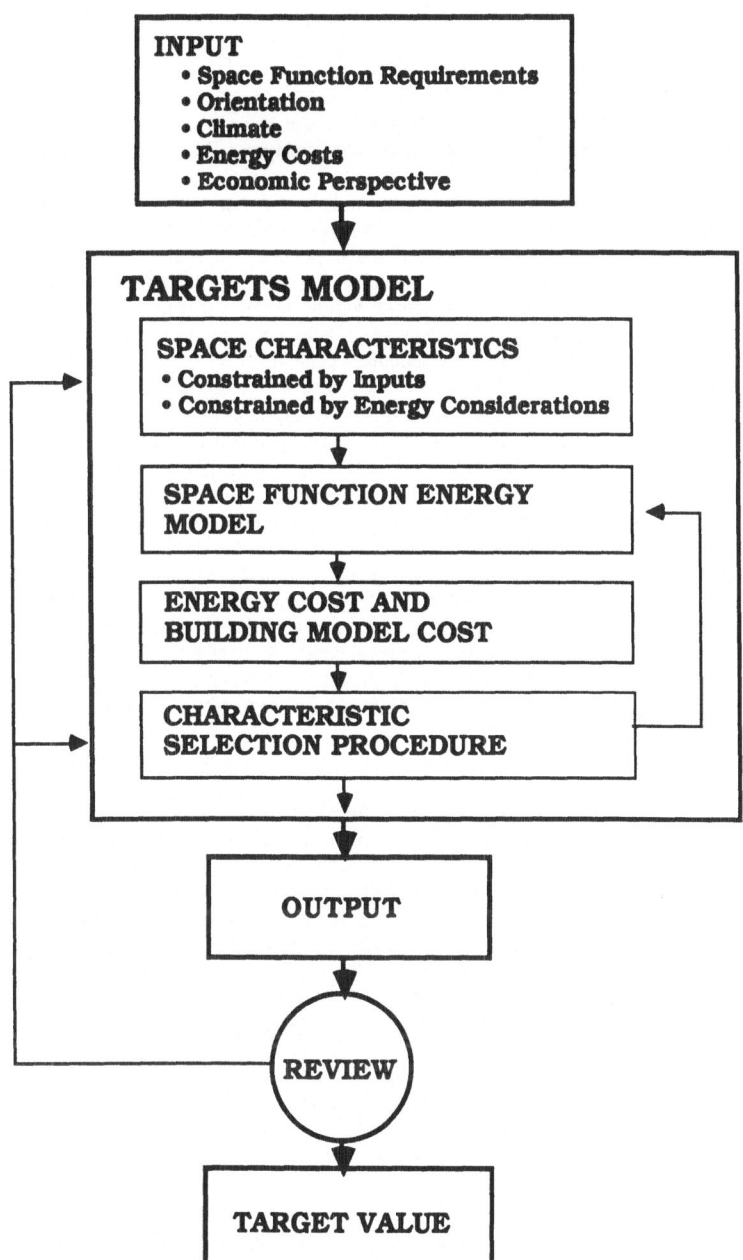

Figure 1. Targets Model — Procedure for Developing Targets

An initial value and a range of acceptable variation around this value will be defined for each envelope and system component of the space that directly impacts the energy required and/or energy costs and building costs . For example appropriate glazing will be selected based on the input to this element . The extent of the glazing, if not specifically defined by the input, will be set and glass characteristics such as shading coefficient and visible transmittance selected. The Input and Space Function Characteristic then define the thermal, solar, and internal load input to the energy analysis. The space function characteristics are then transferred to the Energy Analysis.

Energy Analysis Two approaches to developing a Target energy analysis procedure are under consideration. The first grows directly out of recent efforts to revise the ASHRAE / IES building energy standard (Standard 90). The other grows out of the expectation that there will be a significant improvement in the ability to quickly and inexpensively simulate building performance on micro computers in the next few years. The first approach would develop an integrated set of correlation models based on the results of numerous simulations of building performance using one or another of the state-of-the-art energy analysis program. The second approach, referred to as a compact simulation approach, would be based on a modification of one of the state-of-the-art energy simulation methods to provide the basis for setting targets. The compact simulation methodology would not be a general energy analysis tool. Its function would be limited to providing a mechanism for accommodating design constraints in a way that may be difficult to achieve in the regression approach. It is expected that the compact simulation would be considerably simpler to use than a full scale hourly energy analysis program. It will be designed to run conveniently on the desk top computer hardware that can be expected to be available when the project is completed.

Both the correlation and compact analysis approaches have desirable characteristics and some potential liabilities. For example, the correlation approach would likely be easier, and perhaps faster, to use once developed. However, it would be difficult to develop and more importantly difficult to update in the future. A compact simulation model on the other hand would likely take more time to run (particularly on a microcomputer) but should require less time to develop, be much more adaptable to changes in technology and relatively easy to update. It would also be easier to test and verify than a correlation model.

Although the compact simulation approach is most probably going to be the one chosen, the correlation approach provides a simpler means of illustrating the overall structure of the proposed target setting procedure. The difference in description between the correlation and compact simulation procedures is simply a matter of substituting a simulation procedures for the correlation equations.

The general format for the space - function targets is stated in terms of the energy required to meet comfort and illumination needs. The energy required for lighting is independent of the the energy required for comfort control. However the methodology provides a means to account for the interaction of the space loads which result from lighting and the energy required for comfort control. For example the energy needed, on an annual basis, to maintain comfort by cooling can be expressed as a function of the space loads in the following way:

Cooling Energy Use = c + (Space Load) • (F_{system}) • $(F_{equipment})$

Where

c	is a constant for a given climate and system selection
F_{system}	is a factor which varies with system type and climate
$F_{equipment}$	is an annual energy use efficiency for the equipment selected
Space Load	is a function of the transmission loads, solar gain and Internal loads in each space

In turn the Space Load is expressed in terms of space function characteristics and F_{system} and $F_{equipment}$ are defined by the characteristics of the system and equipment selected. Each of these terms require some additional explanation.

In research carried out in support of the most recently proposed revision of the ASHRAE /IES Standard (Standard 90) it was shown that the annual energy use required to provide comfort and illumination in an individual space in a building could be defined in terms of a basic set of characteristics of the space enclosure and the HVAC and illumination systems that have been selected to serve the space. This relationship was expressed in terms of a linear equation. For example the energy required for cooling a space was expressed in the form:

$$C = C_t(UA_t) + C_s(ScA_g) + C_i(I_pA_f) + C_v(A_f) + C_m(A_t)$$

where

C is the annual energy use for cooling for a space

U is the overall heat transfer coefficient of the exterior wall

A_t is the area of the exterior wall of the space

Sc is the shading coefficient of glass (and blinds or drapes included)

A_g is the area of the glazed fenestration in space

I_p is the lighting and equipment power density for the space

A_f is the floor area of the space

C_t, C_s, C_i are the regression coefficients that link transmission, solar and internal heat gain to climate characteristics

C_v is a coefficient for the particular climate and orientation to account for ventilation.

C_m is a correction to account for the effects of the thermal mass of the wall.

Some question remains about the best approach for defining these relationships in terms of load, climate and schedule variables. Several approaches have been tested with varying degrees of success. However the main point here is that the variation of annual energy use for cooling and heating can be simply expressed as a function of physical parameters.

HVAC System and Equipment Analysis   The next analysis element provides a means of accounting for the characteristics of the HVAC system and controls. Initially, there was some concern that interaction between the HVAC and lighting systems and inter-space heat transfer would introduce effects that would preclude considering space functions separately. However, a preliminary evaluation of building energy performance simulations carried out in proof-of-concept tests have shown that annual energy use for heating and cooling varies with space load in a predictable way for both single space and multiple space systems. Two sets of simulation using an hourly program (DOE 2.1C), one based on a single system serving five zones and the other separate systems of the same type serving each of the five zones were run. A comparison was made of the results for three distinctly different HVAC system in two very different climates. This comparison indicated that the predicted annual energy use for heating and cooling determined as a sum of the single zone simulations did not differ significantly from that predicted by the five zone simulation. Thus the interactions of the system-space-environment combination do not cause significant inaccuracies when the annual energy use is estimated separately for each space-function and then combined with that of other spaces in an additive fashion as opposed to being defined by a multiple space system analysis.

A second issue was whether or not the magnitude of the space loads would influence the relationship between the space loads and equipment energy use. Again there was some concern that the use of an economizer, varying control strategies and the part-load performance characteristics of the heating and cooling equipment would affect the relationship between annual space load and annual energy use. This issue was examined by varying space loads over a reasonable range in a systematic way, again using an hourly simulation program. The nonlinear results which may be observed when hourly loads on a space and hourly heating and cooling equipment energy use are compared do not appear in the annual results. Annual energy use for each of the three systems in both climates varied linearly with annual space load.

The system and equipment performance factors were derived from the building simulation output. The system factor F_{system} was derived by taking the ratio of the coil load (the heating energy added to or the cooling energy extracted from the air stream) to the space load:

$$F_{system} = (\text{Coil Load}) / (\text{Space Load})$$

The results of the proof-of-concept tests indicated that F_{system} was constant with loads varying over a 3 to 1 range for the systems and climates examined. The Equipment factor, $F_{equipment}$ was defined in a similar fashion:

$$F_{equipment} = (\text{Energy Use}) / (\text{Coil Load})$$

Again the factor $F_{equipment}$ was essentially constant over the range tested. The conclusion is again that the magnitude of the load, or its pattern of variation over time, does not introduce non-linear effects when the comparisons are on an annual basis. The final step in the definition of these factors will be to express them as explicit functions of system and equipment characteristic. This has not yet been accomplished. The initial investigation did however shown that it is possible to again aggregate coil loads for individual spaces to system loads for whole building targets.

The most important result of the initial proof-of-concept energy analysis tests has been that it is possible to define the energy needed annually to serve an entire building in terms of the energy needs of the individual spaces within the building. Thus the space-function approach can adequately represent the energy use patterns of a large multizone building with a central HVAC system. The dynamics of space and systems interactions which may influence short term energy use patterns are damped when integrated over the period of a month or a year.

Once the energy analysis has been completed the space function characteristics and energy use results are passed to the Energy Cost and Building Cost element to determine the economic consequences of the characteristics selected within the Targets Model.

Energy Cost and Building Cost Analysis

In developing the Target procedure it has been recognized that a balance between energy and non-energy considerations is necessary . A design that simply minimizes energy use may not be compatible with other design requirements and thus unacceptable to the owner or occupant. Most often cost and economic considerations will prevail if there is a conflict. It is therefore essential that economic efficiency be considered in setting targets. The feasibility of including economic considerations in the target procedure was again tested by means of a proof-of-concept economics model with an encouraging result. This model included building and energy cost data, a preliminary building energy model, and economic optimization routines. The cost criteria selected for the analysis is modified total owning and operating cost for the building (TOOC). The modification occurs in that only those building costs, both first cost and annual cost, that effect or are effected by energy use will be considered.

First Cost   First cost considerations are introduced into the proposed Target procedure through the use of generalized information about the costs of energy related building components or features. Costs for glazing systems, insulation, HVAC equipment and so forth, drawn from available sources of building cost data, will be expressed in algorithmic terms. These will be combined with the algorithms that describe the impact of changes in building components on peak loads, and thus on equipment sizing, to estimate cost impacts. For example the use of high performance glass will increase the cost of the building envelope but reduce peak cooling loads. The reduction in peak cooling loads allows the selection of lower capacity cooling equipment reducing costs. In developing the cost algorithms the relationship between component cost and component performance will be treated as continuous functions. This is obviously not representative of the real case as both cost and physical characteristics vary in a discreet rather than continuous way. However as the procedure is concerned with defining relative building costs, not absolute costs, this approach provides reasonable results. This concept will be tested further in future work.

Annual Costs   The annual energy costs used in the analysis procedure will be based on local energy costs and local utility rate schedules. The cost and rate data will be brought into the analysis through the input block described earlier. Other annual costs, such as maintenance or replacement costs, will be included as averages for the annual period. It will probably not be possible to treat these costs as discrete expenditures at specific points in time. However the use of annualized costs over the life cycle cost analyses period should not limit the effectiveness of the analysis.

Modified Total Owning and Operating Cost   The modified total owning and operating cost has been selected as the criterion for cost effectiveness in the Target procedure. Because some costs only occur once while others are recurrent, standard economic analysis techniques that account for the time value of money will be used in combining all cost elements. The minimum Total Owning and Operating Cost (TOOC) will be used as the criteria in selecting the combination of unconstrained space function and system characteristics to be used in defining the design energy target. This will assure that the target level specified can be reached in a cost effective way. This selection process is described in the next section.

Characteristic Selection Procedure

A structured characteristic selection procedure will be developed to define the unconstrained space and system characteristics that are required in the energy and cost analysis segments of the procedure. Those characteristics constrained by specific design criteria or overriding cost and economic considerations are set in the input block. Those characteristics that are not constrained will be selected on the basis of a balance between energy and related cost considerations using the total owning and operating cost criteria. A rudimentary proof-of-concept characteristic selection model has been developed and tested. This procedure, although far from a final version, has demonstrated that the process is feasible. The range of characteristic values available in this selection process will be limited to ensure that the energy related physical characteristics selected are appropriate for the function served by the space. This portion to the program is in a sense an 'expert system' based primarily on input from design and cost analysis professionals.

The first pass through the energy and cost blocks was based on a preliminary selection of the characteristics needed in these two analyses. The results of the characteristic selection procedures in this block are now fed back into the energy and cost analysis blocks. The energy and cost analyses are repeated to refine the selection of the unconstrained function characteristics until a reasonable balance of energy use and cost effectiveness is achieved. Once this balance has been achieved and the TOOC criteria has been met, the target for the space function is set.

It is important to note that this characteristic selection process will not specify the actual physical parameters to be used in the design. It will only select a one possible set of parameters that can be used in establishing a design target that accommodates both the external constrains on the project design and basic energy concerns. The designer need not even be aware of the characteristic values used in establishing the Target. Once a reasonable target has been set designers are free to use any set of strategies they may wish to choose to meet the target criteria. Thus the proposed procedure provides a building and site specific, custom target without prescribing specific design options.

Output and Review

The final step in the process is to sun the targets for each function (on the basis of the area associated with each function) to obtain a whole-building design energy target. This result, along with an echo of the constrained inputs is printed as output. The output then passes through a final reviewed for consistency and a final building target is provided. The structure of the review block has not yet been defined.

Using the Targets

Developing a Target. The steps in deriving a whole-building energy target, using the Targets Model procedure would be:

- Specify functions to be housed in the building.
- Identify those function characteristics that directly or indirectly influence energy use and are dictated by, or are constrained by non-energy design considerations.
- Specify those function characteristics which influence energy use and are not constrained by design considerations other than energy efficiency.
- Define configuration (size, volume, location - perimeter or interior) and the characteristics of enclosing surfaces (fenestration, orientation and thermal mass) for each function.
- Define thermal comfort, illumination and ventilation requirements for each function.
- Define energy requirements for the comfort, illumination and ventilation needs for each function. There are several steps in estimating the energy requirements. They are:
  - Define the heating and cooling loads in terms of envelope and function characteristics and operating schedule (annual space loads).
    Annual Load = (Transmission Loads+Solar Gain+Internal Loads)
  - Define the HVAC and illumination system performance characteristics for each system (annual system factor)
  - Define equipment performance characteristics (annual equipment factor). For example for the cooling component of the target:
    Cooling = (annual load) x (systems factor) x (equipment factor)
    where the annual cooling loads include consideration of all configuration, envelope and schedule characteristics.
- Multiply the energy used per square foot for each function by the floor area associated with that function.
- Evaluate the energy cost and building cost consequences of the

selected characteristics for each function. Revise and repeat as
necessary to achieve a balance between energy use and cost
effectiveness.
- Sum the energy use-area products for each of the various functions
  to obtain the whole building target.

SUMMARY

The end product of the Targets Model is thus a whole-building energy performance target that can be
tailored to a specific building, rather than being tied to a generic building description that may or may not
be a suitable reference point for a particular building being designed. This approach directly resolves a
number of the problems that have plagued previous efforts in that it will:
- provide information pertinent to specific projects
- provide information at the appropriate stage of design - a point at which
  the design can still be conveniently modified to accommodate energy concerns
- provide this information in a readily usable format, one that requires
  relatively little effort on the part of the user

It is intended that the design targets make a positive contribution to design rather than being an added
burden.

ENERGY AND ECONOMIC SAVINGS FROM NATIONAL
APPLIANCE EFFICIENCY STANDARDS IN THE U.S.

HOWARD S. GELLER

AMERICAN COUNCIL FOR AN ENERGY-EFFICIENT ECONOMY
WASHINGTON, DC

1. BACKGROUND

In March 1987, President Reagan signed into law the National
Appliance Energy Conservation Act (1). The Act contains minimum
efficiency standards for new residential appliances, heating, and
air conditioning equipment (hereafter referred to as appliances).
The standards legislation was supported by a coalition of over 40
national organizations in the U.S. including organizations
representing appliance manufacturers, environmental and
conservation groups, and the electric utility industry.

Adoption of the national standards was prompted in part by the
passage of appliance efficiency standards in a number of states in
the absence of national regulations. Appliance manufacturers
decided to support national standards rather than face a multitude
of state requirements. The national legislation makes it
difficult for states to adopt their own standards in the future.
The basic rationale for adopting appliance standards in the U.S. is
that appliance manufacturers otherwise produce and consumers
purchase products with efficiencies far below economically
justified and socially desirable levels (2).

The national standards are minimum efficiency (or maximum
energy consumption) requirements that apply at the point of
manufacture. The standards also apply to appliances imported
into the U.S. Product performance is judged using the test
procedures currently used for labeling the operating cost or
efficiency of appliances in the U.S. Table 1 shows the standards
requirements, the dates they become effective, and the
efficiencies (or energy consumption levels) of typical products
produced in 1985. The Act provides for periodic review and
revision of the standards in order that the requirements do not
become outdated.

The appliance standards are relatively stringent. For most
products, 70-90% of the models offered or.produced in 1986 won't
qualify to be sold when the standards take effect. Consequently,
the standards will have a significant impact on residential energy
consumption in the future.

This paper discusses the energy and economic impacts that can
be expected from the standards from both the utility and the
consumer perspectives. The broad implications of the standards
for utilities and energy planners in the U.S. as well as foreign
nations is also addressed. This paper summarizes a more detailed
study regarding the energy and economic savings from the national
appliance standards (3).

A. T. De Almeida and A. H. Rosenfeld (eds.), Demand-Side Management and Electricity End-Use Efficiency, 485–496.
© 1988 by Kluwer Academic Publishers.

2. METHODOLOGY

This analysis includes the following products: refrigerators (R/F), freezers (FR), room air conditioners (RAC), central air conditioners and heat pumps (CAC and HP), electric water heaters (EWH), gas water heaters (GWH), gas furnaces and boilers (GF), and gas ranges (GR). Oil heating equipment, direct heating equipment (space heaters), dishwashers, clothes washers, and swimming pool heaters are included in the legislation but not in this analysis because savings for these products are difficult to estimate and are unlikely to be as large as for the other products. The analysis only covers use of heat pumps for cooling because it is difficult to estimate heating savings, e.g., data on the average heating efficiency of heat pumps are unavailable.

The analysis considers the energy use, first cost, and operating cost of products sold between 1986 and 2000. National product sales are projected year-by-year along with the anticipated energy use of typical models produced under a "marketplace scenario" (i.e., assuming national standards had not been adopted) and a "standards scenario". Improvements in efficiency in the marketplace scenario are based on estimates provided by the industry associations representing appliance manufacturers where available.

Table 2 shows the estimated energy use of typical products sold in 2000 in the marketplace and standards scenarios, along with the expected average energy use of products sold in 1987. The values are shipment-weighted averages for different classes of products and are based on typical operating conditions in the U.S. (3). It is assumed that operating characteristics such as daily hot water use or average space heating and cooling loads remain constant over the time frame of interest.

When the standards go into effect, there will naturally be some distribution of shipments whose average efficiency or energy use will surpass the requirements. Table 2 includes the assumed resulting energy consumption levels in the year the standards take effect (typically 1990 or 1992). In all cases besides gas ranges, there is a 5-10% "margin" relative to the ceilings imposed by the standards. After the standards take effect, the energy use of most new products is assumed to gradually decline to the values shown for 2000 in the standards scenario.

For refrigerators and freezers, it is highly likely that the 1990 national standards will be upgraded through a rulemaking by the U.S. Department of Energy. This is necessary in order to avoid having California and possibly other states implement tougher state standards for these products in 1993. This is possible because California had adopted such standards prior to the federal legislation. It is uncertain what the outcome of this federal rulemaking will be -- conservation advocates will be arguing for tightening the standards substantially, while manufacturers will attempt to limit any tightening. In this analysis, it is assumed that the national standards are upgraded to the California level effective in 1995.

Savings in peak summer demand as a result of the standards are in included as part of the analysis. For each electrical product, a particular peak-to-average load factor or diversification factor is used to convert electricity savings to peak demand savings (3). The factors are based on peak demand occurring between 2:00-6:00 P.M. during the summer (4). Thus, the estimates of peak demand savings from the standards are the coincident national savings at this time.

The analysis includes the dollar savings for consumers over the lifetime of products sold by the year 2000. The economic analysis is done in terms of constant 1985 dollars using a 5% real discount rate for equipment and energy costs in the future. The extra first cost for more energy-efficient models is estimated based on a constant cost increase per unit of energy savings. Assumptions regarding extra first cost are derived from engineering studies by U.S. DOE, the California Energy Commission, and others (3). Operating savings are based on the average residential energy prices in 1985 -- $0.078/kWh for electricity and $6.06/MBtu ($6.39/GJ) for natural gas (5). It is assumed that these energy prices remain level in constant dollars.

3. ENERGY SAVINGS RESULTS

Table 3 summarizes the energy savings results, showing the total electricity and energy savings expected in 2000 as well as the aggregate savings over the lifetime of products sold during 1988-2000. Electricity savings are presented in terms of end-use (i.e., TWh and peak MW of demand), while the total energy savings values refer to primary electricity use (11,500 Btu/KWh).

It is seen that by 2000, the appliance standards are expected to reduce electricity consumption by 51.3 TWh/yr with peak summer demand reduced by 21,100 MW. Water heaters account for 39% and refrigerators 26% of the expected electricity savings. CACs and HPs account for the majority of the peak demand savings. The standards on all cooling products represent 78% of the total peak demand reduction expected in 2000.

To place the savings estimates in perspective, residential customers in the U.S. consumed 790 TWh in 1985, with a summer peak demand of approximately 175,000 MW (6). The anticipated reduction in peak demand in 2000 due to the appliance standards is equivalent to nearly 10% of the growth in national peak demand forecast by the utility industry during 1986-2000 (7). The significance of the savings would be even greater if the industry forecast is overestimated, as has been the case in the recent past.

Regarding overall energy savings (electricity and fuel), it is estimated that the standards will lower residential energy use by 917 trillion Btu (0.97 EJ) in 2000. This equals about 6.1% of residential energy consumption at the present time in the U.S. Over the lifetime of products sold during 1986-2000, it is estimated that the standards will lower residential energy use by 14.3 quadrillion Btu (15.1 EJ). This is equivalent to over 18 months of energy imports at the U.S.'s current net import rate. Electrical products account for 64% of the total energy savings in 2000 while gas products account for 36%.

The result that water heaters provide more energy savings in 2000 than any product type is due in part to the expectation that there would be very limited improvements in water heater efficiency if standards weren't adopted (see Table 1). This projection is based on trends during the past decade which showed minimal improvement in the average efficiency of new water heaters (8).

4. ECONOMIC SAVINGS RESULTS

It is possible to estimate the investment in utility plant that can be avoided as a result of the appliance standards. First, it is assumed that the electricity savings due to standards on R/Fs, FRs, and EWHs are baseload savings, that baseload capacity costs $1500/KW, and it operates 65% of the time. Second, it is assumed that the savings due to standards on CACs, HPs, and RACs are peak demand savings, and that peak capacity costs $500/KW. Also, it is assumed that there are 8% T&D losses. Using these assumptions, the calculation of the potential avoided investment is illustrated in Table 4. It is seen that the potential reduction in power generation investment alone is $19 billion by 2000 (in undiscounted 1985 dollars). In addition, utilities will save on investments in power transmission and distribution facilities.

Table 5 shows the economic savings from the perspective of consumers who will consume less energy as a result of the standards. The savings are presented in terms of the reduction in annual operating cost in 2000 as well as net lifetime savings. The latter is the operating savings over the lifetime of products sold by 2000 minus the estimated extra first associated with the more efficient products. The benefit-cost ratio is the value of lifetime operating savings divided by the extra first cost for consumers. All values are expressed in 1985 dollars, without accounting for changes in energy prices as a result of the standards.

Table 5 shows that the standards will save consumers about $3.8 billion per year by 2000. The net economic savings for consumers over the lifetime of products sold during 1986-2000 is $26.3 billion. This is nearly $300 per household. EWHs provide the most net economic savings (31% of the total), followed by GWHs (22% of the total) and R/Fs (20% of the total). The overall benefit-cost ratio associated with the standards is 3.0. The benefit-cost ratio is greater than 1.0 for all product types, ranging from 1.2 for CACs to 8.7 for GWHs. The relatively low benefit-cost ratio for air conditioners is due to the substantial first cost premium associated with increasing the efficiency of this product.

Net economic savings and benefit-cost ratios are based in part on the first cost premium for more efficient products as determined in studies during the early 1980s (3). However, manufacturers claim the extra first cost will fall as more efficient products become the norm once the standards take effect (9). If so, greater savings will occur and the standards will be even more cost effective. If it turns out that there is no first cost premium associated with the standards, consumers would save nearly $40 billion.

It should be remembered that all savings estimates are based on the standards scenario compared to a hypothetical marketplace scenario. In the latter, it is assumed that some efficiency improvements would have occurred "on their own". If the standards scenario is contrasted with the efficiencies of products sold at the present time, the energy and economic savings would be much greater. Also, this analysis is conservative because some products have been left out, and because refrigerators and freezers are the only products for which an increase in the initial standards levels is assumed. The Appliance Standards Act requires that all standards levels be reviewed twice during the 1990s, thus other revisions are possible.

5. IMPLICATIONS OF THE APPLIANCE STANDARDS WITHIN THE U.S.

The national appliance standards will provide substantial energy savings at virtually no cost to the government or utilities within the U.S., i.e., these organizations do not need to operate conservation programs in order to obtain the demand reductions. Furthermore, by having adopted the standards, there is much less uncertainty about future levels of appliance efficiency and energy consumption. Energy analysts and planners no longer need to guess whether manufacturers and purchasers are going to stay with the current generation of appliance technologies, or shift to more efficient equipment.

It is important that utilities as well as state and federal energy planners factor the standards into their forecasts and energy plans. This is best done using end-use models that account for energy use on a product-by-product basis including changes in the energy intensity of new models from year-to-year. Without end-use-based analysis and forecasting, the benefits of avoiding investments in power generation may not be realized, and utilities might construct unnecessary power plants. The fact that appliance standards will result in a substantial changes in new product efficiency makes it all the more important that utilities explicitly consider the impacts that the standards will have on their service areas.

Another issue for utilities in the U.S. is whether or not it makes sense to continue efficiency incentive programs given that the standards have been adopted. Many utilities in the U.S. now pay rebate incentives to consumers who purchase more energy-efficient appliances, air conditioners, etc. [10]. Whether or not utilities should continue these programs depends on their need for energy savings over the short run (before the standards take effect) and the range of efficiencies available in the marketplace once the standards become effective. The efficiency of new products is likely to improve on its own prior to the standards becoming effective as manufacturers introduce new models that comply with the standards and drop less efficient models. If there is still a wide range of efficiencies available after the standards take effect, it might be feasible to offer rebate incentives to consumers who buy very efficient models. Of course, it would not make sense to offer rebates at efficiency levels close to the standards.

6. INTERNATIONAL RAMIFICATIONS OF THE U.S. APPLIANCE STANDARDS

What are the implications of the U.S. appliance standards on other OECD countries? The standards legislation permits U.S. manufacturers to export models that do not meet the U.S. standards, but any product offered for sale in the U.S. (locally or foreign produced) must meet the efficiency standards once they take effect.

International impacts from the U.S. standards could be limited because the U.S. appliance industry has a local orientation. The size and features of U.S. appliances often differ from those in Europe and the Far East, thereby limiting appliance trade between the U.S. and other OECD countries. Most major U.S. appliance manufacturers export less than 5% of their output and much of this is destined for particular markets, e.g., Canada, Mexico, and U.S. employees overseas (11). U.S. manufacturers claim they will not continue producing less efficient appliances in small quantity for export once the standards become effective. Thus, those foreign markets served by U.S. producers should benefit from the U.S. appliance standards.

The U.S. appliance market, unlike electronics equipment and automobiles, is still dominated by U.S. companies. This is changing, however, as international mergers occur, production of components and in some cases finished products moves overseas, and a growing number of foreign producers enter the U.S. appliance market. For example, Japanese companies have gained over a 5% share of the U.S. room air conditioner market in the past few years and General Electric Co., the largest U.S. air conditioner manufacturer, is in the process of obtaining most of its units from Japan (12).

It is not easy to determine the stringency of the U.S. standards relative to the efficiency of appliances produced in other countries because different efficiency test procedures are used in the U.S., Europe, and Japan. The limited data that are available reveal that Japanese companies generally are selling room air conditioners of above average efficiency in the U.S. (13), while the efficiency of their larger refrigerators appears to be comparable to that of U.S. models (14).

Those foreign manufacturers making advanced energy-efficient products might gain an opening into the U.S. market due to the efficiency standards. On the other hand, U.S. producers are expected to engage in considerable R&D and product innovation in satisfying the standards. If foreign manufacturers do not follow suit, their products could fall behind in efficiency and be at a disadvantage when competing in this increasingly international market.

Adopting minimum efficiency requirements in other countries could stimulate manufacturers in these countries to innovate and help them to compete internationally. In addition, consumers and utilities would benefit from the sale and use of more efficient and economical appliances. The Canadian federal government is already considering adoption of the U.S. appliance efficiency standards for these reasons (15).

Adopting national appliance standards in Europe is complicated by the high degree of inter-country sales and the free-trade agreements. Nonetheless, it is suggested that individual countries and/or the EEC consider the possibility of adopting minimum appliance efficiency standards. Naturally, such standards should be tailored to the manufacturing capability, energy and economic conditions, consumer preferences, etc. in particular countries.

REFERENCES

1. "National Appliance Energy Conservation Act of 1987",
   P.L. 100-12, March 17, 1987 (42 USC 6291).
2. H. Ruderman, M. Levine, and J. McMahon, "Energy-Efficiency
   Choice in the Purchase of Residential Appliances", in
   W. Kempton and M. Neiman, eds., Energy Efficiency: Perspec-
   tives on Individual Behavior, American Council for an
   Energy-Efficient Economy, Washington, DC, 1987.
3. H.S. Geller, "Energy and Economic Savings from National
   Appliance Efficiency Standards", American Council for an
   Energy-Efficient Economy, Washington, DC, March 1987.
4. H.S. Geller, et al., "Residential Conservation Power Plant
   Study Phase I - Technical Potential", report prepared for
   Pacific Gas and Electric Co. by the American Council for an
   Energy-Efficient Economy, Washington, DC, Feb. 1986.
5. "Monthly Energy Review", DOE/EIA-0035(86/03), Energy
   Information Administration, U.S. Department of Energy,
   Washington, DC, March 1986.
6. A.H. Rosenfeld, "Shifting Peak Power: At the Meter, Beyond
   the Meter, and at the Checkbook", in C.B. Smith, T. Davis,
   and P.W. Turnbull, eds., Meeting Energy Challenges,
   Pergamon Press, New York, 1985.
7. "35th Annual Electric Utility Industry Forecast",
   Electrical World 198 (9), Sept. 1984.
8. Personal communication with Mr. Jack Langmead, Gas
   Appliance Manufacturers Association, Arlington, VA,
   July 1986.
9. See statements on behalf of the appliance industry
   associations during hearings on appliance standards,
   Subcommittee on Energy Conservation and Power, U.S. House
   of Representatives, Sept. 10, 1986 and Subcommittee on
   Energy Regulation and Conservation, U.S. Senate, Sept. 16,
   1986.
10. "A Compendium of Utility-Sponsored Energy Efficiency
    Rebate Programs", American Council for an Energy-Efficient
    Economy and the Consumer Energy Council of America, to be
    published by the Electric Power Research Institute, 1987
    (forthcoming).
11. Sterling Hobe Corporation, "Comparative Analysis of U.S.
    and Selected Foreign Household Appliance Industries", U.S.
    Dept. of Energy, Building Equipment Division, Washington,
    DC, Oct. 1984.
12. H.S. Geller, "Energy Conservation R&D, Innovation, and
    Industrial Competitiveness: The Case of Household Techno-
    logies", American Council for an Energy-Efficient Economy,
    Washington, DC, Jan. 1986.
13. "1987 Directory of Certified Room Air Conditioners",
    Association of Home Appliance Manufacturers, Chicago, IL,
    March 1987.
14. A.K. Meier, "Energy Use Test Procedures for Appliances:
    A Case Study of Japanese Refrigerators", LBL-22708,
    Lawrence Berkeley Laboratory, Berkeley, CA, 1987.

15. "Appliance Efficiency Information Base", report prepared by Marbek Resource Consultants Ltd., Ottawa, Canada for Energy, Mines and Resources Canada and the Ontario Ministry of Energy, May 1987.

Table 1

NATIONAL APPLIANCE EFFICIENCY STANDARDS IN THE U.S.

Product	1985 average efficiency (1)	Standard level (2)	Year standard takes effect
Refrigerators	1100 kWh	976 kWh	1990
Freezers	790 kWh	671 kWh	1990
El. water heaters	0.836 EF	0.884 EF	1990
Room AC	7.7 EER	8.6 EER	1990
Central AC (3)	8.6 SEER	10.0 SEER	1992
Gas furnace	0.74 AFUE	0.78 AFUE (4)	1992
Gas water heaters	0.494 EF	0.544 EF	1990
Gas range (5)	--	--	1990

1. 1985 shipment-weighted efficiency is expressed in terms of annual electricity use for refrigerators and freezers. For other products, the conventional unit of efficiency is used. EF is the energy factor rating for water heaters, a measure of overall efficiency assuming 64 gallons (242 liters) of hot water use per day. EER is the energy efficiency ratio for room air conditioners, expressed in terms of Btu/hr of cooling output per watt of power input. SEER is the seasonal energy efficiency ratio for central air conditioners, also expressed in terms of Btu/hr per watt. AFUE is the annual fuel utilization efficiency for central heating equipment, a measure of the seasonal space heating efficiency.

2. The standard level is the average for all product classes. It is given in terms of the maximum electricity use for refrigerators and freezers, and minimum efficiency for the other products.

3. The central air conditioner standard applies to split systems; the minimum standard for package units is 9.7 SEER effective in 1993.

4. The gas furnace standard is based on the isolated combustion air test, which is equivalent to about an 0.80 AFUE rating with the test procedure currently used by the furnace industry association.

5. The gas range standard bans the use of pilot lights in ranges and ovens having an electrical supply cord.

Table 2

ASSUMED UNIT ENERGY CONSUMPTION (UEC)
VALUES IN THE STANDARDS ANALYSIS

Product	Estimated 1987 UEC	Resulting UEC with standards (1)	Market case UEC in 2000	Standards case UEC in 2000
	-----------	(KWh/yr or	GJ/yr)	------------
Refrigerators	1070	900	830	620
Freezers	760	620	580	430
El. water heate	4290	3740	3960	3530
Room AC	1000	830	860	750
Central AC	2910	2400	2520	2210
Gas furnace	63	55	58	53
Gas water heaters	28	24	27	23
Gas range	6	5	6	5

1. The assumed UEC in the year the standards take effect. UEC
is the average unit energy consumption of products
manufactured in a particular year.

Table 3

ENERGY SAVINGS IN 2000 FROM THE
APPLIANCE EFFICIENCY STANDARDS (1)

Product	Electricity savings (TWh/yr)	Peak capacity savings (MW)	Total energy savings (EJ)	Lifetime energy savings (EJ)
Refrigerators	13.4	1,785	0.163	3.08
Freezers	2.5	323	0.030	0.62
El. water heaters	19.9	2,456	0.242	3.14
Room AC	4.3	4,573	0.052	0.79
Central AC	11.2	11,955	0.136	1.62
Gas furnace	--	--	0.124	2.86
Gas water heaters	--	--	0.191	2.49
Gas range	--	--	0.029	0.53
TOTAL	51.3	21,092	0.967	15.12

1. Total energy savings accounts for electricity on a primary
basis (12.1 MJ per kWh). Lifetime energy savings refers to
the savings over the life of products sold during 1988-2000
with electricity valued on a primary basis.

Table 4

CALCULATION OF POTENTIAL AVOIDED INVESTMENT
IN GENERATING CAPACITY BY 2000

1. Avoided Baseload Capacity, based on electricity savings from
   standards on refrigerators, freezers, and water heaters.

   35.8 TWh/yr x (8760 hours x 0.65)$^{-1}$ x 1.08 x \$1500/KW
   = \$10.2 billion

2. Avoided Peak Capacity, based on peak demand savings from
   standards on central air conditioners and heat pumps and
   room air conditioners.

   16,530 MW x 1.08 x \$500/KW = \$8.9 billion

Total Potential Savings = \$19.1 billion

Table 5

ECONOMIC SAVINGS IN 2000 FROM NATIONAL
APPLIANCE EFFICIENCY STANDARDS

Product	Annual operating savings (million \$)	Net lifetime savings (1) (million \$)	Benefit-cost ratio
Refrigerators	634	5394	3.37
Freezers	119	1261	5.88
El. water heaters	1000	8237	8.14
Room AC	217	777	1.53
Central AC	553	788	1.19
Gas furnace	449	2872	1.90
Gas water heaters	708	5879	8.67
Gas range	110	1105	7.09
TOTAL	3790	26,313	3.01

1. Net lifetime savings are the operating savings minus the extra
first cost over the lifetime of products sold during 1988-2000.

SOCIOLOGICAL AND PSYCHOLOGICAL BARRIERS TO ELECTRICITY SAVINGS

JORGEN S. NORGARD* and BENTE L. CHRISTENSEN

*Physics Laboratory III, Technical University of Denmark
DK-2800 Lyngby, Denmark.

1. INTRODUCTION

For about a decade, substantial technical electricity savings have been known to be favourable from any **rational** point of view. Despite this, the savings are not readily implemented. Gradually some of the necessary efficient technology is being introduced, but there is a long way to go before a reasonable balance is reached between the investments in electricity savings and electricity supply (see reference (1)).

The period during which most of the electric equipment was introduced was also a period with a perception of unlimited resources. This might have contributed to shaping not only the technology, but also the mind of decision-makers and the structure of societies towards a non-rational favouring of supply investments over conservation investments.

Since there seems to be few technical and economical obstacles on the conservation path, we will focus on some sociological and psychological barriers as experienced during 15 years of work on electricity conservation. We will not analyze here the economic and institutional structures, even though they are important factors. We will rather go one step further and focus on the psychological barriers held by the individuals involved in shaping the energy future. Many of these barriers are non-rational and hence often not obvious to people in an industrialized society. From a natural science point of view, such barriers to electricity savings are, therefore, difficult to "prove". Nevertheless, we find it extremely important to identify such possible barriers. We hope this paper can help make individuals aware of these unconsciously held barriers against conservation, and thereby open-up a discussion on how to break through them on the way towards sustainable societies.

People's behaviour is closely associated with the roles they play in society. Therefore, we will describe the barriers as we perceive them as typical to various groups of decision-makers. Firstly, however, we will outline the chain of concepts involved in converting the natural resources into satisfaction of human needs and wants.

2. ENERGY CHAIN

Figure 1 illustrates the major links in the chain which leads from basic means to ultimate ends - from natural resources to satisfaction of human needs and wants.

In the case of electricity, an extensive chain of technology first converts the primary energy sources like coal, hydro, wind or gas into electricity to be sold to the con-

A. T. De Almeida and A. H. Rosenfeld (eds.), Demand-Side Management and Electricity End-Use Efficiency, 497–503.

sumers. At the consumers, including industry, service institutions etc., a chain of end-use technology converts electricity into some useful physical and quantifiable energy service such as refrigerated storage volume, washing of clothes, cooking of food etc. These services in turn, provide fresh food, clean clothes and warm meals etc., which are physical but not easy to quantify. Finally, these services give people non-physical pleasures of satisfying their needs and wants.

ENERGY CHAIN

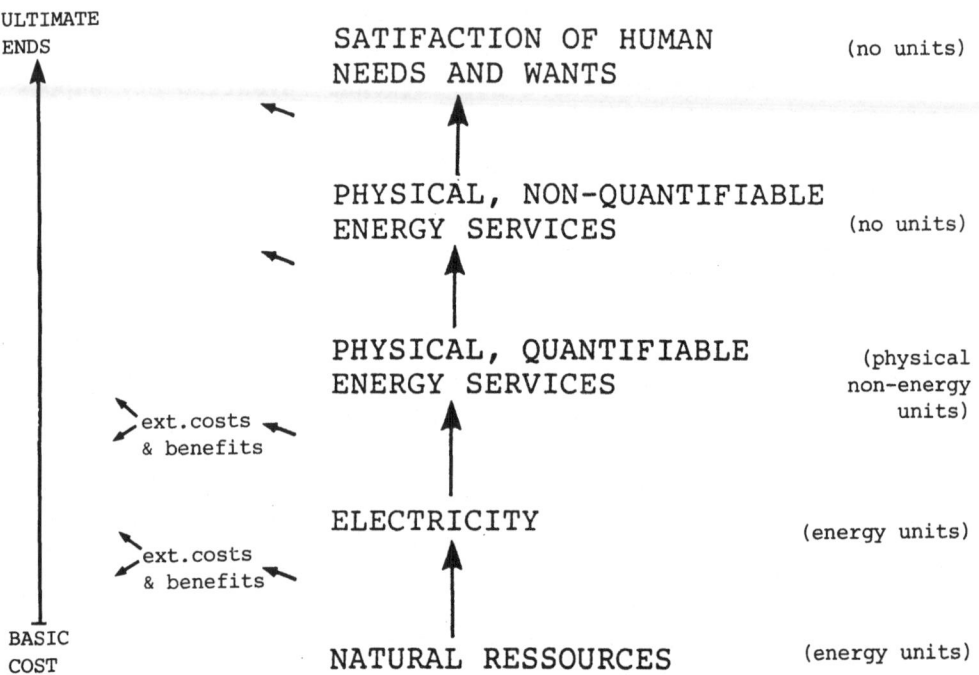

FIGURE 1. Illustration of the chain of concepts involved in the conversion of natural resources into human satisfaction.

2.1 Overall efficiencies
A general definition of efficiency is:

$$\text{Efficiency} = \text{Benefits}/\text{Costs},$$

where benefits and costs should be interpreted very broadly as all desired output from the system and all its undesired consequences, respectively.

In the energy chain in figure 1, the ultimate benefits are the non-physical energy services of satisfying needs. The costs are not only the depletion of natural resources, but also the capital and labor costs plus the environmental costs and other non-quantifiable costs associated with the conversions.

The potentials for saving electricity without human sacrifice are mainly in the links above "electricity" in Fig.2. This involves improvements in the efficiencies of end-use technology, of social structure, and of individual lifestyle. The potential for savings outlined in (1) only consider efficiency improvements in the end-use technology.

3. ENGINEERS

Obviously engineers and other technologists play a dominating role in shaping the energy systems we rely on. We will suggest some psychological reasons why still only relatively few engineers and engineering students are attracted by the large technologial opportunity in electricity conservation.

3.1 Prestige and glamour

From a traditional engineering view-point the electricity saving end-use technology has usually less glamour and less prestige compared to the building of large power plants, windmills or other spectacular electricity supply systems.

The engineering skills required for developing electricity saving systems have a more holistic and inter-disciplinary character than the traditional engineering expertise. These skills have only recently emerged and begun to be recognized and estimated. The electricity saving technologies can be equally or more sophisticated in their modest and integrated way as are the dominating, large-scale supply technology.

3.2 Measurable achievements

With their natural science background, engineers generally prefer exact definitions and achievements which can be measured. Total efficiency in the energy chain from coal to human satisfaction, as shown in Fig.1, cannot be defined precisely in numbers.

The efficiency of the links in Fig.1, which convert coal into electricity, can be measured precisely, even in a dimensionless number. Enormous efforts have been successfully devoted to improving this efficiency from 10%, 20%, etc., to its present level above 40%.

Efficiency of the next step, the end-use technology, can be expressed quantitatively as, for instance, "litres of $5^{\circ}C$ refrigerated volume per kWh electricity annually" or "kg clothes washed per kWh electricity". Such efficiencies cannot be added up to one total efficiency of, for instance, a household. Such difficulties in measuring the achievements might make them less attractive to technologists, leaving huge potentials for savings behind.

If we go further up in the chain in Fig.1, the efficiencies are equally important, but they become impossible to qualify. They are, by and large, ignored by technologists.

4. ECONOMISTS
4.1 External costs and benefits

Traditional economists tend to ignore the external costs and benefits when they evaluate the basis for decisions. In energy planning, these externalities can typical be natural environment, air quality, risk of accident, supply security and independence. Electricity savings are almost always superior to electricity supply in these external costs and benefits, and when they are ignored, the advantages of choosing **savings** are consequently underestimated (see Fig.2).

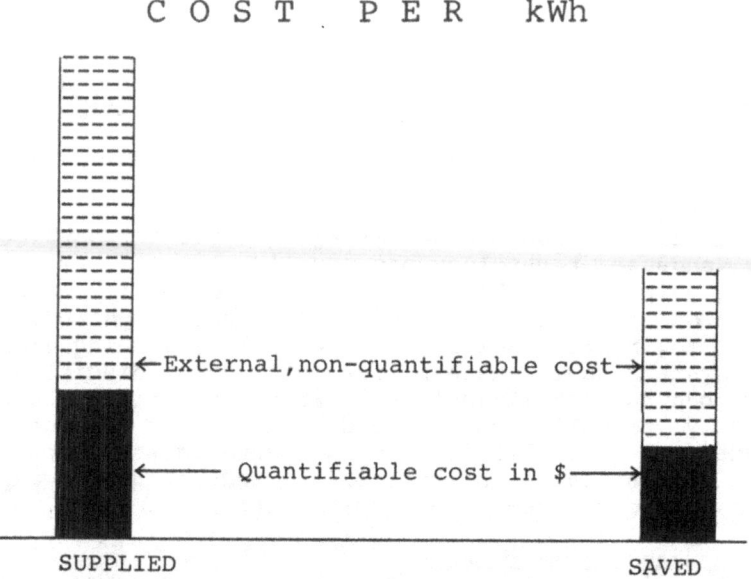

C O S T   P E R   kWh

←External,non-quantifiable cost→

←――― Quantifiable cost in $―――→

SUPPLIED
kWh

SAVED
kWh

FIGURE 2. Qualitative illustration of the costs of supplying and saving one kWh of electricity.

The reason why economists tend to ignore the external costs and benefits in the fundamental evaluation seems to be that they - or maybe rather the politicians - prefer hard precise facts, that is, dollars and cents, so nothing is left to personal judgement and responsibility. The external costs and benefits cannot - and we believe should not - be squeezed into a narrow frame of dollars and cents.

4.2 Efficiency and GNP

Electricity savings will, to a very large extent, come out favourable to supplying electricity, even in pure dollars and cents, that is, even neglecting the external costs and benefits described above, as illustrated in Fig.2. Still, however, traditional economic forces often choose the more expensive solution, namely, to supply more electricity, rather than to save.

Most politicians, and hence most economists, seem to consider economic growth as a basic value in society. For that reason they tend to prefer solutions which contribute the most to economic activities. Electricity end-use savings are generally simple and cheap and therefore do not contribute so much to GNP. When economists do appear sometimes to favour electricity savings, it often turns out that they see a chance for extra economic activities by producing (and maybe exporting) the new efficient technology. This usually implies a forced scrapping and replacement of less efficient technology, which makes sense in some cases, but often results in waste of resources, including energy.

True energy savings can of course be considered by economists as an option for providing other goods and services to the consumers. However, in many industrialized countries, the basic economic necessities are satisfied for most people and saturation is approaching (2). In such cases, the emphasis on economic growth tends to favour inefficiencies in satisfying needs and wants. The choice of the most costly solutions will be sought justified in one way or another.

A true preventive environmental measure consists of the introduction of electricity efficient technology in the course of the normal rate of replacement (or maybe slower). This will most likely reduce GNP but leave the nation and its citizens better off.

5. POLITICIANS

Politicians in highly industrialized nations become very dependent on the advices from experts in technology and economy. This dependence is a threat to true democracy and is in itself a reason for choosing more simple solutions like those represented by conservation technology.

Because of the politicians dependence on the engineers and economists, they are subjected highly to the attitudes that dominate these professions as described above.

5.1 The employment catch

In periods with high unemployment like presently, politicians are very aware of the employment implications of the various options and choices. Indeed, electricity conservation activities can often provide many job opportunities. Overall, however, there will be less employment from saving one GWh than from producing one. This is why the savings are cheaper than the production of electricity. Whether the employment occurs within the nation or abroad is what really concerns politicians, since this determines the activity's contribution to the national balance of trade.

Generally, employment problems should be solved by sharing the work, that is, by regulating working hours, rather than by initiating wasteful production. The employment catch is, of course, a variant of the high preference for economic growth.

5.2 The monumental aspect

For politicians - maybe especially local politicians - it seems important to establish a symbolic monument of the political achievement, such as a new school, a theater, or a local district heating plant. Generally, energy conservations do not manifest themselves in a similar photogenic way, suitable for TV, newspapers, etc. This might subconsciously give the con-

servation options a lower priority in the mind of the traditional politician.

6. CITIZENS
The ultimate decision-makers, the citizens, also constitute
the ultimate purpose of the whole energy systems. Nevertheless,
they also seem to make non-rational choices.

6.1 Lack of information
In the complicated technological systems of the industrialized countries, citizens are not able to get a clear picture
of the choices available for saving energy, for instance,
electricity. Lack of information about these options for electricity savings explains why these ultimate decision-makers,
the consumers, do not even make the energy choices that seem
best for themselves as individuals. Energy conservation and its
environmental benefits are, however, most of all a social responsibility, but still the citizens are short of information
about the wide range of options. Consequently, they cannot ask
the politicians for the appropriate energy policy.

6.2 Collective activities
Many industrious people have done something about the energy problems in Denmark. But they usually have preferred collective energy systems like a district heating system or a shared
wind-mill. Systems like these have a social purpose by providing opportunities to share something and get together.
Energy saving activities such as insulating your home or
choosing proper equipment for your shop, do not require cooperation with neighbours, even though such a collective organized
effort could promote the activities to save energy.
Furthermore, when a collective energy supply system is
established, the incentive to save energy diminishes. A joint
owner of a supply system is often seized by the business mind
of "selling more". Reducing the consumption and hence the
supply, will often also make the supply system less profitable
or increase the unit price of the energy delivered.

6.3 Neighbour-effect
If an individual citizen decides to do something about the
energy problems, the choice will often be some technology which
is visible. In our competitive society, it is important to
"impress" neighbours and other people by, for instance, installing a solar water heater on the roof or build a wind-mill.
Compared to such supply systems, the electricity savings technologies, like better ventilation system and refrigerators, are
almost invisible, which ironically are also to their advantage
from an environmental point of view.

6.4 Private economic consideration
Electricity consumption is usually distributed over a dozen
different uses or more, each of which does not amount to anything which seems worth much consideration. This is the case
especially in rather wealthy countries like Denmark.

It is also important to notice that a consumer, who invests in some durable goods like a washing machine, a sofa or a wood stove, does not suddenly apply rational economic considerations just because some of the outputs are measurable like heat or electricity savings. The decisions are made on the basis of a whole pattern of values, some of which might be rational from an economic point of view.

7. CONCLUDING REMARKS

We are well aware that this paper does not in any way cover the whole spectrum of psychological barriers to electricity savings activities. One general and simplified conclusion could be that the cheapest, simplest and most durable solution to the energy problems – energy conservation - is often rejected exactly because it is cheap, simple and durable!

Some basic changes in our attitudes and values seem necessary in order to overcome the barriers on the true energy conservation path.

REFERENCES

1. Norgard, J.S, and T. Guldbrandsen: "Potentials for Technical Electricity Savings Using Known Technology". In this proceedings from NATO Advanced Study Institute in Povoa do Varzim, Portugal, 20-31 July 1987.
2. Norgard, J.N. and B.L. Christensen: "Individual Attitudes in Scandinavia Point Towards a Low-energy Saturated Society. Vol.F, Proceedings from ACEEE 1984 Summer Study of Energy Efficiency in Buildings, Santa Cruz, California, August 1984.
3. Christensen, B.L., and J.S. Norgard: "Social Values and the Limits to Growth", Technological Forecasting and Social Change, 9, p. 411, 1976.

THE POLITICAL AND ECONOMIC IMPLICATIONS OF AN INTENSIVE ELECTRICITY
CONSERVATION PROGRAMME

C.H. Davies
Central Electrical Generating Board, England

INTRODUCTION

1.   There is clearly a difference between the
policies being followed in Europe and those being
followed in North America by electricity supply
utilities with respect to the promotion of energy
conservation. In North America many utilities are
engaged in active policies to promote increases in the
efficiency in use of electricity by their customers.
These policies take a number of different forms
including the provision of free energy audits, low or
zero interest loans to customers to improve their
energy efficiency and subsidies to help customers
replace less efficient electricity using equipment
with more energy efficient equipment.

2.   In Europe in general, while most utilities
have had policies aimed at the encouragement of energy
efficiency, most have not engaged in the active support
of energy and electricity efficiency improvements to
the point of subsidy as has been the case in North
America. There are many reasons for this difference.
These include, for example, government policy and the
legal framework within which utilities operate. Thus,
in 1978, President Carter's National Energy Policy
which included the National Energy Conservation Policy
Act (NECPA) required utilities to offer energy audits
to residential customers, provided for weatherisation
grants for low income families, loans for energy
conservation investments and set new thermal standards
for buildings. In addition, the legal framework within
which utilities operate, particularly pressures from
the regulatory authorities, have encouraged utilities
to finance energy conservation schemes. Sometimes this
has been via the positive encouragement of such
schemes; sometimes they have been forced upon utilities
because of the unwillingness and/or time taken by
regulatory commissions to approve rate increases
required to finance new power station investments.

3.   However, this paper is going to concentrate
on the questions which arise in this context for a
utility which is currently not actively engaged in
these activities. In essence, I am going to consider
the problems which I see have to be solved before an
active policy of this sort could be implemented.

505

A. T. De Almeida and A. H. Rosenfeld (eds.), Demand-Side Management and Electricity End-Use Efficiency, 505–517.
© 1988 by Kluwer Academic Publishers.

Demand Forecasts

4.   The starting point for my analysis has to be
an assessment of the effect on the demand for
electricity of an active conservation policy. In the
first instance, therefore, we have to be able to assess
the position with respect to the efficiency of energy
use and electricity use in the absence of an active
conservation policy. Such an estimate can then form
the benchmark from which the effects of an active
policy can be estimated.

5.   In our case, forecasts of electricity demand
which are used for planning purposes are made on such a
basis, i.e we do not assume in our forecasts that the
ESI in England and Wales will be actively subsidising
energy improvements by customers. However, this does
not mean that we do not expect significant increases in
the efficiency with which electricity is used. Neither
does it mean that efficiency increases are not treated
seriously in our forecasting process nor that they do
not have a significant impact on future electricity
demand. It does, however, mean that the efficiency
improvements included in our forecasts arise indirectly
from our general policies of providing customers with
information on efficiency in use and on the technology
available and by assessing the reaction of customers to
this information, given the expected overall economic
background, the general level of energy prices and our
own tariff policies.

6.   In our current forecasts of electricity
demand to the year 2000, our assumptions about the
improvements in electricity's efficiency in use have
reduced our forecast by almost 10% compared with the
continuation of current practice. In more general
terms, these forecasts also point to a decline in the
primary energy ratio in the UK (primary energy demand
per unit of constant price Gross Domestic Product) of
25% and the electricity intensity of output of 8%.
These are significant reductions and ones which any
active conservation or efficiency policy must exceed
before any effect on demand can be taken into account.

7.   The ability to estimate the effect on demand
of an active conservation policy will depend on the
extent to which the components of the base or benchmark
forecasts are affected by the policies. In our demand
forecasts, covering a period of up to about 10 years
ahead, as I have already referred to, efficiency in use
estimates are made at a very detailed level and are a
very important part of the forecasting process.

8.   This is a reflection of the general approach
which has been adopted to demand forecasting which
involves a high degree of disaggregation. The
following table illustrates the component parts of a
ten year ahead electricity demand forecast.

Domestic Sector

> Space Heating
> Water Heating
> Cooking
>
> Appliances
>   Refrigeration
>   Washing Machines
>   Tumble Dryers
>   Dish Washers
>   Irons
>   Vacuum Cleaners
>   Kettles
>   Televisions
>   Lighting
>   Other

Industrial Sector

Industries	Usages
Paper and Board	
Printing and Publishing	
Iron and Steel	Process Heat
Chemicals	Motive Power
Engineering	Electro Chemical
	Processes
Non-Ferrous Metals	Space and Water
	Heating
Textiles, Leather and	Other
Clothing	
Food, Drink	
and Tobacco	
Mineral Products	
Energy Industries	
Other Industries including	
Construction	

Commercial Sector

Categories	Usages
Shops	Space Heating
Offices	Water Heating
Education	Inside Lighting
Hotels	Outside Lighting
Transport and	Catering
Communications	Air Conditioning
National and Local	Miscellaneous
	Appliances
Government	
Health	
Other Premises	

9.   Demand for electricity and for other fuels is assessed for each of the component parts of the forecast. So for example, we will have an assessment, for a given economic and fuel price background, for the amount of electricity used for motive power in the chemicals industry or for the amount of gas used for space heating in shops.

10.   For each of these components, therefore, there will be an assessment of the efficiency with which electricity and other forms of energy are used. The sources of these assessments are varied and rely both on assessments of current and future technology and the extent to which 'best practice' will be implemented. In addition, for many uses, particularly for space heating, account has to be taken of the current and future state and type of the stock of buildings in which the equipment will be operating.

11.   Clearly, there are a very large number of judgements which have to be made and many if not all of them have to be made on the basis of a limited amount of information concerning current practice and future trends. Nevertheless, it is our experience that this type of forecasting framework is of considerable practical value. While such a forecasting approach is intensive both in terms of the amount of time and effort to produce it and in terms of its data requirements, it has particular advantages compared with previous more aggregate methods of forecasting. Foremost amongst these is the ability to show precisely how the overall level of demand is built up and to expose detailed estimates concerning market shares and of course efficiency improvements to detailed scrutiny. In terms of the performance of this approach since its adoption in 1980, the electricity demand forecasts made at that time for 1986/87 showed an error of about 2%.

12.　　　　It is necessary therefore in the first
instance to be able to assess the effect of an active
conservation policy on the efficiency with which
electricity (or energy in general) is assessed at this
detailed level.　Given the broad nature of many of the
judgements and assessments which have to be made, this
is a difficult process and one involving large elements
of judgement.　Nevertheless, it can be argued that it
is no more difficult than the process by which the
initial judgements are arrived at, and therefore cannot
be ruled out as a matter of principle.　However, it is
also the case that the introduction of such a policy
and therefore the need to make the assessments of its
effects introduces one further uncertainty into the
demand forecasting problem.

13.　　　　However, this of course is only the first
part of the problem in that an increase in efficiency
of use cannot be directly translated into a reduction
in demand.　In particular a number of factors would
have to be assessed before an estimate of the effect on
demand could be made.

> i) the extent to which standards improve or the
> amount of use increases as a consequence of
> increases in efficiency.　Evidence from the
> USA suggests that this could reduce the
> original savings by up to 40%.(1)

> ii) the effect on the relative competitive
> positions of various energy forms.　If
> electricity becomes more efficient relative
> to other fuels, this may encourage fuel
> substitution and a consequent increase in
> sales.

> iii) the extent to which the higher level of energy
> efficiency arising from an active
> conservation policy represents a bringing
> forward in time of increases which would have
> occurred anyway as a consequence of normal
> market pressures.　Once again there is
> substantial evidence from the USA which
> suggests that conservation programmes largely
> accelerate conservation investments rather
> than induce investments which might
> not have otherwise taken place.(2)

iv) in the case of electricity the extent to
which reductions in electricity sales
actually affect maximum demand. In the UK
for example over two thirds of domestic space
heating is undertaken on off peak tariffs.
Increased efficiency in use of electricity
for this purpose would have little effect on
peak demand.

14.   A further difficulty in making these
assessments, and particularly, in assessing the effect
of a conservation programme on efficiency in use is the
extent to which efficiency in use increases may be
determined by factors other than the conservation
programme. Some efficiency improvements will be
greatly influenced by the overall economic background.
Fast economic growth leading to a more buoyant economy
and therefore to a more rapid turnover in the capital
stock of buildings and equipment may be of considerably
greater importance in determining the rate of
efficiency improvement than the energy conservation
programme. A recent study conducted by Pacific
Northwest Laboratory (PNL) for the US Department of
Energy (2) suggested that consumers have made
substantial investments in energy efficiency since the
early 1970s. However, while some of these could be
attributable to government policies (e.g tax subsidies,
efficiency information, appliance standards) and
utility programmes, the evidence suggested that
consumers had frequently acted on their own.

15.   In summary, therefore, it is going to be very
difficult to assess the effect on demand of an active
conservation policy. First, this is because of the
adjustments which would have to be made to estimates
themselves which are only based on using broad
judgements; secondly because of the relationship between
standards and efficiency; thirdly, because of the
effects of efficiency on market shares and finally
because of the extent to which the active policy may
only involve a bringing forward of increases. The US
Department of Energy in its Report to the President (3)
in March 1987 summarised these problems by pointing out
that

"Careful planning and research by individual utilities are needed to develop and implement cost-effective programs. It is particularly important to distinguish between the amount of market-generated conservation that would occur independent of utility programs and the incremental amount attainable through such programs. Without this distinction, the amount of conservation attributable to utility efforts is almost certain to be overestimated, and its costs underestimated"

16.   I recognise that these problems are not insuperable and that considerable work has been undertaken in these areas. However, in the context of the particular forecasting approach which we have adopted and which in my view is best suited to the circumstances of the market that we face, I see that the problems are likely to be considerably greater. A more formal approach to the demand forecasting problem may be more amenable to assessing the impact of an active conservation policy than is our own.

The Economics of Conservation Policy

17.   So far this paper has only considered the very real problems in my view of assessing the effect of an active conservation policy on demand. These problems are severe but if we were to assume that they could be overcome, at least to an extent that the efficiency improvement assessments do not lead to an increase in the uncertainty which is inevitably attached to a ten year ahead demand forecast, then the next point to consider is how to assess the economic merit of such a policy.

18.   In concentrating so far on the question of our ability to assess such effects, the question of the costs of such a policy has been ignored. In order to undertake an economic assessment, it is necessary to cost such policies. Clearly, in my view and consequent upon the problems set out above, there will be considerable difficulty in quantifying the link between expenditure by the utility on promoting energy efficiency and the reduction in demand resulting from it.

19.   In this context a particular problem is posed
by the need only to consider as a benefit of the policy
the additional increment in energy efficient investment
brought about by the active policy. If one considers
for example a policy of actively supporting the
installation of high efficiency lighting in the
commercial sector. It is clearly inappropriate to
consider all recipients of support as effectively
providing a return on the investment. Many of the
customers would have chosen to install high efficiency
lighting without the subsidy. This number in effect is
set by the base or benchmark forecast. Yet, it is
likely to be the case that all customers who undertake an
investment in high efficiency equipment will be in
receip of a payment. Under these circumstances, which
are likely to be typical of an active conservation
policy, considerable care would have to be taken to
identify the increment in efficiency improvement
together with the total cost of the scheme.

20.   However, if it is assumed that this can be
done, an economic assessment would require this cost to
be assessed against the benefits if any, of lower
demand. From the utility's point of view, it must be
borne in mind that kWh not supplied because of its
policy to promote efficiency in use not only involves a
reduction in cost (i.e the saving in not having to
produce that kWh) but also a reduction in revenue from
not selling that kWh. The utility's interest is
therefore served if the saving in cost is greater than
the loss of revenue. That difference, if there is any,
is the amount of money available to pay for the energy
efficiency programme.

21.   This requirement leads to a further
complication in the analysis, namely a need to identify
the particular loads which are reduced as a consequence
of the energy efficiency programme, a kWh not supplied
at peak may be of considerable value to the utility as
defined above but a kWh saved at off peak times may be
of little or no value.

22.   However, if as may be necessary, the savings
in kWh terms are merely assumed to be representative of
total sales, then the extent to which savings to the
utility will occur will depend on how its marginal
costs at some date in the future relate to its prices.
On this basis, the extent to which costs and prices may
differ in the short run and in the long run depends
considerably on the basis on which prices are charged.
In our case, with marginally orientated tariffs (whilst
also working to an overall requirement set in terms of
a rate of return on the replacement cost of assets),
any gap between costs and prices is likely to be small
and the incentive for a conservation policy similarly
limited.

23.   In this context, however, the question which
must arise is what is the utility's interest? In
essence, what is being said above is that unless
overall costs are reduced relative to revenue, or at
least remain unchanged, consumers in general will have
to bear an increased burden in the form of higher
electricity prices as a consequence of the policy.
Clearly, a customer who is in receipt of a subsidy will
benefit. In general, one would expect the efficiency
improvement to be worthwhile for a customer, even if he
paid for it himself. With part or all paid for by the
utility it will clearly be even more worthwhile. But
what about the situation of the customer who has
already undertaken such an investment at his own
expense, prior to the implementation of the policy. In
the circumstances where prices rise as a result of the
implementation of such a policy, he will be worse off
while the customer who has only undertaken the
investment as a result of the policy may be better off
i.e his gain is greater than the effect of higher
electricity prices.

24.   There may well be circumstances where
electricity prices rise as a consequence of such a
policy but the total benefits accruing to customers in
receipt of the subsidies may well be greater than the
costs. Under these circumstances, it could be argued
that a utility should proceed with an energy efficiency
policy. In my view, these arguments are not
supportable in that they involve a redistribution of
income from those customers not in receipt of the
subsidies to those who are. Decisions on
redistribution of income are really appropriate to
Government and not to electricity utilities.

25.   In practical terms, the relationship between marginal costs and average costs (assuming the overall price is determined by average costs) will give a broad indication of the opportunity to benefit all consumers via such a policy. In circumstances where a utility has excess capacity, marginal costs are likely to be lower than average costs, for the period for which the surplus is expected to exist. Under these circumstances it is implausible that an active conservation policy could pay from the utilities point of view. The only exception would be if the plant at the margin had very high costs of generation. This would be the case if for example oil prices were high relative to say coal prices and if the utility had a preponderance of coal plant but oil capacity was at the margin.

26.   Leaving aside the possibility of excess capacity, which in any case one would expect to be a short run phenomenon, the benefit to the utility will be determined by the relative costs of putting new capacity on the system relative to the cost of meeting demand with existing capacity.

27.   In our case at present, 60% of our plant is coal fired and about 10% is nuclear. New main generating plant on our system would be nuclear or coal. In the case of nuclear, the cost of this capacity in terms of its overall effect on system costs is likely to be low. However, it is probably inappropriate in our case to consider nuclear plant to be marginal on our system for the foreseeable future. On this basis, it is probably necessary to consider the marginal costs as being determined by the cost of putting new coal fired capacity onto our system. This coal plant is, as I have already indicated, depreciated at its replacement cost value. On this basis the long run marginal cost on our system based on building new coal plant is not likely to be substantially higher than current average costs.

28.   The difference between the costs of meeting an increment in demand (or alternatively the benefit arising from not having to meet an increment in demand) is not likely to be very different from current costs and prices. This does not of course mean that electricity prices may not rise in the future but this would be as a consequence of rising coal prices and not as a consequence of having to meet increase in demand. On these a priori grounds, the benefits of an active conservation programme for our system are likely to be very limited.

29.　　　Clearly, under other circumstances, long run marginal costs of new capacity can be considerably higher than current average costs. In this case the benefits of a lower demand for a utility can be large. Certainly, this would appear to be the case for many North American utilities, particularly where prices are based on the historic rather than replacement cost of the utilities' assets.

30.　　　In summary, therefore, to assess the overall economics of an active conservation programme, it is first of all necessary to carefully cost the programme. This means assessing against the expected incremental increase in efficiency the total cost of the programme. A significant part of this cost is by definition likely to be wasted in providing incentives to customers who could reasonably be expected to make that decision anyway. Having done this considerable care has to be taken to differentiate between the benefits which accrue to the utility (or more precisely to electricity consumers as a whole), to the consumer who is in receipt of the subsidy and to the nation as a whole. It is likely that the benefits for these groups will be rather different. In my judgement it is likely in most instances that the benefits to the utility will be the smallest and are quite likely to be negative while those to the consumer in receipt of the subsidy are likely to be greatest.

Conclusions

31.　　　The title that I was given for this paper was the political and economic implications of an intensive electricity conservation programme. I have perhaps dealt with the topic in an indirect fashion through looking at the sort of factors which have to be, in my view, considered in contemplating such a programme.

32.　　　The main problems that I see in such programmes are:

　　i) of measuring the incremental increase in efficiency in use arising from the programme

　　ii) of converting this assessment into an effect on demand taking into account changes in use arising from the conservation savings

　　iii) of costing the reduction in demand and of assessing how much expenditure is necessary to obtain a certain level of demand reduction

    iv) of identifying the benefits of the active
        conservation policy and the recipients of
        these benefits

33.   The economic implications of an intensive
energy conservation policy will of course depend on the
outcome of such an analysis. What in my view is clear
is that a pursuit of such a policy without undertaking
the above sort of analysis and being sure that the
results are soundly based could have a number of
unfortunate consequences.

    i) The first is that the policy itself does
      not achieve its desired ends. This is not to
      say that the energy market with respect to
      conservation works efficiently but rather
      that the policy itself may not be sufficient
      to overcome the inefficiencies of the market.
      Under these circumstances at the least the
      policy would be a waste of money and at the
      worst could lead to insufficient generating
      capacity being installed.

    ii) Secondly, that the policy may have
       redistributive effects through benefitting
       one group of consumers at the expense of
       others. While these redistributive effects
       may not be all that serious it is my view
       that it is inappropriate for a utility to
       follow policies which have this effect.

    iii) Thirdly, such a policy adds to the
        uncertainty of demand forecasting which
        itself could affect capacity requirements.

34.   On the positive side of course such policies
could lead to substantially more efficient use of
resources. I happen to believe in the UK at least this
is a highly implausible outcome.

Bibliography

(1)   Frederick D Sebold and Eric W Fox,
          "Realised Savings from Residential
          Conservation Activity", The Energy Journal,
          Vol 6, No 2, April 1985, pp 73-88.

(2)   E Hirst, et al, "Evaluation of the BPA
          Residential Weatherisation Program", Oak
          Ridge, TN: Oak Ridge National Laboratory,
          ORNL/CON - 180, June 1985.

Pacific Northwest Laboratory, "A Retrospective Analysis of Energy Use and Conservation Trends: 1972-82, prepared for the US Department of Energy, June 1985.

(3)   US Department of Energy, "Energy Security: A Report to the President of the United States", US Department of Energy, March 1987.

INTERNATIONAL AND NATIONAL APPROACHES TO END-USE TECHNOLOGY DEVELOPMENT

CONNIE SMYSER

INTERNATIONAL ENERGY AGENCY, ORGANISATION FOR ECONOMIC RESEARCH AND DEVELOPMENT

1. INTRODUCTION

The IEA has recently completed the first of two rounds of reviews of Member countries' research, development, demonstration and dissemination (RDD&D) programmes for end-use technologies. In this first round, six countries were reviewed in one of three "focus sectors": Buildings, Industrial, or Transportation. For some of the smaller programmes, the whole programme has been or will be reviewed. Table 1 shows the countries and focus sectors for both rounds of reviews.

During the first round a questionnaire was developed for structuring the information provided by each country on their programme, one-week site visits were conducted by IEA Secretariat staff and at least two Rapporteurs from other Member countries, and individual country reports were prepared (which will appear in monograph form in the Autumn, 1987).

TABLE 1

PROGRAMMES AND SUBSECTORS IN IEA END-USE TECHNOLOGY THEMATIC REVIEW

COUNTRY UNDER REVIEW	FOCUS SECTOR
1986/87 Review Round	
Austria	Buildings
Greece	Whole Programme
Italy	Transport
Japan	Industry
Netherlands	Industry
Sweden	Buildings
1987/88 Review Round	
Canada	Buildings
Ireland	Whole Programme
Norway	Buildings
Portugal	Industry
Spain	Industry
Switzerland	Industry
United Kingdom	Industry
United States	Transport

The second round will consist of country reviews and site visits similar to the first round except that it will concentrate attention on substantiating the issues that were developed in the first round in order to support better the conclusions and recommendations that will be made to

A. T. De Almeida and A. H. Rosenfeld (eds.), Demand-Side Management and Electricity End-Use Efficiency, 519–532.
© 1988 by Kluwer Academic Publishers.

the IEA's Committee on Research and Development (CRD) and Member countries for future activities. The number of programmes in each sector that will have been reviewed by the end of the second round will be: four Transportation programmes, six Buildings, and eight Industry.

The remainder of this paper will provide a brief background on how the IEA came to be studying this subject. It will then present the findings of the first round and the issues which have been identified for clarification and expansion in the next round of reviews. An overview of the specific technologies which were reviewed in the country visits or in material provided by the countries will be provided as well. However, because this course is oriented primarily toward electricity efficiency, findings and issues arising out of the transport sector review, which cannot be generalised to all sectors, will be omitted. Finally, there will be a discussion of the implications for electricity efficiency and demand-side management.

2. BACKGROUND

For the past several years, the CRD has undertaken thematic reviews of specific areas of energy RD&D. The two primary purposes of these reviews have been 1) to establish a current view of IEA-wide developments in important technology areas, including both progress in the technology and impediments to its further development and implementation, and 2) to understand the different country approaches to development and implementation of the technology and to assess the effectiveness of their RD&D programmes. The first two thematic reviews focussed on synthetic fuels (1985) and the clean use of coal (1986). In May the CRD approved the examination of the energy end-use technologies in the next review cycle.

Two issues which began to assume greater importance within IEA countries during this time, the sharp decline in oil prices and the increased concern over the environmental effects of conventional electrical generation technologies, were made one focus of the review because of their potential effects on the direction of policies and programmes in end-use technology RD&D. Furthermore, the IEA had at approximately the same time completed an energy conservation policy study [1] which has already been discussed in this course by David Jones of the IEA. This work was therefore made the basis from which the reviews proceeded.

3. MAIN ISSUES AND FINDINGS AFFECTING END-USE TECHNOLOGY POLICY
3.1. Which Technologies Are Considered End-Use?

While this may seem a rather basic question, what is considered an end-use technology varies considerably among the countries studied. This review, therefore, necessarily uses a very broad definition encompassing:
-   all energy-efficiency technologies;
-   the "end-use" renewable-energy technologies;
-   fuel-switching technologies; and
-   technologies improving energy efficiency in the electricity transformation sector.

Such a broad range can lead to difficulties in comparisons of whole programmes, such as this review attempts, and to possible confusion in efforts to transfer information among countries, e.g. about programme results or funding levels. Despite these potential problems, it was considered inadvisable to narrow the focus of the study because any exclusions would artificially constrain understanding how end-use technology programmes are structured. Rather it was found to be more important to describe what is considered end-use technology in each case and to use the thematic review as a means of disseminating information on what is included in each programme and why.

3.2. What Government Activities Are Included in End-Use Technology Development Programmes?

To be successful in end-use technology RD&D, more than for many other types of energy technologies, dissemination of results has been found to be inextricably linked with the development and demonstration phases. Therefore, many programmes are designed to foster integration of the research, development and demonstration efforts with dissemination activities. To accomplish this, governments have been using more or less the same structure in their end-use technology development programmes as they do in their RD&D programmes for other energy technologies and adding a dissemination (or technology-transfer) function, in essence a third "D", in the attempt to make an integrated RDD&D programme.

The concept is that of building a programmatic "bridge" to move a technology through the different stages of development until it achieves successful market introduction where each supporting structure (policy, policy-making, research, development and/or experiment, demonstration, market introduction and practical use) is necessary to "bridge the gap" between policy and implementation. Forming such a bridge to ensure or assist commercialisation is, in fact, the governmental response to results of numerous analyses of conservation-technology uptake which have shown that there are sometimes large institutional or market barriers which impede optimum utilisation.

To be effective this concept requires a wide array of activities primarily in the dissemination stage to handle the range of technologies and end-use sectors. However, there are in fact widely differing national policies on what government programmes should include and what should or can be left to the private sector, i.e. exactly what is considered an appropriate role for government in technology development versus market introduction activities. That having been said, there were a number of government activities highlighted in the reviews which are worth describing here in a little more detail (without taking a position of their appropriateness for all national situations).

Practically all governments have added advisory bodies containing experts from the benefiting industry and other technical experts. These bodies are supposed to provide a "real world" assessment of a technology's chances and to bring interested parties into the development as proponents early in the process. Therefore, on the one hand, having such boards and advisory groups is useful to ensure both an acceptable direction to the RD&D programme and the timely application of results. As always, however, the complication added to the planning and decision-making process can offset some of the benefits. Many countries have far too many advisory bodies, while others could increase this sort of outside involvement. The balance between too few and too many needs to be carefully handled.

Several countries, such as the Netherlands and Japan, have created separate organisations which specifically handle technology transfer and dissemination of all of the energy technologies and systems developed in the government's RD&D programmes. It should be noted that such an organisation is not only an information centre; rather, the function is that of an "acquirer" (the Netherland's terminology) of demonstrable technologies and innovations and a project catalyst/manager which insures that these new technologies find appropriate applications. The catalysis can take such forms as loans of venture capital or even entering into joint-ventures. Another function of such an organistion can be working with municipal governments (and their respective utilities) to develop long-term energy plans incorporating technological changes available and likely to be available during their planning horizon.

Japan has also developed other innovative means to achieve technology transfer, among them are the various measures administered by their Energy Conservation Center (ECC). The ECC is quite typical of the vast number of non-profit organisations (foundations) set up jointly by the government and the industrial sector in order to facilitate the dissemination of new technologies. One interesting measure administered by ECC is the designation of "energy management factories". Any industrial plant exceeding a certain yearly energy use become a designated energy management factory. The owner (manager) of the factory is then required to appoint a certain number of energy managers in charge of the rationalisation of energy use. Such energy managers must be selected from those who have passed an "Energy Management Training Course" which is given by the ECC. With the help of these energy managers, ECC has been able to facilitate the dissemination of information on new energy end-use technologies to many large and medium industrial companies.

For programmes in the buildings sector, when market introduction is left up to the construction industry, there is often little use of new technologies which might be seen as added costs without added benefit to the builder. In this sector, development and introduction of building codes (to be effective, necessarily supported by benefiting-industry and/or government RD&D) has been used by some countries to ensure market introduction. In the Netherlands, for example, the government's RD&D efforts to develop construction methods and technologies that would meet a prescribed level of energy consumption allowed them first to define what could be required in building codes that would both save energy and not cost the consumer more than conventional construction techniques and subsequently to enact changes to their building codes reflecting this state of the art.

In the industrial sector, like many countries, Japan recognized another "structural weakness" in the bridge where important, very large, risky and often cross-cutting technologies may not be taken up by private-sector R&D efforts because of the financial commitment and risk involved. To overcome this barrier, Japan first emphasizes an early conceptual phase where the government performs a combination of market and technology potential assessment as an initial input to the decision for the government to pursue development of a certain technology, thus providing preliminary assurance that commercialisation later will have a high probability of success if the development is successful. Only projects passing this first assessment become government-sponsored RD&D projects under its "Moonlight Project." Then, depending on the likelihood of a technology developing into a commercial product within a near- to mid-term timescale, projects become part of either the Leading and Basic Technologies Project or the Large-Scale projects. The latter distinction puts a project on a "fast-track" for development with ambitious goals and milestones to meet and often entails setting up a engineering research association of industries capable of supporting some aspect of the product development with 100% government funding. Leading and Basic Technologies, which have long-term potential but which need consistent government support for progress to the point where a commercial-product potential can be envisaged, are likely to be developed with full government funding at the appropriate national laboratories.

Many of the engineering research associations performing parts of the R&D work within the Large-Scale Projects are temporary, "cooperative" research institutes. Most of the researchers at such institutes are detailed from the laboratories of the member companies of the engineering research associations. When the particular Large-Scale Project has been completed, the researchers at the "temporary" institute will return to their employers. In case the association is not setting up a temporary institute, the association will contract out the R&D work to its members and sometimes parts of the R&D work will be performed by a national laboratory.

The Netherlands adapted its buildings-sector programme structure to its newer industrial-sector programme by the addition of sectoral audits as the first step of the programme planning phase. The audits consisted of an assessment, backed up by numerous on-site visits of the major end-uses in the sub-sector, of the potential for savings with both operational and technological modifications to those major end-uses. The RDD&D programme was based on these audits and then carried out via a series of one-of-a-kind demonstrations of new technologies and process modifications (as well as the operational and maintenance changes identified) in a combination of contracts for some research and tax incentives for other aspects of the industries' innovative R&D investments. The subsequent publicity on the success of the earlier audits has brought new sub-sectors into the programme.

Some governments are trying to push priority technologies into commercial markets before they would be considered "ready" for introduction, based solely on economic competitiveness with the alternatives. Examples are the Austrian and Swedish programmes for introduction of heat pumps in the residential retrofit market when their cost can presently be twice the cost of the most expensive competing "conventional" technology. This is being attempted through programmes of consumer information, financial incentives, and direct contact with the interested contractors. The public message takes advantage more of the consumers' environmental consciousness than of the economic aspects (for obvious reasons). It remains to be seen if this type of government effort (going back to the bridge analogy: a much strengthened section of bridge compensating for a weaker section) can overcome the economic problems of an otherwise desirable technology.

Each technology or new component is recognized as having different characteristics that need to be tried on the targeted user, yet none of the programmes studied specifically included behavioural or market research as a formal component. However, in numerous specific projects, references were made to the use of such research as targeted studies of market potential on an as-needed basis, and many programmes show the result of careful market and/or behavioural research. Market research is, however, a much more common occurence than behaviorial or social research. Such research has been incorporated at several points in the development process: at the initial concept stage to identify basic barriers to the concept and at the final demonstration stage to identify additions that would make the product more "user friendly." Programme managers have thus begun to acknowledge the increasing need for such research to help the dissemination of technologies when they are developed but also need to explain and document this need effectively to their funding agencies.

The means that governments use to assure dissemination are therefore varied and evolving. No one method would necessarily work for all technologies or sectors. Therefore considerable ingenuity is needed to develop the structure that works best for the technology under development and the sector targeted. Information on the different means that have been used with some success is beginning to be disseminated among governments, but very little analysis is available on what works best or what factors affect the way the programme should be structured. More work is needed in this area, and much additional information will be sought and analysis done on this in the IEA's next round of reviews.

3.3. What is the Status of Government Funding for End-Use RDD&D?

Fortunately, there seems to be a general resistance to reducing budgets for end-use technology RDD&D even where overall energy R&D budgets are being reduced. Nevertheless, governments have been attempting to make their funding more effective. For example, there have been efforts to improve the co-ordination of financial-aid mechanisms with the RD&D programmes, but additional linking could be effective. The review team recommended that Sweden start tying its extensive financial aid more to the highest-priority technologies. Italy has implemented a special law encouraging conservation and renewable energy technology which is tied to payback and oil-saving criteria. The Netherlands has been examining how much aid goes to projects where there is no "interested industry" involvement and has been trying to minimize this and to encourage a larger industry contribution, even in its longer-term research. Finally, many countries like Japan are recognizing and acting on the need for more international collaboration as a way to stretch RD&D budgets at a phase when some large projects are requiring more capital-intensive prototype plants.

Funding requirements for dissemination tend to be greater than policy makers and programme managers would expect for technologies which are considered technologically "ready" for market introduction. Indeed, comparison of some budgets for RD&D versus those for dissemination activities shows that a strong dissemination programme can cost as much or more than the development programme for the same technology. Dissemination activities which often have no tangible product in themselves are therefore much harder to justify for expenditures or to evaluate either for cost-effectiveness or for the degree to which the activity itself was the cause of the observed effect. The need for adequately justifying the relatively large proportion of funding for these stages should be recognised and incorporated into programme plans.

3.4. How Are Governments Incorporating The Effects Of Technologies Under Development Into Their Energy Supply Situations?

All of the governments reviewed list conservation (or similar concepts such as rationalisation of energy use or end-use efficiency) as a major energy policy goal. All of the governments view end-use technology development through government and industry RD&D, and sometimes dissemination of technologies, as a key component of their energy conservation programmes. However, not all of them have actually established the potential for the achievement of their national conservation efforts and even fewer either have quantified the contribution which will come from each of the technologies they or industry are developing or specifically established the link between energy supply requirements and the potential effects of these new technologies.

Numerous attempts to quantify technological potential are being made which can be characterised as one of three basic approaches: 1) a "top-down" approach of setting goals for technologies to be developed in government programmes, 2) a "bottom-up" approach of estimating the probable contribution of a technology being developed by government and/or industry, or 3) a systems-analysis approach which tries to incorporate both the bottom-up and top-down approach. At least one country reviewed provides illustration of attempts to do the first. The Netherlands has developed ambitious energy conservation targets by year at for each sector; although, it has not yet related how specific programme components of its end-use technology RDD&D programme will contribute to these targets.

Most countries reviewed, however, such as Austria or Sweden, work largely from the "bottom-up" approach which resembles a traditional technology RD&D planning process. They target the technologies with greatest potential for efficiency improvements, either in certain applications or end uses which are major sources of energy demand in aggregate and/or those which have large potential for efficiency improvements. For example, some of the institutes involved in Sweden's RDD&D programme have made attempts to develop a range of contributions predicted to result from groups of technologies.

Because neither the top-down or the bottom-up approach can capture such interactions and the range of outcomes, certain governments have recognized and begun to incorporate a "systems approach" to integrated supply and demand planning. To assist their efforts, comprehensive and highly sophisticated simulation models have been necessary, as have certain supporting techniques. For example, the Lawrence Berkeley Laboratory in the United States has been developing the concept of "conservation supply curves" for various new technologies. [2] This work is being utilised by various utilities in the United States in their supply planning.

Japan has developed a systems approach at their Electrotechnical Laboratory under the Moonlight Project of the Agency of Industrial Science and Technology. Although acknowledging that a quantitative estimate of contribution of new energy technologies to energy conservation has a speculative nature, Japan finds such estimates informative in planning energy research and development projects especially when it is carried out with a sensitivity analysis. The projection of future energy demand and supply balances have been analysed using the MARKAL model (which is well known among the IEA participating countries as a fruit of the Energy Technology Systems Analysis Project.)

Using economic analysis techniques, some governments have also started to evaluate the "robustness" of end-use technologies under different energy price scenarios for inclusion in their planning. [3] This essentially provides a range of scenarios of the worst- to the best-case contribution of the conservation technologies being developed and introduced for a range of supply requirements. (It also allows them to apply a slightly different criterion to their setting of technology priorities, that of cost savings potential, which for some governments is becoming a more relevant criterion than energy savings. But see the discussion later on the application of "multi-criteria" to project selection.) Governments can then determine the level of technology development effort required under the most probable case and can later evaluate the progress in technology development and diffusion in order to determine if the conservation supply goals are being met.

Because conservation can be considered a form of energy supply which eliminates the need for some quantity of new energy supplies, better integration of supply planning and the predictable results of new end-use technologies is advisable. Lack of quantified targets makes it impossible to determine if the technologies being developed add up to an overall level of effort that is "adequate". Despite difficulties in knowing the present and future structure and interactive nature of end-use and uncertainties of energy prices and ancillary economic factors, the maximum potential effects of new technologies should be quantifiable and the quantification useful. This is especially important where conservation will be relied on to reduce the amount of new supplies of energy needed, such as is presently the aim in Sweden or Japan, and thus where it is most important to have a quantitative and reliable estimate of the contributions of conservation and particularly of new technologies being developed.

3.5. How Are Government End-Use Technology Programmes Affected By Other Government Policies?

Important, but not necessarily energy-related, government goals and policies, in such areas as industrial development (and the related development of exportable technologies), environment, housing or transportation, can largely affect the direction and focus of end-use technology RDD&D programmes. For example, most governments subscribe to the idea that development of technologies "indigenously" (as opposed to importation) leads to greater spin-off economic benefits. Some countries, such as the Netherlands and Japan, not only recognize but strongly articulate the connection between energy policy and industrial policy (or technology policy) and attempt to maximise the benefits from the relationship in an integrated energy-technology development policy covering all energy technologies. Decisions to emphasize development of certain end-use technologies are, therefore, predicated on the markets available locally and often internationally and the capability of the country's industry and/or the government's research facilities to handle the development.

Such focusing on markets is entirely necessary to ensure there will be a place for the developed technology, but could lead to inappropriate technology development, especially if the market assessment does not take account of similar plans in private industry and other countries, both private and public. Given that market assessment is a skill relatively new to government planners and that gauging international competition is tricky even for industry, it was found that care should be taken to assure that these market assessments are realistic particularly where the result might be to promote a technology before one with higher priority for the national energy future.

Some governments find their end-use technology development almost completely dominated by non-energy policies and goals, perhaps to the detriment or reduction of the end-use technology RDD&D programme's effectiveness. In the case of Sweden, environmental and social policies dominate the energy goals in the building sector to the extent that the energy conservation potential of end-use technology development decisions is secondary to these other concerns. Countries with relatively small RDD&D programmes, such as Greece, sometimes find their programmes dominated by regional development policies and outside interests in end-use technology development. This is especially true where there are attractive funding opportunities available for certain types of projects, such as foreign industrial development projects or the EEC's technology development programme.

Even emphasis on other important energy goals, such as diversification, fuel switching, or utilization of indigenous resources can affect the direction of an end-use technology RDD&D programme. For example, in the Netherlands, large supplies of indigenous natural gas induce planners to orient towards development of technologies, such as fuel cells, which use natural gas. In fact, this motivation, added to the desire to couple industrial policy with energy policy previously mentioned, leads the Netherlands to a strong commitment to fuel cells that goes far beyond the probable contribution of the fuel cell to the Dutch energy-supply future. Austria's end-use programme provides a different example of combined energy goals: biomass-combustion technology development which not only assists diversification to an indigenous resource but achieves fuel-switching as well.

The previous examples highlight fortuitous concurrences of energy policy goals, but often there can be conflicts. Sweden provides an additional example in their heat-pump development and dissemination programme which would add to electricity-supply requirements when future electricity supplies are uncertain. Clearly the ability to connect a number of important government policies to any activity <u>should</u> strengthen the governmental support for the activity. This can be especially important in times of lower oil prices when government support for end-use technology development might be being re-directed or when RDD&D budgets in general are being reduced. In fact Hagler, Bailly and Co.[4] found that a number of the governments it studied were tending to focus more on those projects which could be justified as "multi-criteria" projects. (They found such additional criteria applied as improved industrial process cost or environmental effects.) Nevertheless, such coupling must be done so that the technologies' energy-saving merits are not obscured in the short run (which could possibly lead to a shortfall in the technologies having the most suitable and cost-effective energy-saving potential in the long run). Further information and analysis is needed to determine if this is in fact occurring, and this issue will therefore be the subject of further investigation in the second year of this thematic review.

3.6. Government/Private Industry Relationships in End-Use Technology Development

Many of the end-use technologies being developed have mid- to near-term development times, and it is generally conceded by government researchers that industry develops most of the product ideas and spends far more than government for research in the end-use technology area. It is therefore important to understand what the industry is doing in end-use technology development, how it relates to what governments are doing and what means are used to achieve technology transfer between them.

In this way, government can plan programmes which are neither duplicative nor leave gaps in important research. However, the broad range of end-use technologies (especially the more generic research, such as in new materials, and cross-cutting applications) makes it particularly difficult for governments to keep abreast of private industries' efforts.

The type of research <u>not</u> generally tackled by industry is the high-risk, long-term projects (as mentioned previously) because industry sees higher profits in shorter-term, lower-risk projects. Even the latter may not be taken up by industry if the industry is fragmented or immature, and because the end-use industry is relatively immature (often 10-15 years at most), there are still some areas of fragmentation.

Recently, however, industry has even begun to be a leader in RD&D in markets where it has not traditionally been active if the market is seen as large enough and the product seen as near-term enough. For many European manufacturers and for Japan, for example, aiming at an international market may be the only way that product development can be justified. An example is a special niche found by Sweden's building-sector industry, where so much of the housing is pre-fabricated and therefore able to be exported that the market is far larger than the housing creation rate in Sweden could promise. Some non-energy technology innovations have had spin-off energy-saving benefits. The best example is the "smart building" (microprocessor) technology which often has more consumer appeal for its building management capabilities but nevertheless saves considerable energy and can be coupled with load-leveling capabilities.

Nevertheless, with the exception of these examples, the buildings sector remains one where private industry efforts have lagged primarily because of the still-fragmented nature of the industry and the relative lack of market pull by the consumer. Government regulations (product standards, specifications and testing and building codes) have already been mentioned as a key component of many building sector programmes. Performance-based building codes can be particularly effective in stimulating private sector ingenuity. Targeting associations of manufacturers, architects, and builders has also been effective, as has the intervention of utilities to inform builders on how to use energy efficiency as a selling feature in a building. Shared-savings programmes are also having the effect of energising private-sector buildings RD&D, especially in the retrofit of large multi-occupant buildings.

Participation by industry in government's end-use RDD&D has not only been recognized by most of the reviewed countries as being very necessary, especially for those technologies which are considered to be near-term development prospects, but is also felt to stimulate private sector ingenuity if properly handled. For example, in almost all areas of energy RD&D (except for nuclear) the Dutch government attempts to have a large proportion of industry involvement in first-of-a-kind demonstrations which will have high replicability-potential. By doing this, it is expected that the programmes will benefit from better assessment of market potential by including commercial risk assessment, more assurance that the developed product will indeed be used, and maximization of technological innovation.

Ways to involve industry in government projects include reducing industry's "opportunity costs" by contracting directly with industry for development of some component or all of a certain technology, as is done in the Netherlands, Japan, and Sweden. Industry advisory boards to government programmes have already been mentioned as have Japan's project-specific research institutes which ensure a highly-integrated involvement of the diverse set of industries required for the more complicated development projects as well as two-way information flow between industry and government research efforts. The Netherlands requires that a certain portion of some targeted RD&D projects have a minimum amount of industry funding, even if the funding recipient is the government research laboratory which will ultimately perform the research.

Japan also has developed an approach to contracting with industries for product development which involves competition between two or more companies to develop a product for a specific application. The company which "wins" is then guaranteed the rights to produce and market the product provided that part of the development costs are paid for by the company itself and the government funding has been given as a "conditional loan". In case all of the contracted development work is funded by a government grant, the results (patents etc.) belong to the government. Any Japanese company is then allowed to bid for licences. The company which performed the development work will have to make the product in competition with other companies.

Alternatively, there is little evidence that governments are utilising the marketing and technical expertise and privileged position of electric, gas and district heating utilities to full advantage, especially in the last phases of demonstration and dissemination of technologies. In addition, utilities themselves should be recognised a part of the market to which end-use technologies should be directed. Load-leveling devices are a prime example as they are attractive only to utilities (unless a time-of-use or similar tariff has been instituted.)

The importance of utilities (many of which are private companies) to the area of end-use technology RDD&D should not be underestimated, especially because of their close connection and knowledge of the end-user, as well as their direct interest in energy supply and demand. Much can be learned from evaluating their past efforts, and they can be valuable research partners. In the upcoming round of reviews, there will be increased emphasis in analysis of the integration of government and utilities' efforts and encouraging utilities' involvement in end-use technology development.

Finally, it should be noted that both informal and formal technology transfer occurs routinely between and among private industry research efforts in this field even on an international level. Government efforts to foster this exchange, especially on the international level, nevertheless seem to fall far short of expectations. A further area of emphasis for the next round will therefore be how governments could improve their methods for obtaining industry involvement in end-use technology RDD&D.

The foregoing discussion suggests that governments efforts logically fall into the area of:
- gap filling,
- product testing and performance rating, product specification, and development of building codes.
- long-term generic and/or cross-cutting research
- co-operative applied research.

Herein lies one of the biggest problems. A gap-filling approach tends to be fragmented and hence could appear to lack dynamism or clear direction; activities such as developing product standards are necessary but actually ancillary to the development work; and generic work often tends to be the first hit by budget cuts and the hardest to justify in times of unstable oil prices. Such problems will be magnified in smaller government programmes. The final role of cooperative applied research may be the most viable role for them.

4. TECHNOLOGIES BEING DEVELOPED WHICH AFFECT ELECTRICITY END-USE EFFICIENCY

This section will briefly describe a range of technologies being developed by one or more of the countries reviewed in the first round. It is not meant to be comprehensive, but rather to indicate the types of activities which will affect ultimately electricity efficiency and the demand-side management programmes of governments and utilities.

4.1. Electricity Transformation Sector Technologies

As was pointed out at the beginning of this paper, some countries view their electricity transformation sectors as part of their industrial end-user sectors. While this might seem intuitively difficult to accept, a convincing argument can be made when the technologies themselves are considered because most being developed would be appropriate for non-electricity transformation sector applications across one or more sectors. Numerous examples below will make this apparent.

For example, Japan and the Netherlands are developing short- and long-term electrical energy storage technologies. In the Netherlands, the types of large scale electricity storage systems being investigated are underground storage, storage reservoirs, and low pressure air systems. Japan is actively developing large scale superconducting underground magnets for long-term storage of electricity. For shorter term, Japan is developing several advanced types of batteries whose primary application would be smaller scale storage for load leveling purposes.

The significance for utilities and their demandside management programmes aimed at load shifting is clear. These technologies would tend to obviate the need for such programmes in the long term or relegate them primarily to stop-gap measures between the construction of increments of electricity storage capacity, much as today some demand-side managment programmes are used between additions of increments of electricity capacity.

Substantial work is being conducted in a number of countries on several types of fuel cells. Commercialisation activities to promote phosphoric-acid types are just beginning with such activities as workshops and demonstrations to occur in the near future. Several other types of fuel cells, molten-carbonate and solid-oxide types, are envisioned to be ready for similar commercialisation activities in the middle 1990's. They are expected to have significant advantages over conventional generation techniques, including lower environmental effects and higher efficiency because the waste heat will be able to be used on site.

Work also continues on development of more reliable and more efficient central heating and power (CHP) units which in some localities are known as cogeneration units. These efforts, combined with related activities to remove barriers to cogeneration dissemination, such as improving gas tarifs, promoting more attractive wheeling arrangements, and assistance in identifying matched thermal and electricity loads, should ensure that CHP makes a much larger contribution to electricity supplies in the future.

These two decentralised electricity-generation technologies could have tremendous effects on the demand-side structure as more and more businesses and industries become, in effect, their own power producers. Again demand-side management programmes which presently are most often operated by utilities may no longer be very effective or justifiable when there are large classes of customers no longer requiring the utilities' services as end-users. A rethinking of the most appropriate format to obtain energy efficiency, which will still most certainly be a primary goal of governments concerned with energy security, may be necessary.

Japan's and other countries' work on advanced gas turbines will have repercussions, albeit rather long term, on electricity efficiency as well. While their present work is oriented almost exclusively towards very efficient large-scale generating turbines, perhaps ultimately capable of utilising coal gas, the spin-off benefits, such as from the research into new materials and turbine designs, will substantially assist efforts to build smaller-scale, and even packaged, CHP units cost-effectively.

It is probably worth mentioning one other research effort in the electricity transformation sector in Japan where technology is being developed to enable waste heat utilization from underground electricity transmission lines. Again this type of effort to reduce the losses associated with electricity generation and transmission make electricity more viable as a fuel supply. The possible effect on the structure of the demand side of the equation is obvious but the extent and timing of such effects, especially in light of the competition from decentralised generation equipment, is difficult to guage.

4.2. Heating Systems Developments

There is also substantial RDD&D being conducted on heating, ventilation and air conditioning systems. Among those countries studied, Austria, Japan, Netherlands and Sweden all have very active programmes in this area. Two primary foci are heat pumps and waste heat utilization.

Work on heat pumps is proceeding on several fronts. The Netherlands is concentrating largely on the development of a 40 kW gas-driven absorption heat pump. Sweden and Austria are endeavoring to reduce the cost of electrically driven compressor-type heat pumps and in addition are developing heat pumps for special applications, such as hot tapwater, and with a variety of heat sources, such as waste water or solar heat. Japan, again taking a slightly longer-term perspective, is developing a "Super Heat Pump" with chemical heat storage capabilities which will provide some load-leveling capacity in addition to the normal coefficients of performance for heat pumps.

In their attempt to increase efficiencies, numerous components are being developed which can have much wider applicability than heat pumps. For example, the Netherlands has developed a vacuum-freeze evaporator for large heat pump applications which allows the use of the latent heat of ice-formation as a heat source. Japan has recently had a major breakthrough with its development of advanced electrohydro-dynamic heat exchangers which use an electric voltage to increase dramatically the efficiency of heat exchange at the surface of the heat exchanger.

Japan's work on the Stirling engine also promises wide applicability, such as in household air-conditioning, energy recovery systems, and total energy systems, as well as construction machinery, automobiles and shipping. Austria is developing wood and other biomass furnaces as another heating system alternative which may be combined with heat pumps to reduce their peak load effects.

Even reviewing just one country's programme in the space conditioning area leads to questions of overlapping or competing technologies for a limited number of applications and markets. Reviewing several countries compounds the impression. It is becoming impossible for researchers to keep abreast of the developments in other countries in overlapping research. Attempts to do so are being made and will be described in the final section of this paper on international cooperation.

This sketchy overview of a few activities in development of technologies for more efficient space conditioning and advances in energy storage should convince the reader that there is a high probability of substantial advances in the efficiency of energy, and many cases electricity, use in this area. The timing of the effects is somewhat unknown. Even the overall direction, i.e. whether towards or away from greater electricity use, is unpredictable at this stage. What is clear is that energy-supply planners as well as demand-side-management programme planners should expect a changing field. There is in effect no one technological fix possible, nor does just one appear to be desirable from the standpoint of the energy policy makers in the respective countries. Where the such decisionmaking and the resulting technological innovations will affect these planners and their own programmes or capital improvement plans, it is imperative that maximum cooperation occur.

5. INTERNATIONAL CO-OPERATION AND OPPORTUNITIES FOR FURTHER ENHANCED COLLABORATION

The IEA is already a forum for international collaboration most notably in the form of its Working Party structure and its Implementing Agreements and Annexes. A recent IEA publication describes the numerous end-use collaborative projects and the countries participating.[5]

In addition the IEA Heat Pump Centre and Air Infiltration and Ventilation Centre provide members with information exchange services, etc. Additionally the IEA has a number of activities planned, such as the workshops on separations, tribology, and fuel cells and a Center for Analyses and Dissemination of Demonstrated Technologies (CADDET) which will further augment international exchanges of information.

The question of the optimum forum, form and content for international collaboration arises because there are always ways of improving a process, no matter how good. Despite the benefits of these present and planned IEA activities, there are some limitations inherent in a large proportion of end-use technologies which make them inappropriate for the present IEA collaborative project structure. These are technologies with a very high commercial content, i.e. those that are very near to commercialisation, and those which are highly constrained by national or local particularities, such as urban transport systems. More work will be needed to develop the best way to identify duplication of end-use RD&D,

the ways that international collaboration can help to eliminate duplication, and of what additional collaborative activities might be necessary.

REFERENCES

1) Energy Conservation in IEA Countries, International Energy Agency, Paris, France, 1987

2) "Supply Curves of Conserved Energy: A Tool for Least-Cost Energy Analysis", Alan Meier and Anthony Usibelli, Energy Technology XIII, Proceedings of the Thirteenth Energy Technology Conference, March 17-19, 1986, Government Institutes, Inc. Washington D.C.

3) "Government-Sponsored Industrial Energy Conservation Research, Development, and Demonstration in Europe, A Review of Programs in the Federal Republic of Germany, France, Sweden, the United Kingdom and the European Community [sic]" prepared for the Office of Industrial Programs, US Department of Energy, Hagler, Bailly and Co, Washington D.C. January 1987.

4) Ibid.

5) Collaborative Projects in Energy Research, Development and Demonstration, A Ten Year Review, 1976-1986, International Energy Agency, OECD, Paris, 1987.

VII. Case Studies

ELECTRICITY CONSERVATION IN BRAZIL: POTENTIAL AND PROGRESS

HOWARD S. GELLER, JOSE GOLDEMBERG, ROBERTO HUKAI,
JOSE ROBERTO MOREIRA, CLAUDIO SCARPINELLA, MAMIRO YOSHIZAWA

American Council for an Energy-Efficient Economy,
University of Sao Paulo, the Sao Paulo Energy Company

1. INTRODUCTION
1.1. Background

Total electricity consumption in Brazil grew from 38 TWh in 1970 to 175 TWh in 1985, an average growth rate of 10.7%/yr (1). Figure 1 shows rates of growth in electricity demand and electricity demand per unit of GDP during this period. Overall, the electricity intensity of the Brazilian economy increased 82% between 1970 and 1985. High growth was caused by economic expansion, the relatively low level of absolute electricity consumption (1220 kWh/capita/yr as of 1985), the development of plentiful hydroelectric resources throughout the nation, and efforts to substitute electricity for oil products.

The rapid expansion of hydroelectric facilities compounded with the economic recession of 1981-83 led to excess generating capacity in the industrialized Southeast region of the country in the early 1980s. In response, utilities temporarily offered very low-priced power for the substitution of oil products in industrial boilers and furnaces. This effort accounted for a large portion of the growth in power demand in 1983-84. Inexpensive power for oil substitution was discontinued at the end of 1985.

Power sector planners anticipate that electricity demand will continue to grow at a relatively high rate during the remainder of the century. Table 1 shows the official demand forecast completed in 1985 by Eletrobras, the national holding company and planning authority for the electricity sector in Brazil (2). National electricity consumption is projected to increase at an average rate of 7.4%/yr during 1985-90, 6.2%/yr during 1990-95, and 5.6%/yr during 1995-2000.

Table 1 also shows estimated generating capacity requirements through 2000, assuming a slight improvement in the average capacity factor and a slight reduction in line losses over time. It is estimated that 95 GW of installed capacity will be needed by 2000, 2.3 times the level in service in 1985. Since there was approximately 25 GW of capacity under construction in 1985 in addition to the 42 GW in operation, the official forecast calls for starting and completing an additional 28 GW by 2000.

Table 2 shows the prevailing electricity tariffs in Brazil as of December, 1986. The tariffs, including taxes, are typically about US $0.03/kWh for industrial customers and US $0.05/kWh for residential customers. These relatively low rates have declined in real terms in recent years, and are not adequate to cover the operating costs, debt service, and capital investments by the

A. T. De Almeida and A. H. Rosenfeld (eds.), Demand-Side Management and Electricity End-Use Efficiency, 535–558.
© 1988 by Kluwer Academic Publishers.

power sector. Consequently, the government intended to raise average tariffs 15%/yr in real terms (i.e., above inflation) in 1986 and 1987, with similar real increases expected through 1989 (3). However, national economic policies led to a freeze on tariffs throughout much of 1986.

1.2. Why Pursue End-use Efficiency?

Improving the efficiency with which electricity is used is desirable for a number of reasons:

1) Expanding electricity supply is highly capital intensive.

Although hydroelectric power is a renewable and plentiful resource in Brazil, it is not inexpensive to harness. At the present time, new hydroelectric capacity costs about $1500 per installed kW (4). Associated investments in transmission and distribution raise the capital cost to over $2000/kW. Adding nearly 53 GW of new generating capacity during 1985-2000 costs over $106 billion (undiscounted). This is nearly equal to Brazil's foreign debt as of late 1986.

2) Increasing end-use efficiency is less costly than increasing electricity supply.

The investment associated with serving a kW of new firm demand with Brazil's hydro-based system is typically at least $3000, taking into account the average load factor and typical T&D losses (4,5). As will be shown in the next section of this paper, increasing the efficiency of electricity use costs much less than expanding electricity supply when the two options are evaluated on an equivalent basis. Thus, investing in electricity conservation is a more rational allocation of financial resources for the nation as a whole.

3) There is massive potential for increasing the efficiency of electricity use in a cost-effective manner.

Numerous technologies are already available in Brazil for increasing efficieny and lowering electricity use. Other products, now manufactured in industrialized nations, could be made in Brazil if so desired.

4) Adopting more efficient end-use equipment holds down electricity bills and can lead to improved standards of living.

Increasing end-use efficiency can moderate the impact of rising tariffs during the late 1980s. Also, reducing the pace of supply system expansion should lower electricity costs in the future. Keeping costs and tariffs as low as possible are an important mechanism for expanding the level of energy services and well-being among poorer segments of the population (6).

5) Increasing the efficiency of end-use equipment can strengthen Brazil's growth and export potential.

The production and use of energy-efficient equipment is rapidly advancing in industrialized nations. If similar developments are not made in Brazil, it could become stuck with obsolescent products and factories. This in turn would restrict economic growth and hinder Brazil's efforts to maintain a high level of exports.

6) **Slowing the rate of electricity supply expansion has positive environmental impacts.**
Providing 1000 MW of hydroelectric capacity typically inundates 65,000-200,000 hectares of land, often with disruption of settlements as well as loss of natural resources. Coal-fired and nuclear power present air pollution, safety, and waste disposal problems. Increasing end-use efficiency, on the other hand, causes little or no environmental degradation.

1.3. **How electricity is used in Brazil**
Before considering the opportunities for greater end-use efficiency, it is important to understand how electricity is used. Figure 2 presents estimates of the major electricity end-uses in 1985.

The industrial sector accounts for about 55% of total electricity consumption in Brazil. The iron and steel, non-ferrous metals, chemical, and food and beverage industries together represent 54% of all industrial electricity demand (1). Motors account for the majority of industrial electricity consumption, followed by thermal processes (7,8).

The residential sector and the commercial/public services sector each contribute about 20% of total electricity demand in Brazil. In the residential sector, it is estimated that refrigerators account for about 32% of electricity use, followed by water heating and lighting which each account for about 25% of sectoral demand (4). In the commercial/public services sector, lighting is estimated to account for about half and motors one third of the total (8).

The regional disaggregation of electricity use in Brazil is shown in Figure 3. Like income and population, electricity use is geographically skewed. Per capita consumption in the Southeast region is about 50% greater than for the nation as a whole. Per capita demand in the Northeast, North, and Center-West regions, on the other hand, is about half the national average.

2. SAVINGS POTENTIAL
Even though per capita electricity consumption in Brazil is low compared to that in industrialized nations, opportunities for efficiency improvement abound. This section of the paper summarizes and updates a detailed study of the electricity conservation potential in Brazil completed in 1984 (4).

2.1. **Industrial sector**
Motors are the primary end-use in the industrial sector. It is possible to obtain direct electricity savings through the use of more efficient motors as well as indirect savings through better matching of motor output to loads. The industrial mix and the rates of growth of different industrial subsectors also have an important effect on electricity demand.

2.1.1. **More efficient motors.** Studies in the U.S. show that larger motors dominate in terms of total electricity usage, with motors greater than 50 HP accounting for nearly two-thirds of total electricity consumption by motors (8). Similar conditions are expected in Brazil.

Larger motors are already relatively efficient; standard models greater than 10 HP have efficiency ratings above 85% while models above 100 HP have ratings of 90% or greater (9). Small efficiency improvements are available, however, through the purchase and use of "energy efficient" motors. Manufacturers in Brazil produce high quality motors that are typically 4-5% more efficient than standard motors at smaller sizes (< 20 HP) and 2-3% more efficient at larger sizes (4).

Assuming the relative share of electricity consumption by motors remains constant between 1985 and 2000, improving the efficiency of motors in the commercial and industrial sectors by only 3% could save an estimated 5.8 TWh by the year 2000.

More efficient motors have a greater first cost due to the use of better quality steel and other improvements. It is estimated that more efficient motors cost approximately 25% more than standard motors in Brazil, with an extra cost of about $350 for a 100 HP motor (4). For new installations, the simple payback period is generally less than three years and the internal rate of return on the extra first cost is in excess of 40%/yr (4). It is also cost-effective for consumers to replace existing motors with more efficient models at the time an older motor needs to be rebuilt.

2.1.2. Motor speed controls. For many motor applications, there are substantial energy losses during part load operation. When the load on most motor-driven pumps, compressors, blowers, etc. is below the rated value, the motor drive is normally operated at constant speed and a throttle valve or damper is used to reduce the flow rate. There are large pressure and energy losses across the throttle or damper.

Various technologies for motor speed control and energy conservation during part load operation are available, including eddy current couplings, D.C. motors with a converter and voltage controller, A.C. variable speed drives (10). Variable speed drives (VSDs) are becoming increasingly popular in industrialized countries as a result of decreasing prices, improved reliability, and high efficiency. In Brazil, a few companies manufacture VSDs (4).

The energy savings resulting from use of a VSD depends on the specific application. One study in the U.S. estimated that the savings potential is typically 20-30% for motors powering industrial pumps, compressors, blowers, and fans as well as commercial air conditioning and refrigeration equipment (11). Together, these applications account for 30% of electricity use in the U.S., leading to an estimated savings potential of 7.3% of all electricity consumed there.

Assuming a similar savings potential in Brazil, full implementation of VSDs by the year 2000 could reduce baseload electricity demand by about 30 TWh. A reduction in electricity consumption of this magnitude in the commercial and industrial sectors in 2000 could obviate the need for nearly 6.6 GW of installed generating capacity.

In Brazil, VSDs on the order of 100 HP cost approximately $270/HP (4). This is 50-100% greater than equipment costs in the U.S. With electricity typically costing industrial customers $0.023-0.038/kWh, motor usage as well as energy savings have to be very high in order for VSDs to be economically viable as a conservation investment in Brazil (4).

On the other hand, VSDs are economical from the social perspective when the cost of saving a kW of demand is compared to the marginal cost for electricity supply. This relationship is shown in Figure 4, assuming that there is a 20% electricity savings using a 100 HP VSD. The graph shows the net present value of investments for 50 years for both VSDs and new electricity supply. Even with the relatively high cost for VSDs in Brazil, the capital cost for reducing demand is about half that for supplying demand with hydroelectric power at accepted discount rates of 10-15%.

2.1.3. Industrial structure. Figure 5 shows the electricity intensity and value added of different industries in Brazil. Electricity use per unit of value added for non-ferrous metals is over ten times greater than for the food and beverages or chemical industries. Non-ferrous metals accounted for 17% of industrial electricity use but just 1% of industrial value added in 1980. Furthermore, the seven most electricity-intensive industries account for 52% of industrial electricity use but only 13% of industrial value added.

The shift away from basic materials processing toward fabrication, finishing and service activities has significantly lowered energy demand in OECD nations. In the U.S., for example, it is estimated that nearly half of the 32% reduction in industrial energy intensity during 1973-84 is due to this structural shift (12). Restructuring economic development away from electricity-intensive industries could be an important means for containing growth in electricity consumption in Brazil.

2.2. Residential sector

There is considerable potential for increasing end-use efficiency in a cost-effective manner in all three major end-use areas -- refrigerators, water heating, and lighting.

2.2.1. Refrigerators. The national housing survey showed that 50% of the households in Brazil had refrigerators in 1980 (13). It is estimated that the refrigerator saturation reached 60% by 1985.

Single door refrigerators with volumes of 200-350 liters are the predominant type. Standardized energy consumption tests conducted in Brazil beginning in 1986 show that single door refrigerators in this size range consume 380-660 kWh/yr (14). Two-door refrigerators are more electricity intensive and are gaining in popularity. Data from one manufacturer show that two-door refrigerators 350-450 liters in capacity typically consume 1300 kWh/yr (4).

Energy-efficient refrigerators produced in industrialized countries consume considerably less electricity than the refrigerators produced in Brazil. The best European and Japanese single door models consume 180-300 kWh/yr (15). The best two-door model produced in the U.S. in 1986 is a 500 liter model that consumes only 750 kWh/yr, 42% less than smaller Brazilian two-door models (15). Interestingly enough, this model includes an efficient compressor manufactured in Brazil for export. The efficient compressor is not used in any refrigerators sold within Brazil.

Without any improvement in efficiency, refrigerators could consume 25 TWh by the year 2000. This is about 2.4 times their estimated consumption in 1985. However, if the efficiency of new refrigerators is improved to the level of the best models now produced worldwide (about a 60% reduction in electricity consumption compared to current Brazilian models), electricity demand for this end use could be held to about 10 TWh in 2000.

Experience in other countries has shown that high efficiency refrigerators turn out to be very cost-effective for consumers. In the U.S., for example, highly efficient models typically cost about 10-15% more than models of average efficiency. This leads to paybacks on the extra first cost of under three years and internal rates of return in excess of 40% (16). Producing and selling more efficient refrigerators should also be very cost-effective in Brazil, both for consumers and for society (4,5).

2.2.2. Hot water heating. The most common domestic hot water heating device in Brazil is the chuveiro, a point-of-use shower water heater. Typical chuveiros have a power rating of 3000-5000 watts. A field study conducted in metropolitan Sao Paulo showed that chuveiros typically consume about 800 kWh/yr (17). However, chuveiros are heavily used during the peak demand period (6-9 PM). They are estimated to account for over 50% of the residential peak demand in metropolitan Sao Paulo (17).

Storage water heaters are used only in wealthier households in Brazil. Based on the Sao Paulo survey, it is estimated that the typical storage water heater consumes 3200 kWh/yr and presents a diversified average demand of about 600 watts during the daily peak period. Point-of-use water heaters are more efficient than storage water heaters because there are no heat losses in storage and distribution. However, both types use electric resistance heating, which is inherently less efficient than using a heat pump.

Small heat pump water heaters (HPWHs) are produced for domestic water heating in North America and Europe. They typically provide a 50% reduction in electricity consumption compared to electric resistance water heaters (16). In the U.S., HPWHs cost $700 or more. However, they are economical for consumers when considered in terms of life-cycle cost (16).

In Brazil, private companies are developing HPWHs for export as well as the domestic market. One company estimates that they will be able to produce a small heat pump unit that will cost $200 in addition to the storage tank (18). A HPWH at this price should be cost-effective in households where a storage water heater is used. HPWHs are especially attractive in tropical countries because their efficiency increases as the ambient temperature rises.

It may even be practical to replace chuveiros with storage HPWHs, although this may result in other benefits besides kWh savings. If the heat pump unit was efficient enough (COP of 3.0 or greater) and the tank is well-insulated, the level of hot water supply typical of storage water heaters (approximately 180 liters/day) could be provided without consuming more than 1000 kWh/yr. Furthermore, if a HPWH is used along with a storage tank, the heat pump could be automatically switched off during the peak demand period for load management purposes. Simple timer controls are already available for switching on and off storage water heaters in Brazil. Lowering peak demand in this manner would be of value since utilities need to invest about $1000 for supplying power to each chuveiro in operation.

2.2.3. Lighting. An analysis of residential electricity consumption in Brazil in 1975 estimated that lighting accounts for nearly 25% of total residential use (19). This corresponds to 340 kWh/yr in electrified households as of 1982. However, the survey completed in metropolitan Sao Paulo estimated that lighting consumes only 250 kWh/yr on the average (17). Incandescent bulbs account for almost all domestic lighting.

Large savings in electricity demand are possible through the use of compact fluorescent lamps. These lamps use 60-75% less power than incandescents bulbs for the same level of light output. In addition, compact fluorescent lamps provide "warm" light and last 5-10 times longer than incandescents.

Two major lighting manufacturers in Brazil (Philips and Osram) started marketing "PL" type compact fluorescent lamps in 1986 (20). The PL lamp has a separate tube and base/ballast. Although the PL lamp is not as convenient as an integrated lamp that screws into an ordinary socket, it is easier to produce and it has the advantage that tubes can be replaced when they wear out.

Compact fluorescent lamps typically cost $10-20 in industrialized countries, compared to less than $1 for incandescent bulbs. But the fluorescent lamp is often cost-effective for consumers on the basis of life-cycle cost (16).

The first PL lamps sold in Brazil cost about $20, with about 60% of the cost for the tube and about 40% for the base and ballast (20). For higher use residential consumers paying a marginal electricity price of $0.08-0.09/kWh as of December, 1986, the PL lamp provides a rate of return of more than 17% per year if the lamp is used at least three hours per day. Like other conservation technologies, compact fluorescent lamps are economical for society as a whole when the cost for reducing demand is compared to the marginal cost for serving demand (5).

Residential lighting could consume 16.5 TWh by the year 2000 asssuming that lighting accounts for 20% of residential electricity demand that year. It should be possible, however, to reduce this electricity demand by 50% through the widespread use of fluorescent lamps. Savings of this magnitude would obviate the need for about 1.9 GW of generating capacity based on expected capacity factors and T&D losses to low-voltage consumers.

2.3. Commercial and Public Service Sector

The commercial and public services sector accounted for about 20% of national electricity demand in 1985. The end uses of indoor lighting, outdoor lighting, and air conditioning are addressed below.

2.3.1. Indoor lighting. There are a variety of ways for reducing electricity consumption for lighting in commercial buildings, including:

o reduced illumination levels
o use of more efficient lamps
o use of more efficient ballasts
o use of improved light fixtures
o use of lighting control systems.

Standard two-lamp fluorescent fixtures emit about 6000 lumens of light while consuming 90-100 watts (4). Better lighting systems produced in industrialized countries consume about 75 watts, while the most efficient systems consume about 60-65 watts. The latter is achieved using a high frequency electronic ballast and improved lamp tubes. Energy-efficient fluorescent lamps are starting to be sold in Brazil (20), and one company began producing electronic ballasts on a pilot scale in 1986.

In the area of sensors and controls, automatic controls for daylighting, computerized switching systems, and occupancy sensors are available in industrialized countries. These devices can cut lighting power consumption by 30% or more, in addition to the savings achieved through more efficient lamps and ballasts (4). Simpler automatic lighting controls are available in Brazil.

Combining measures, it is possible to reduce the electricity consumption for lighting in commercial buildings by 60% or more (21). Assuming that indoor lighting continues to account for 30% of the electricity consumption in the commercial and public services sector, this end-use could account for about 25 TWh of demand by the year 2000. However, if electricity consumption can be reduced by 60% through efficiency improvements, the direct savings in 2000 would equal 15 TWh. In addition, air conditioning requirements would also be reduced, leading to further electricity savings.

Economic analysis shows that with the relatively high tariffs paid by commercial customers, the various lighting conservation measures are cost-effective for consumers (4). Paybacks are on the order of three years or less.

2.3.2. Air Conditioning. Air conditioning is an important end-use in the commercial sector in certain parts of Brazil. In Rio de Janeiro, for example, air conditioning represents a growing fraction of peak electrical demand.

Motor speed controls were mentioned previously as a means for reducing electricity use in commercial HVAC systems. There are other savings opportunities as well, including "economizer cycles" (i.e., ventilation with outdoor air when ambient conditions are suitable), better architectural design in new buildings, and better window shading in existing buildings.

Room air conditioners sold in Brazil are not efficient by international standards. Data from one major manufacturer shows that in the size range of 7,000-18,000 Btu/hr, the energy efficiency ratio (EER) is only 6.4-7.4 Btu/Wh (22). For comparison, the average EER of room air conditioners sold in the U.S. in 1985 was 7.7, and models are available with EERs as high as 11.5 (23).

One major Brazilian producer is assembling efficient room air conditioners in Brazil for export only. These models include rotary compressors imported from Japan and have an average EER of 9.0 at capacities below 12,000 Btu/hr (22). The export models consume 20-25% less electricity than domestic models.

2.3.3. Outdoor lighting. Outdoor lighting is estimated to account for about 4% of total electricity consumption in Brazil. In Sao Paulo state, mercury vapor lamps predominate although less efficient lamps still accounted for about 10% of the installations in 1983 (4). Incandescent and other inefficient types of street lamps are more common in less developed regions of the country.

High pressure sodium (HPS) lamps are starting to be produced in Brazil for street lighting. HPS lamps consume about 40% less power than mercury vapor lamps with slightly greater light output (e.g., a 150 W HPS lamp is used instead of a 250 W mercury vapor lamp). Compared to incandescent lamps, HPS lamps consume 80% less power for the same light output.

If street lighting continues to account for 4% of national power consumption, this end-use would represent 16.8 TWh of electricity demand in 2000 based on the 1985 forecast. Fully implementing HPS lamps could save about 6.7 TWh of demand in the year 2000.

Economic analysis shows that even with the low tariffs for public illumination, HPS lamps are cost effective for municipalities in Brazil either for new installations or for existing lights when lamp replacement is needed (4). Increasing tariffs and anticipated reductions in the cost of HPS lamps will make this conversion even more attractive.

2.4. Overall savings potential

Table 3 shows the potential electricity demand and the potential savings in the year 2000 in six major end-use areas. It is estimated that these end-uses will account for about two-thirds of total electricity consumption in 2000 without efficiency improvements beyond those incorporated in the 1985 Eletrobras forecast. The potential for additional savings, 83.3 TWh, is almost 20% of the total electricity demand projected for 2000. This savings is provided entirely through technologies that are technically and economically feasible and, in many cases, already available in Brazil. Additional savings may be possible in other areas such as water heating and air conditioning.

Reducing electricity demand in 2000 by 83 TWh could eliminate the need to construct nearly 19 GW of power capacity, assuming an average capacity factor of 56% and 10% T&D losses. This is equal to about two-thirds of the new capacity that must be completed by 2000 (that is not already under construction) according to the 1985 forecast. Avoiding 19 GW of new generating capacity would save utilities from having to invest at least $38 billion (in undiscounted 1985 dollars). The required investment in greater end-use efficiency, on the other hand, is estimated to be on the order of $8 billion (4).

3. PROGRESS AND CHALLENGES

Awareness of the need to use electricity more efficiently is rising in spite of continuing growth in electrical intensity in Brazil. An increasing number of energy-efficient products are reaching the marketplace and are starting to be implemented. In addition, policies and programs to stimulate more efficient electricity use are underway. These technical and policy developments are summarized below.

3.1. Technological developments

Table 4 shows the status of various electricity conservation technologies in Brazil. Manufacturers in Brazil are producing energy-efficient fluorescent lamps, high pressures sodium lamps, energy-efficient motors, variable speed motor drives, and lighting control systems. These products were being marketed as of 1986, although sales in most cases are still limited.

Heat pump water heaters (HPWHs) are manufactured and installed to a limited extent for water heating in institutions such as restaurants, hotels, and sports clubs. Local manufacturer also are developing smaller HPWHs for household use. Prototype models were built and tested in 1986 (18).

As mentioned above, relatively efficient room air conditioners are being assembled in Brazil for export. A 300% tariff on imported rotary compressors inhibits their sale and use within the country (the tariff is waived if the air conditioners are exported). Some manufacturers have expressed an interest in producing rotary compressors within Brazil, but the required captital investment and other factors have prevented this from occurring so far (22).

A number of electricity conservation technologies for commercial buildings are under development. Prototype solid-state electronic ballasts for fluorescent lamps have been produced at the University of Sao Paulo. One private company has started making these energy-efficient ballasts on a limited basis.

Prototype reflective light fixtures have been produced in small quantities using mirrors for light reflection. However, these light fixtures are heavy and difficult to maintain. Use of aluminum film or anodized aluminum for light reflection, as is occuring in industrialized countries, is under investigation in Brazil.

3.2. Policy developments

Until recently, there has been a lack of concrete policies and programs to stimulate more efficient electricity use in Brazil. This is changing as awareness concerning the potential for cost-effective electricity conservation increases and new conservation technologies become commercially available.

The most significant policy initiative is the national electricity conservation program begun in late 1985 (24). This program, known as PROCEL, was established by the federal government and is funded mainly by the federal government. Institutionally, PROCEL is based at Eletrobras and is implemented in cooperation with other electric utilities and government agencies.

Table 5 lists the major objectives and activities of PROCEL. This comprehensive program is intended to institutionalize electricity conservaton through technology development and testing, market analysis, financial incentives, educational and training activities, regulatory measures, and institutional development. By the end of 1986, PROCEL could point to a number of accomplishments.

First, electricity consumption testing and labeling for domestic refrigerators was started. The testing facility was established in 1985-86 at the national laboratory for the electricity sector. A standardized test procedure was agreed upon and testing of single door models was completed in 1986 (14). Energy consumption labels will appear on refrigerators beginning in 1987. Also, a similar program for testing and labeling room air conditioners was announced.

Second, a national campaign to replace remaining incandescent street lights with mercury vapor and high pressure sodium lamps is underway. In the first phase begun in 1986, 210,000 lamps were replaced with financial support of approximately $5 million from PROCEL (25). This campaign will continue for a number of years.

Third, a new financing program for energy conservation projects was set up within the national development bank (26). Financing is available to industries, commercial building owners, and public utilities for electricity as well as fuel conservation projects. Loans are provided with 5-7% real interest, considerably below market interest rates.

Fourth, in the area of electricity planning, an explicit examination of conservation potential and establishment of savings goals are included in the most recent national electricity plan (known as the Plan 2010). The preliminary conservation targets are for savings of 41 GWh by 2000 and 88 GWh by 2010 (27). The latter approximately equals the conservation potential identified above in six major end uses as of 2000 (see Table 3).

In addition to the activities associated with PROCEL, some utilities are conducting their own conservation projects. The Sao Paulo utility, for example, is demonstrating and monitoring the performance of high pressure sodium lamps, electronic lamp ballasts, and reflective fixtures. The utility also conducted field tests of a prototype household heat pump water heater.

3.3. Challenges for the future

Many hurdles still need to be overcome before the overall efficiency of electricity use in Brazil improves. Relatively low electricity prices, lack of information, a shortage of capital on the part of businesses and households, and other institutional barriers inhibit greater energy efficiencies. While PROCEL is creating a number of educational, financial incentive, and other programs to overcome these barriers, much more needs to be done.

The relatively low electricity tariffs particularly for industrial customers present a strong disincentive to conservation investments. In March, 1986, tariffs were frozen at levels about 30% lower than those in 1981 (corrected for inflation). A 10% increase in industrial tariffs and a 20-40% increase in residential and commercial tariffs in November, 1986 did not fully make up this difference. Further tariff increases, while unpopular, are needed both to stimulate greater efficiency and to assist the financially-troubled electricity sector. Although official policy calls for increasing the average tariff 14% per year during 1987-89 (3), whether or not this occurs remains to be seen.

The conservation programs initiated under PROCEL need to expanded and effectively implemented. Experience in Brazil has shown that well-meaning conservation programs do not necessarily have the desired impacts. For example, a government program implemented during the early 1980s for financing investments in oil conservation and substitution on the part industries had mixed results. Poor program organization, inconsistent pricing policies, and the economic recession limited the response and program impact (28).

Certain technological gaps also need to be filled. For example, more energy-efficient air conditioners and domestic refrigerators are urgently needed. For products such as motor speed controls and heat pump water heaters, cost reductions and/or technical improvements would be valuable. The federal government may have to take a more active role in prodding equipment manufacturers in these areas.

Besides further development of specific conservation technologies and programs, some fundamental changes in perspective are required. Utilities in Brazil are accustomed to building power plants, providing power in a reliable manner, and promoting electricity use. While traditional activities such as building power plants and transmission lines must continue, end-use efficiency has to receive equal political and financial support. In essence, Brazil's energy authorities and utilities need to make a commitment to both demand-side and supply-side investments with the overall objective of providing "least cost energy services". Although some energy planners and utility officials acknowledge this, it is by no means widely accepted.

One specfic challenge relates to the structure of the electric power industry in Brazil. A few large utilities are responsible for financing and building generating plants. These utilities have the strongest incentive for reducing growth in power demand, but they are not in direct contact with consumers. A larger number of power distribution companies have direct contact with consumers, but less incentive to stimulate conservation. These utilities must become involved in PROCEL in order to bridge the gap between the generating companies and consumers. In the U.S., the Tennessee Valley Authority is in a similar situation and has been successful in implementing its conservation programs through local power distributors (29).

Yet another challenge relates to the overall direction of industrial development in Brazil. Decisions that affect the rates of expansion of different industries can have a substantial impact on electricity demand. In particular, reducing or eliminating subsidies for the aluminum and other non-ferrous metal industries could lead to lower electricity demand with negligible impact on economic growth.

4. CONCLUSION

Development of hydropower resources and increasing electricity consumption were key factors propeling economic growth in Brazil during the past twenty years. Now, however, continued rapid expansion of electricity consumption and supplies poses a severe economic strain. In addition, the negative environmental impacts of conventional power plants are receiving greater attention.

Increasing the efficiency of electricity use is one attractive alternative to "business as usual" electricity supply expansion. Vigorously pursuing conservation could greatly reduce the need for new power plants over the next 15 years. For example, the estimated savings potential in six end-use areas is 19 GW, nearly 20% of the power capacity required in 2000 according to current forecasts. Moreover, efficiency investments are cost-effective on a life-cycle basis and are much less capital intensive than expanding electricity supply.

The energy efficiency strategy involves no reduction in economic growth or energy services. In fact, by providing these services in the most cost-effective manner, economic output and standards of living can rise more rapidly.

Although the overall electrical intensity of the Brazilian economy is still increasing, technologies and policies to stimulate more efficient electricity use are moving forward. A host of energy-efficient products are already available. The federal government and utilities are proceding with a comprehensive national program involving technology development, financing, standards, incentives, education, and institutional development.

In spite of these encouraging signs, more efficient use of electricity is still a new endeavor in Brazil. A number of important steps still need to be taken including raising average tariffs, further development of more efficient equipment and conservation programs, more active involvement of utilities in promoting and investing in end-use efficiency, and removal of subsidies for highly electricity-intensive industries. If progress along these lines continues, end-use efficiency improvements could become an important "electricity source" in Brazil and serve as an example for developing countries throughout the world.

REFERENCES

1. Balanco Energetico Nacional, Ministerio das Minas and Energia, Brasilia, 1985.
2. "Relatorio Anual 1985", Eletrobras, Brasilia, 1986.
3. "Plano de Recuperacao do Setor de Energia Eletrica", Eletrobras, Brasilia, 1985.
4. H.S. Geller, "The Potential for Electricity Conservation in Brazil", CESP, Sao Paulo, 1984.
5. J. Goldemberg and R.H. Williams, "The Economics of Energy Conservation in Developing Countries", PU/CEES Report No. 189, Center for Energy and Environmental Studies, Princeton University, Princeton, NJ, 1985.
6. J. Goldemberg, "An End-use Energy Strategy of Latin America", paper presented at the OLADE Meeting of Energy Ministers, Montevideo, 1985.
7. Pesquisa de Consumo e Desempenho Energetico na Industria 1981", Conselho Nacional de Petroleo, Brasilia, 1985.
8. "Penetracao da Eletricidade Avalicacao do Potencial Area CESP/1981", Divisao de Desenvolvimento de Mercado, CESP, Sao Paulo, 1981.
9. "Classification and Evaluation of Electric Motors and Pumps", DOE/CS-0147, U.S. Dept. of Energy, Washington, DC, 1980.
10. B.L. Jones and J.E. Brown, "Electrical Variable-Speed Drives", IEE Proceedings, Vol. 131, Pt. A, No. 7, 1984.
11. "Impact of Advanced Power Semiconductor Systems on Utilities and Industry", EPRI EM-2112, Electric Power Research Institute, Palo Alto, CA, 1981.

548

12. M. Ross, E.D. Larson, R.H. Williams, "Energy Demand and Materials Flows in the Economy", Energy, the International Journal (forthcoming).
13. Anuario Estatistico do Brasil, FIBGE, Rio de Janeiro, 1982.
14. "Energy Conservation in Home Appliances", Informe, No. 20, CEPEL, Rio de Janeiro, April-June, 1985. Also, "Etiquetas para refrigeradores", Boletim No. 1, PROCEL, Eletrobras, Rio de Janeiro, 1986.
15. H.S. Geller, End-Use Electricity Conservation: Options for Developing Countries, Energy Dept. Paper No. 32, World Bank, Washington, DC, 1986.
16. H.S. Geller, Energy Efficient Appliances, American Council for an Energy-Efficient Economy, Washington, DC, 1983.
17. G.M.G. Graca and A. Barghini, "Uso de Eletricidade no Setor Residencial da Cidade de Sao Paulo", paper prepared for the Special Workshop on the Rational Use of Energy: an End-Use Oriented Energy Strategy, Sao Paulo, 1985.
18. Personal communication from Oswaldo de Siqueira Bueno, Director of Engineering, STARCO, Sao Paulo, 1986.
19. M.C. Arouca, F.B.M. Gomes, L.P. Rosa, "Estrutura da Demanda de Energia no Setor Residencial no Brasil", COPPE, Universidade Federal do Rio de Janeiro, 1983.
20. Personal communication from Isac Roizenblat, Director of Engineering, Philips do Brasil, Sao Paulo, 1985.
21. C.L. Robbins, K.C. Hunter, N. Carlisle, "Energy and Economic Efficiency Alternatives for Electric Lighting in Commercial Buildings", SERI/TR-253-2574, Solar Energy Research Institute, Golden, CO, 1985.
22. Personal communication from Jorge Marques, General Manager, Springer Carrier, Canoas, Rio Grande do Sul, Brazil, 1985.
23. "The Most Energy-Efficient Appliances", American Council for an Energy-Efficient Economy, Washington, DC, 1986.
24. "Programa Nacional de Conservacao de Energia Eletrica (PROCEL)", Ministerio das Minas e Energia e Ministerio da Industria e do Comercio, Brasilia, 1985.
25. "Substituicao de Lampadas", Boletim No. 1, PROCEL, Eletrobras, Rio de Janeiro, 1986.
26. "Criado um Programa Especial de Incentivo", Boletim No. 2, PROCEL, Eletrobras, Rio de Janeiro, 1986.
27. Resolucao de Grupo Coordenador de Conservacao de Energia Eletrica - 06/86, Diario Oficial, Secao 1, 19273, Dec. 19, 1986.
28. A. Behrens, "Brazil's Industrial Energy Conservation Programme", Energy Policy, Oct. 1985. Also, A. Behrens, "Uma Avaliacao do Programa CONSERVE/Industria", INPES/IPEA, Rio de Janeiro, April 1985.
29. "Energy Conservation Program", Division of Conservation and Energy Management, Tennessee Valley Authority, Knoxville, TN, May 1984.

Table 1

1985 ELECTRICITY DEMAND FORECAST AND CAPACITY REQUIREMENT (a)

Year	Electricity Demand (TWh)	T&D Loss (%)	Capacity Factor (%)	Capacity Required (GW)
1980	114.5	12.6	49.0	30.5
1985	165.9	11.2	51.3	41.6
1990	237.5	10.7	53.0	57.3
1995	320.1	10.3	54.5	74.7
2000	420.7	10.0	56.0	95.3

(a) The electricity demand forecast was taken from the 1985 Eletrobras annual report (Reference 2). The forecast does not include capacity or electricity consumption provided by self-producers. Capacity factor is the ratio of gross electricity production to potential production if installed capacity was operated continuously. T&D losses and capacity factors are actual values in 1980 and 1985, estimates for 1985-2000.

Table 2

ELECTRICITY TARIFFS IN BRAZIL - DECEMBER 1986 (a)

SECTOR	RATE ($/kWh)
Industrial (b)	
88-138 kV	0.023
2.3-13.8 kV	0.038
Residential	
< 30 kWh/mo	0.013
30-100 kWh/mo	0.048
100-200 kWh/mo	0.053
200-300 kWh/mo	0.083
> 300 kWh/mo	0.092
Commercial	
low voltage	0.091
Public services	
street lighting	0.018

(a) Tariffs were converted to dollars at the official exchange rate prevailing in mid-December, 1986. The tarrifs include all government taxes.

(b) Industrial tariffs include both kW and kWh charges and are based on the average load factors in each voltage class.

Table 3

ELECTRICITY DEMAND AND CONSERVATION
POTENTIAL IN THE YEAR 2000

End-use area	Current Forecast (TWh)	Savings Potential (%)	Savings Potential (TWh)
Industrial motors	164.8	20	33.0
Domestic refrigerators	24.7	60	14.8
Domestic lighting	16.5	50	8.2
Commercial motors	28.0	20	5.6
Commercial lighting	25.0	60	15.0
Street lighting	16.8	40	6.7
TOTAL	275.8	--	83.3

Table 4

STATUS OF MORE ENERGY-EFFICIENT ELECTRICAL EQUIPMENT
IN BRAZIL, DECEMBER 1986

End-use	Commercially available	Produced for export	Prototype available	Under development
Efficient residential refrigerators				x
Compact fluorescent light bulb	x			x
Heat pump water heater		x	x	
Efficient room Air conditioner		x		
Efficient fluorescent lamp tubes	x			
Electronic lamp ballast			x	x
More efficient motors	x			
Variable frequency motor drives	x			

Table 5

BROAD OBJECTIVES OF THE NATIONAL ELECTRICITY CONSERVATION
PROGRAM STARTED IN NOVEMBER, 1985

I. Technology development

   o Support development of more efficient refrigerators,
     lighting products, heat pumps, control systems, etc.
   o Demonstrate energy efficient equipment

II. Market analysis

   o Study electricity use and savings potential in the
     commercial and industrial sectors
   o Incorporate energy use-related questions in the national
     household survey
   o Evaluate savings potential and applications for variable
     speed motor drives

III. Program and policy development

   o Develop equipment test procedures and labeling programs
   o Develop minimum efficiency standards for equipment and
     buildings
   o Study and propose financing programs for investments in
     more efficient equipment by consumers and manufacturers
   o Study and propose tax incentive, rebate or other
     incentive programs
   o Develop manuals, consumer guides, and other educational
     materials
   o Develop training programs and professional courses for
     commercial and industrial energy managers

IV. Institutional development

   o Assist utilities interested in conservation R&D,
     demonstration, incentives and promotion
   o Support development of diagnostic techniques and audits
   o Investigate the possibility of standardizing voltage
     levels among utilities
   o Develop methodologies for social cost-benefit analysis
     and conduct studies where appropriate
   o Investigate and support technology transfer

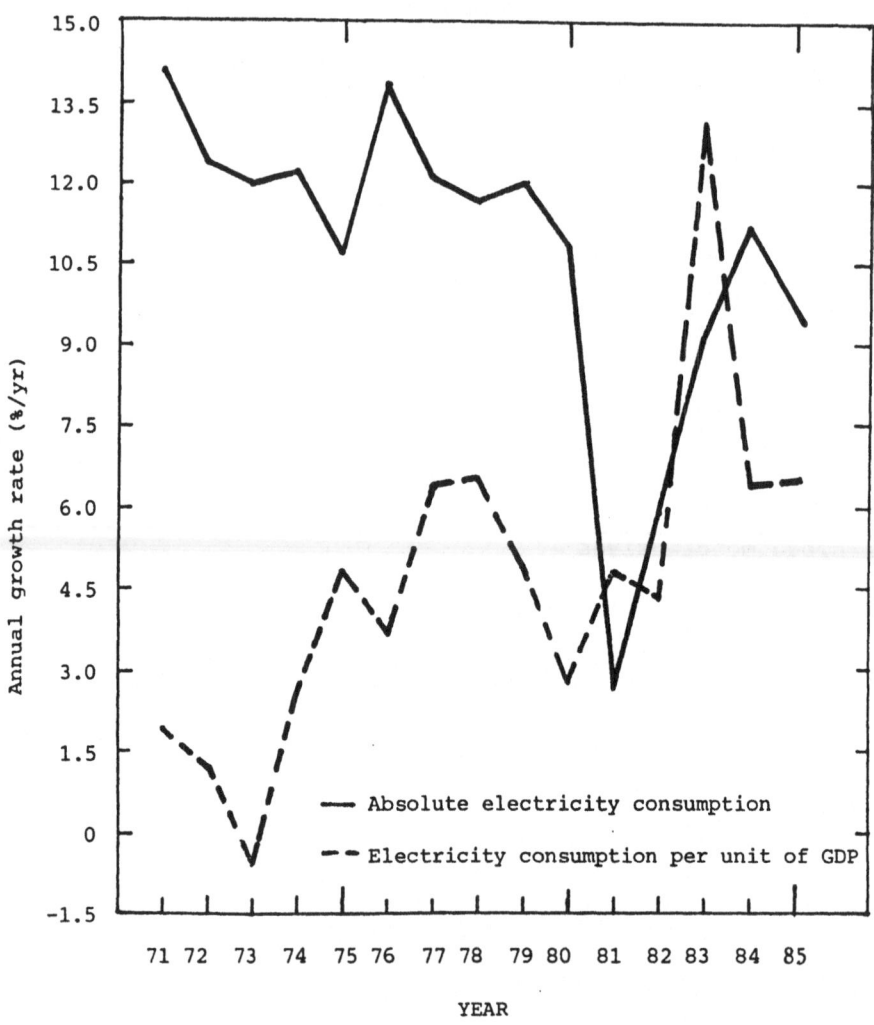

Figure 1 - Trends in Electricity Consumption in Brazil

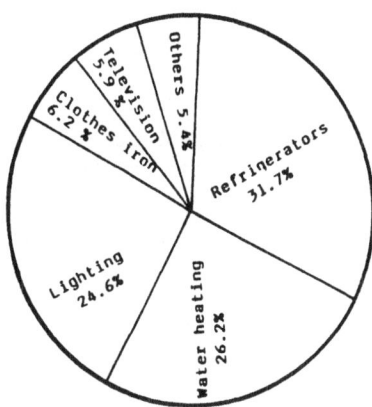

Residential - 32.9 TWh

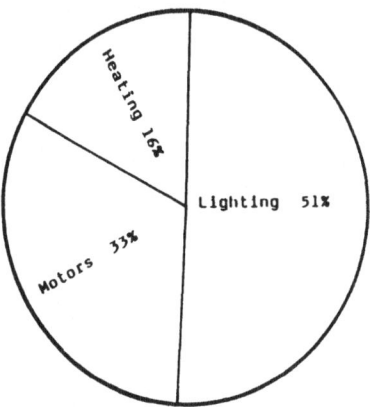

Commercial/public services - 33.5 TWh

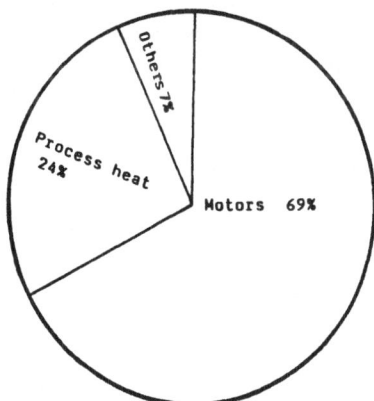

Industrial - 91.4 TWh

Figure 2 - Electricity demand in Brasil by end-use
and sectors (Totals apply to 1985)

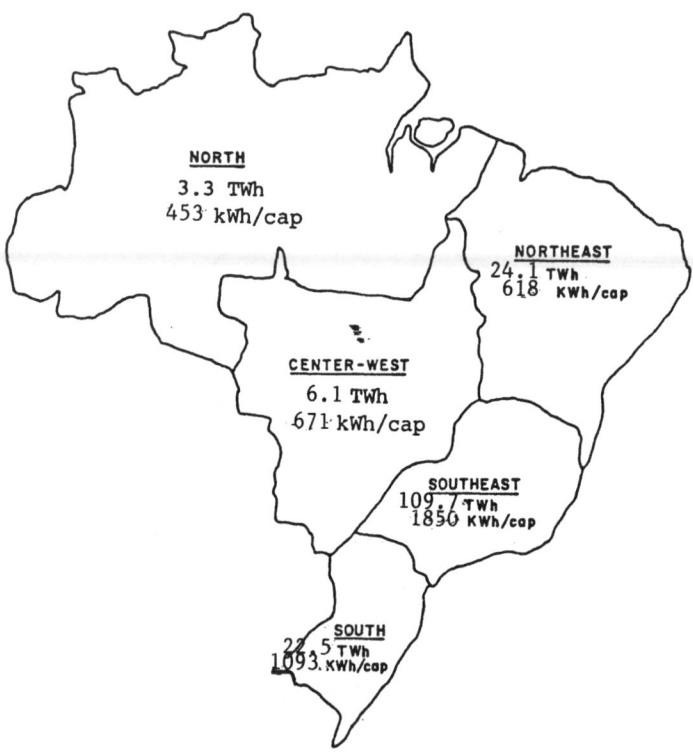

BRAZIL
165.9 TWh
1,220 KWh/cap

NORTH
3.3 TWh
453 kWh/cap

NORTHEAST
24.1 TWh
618 KWh/cap

CENTER-WEST
6.1 TWh
671 kWh/cap

SOUTHEAST
109.7 TWh
1850 KWh/cap

SOUTH
22.5 TWh
1093. KWh/cap

Figure 3 - REGIONAL ELECTRICITY CONSUMPTION IN BRAZIL IN
1985

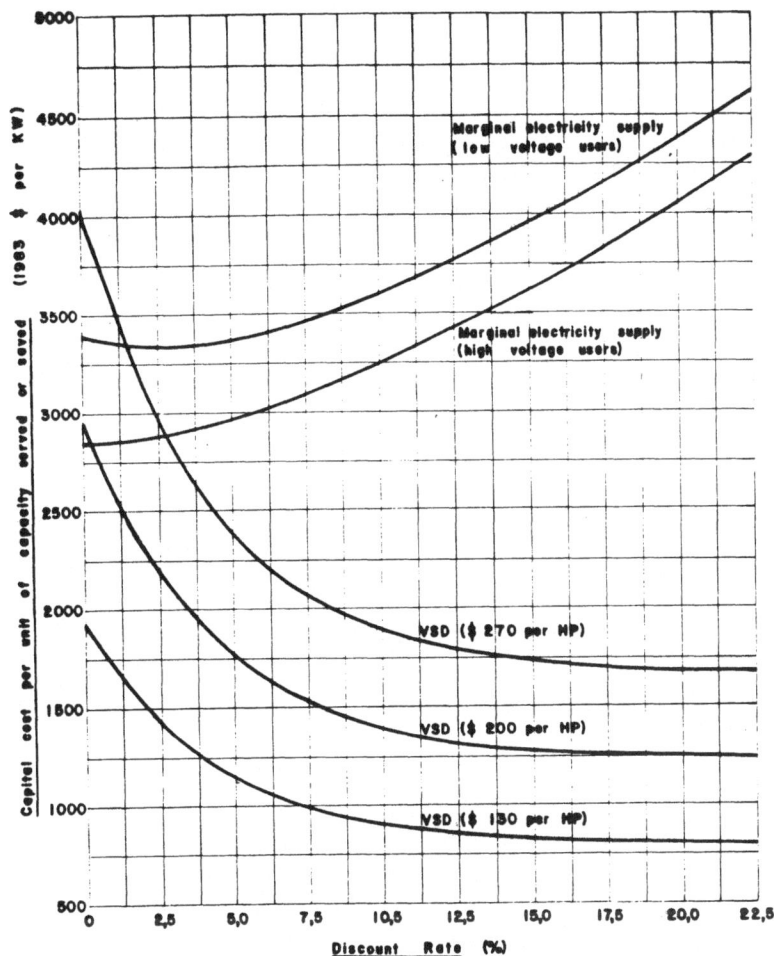

Figure 4 - <u>Comparison of the Capital Costs for Electricity</u>
<u>Supply and Demand Reduction Using Variable Speed</u>
Motor Drives

    ---- present value of investments over 50 years
    ---- 20% electricity savings assumed with 100 HP
         variable frequency drive

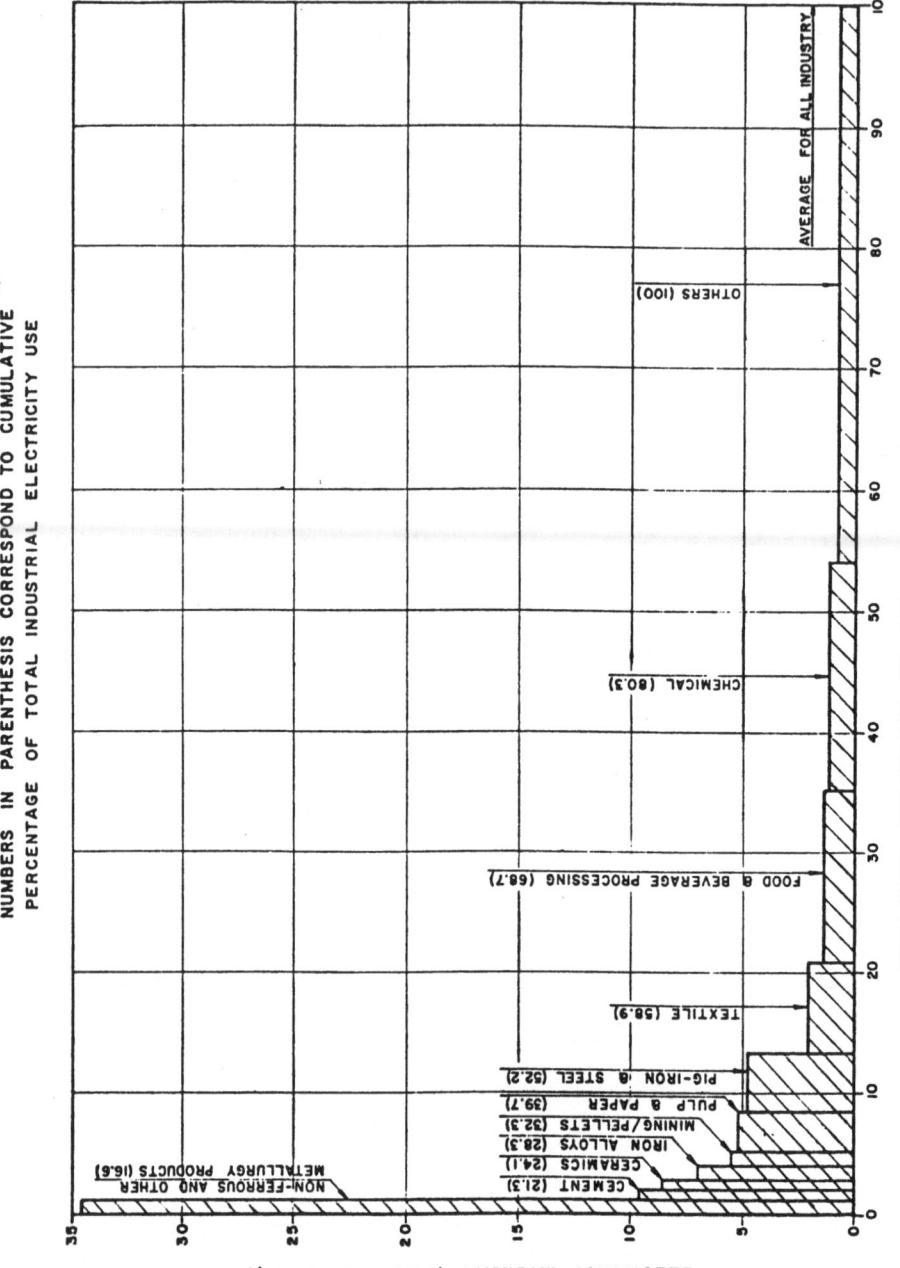

Figure 5 - The Electricity Intensity of Industries in Brazil, 1980

ELECTRICITY CONSERVATION IN DEVELOPING COUNTRIES - A CASE STUDY FOR PAKISTAN

Anibal T. De Almeida
Dep. Eng. Electrotecnica
Universidade de Coimbra

1. INTRODUCTION
1.1 Historical Perspective

Electricity consumption in Pakistan has been growing at a fast rate, at around 13% annually during the past twenty five years, reaching 17600 GWh in 1985 (1,2). During the last 10 years, the rate of growth of electricity consumption slowed to 11.2% but without any tendency for further decrease in the last five years. These high growth rates are associated with a low electricity consumption of 180 KWh per capita, a fast population growth of 3%/year, and stable economic growth with an average increase in gross domestic product (GDP) of 5.8% over the last twenty five years (2). The ratio of electricity growth over GDP growth was 2.2 for the same period, reflecting a situation common to developing countries, where industrialization, household electrification and large scale electrical irrigation are responsible for high growth in electricity intensity.

Electricity sector development is highly capital intensive, requiring large investments for the expansion of the generation, transmission and distribution facilities. This effort, with a foreign exchange component of around 40%, places a heavy burden on the national economy and competes with other important sectors (i.e., health, education, transport housing, telecommunications, agriculture, industry) for the allocation of scarce financial resources. The annual investments in the electric power sector, required over the next four years, amount to over 4% of the GDP. Electric utilities in Pakistan have not been able to keep up with the fast growth of electricity consumption. The expansion in installed capacity, presented an average growth rate of 10.6% during the last ten years. There is a widening demand-supply gap which has led to widespread load shedding in recent years. The monthly percentage of peak load shed ranges from 4% in the summer months to 20% in January. As more and more activities in society depend upon the availability of electricity, load shedding can have a serious impact in all sectors of the economy. The monthly fluctuations in load shedding are mainly due to the the seasonal variations in generation from hydroelectric power plants, which were responsible for 52% of the energy generated in 1984-85 (2).

Electricity losses (electricity generated minus electricity sold) are significant in Pakistan. Besides the technical losses in generation auxiliaries, transmission, transformation and distribution, there are relevant non-technical losses (electricity theft and faulty metering). Electricity losses in the largest utility have been brought down from 37.6% in 1976-77 to 26.7% in 1984-85. The latter value includes 15.2% in distribution, 9.4% in transmission, and transformation and 2.1% in auxiliaries.

The province of Sind, which includes the Karachi industrial area, presents the highest per capita consumption of 256 KWh. This is a modest amount when compared with developed countries where the per capita consump-

A. T. De Almeida and A. H. Rosenfeld (eds.), Demand-Side Management and Electricity End-Use Efficiency, 559–584.
© *1988 by Kluwer Academic Publishers.*

TABLE 1.1

DEMAND AND ENERGY FORECAST
(ACCELERATED RURAL ELECTRIFICATION PROGRAMME)

YEAR	DEMAND (MW)	ENERGY GENERATED (GWH)	LOSSES %
1986–87	5361	29446	25
1987–88	5920	32516	24
1988–89	6531	35873	23
1989–90	7205	39578	22.5
1990–91	7909	43347	22
1991–92	8625	47380	21.5
1992–93	9404	51658	21
1993–94	10285	56504	21
1994–95	11258	61845	21
1995–96	12318	69672	21
1996–97	13477	74043	21
1997–98	14747	81021	21
1998–99	15903	87373	21
1999–2000	17150	94227	21
2000–2001	18439	101609	21
2001–2002	19943	109580	21
2002–2003	21507	118173	21
2003–2004	22978	126262	21
2004–2005	24551	134904	21
2005–2006	26232	144143	21

Source: (1,2)
Notes: Average growth rate of demand and energy generated are 9.7%
and 10.3% respectively for the period 1984-85 — 1999-2000

tion is between twenty to forty times that value. The rural province of Baluchistan shows the lowest per capita value with a meager 100 KWh.

1.2. Future Trends

Due to the factors already mentioned, it is inevitable that the electricity sector continues to expand at a vigorous rate. Additionally, the Government of Pakistan has set a target of electrifying 90% of the villages by January, 1990 (2). This target combined with plans to eliminate load shedding by January, 1990 imply an investment in the electricity sector of 116 billion Ruppees * of which 42 billion are foreign exchange requirements. On an annual basis, the foreign exchange requirements of about 10.5 billion Ruppees per year represent about 20% of export earnings in 1985-86.

Table 1.1 presents the forecasts for electricity supply and demand during the next two decades. The average growth rate for electricity generated is 10.3% and for peak demand is 9.7%. These forecasts mean that by the year 2000, electricity supply will be almost four times the value reached in 1985.

Besides the economic and financial effort associated with the expansion of the electricity sector, there is additionally an increasing percentage of electricity generation obtained by burning valuable fossil fuels, part of which are imported placing an additional burden on the economy. Energy conservation is recognized as one of the least expensive and fastest ways to reduce the expansion of electricity supply (3). Because of the reduced dependence on foreign resources, electricity conservation can also improve national self-sufficiency. The reduction of the supply requirement is achieved not through a restriction in economic growth, but through enhanced energy productivity. This means the production of the same level of goods, services and amenities with reduced energy requirements.

Figure 1.1 shows the sectorial distribution of electricity consumption in Pakistan during the last five years.

The industrial sector is the largest consumer, 36 % of the total, with a growth rate of 8.8%. The commercial sector in Pakistan is defined in a restrictive way including only private non-residential buildings. In this study, the commercial sector is defined in a broader way which includes both private and governmental non-residential buildings, as well as street lighting. This combined sector ("Commercial" plus "Other/Government" in Figure 1.1) is basically the services sector. It is responsible for 19.% of electricity consumption, with an average growth rate of 14% during the last 5 years. The sector "Other/Government" also includes the electric railways, which account for only 2% of electricity consumption in that sector.

Due to government policy, electricity rates are quite different from sector to sector, with the industrial and commercial sectors subsidizing the domestic and agricultural sectors. During 1984-85, the main utility charged the following average prices (1,2):
- Industrial Sector   0.635 Ruppees/Kwh
- Commercial Sector   1.077   "
- Residential sector 0.409   "
- Agricultural sector 0.353   "

The prices shown for the industrial sector include average demand charges. With the exception of the commercial sector, the above prices are substantially lower than the long-term marginal cost (LTMC) of delivered electricity. In recent studies, the LTMC has been estimated to equal 1.1

---

* 1 US Dollar = 16.8 Ruppees, in August 1986

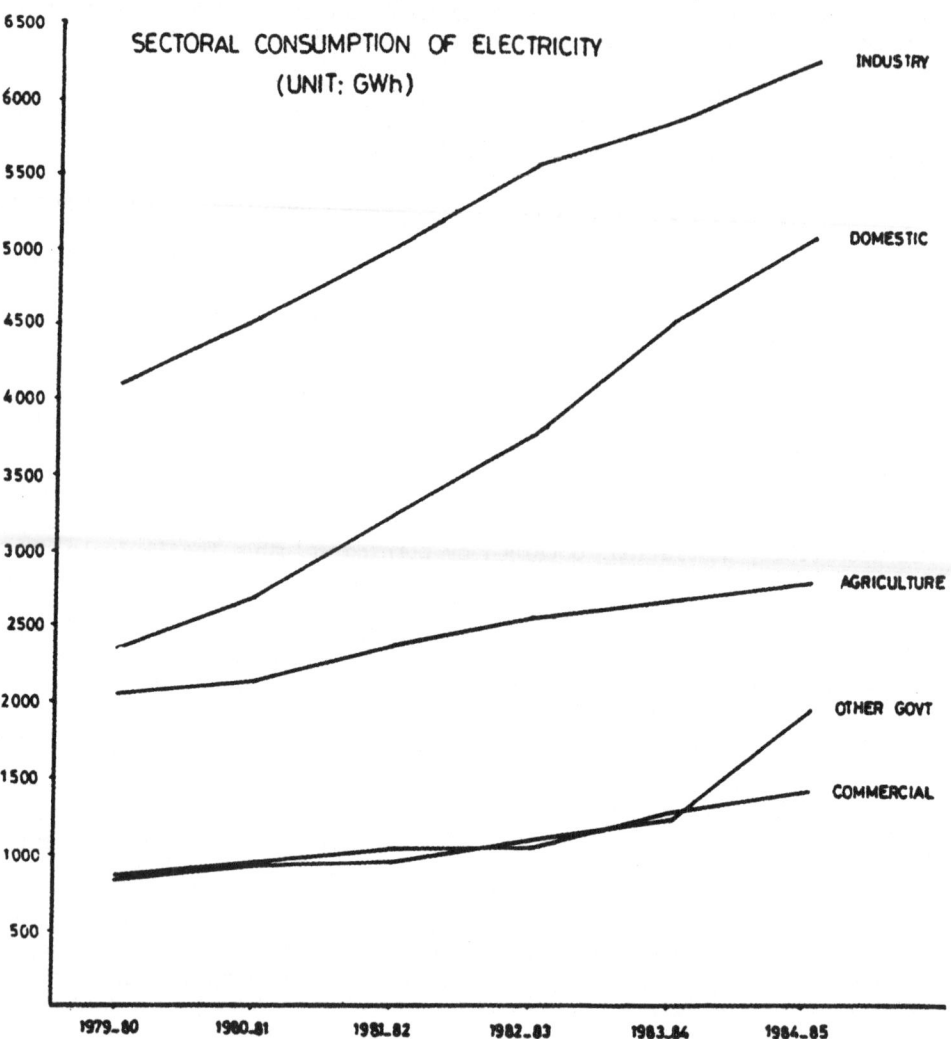

FIGURE 1.1
Source: (2)

Rs./KWh and 1.2 Rs./KWH (1). These prices are typical of projects under consideration by the utilities. Prices based on the LTMC of energy are advised by the World Bank for healthy development of electric utilities (4). LTMC-based prices of electricity send consumers the appropriate signal for choosing energy-efficient equipment and cut wastage, besides increasing the self-financing capabilities of utilities.

There are also demand charges for industrial consumers, in the range of 24-88 Rs./KW/MONTH. These charges are well below the LTMC of power capacity. A study carried out in 1981 (4) estimated the following costs for marginal capacity (including generation, transmission and distribution equipment) :
- At high voltage ....230 Ruppees/KW/month
- At medium voltage................280   "
- At low voltage...................440   "

These values are based on a generation capital cost of 10500 Ruppees/KW. The higher costs at lower voltage levels are due not only to the additional investments required for transmission & distribution, but also to increased losses. Present values of marginal capacity cost should be even higher, at least 40%, to reflect inflation and currency devaluation.

New base load power plants, both thermal and hydroelectric, can cost over 15000 Ruppees/KW (1,2). The biggest power plant under consideration, the Kalabagh dam with 1800MW in the first phase to be completed in the early 1990s, requires an estimated investment of 44 billion Ruppees at 1985 prices. This is equivalent to over 24000 Ruppees/kW.

Taking into account the projected losses of 21% (Table 1.1), a capacity factor of 80% and a reserve margin of 10% , the value of a kW saved at the point of consumption can be worth 1.8 kW of saved base generating capacity. Additionally 1kWh saved at the point of consumption is equivalent avoiding the generation of 1.27 kWh. Considering all the investments required (generation, transmission and distribution), a kW of firm power supplied to the consumer can cost in the range of Rs 40000-50000, depending upon the type of base generation considered (thermal or hydro).

1.3. Purpose of this Study

This study was carried out to investigate the potential electricity conservation in the industrial and commercial sectors in Pakistan. The most promising technologies for electricity conservation were scanned to investigate their suitability in Pakistan. Cost-effectiveness analysis was carried out and the approximate electricity savings potential for the industrial and commercial sectors were estimated.

There is a substantial amount of published data about the electricity supply sector in Pakistan. However there is a scarcity of information relating to the demand side, namely on the following aspects:
-How electricity consumption is distributed by major end-use in each economic sector.
-Percent penetration of different technologies for each major end-use
-Growth rates of major end-use equipment
-Pattern of use of major end-use equipment
-Technical characteristics, with particular emphasis on energy performance and prices, of different types of equipment

The data collection was designed and performed (2) to obtain enough information on the above topics, to allow the investigation of the electricity savings potential. As more detailed data are collected and data bases are built, the estimates provided in this study can be updated.

2. POTENTIAL FOR ELECTRICITY CONSERVATION IN INDUSTRY

A limited survey carried out for this study (2) and other surveys show that motors account for more than 80% of the total industrial load, with lighting normally under 5% and other loads (furnaces, electrolytic processes, air conditioning, etc.) accounting for the remainder. Based on the above sources, the following end-use distribution is utilized:

-Electric motors - 80%

-Lighting - 4%

-Other - 16%

2.1. Electric Motors

2.1.1. <u>Background</u>. The industries which make electricity end-use equipment in Pakistan are either relatively large units with well trained staff or small ("garage") industries normally employing unskilled workers. Large manufacturers produced in 1985 around 22000 motors with a total 230000 HP(2). Motors in the range 6-50 hp account for over 70% of the total horsepower introduced each year.

There are also small manufacturers, mainly the "garage" industries, making an estimated 25000 motors per year in the range below 10 HP. These motors add another 80000 HP per year to the motor market in addition to the 230000 HP produced by large manufacturers. They are built with low quality materials (mild steel and sometimes impure copper) presenting poor efficiency and power factor. These small industries, mainly located in the Karachi and Gujranwala districts, sell their motors 50-60% below the price of the major manufacturers and supply motors to tubewells and other "garage" industries (50-70% of their sales).

Chinese motors are imported on the order of 8000 units per year, adding 120000 HP to the market mainly in the range below 30 HP. Chinese motors are low cost but have similar characteristics (power factor and efficiency) to the motors made by the large domestic manufacturers. A substantial part of the motors produced in the range 10-30 HP are used for agricultural purposes such as tubewells.

Table 2.1 presents a comparison of motor efficiencies at full load for the four main manufacturers and motors from the U.S.. It can be seen that the average performance of Pakistan motors is similar to standard motors in the U.S.. The motors made in Pakistan by the major manufacturers use high quality imported silicon steel.

Table 2.2 presents the results of a survey regarding rewind costs and a comparison with costs of new motors, showing an average ratio of 1 to 3. This is due to savings in materials and also due to the reduced labor costs in Pakistan. Some rewinding methods, which can degrade motor efficiency, are widely used in Pakistan (2). Normally the stator is heated at a high temperature for stripping the windings. Heating the motor above 650F can destroy the interlamination resistance and thus increase the magnetic losses.

2.1.2. <u>Energy-Efficient Motors</u> Energy-efficient motors (EEM), available in the U.S. and other countries, use an improved design as well as larger magnetic and electrical circuits for lower losses. It can be seen from Table 2.1 that if energy-efficient motors were used, there would be an average weighted efficiency improvement of around 4%. Motor manufacturers in Pakistan are receptive to making energy-efficient motors provided there is a market. The price premium over standard motors would be around 25%.

Assuming that there is a 4% average efficiency improvement potential when using EEMs in industry, demand growth can be slowed over time with

TABLE 2.1

COMPARISON OF EFFICIENCIES BETWEEN MOTORS MADE IN PAKISTAN AND IN USA AT FULL LOAD

	MOTOR HP RANGE				
	1-5HP	6-20HP	24-50HP	51-125HP	Above 125HP
Manuf. A	85%	89%	92%	93%	95%
Manuf. B	83%	88%	91%	91%	91%
Manuf. C	83.8%	85.6%	88.9%	90%	-
Manuf. D	82%	88%	89%	-	-
Average Pak.	83.5%	87.6%	90.2%	91.3%	93%
High-eff. (USA)	88.5%	93%	95%	95.4%	96.2%
Standard (USA)	83%	87.5%	90.4%	91.8%	93.3%

Source: Manufacturers A,B,C and D and IEEE. In each range the value of the highest rating available was taken.

TABLE 2.2

COMPARISON OF EQUIPMENT COSTS FOR NEW AND REWOUND MOTORS

MOTOR HP RANGE	NEW MOTOR RS/HP	REWOUND MOTOR RS/HP
<1 HP	1200	400
1-5 HP	700	250
6-20 HP	450	150
21-50 HP	300	125
51-125	350	125
>125	400	125

Source: (2)

the introduction of more efficient models. In the base case (growth at the present average rate of 8.75%), industrial electricity use will increase by 435% between 1985 and 2005. Approximately 4500 MW of additional base generating capacity would be required by the year 2005 to supply around 22000 GWh of new electricity demand from motors. This compares with 17600 GWh of total consumption in Pakistan in 1985.

In the more efficient case, it is assumed that existing standard motors are converted to more efficient models over a twelve years period (1988-2000). When new motors are purchased, EEMs are installed. When a motor needs to be replaced, the old unit is scrapped instead of being rebuilt and a new EEM is installed. The approximate savings potential in the year 2005 is 1060 GWh of electricity consumption and 220 MW of base generating capacity. This is a conservative estimate as the improvements which can be made by replacing motors made by "garage" manufacturers are substantially higher.

If it is assumed that EEMs cost 25% more than standard motors, for a new application, a 50 HP EEM offers a rate of return in the range 55-220% for 2000-8000 hours of annual use, with present electricity prices (Rs 0.63/kWh). At an average 4000 hours of annual usage, and even with current low electricity prices, the rate of return is an impressive 112%.

In the case of existing motors, there is the option of rebuilding a failed motor or buying an EEM. Considering the 25% cost premium of EEMs over standard motors and the costs presented in Table 2.1, investing in EEMs can cost three to four times as much as a motor rewinding. It should be emphasized that the efficiency gain in this case may be greater than 4%, due to the possible degradation of the silicon steel during rebuilding.

Although the additional investment required to use an EEM is now much higher, the rate of return is still in the range of 15-60% for the same 50 HP motor. For an average 4000 hours of annual usage at the present electricity price, the rate of return is 31%.

EEMs are not only cost-effective from the consumer point of view, but also from the utility point of view. The utility can either make new investments to increase available supply (construction of new power plants) or make investments in conservation technologies to reduce demand. It was already mentioned that the widespread use of EEMs could save 220 MW of base generating capacity by the year 2005. Assuming 4000 hours of annual operation, a lifetime of 40000 hours, and a real interest rate of 8%, the investment required for non-diversified demand is only RS 3200/kW in the case of new motors and Rs 10700/kW in the case of replacing old motors. The cost of avoiding 1 kW of firm power is Rs 7000/kW and Rs 23400/kW for old and new motors respectively. These values compare favorably with the cost of supplying firm power to consumers, which is in the range of Rs 40000-50000/kW.

2.2. Speed Controls

Conventional methods of controlling liquid or gas flow, use flow restriction devices such as throttling valves, inlet vanes and outlet dampers. Flow control is necessary either due to process requirements or due to oversizing. Although these methods are simple and require a small initial investment, they are very inefficient, as the power grows with the cube of flow.

A limited survey was carried out for this study to find out the penetration of pumps, fans and blowers in a selected sample of industries (2). Those types of loads were responsible for around 55% of the motor load. The results are similar to the ones obtained in the USA where those loads represent more than 50% of total motor load. The typical savings potential associated with variable speed motor controls is in the range of 25-30%, when applied to pumps, fans, blowers and compressors (5). In Pakistan, full utilization of variable speed motor controls would therefore save about 16% of the total electricity consumption of industrial motors.

The potential impact of variable speed technologies on energy and power requirements is substantial. Their introduction can save over 4000 GWh/year of electricity demand in 2005, which is two-thirds of total industrial consumption in 1985. Additionally 840 MW of base generating capacity would be avoided.

Two technologies, multi-speed motors and adjustable speed drives (ASDs), are particularly attractive for adjusting the speed of industrial motors.

2.2.1. Multi-Speed Motors. Multi-speed motors can operate at different speeds by changing the number of poles, which is achieved by rearranging the windings connections to produce a different number of poles (6). This technology is suitable in situations where the number of different flow rates is limited (e.g.,two or three). Multi-speed motors can be made in Pakistan with a single or double winding with an extra cost of 65% and 95% respectively (2). Single winding pole-amplitude-modulated (PAM) motors are particularly attractive because of being cheaper and of having the same frame size of standard motors of the same horsepower. To change the speed, multi-speed motors require a multipole breaker which adds 10-15% to the motor cost.

In a new application, the use of a multi-speed motor instead of a single speed motor, can eliminate the need and the cost of the flow restriction device (valves,vanes or dampers), and thus reduce its extra cost. Assuming typical energy savings of 28%, the rate of return of using a 50 HP multi-speed motor instead of a single speed 50 HP motor, is in the range 60-500%, for 1000-8000 hours of operation. For a typical 4000 hours of operation and with the 1985 electricity price, the rate of return is 250%

An existing single speed motor can also be replaced by a multi-speed motor if the application warrants this. Assuming 28% savings for a 50 HP motor, the rate of return is in the range of 25-200%, not considering the resale value of the existing motor. For the typical 4000 hours of motor operation, the rate of return is about 100%.

2.2.2. Electronic Adjustable Speed Drives (ASDs)

Electronic adjustable speed drives (ASDs) have the possibility of changing continuously the frequency (and thus the speed of an induction or synchronous motor) in a typical ratio of 50/1. Imported ASD units have been installed mainly to the cement and paper industry in applications where precise speed control is necessary.

The unit price of ASDs is strongly dependent upon the horsepower. This is due to the fact that the control electronics are basically the same in small and large units, the main difference being in the power stage. Table 2.3 shows the unit prices of imported ASDs sold in Pakistan.

During recent years, due to advances in electronics and increasing market size, ASDs have been decreasing steadily in price, presenting at the same time increasing reliability and compactness. Due to their compactness and freedom of positioning, ASDs can be easily applied in new installations or as a retrofit.

The cost-effectiveness of ASDs is dependent on the horsepower of the motor(Table 2.3). Figure 2.1 shows the rates of return for 10, 50 and 125

HP ASD drives, as a function of the motor annual usage, and considering 1985 and marginal prices of electricity. Again 28% average energy savings are assumed and the possible cost savings associated with avoiding the mechanical flow control devices are not considered.

For the 10 HP drive the rate of return is in the range of 2-43% with a payback of 2.3-16 years. Small ASDs drives are not economical, except if marginal electricity prices are considered and the annual motor use is over 6000 hours.

For the 50 HP drive , the rate of return is 7-65% with a payback of 1.5-11 years. In this case ASDs are attractive in applications with annual use above 3500 hours if LTMC prices are considered. With current electricity prices, only annual use above 6500 hours can justify the investment.

Finally for the 125 HP drive, the rate of return is 18-130% with a payback of 0.8-5 years. ASD applications in drives of 100 HP can be attractive for annual use over 2000 hours with LTMC prices, whereas with present prices their use is only economical above 3500 hours.

Although ASDs present a large potential for electricity savings, present prices do not make this investment attractive in some applications. As ASD prices move downwards, the economic thresholds of annual motor use will also decrease accordingly. The attractiveness of ASDs can be improved by the following actions:
- Eliminating the present tax/duties on ASDs, which are in the range of 20-30%
- Local manufacture of ASDs; at least one of the motor manufacturers has the technology and is willing to produce ASDs locally if there is a suitable market.
-Giving financial incentives to customers to partially offset the high initial cost of ASDs.

The last option is pertinent when the potential power savings are considered. It was already mentioned that speed control technologies can displace 840 MW of base generation capacity. If the cost of conservation technologies is considered against the power savings provided, the cost of non-diversified power capacity is the following:
-Rs 36000/kW for a 10 HP ASD
-Rs 23000/kW for a 50 HP ASD
-Rs 12000/kW for a 125 HP ASD

The costs of avoiding firm power are Rs 80000/kW, Rs 50000/kW, and Rs 27000/kW for the above drives, for a typical 4000 hours application.

2.3. Lighting

Lighting represents a small fraction (2-5%) of the total load in a sample of selected industries (2). In some industries, such as textile and packaging, its contribution to the total load can go up to 15%.In accordance with the values obtained in the limited survey, lighting is assumed to represent 4% of the industrial electricity consumption.

The survey carried out in conjunction with this study also shows that fluorescent is the dominant type of lighting with 85% of the total, incandescent and mercury vapor types being responsible respectively for 11% and 4 % of the industrial lighting load.

Although the potential for savings is much less than with industrial motors, the following actions seem worthwhile:
-Replacement of existing fluorescent lamps and ballasts by high efficiency equipment of the same type. The potential savings is over 40%.
-Replacement of incandescent lamps by compact fluorescents or by

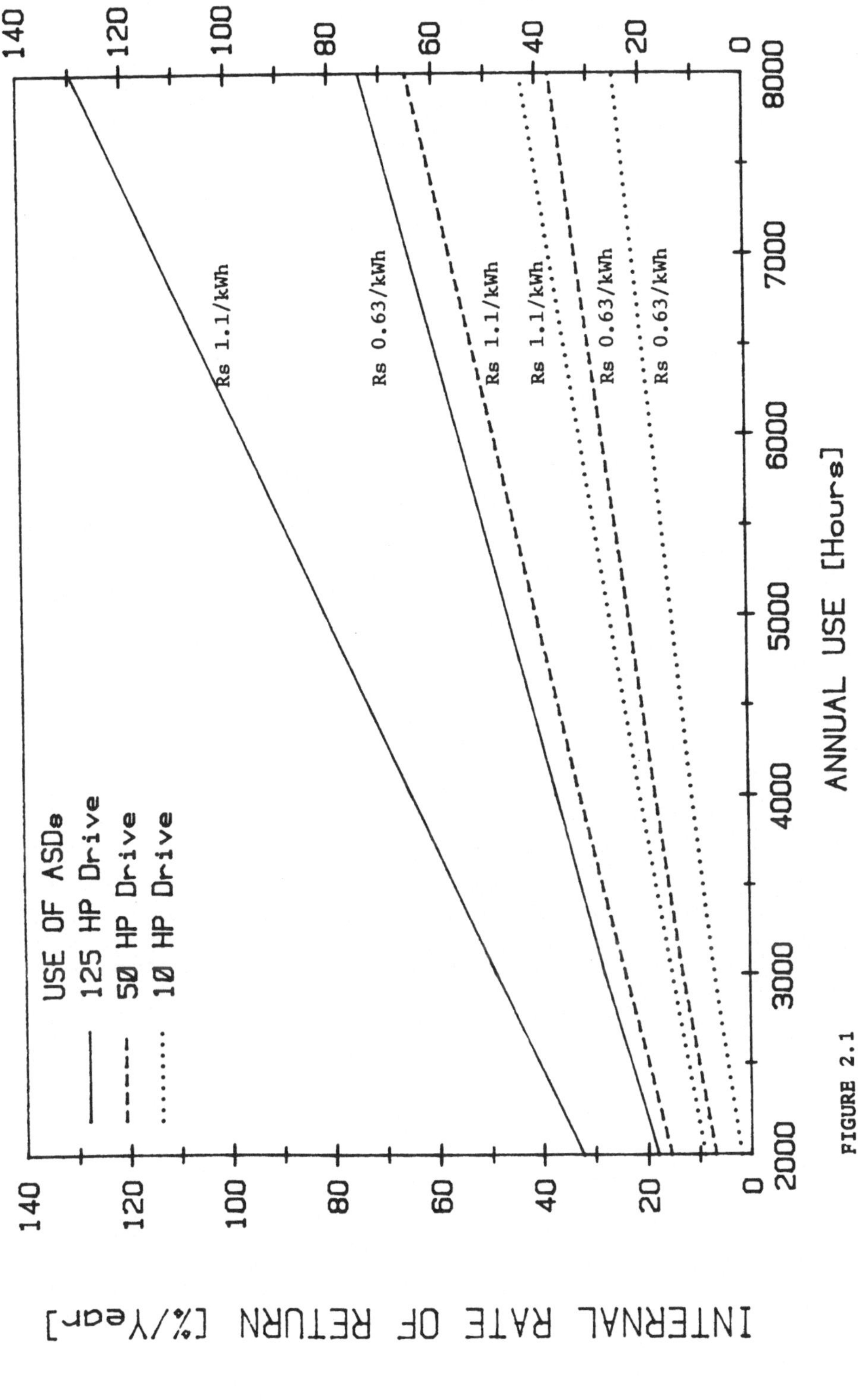

FIGURE 2.1

The figure shows curves plotting INTERNAL RATE OF RETURN [%/Year] versus ANNUAL USE [Hours].

Legend: USE OF ASDs
— 125 HP Drive
- - - 50 HP Drive
······· 10 HP Drive

Curve labels: Rs 1.1/kWh, Rs 0.63/kWh, Rs 1.1/kWh, Rs 1.1/kWh, Rs 0.63/kWh, Rs 0.63/kWh

fluorescent tubes if the former is not available. The potential savings are around 75% of the incandescent lighting load.

-Replacement of high high pressure mercury vapor lamps by high pressure sodium vapor lamps in places where color rendering is not important and with high ceilings (over 6m). The potential for savings are 40% of the mercury vapor lamp load.

The potential impact of the above measures is modest, leading to global savings of 1.6% of the total industrial load. (The impact of the three actions is 1.2%, 0.34% and 0.07% respectively). By the year 2005, the potential industrial lighting improvements represent 535 GWh of electricity savings and 110 MW of avoided base generating capacity.

2.4. Load Management

In Pakistan, the customers with sanctioned loads above 500 kW have a non-diversified demand close to 900 MW (2). With the billing demand charges of 85 Rs./KW, consumers with a peak demand of 200 KW or even less may find economical the purchase of a maximum demand controller if they can shift 10% or more of their peak load to non-peak periods. In general, all non-essential loads (heating and cooling loads, compressors, mills, battery charging) can be transferred to other periods to limit the maximum demand.

In developed countries, most utilities offer the medium and large consumers time-of-use (TOU) rates for energy and peak demand to reflect the different costs of generation and power. Pakistan utilities have not implemented TOU rates, but are receptive to this concept. The introduction of TOU rates will make load management much more attractive as transferring loads outside the peak period can lead to savings in demand charges and also in energy charges. Typically, load management equipment can reduce the peak load between 10 and 20% without disruption of the production. A recent audit of the largest industrial consumer in the Lahore region showed a potential peak reduction of over 20% (2). The prices of demand controllers sold in Pakistan range from Rs. 30000 for simple units, to Rs. 300000 for higher performance units. These latter units, besides being able to monitor a large number of loads and submetering circuits, can perform other functions besides demand limiting, such as load scheduling, duty cycling, optimal start-stop and data logging.

Without TOU rates, the cost of LM equipment has to be recovered through the reduction in the demand charges. For a consumer with 500 kW peak load, having 5-20% potential for peak reduction, the use of an inexpensive demand controller (Rs 30000) can lead to a rate of return of 85-340% .

If an average 15 % peak demand reduction is assumed for consumers above 500 kW, a potential reduction of 130 MW can be achieved nationwide. This value is non-diversified peak at the consumer point in 1985. The projected savings in the year 2005, if the demand grows at the rate forecasted in Table 1.1, can reach 830 MW of non-diversified peak demand. At the generation point, because of the losses and capacity factor, the savings are even larger.

2.5. Power Factor Correction

The average power factor in industry is only 0.75 and in the main utility network is only 0.8 (2). Maintaining a higher overall power factor would reduce line losses and lead to better utilization of power supply facilities. An improvement of 0.75 to 0.95 in the power factor can lead to a 38% decrease in the losses, that is 3.4% of the electrical energy supplied to the industry. If the correction is made from 0.75 to 0.85 the decrease in the losses is only 22%.

TABLE 2.3

PRICES OF ELECTRONIC VARIABLE SPEED DRIVES
IMPORTED BY MANUFACTURER A

Horsepower Range	Price
UP TO 15 HP	RS. 5000/HP
15 - 100 HP	RS. 3000/HP
ABOVE 100 HP	RS. 1500/HP

Source: (2)

TABLE 2.4

POTENTIAL ELECTRICITY SAVINGS IN INDUSTRY IN THE YEAR 2005

	Power Savings Base/Peak (MW)	Energy Savings at Consumption (GWh)	Energy Savings at Generation (GWh) *
E. E. Motors	220	1060	1340
Speed Controls	840	4100	5190
Lighting	110	530	670
Load Management	830 **	----	----
Power Factor	230	1140	1440
TOTAL	1400/830	6830	8640

Notes: - * Includes 21% losses
        - ** Non-diversified peak at consumer point

Consumers with sanctioned loads above 70 kW are subject to a penalty if the average power factor is below 0.85. In the event of a power factor below 0.85, the consumer pays a penalty of one percent increase in the fixed charges for every one percent decrease in the power factor below 0.85.

Power factor compensation costs in the range of 150-300 Rs./KVAR, depending upon the voltage level, leading normally to payback times of less than a year. Although power factor penalty is only charged for values below 0.85, a new pricing policy is being studied for consumers willing to upgrade their power factor above 0.85, up to 1. Additionally, the price penalty paid by consumers with a power factor below 0.85 would increase with the square of the difference between 0.85 and the actual power factor. This pricing policy is already being used in other countries with large transmission and distribution losses.

In the year 2005, power factor correction to 0.95 in the industry can lead to 1440 GWh of electricity savings. Also 230 MW of base generating capacity can be avoided. The cost of avoiding non-diversified power, through the investment in power factor correction from 0.75 to 0.95 is only Rs 2400-4800/kW. This fact coupled with the short payback to consumers, makes power factor correction extremely attractive.

2.6. Aggregate Savings in Industry

Table 2.4 presents the summary of the potential for savings in industry. The potential electricity savings in the year 2005 could reach 6800 GWh, which is greater than present industrial electricity consumption. Additionally, over 1200 MW of base generating capacity and over 800 MW of non-diversified peak demand could be avoided.

Figure 2.2 shows the evolution during 1985-2005 of industrial electricity consumption in the base case and with the conservation alternative. Total electricity savings possible by 2005 amount to around 21% of the projected base case projection.

Speed controls present the highest potential both for electricity and power savings, although their cost-effectiveness may not be attractive yet in some applications.

Load management presents a high potential for peak demand reduction and is cost-effective for medium and big consumers (above a few hundred kWs).

Other technologies have more modest potentials, although they are worth implementing. Energy-efficient motors are attractive, specially in new applications. Power factor improvement and the use of higher efficiency lighting technologies supply additional cost-effective savings.

3. POTENTIAL FOR ELECTRICITY CONSERVATION IN THE COMMERCIAL SECTOR

With the available data, the following end-use distribution was estimated for 1985:
- Air conditioning.............................15%
- Lighting 70%
- Fans and evaporative coolers.................5%
- Other..... ....10%

For further analysis, it is assumed that the rate of growth of electricity consumption in the commercial sector will come down from the present 14% to 12% to be in accordance with the general projections made by the electric utilities in Table 1.1. Although the overall rate of growth for the commercial sector is assumed to be 12%, due to the faster growth of

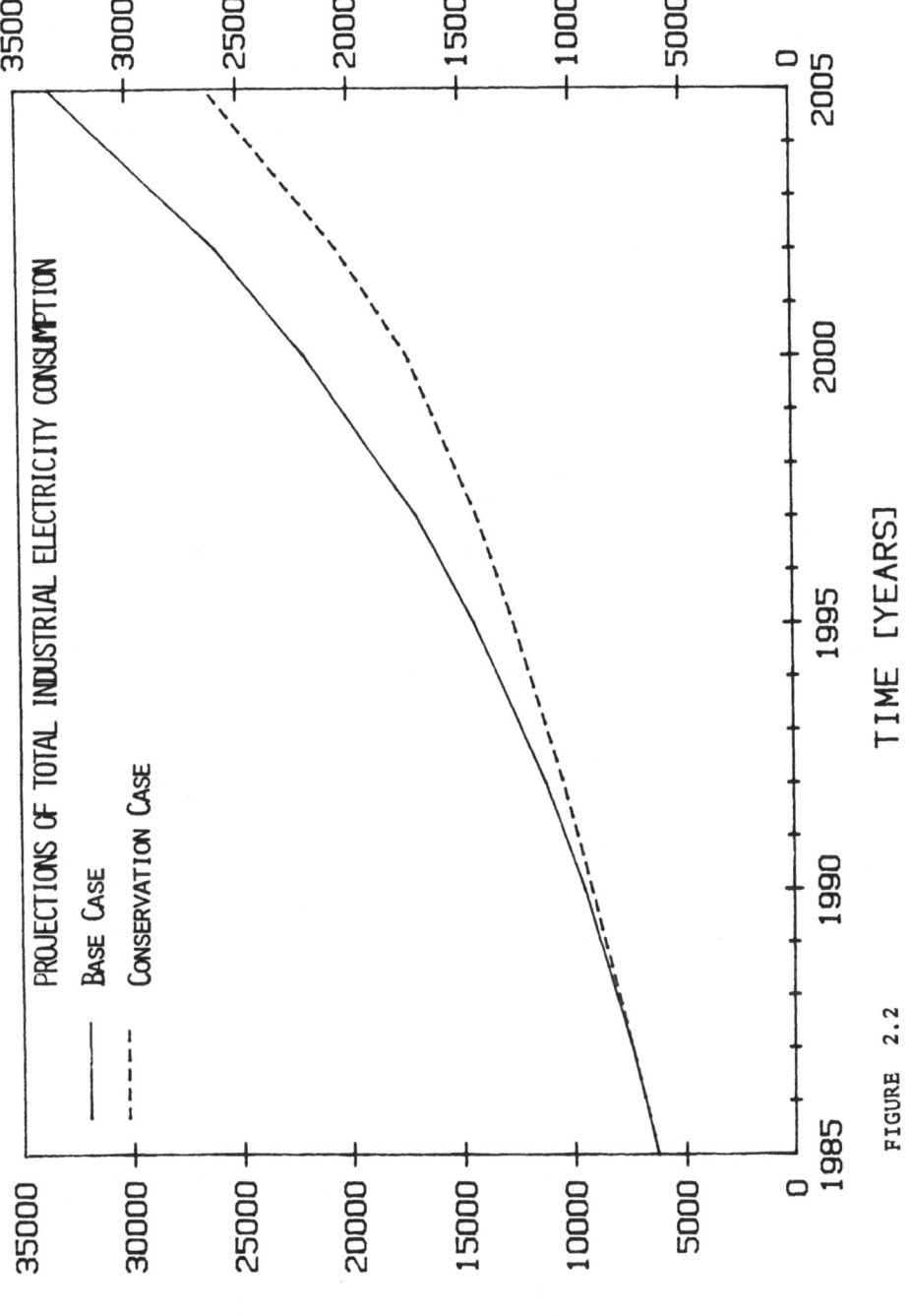

FIGURE 2.2

air conditioning the following distribution is assumed for 2005 :
- Air conditioning............30% (growth rate of 16%/year)
- Lighting....................55% (growth rate of 10.6%/year)
- Fans.......................5% (growth rate of 12%/year)
- Other.....................10% (growth rate of 12%/year)

The rates in the commercial sector are twice the average rates and are based on policy choices. However, these rates cannot be considered too high as they are close to the estimates of the long term marginal cost (LTMC) of energy, Rs.1.1- 1.2/KWh (2). If the fuel adjustment surcharge is included, the range of electricity prices in the commercial sector is Rs 1.25-1.38/kWh. In the following analysis the intermediate value of Rs 1.3/kWh is used.

3.1. Lighting
3.1.1. Background The lamp distribution in commercial buildings was estimated to be the following (2):
- Fluorescent...........................80%
- Incandescent..........................20%

Table 3.1 shows the power, efficacy, lifetime and price of lamps made and sold in Pakistan. Table 3.2 shows the characteristics and CIF prices of high-efficiency lamps whose import is banned in Pakistan. Both thin fluorescents and compact fluorescents can be applied in commercial buildings in new applications or as an inexpensive retrofit. Importation of these lamps is banned at the present time in order to protect domestic manufacturers of conventional lighting equipment.

3.1.2. Fluorescent Tubes. Fluorescent tubes are responsible for an estimated 60% of the electricity consumption of non-residential buildings, being therefore the most important load.

In Table 3.1 it can be seen that the efficiency of the widely used 40W fluorescent lamps sold in Pakistan is in the range of 50-65 lumens/Watt. If the lamps are combined with an average ballast having 10 W losses, the efficiency is in the range of 40-52.5 lumens/Watt.

In other countries , namely in the U.S. and in Europe, the fluorescent lamps have a substantially higher efficiency. In the 1970s "thin" tubes (28 mm diameter versus 37.5 mm in conventional tubes) were introduced with improved phosphors, providing an efficiency close to 100 lumens/Watt. The biggest lamp manufacturer in Pakistan, sells in Europe the TLD 36 tube which uses the same fixtures of the conventional 40 W tubes and provides 96 lumens/Watt (Table 3.2). Even when combined with an average ballast having 10 W losses, the total efficiency is 75 lumens/Watt. If these lamps replace the 40 W tubes now used in Pakistan, the potential savings are 30-47% of the present consumption.

High efficiency tubes use tri-phosphors (rare earth elements), which increase their price 50-100% over conventional tubes. The rates of return are in the range of 64-420%, for an annual use of 1000-8000 hours, considering a price premium of 75%, a lifetime of 8000 hours and average savings of 35%. For a typical 2000 hours of annual use, the rate of return is 125%.

Assuming a 12% growth rate, the projected commercial sector energy demand will be 32400 GWh in the year 2005. High efficiency fluorescent tubes can save over 15% of that energy, that is around 5000 GWh of electricity savings. Additionally, 1000 MW of base generating capacity would be avoided.

The cost of avoiding non-diversified power demand is less than Rs

TABLE 3.1

CHARACTERISTICS OF LAMPS MADE AND SOLD IN PAKISTAN

Type of Lamp/ Manufacturer	Power (W)	Efficiency (Lumens/Watt)	Lifetime (Hours)	Price (Rs)
Incandescent/A	60	11	1000	7
"	100	14	1000	8
Fluorescent/A	20	52 (35)*	7500	35
"	40	65 (52.5)*	7500	37.5
Fluorescent/B	20	42-47	5000	35
"	40	50-60	7500 - 12000	37.5
"	65	50-60	12000	40

Source: (2)

Note: *including 9.5 W losses in the ballast from the same manufacturer

TABLE 3.2

CHARACTERISTICS OF HIGH-EFFICIENCY LAMPS WHOSE IMPORT IS BANNED

Type of Lamp	Power (W)	Efficiency (Lumens/ Watt)	Lifetime (Hours)	Price CIF (RS)
Thin Fluorescent				
TLD 36	36	96	8000	
TLD 58	58	93	8000	
Compact Fluorescent				
PL 9	9 (12.5)*	67 (53)	5000	60
PL 11	11 (15)*	83 (65)	5000	72
SL 13	13*	46	5000	140*
SL 18	18*	50	5000	161*
SL 25	25*	48	5000	178*

Source: Philips Pakistan

Note: *These values include ballast

6000/kW, considering a typical 2000 hours of annual use, a lifetime of 8000 hours and a real interest rate of 8%. Thus the introduction of high efficiency fluorescent lamps should be promoted as soon and as fast as possible due to the large potential for savings, high rates of return and low cost of avoided power.

One lamp manufacturer has the technology to make thin fluorescent lamps in Pakistan and is considering introducing them in 1987. They have not been introduced so far because lamp prices are controlled by the government, which discourages investments in equipment to produce new types of lamps.

3.1.3. Ballasts. The ballasts manufactured in Pakistan present large differences in efficiency and power factor (2). In ballasts for 40W lamps, the difference can reach 12 W.

Again "garage" manufacturers account for a substantial fraction of the ballast market, with prices 50% below that of good manufacturers. Poor quality materials, such as mild steel and even impure copper, are used in these low performance ballasts. The price range for 40 W fluorescent ballasts is 20-60 Rs.

In U.S., high efficiency ballasts are widely sold. Whereas an ordinary ballast used with two 40 W fluorescent tubes has 16 W losses, a high efficiency model used with the same two lamps only dissipates 8 W. The use of a ballast with two lamps instead of one is substantially more efficient as the losses are basically the same. Additionally, most ballasts used in USA are of the quick-start type, which provide a much longer lamp life (up to 20000 hours instead of 8000 hours).

High efficiency ballasts have less electrical and magnetic losses, but as they use more copper and steel, they have a cost premium of around Rs 50 (US$3) over conventional ballasts. These cost Rs 134 (US$8) for a two lamp fixture, in medium quantities in USA.

For further analysis, it is assumed that the Pakistani ballasts have similar characteristics to the ballasts made by the biggest lamp manufacturer (9.5 W losses for a 40 W lamp) and cost Rs 60. If these single lamp ballasts are replaced by high efficiency double lamp ballasts, there would be a reduction in the losses of 11 W with an extra cost of Rs 64.

If high efficiency fluorescent tubes (TLD36 or similar) are used with high efficiency ballasts, the combined efficiency can reach 86 lumens/Watt. This value represents a reduction of energy consumption of 38-43% when compared with ordinary lamps and ballasts made by the larger manufacturers in Pakistan. If ballasts made by the "garage" manufacturers are considered, the savings are even larger.

High efficiency ballasts have the potential to reduce by 8% the total fluorescent lighting load, in addition to the potential 35% savings due to the high efficiency fluorescent tubes. This translates into 1140 GWh of electricity savings by the year 2005 and 230 MW of avoided base generating capacity. Assuming that high efficiency ballasts have a lifetime of 20 years, the cost of avoiding non-diversified power is Rs 5800/kW, again much lower than the cost for new power supply.

In the commercial sector the rates of return are in the range of 22-180%., for 1000-8000 hours of operation. For a typical 2000 hours of annual operation, the rate of return is 45%.

3.1.4. Compact Fluorescents. It was previously estimated that incandescents represent about 20% of the lighting load in non-residential buildings. Compact fluorescents lamps, which fit in the same socket of incandescent lamps , can replace them with substantial savings. It can be seen from

Tables 3.1 and 3.2 that energy savings in the range 70-80% of the consumption by incandescents can be achieved.

For the purpose of analysis it is assumed that 60 W incandescent lamps are replaced by compact fluorescents. PL 9 and SL 13 lamps provide 90% of the light output of the 60 W incandescent and the savings are adjusted by that factor.

Using the lifetimes and prices of Tables 3.1 and 3.2 the rates of return are in the range of 67-360% and 37-176% for the PL9 and SL 13 lamps respectively. For a typical 2000 hours of annual use the rates of return would be 127% and 68%.

If an average 75% efficiency improvement is achieved with the replacement compact fluorescent lamps, the potential savings by the year 2005 are 8% of the total commercial sector electricity consumption. This translates into 2600 GWh of energy and 550 MW of base generating capacity.

The cost of avoiding non-diversified demand, through the investment in compact fluorescent lamps, assuming 2000 hours of annual use and a real interest rate of 8%, is Rs 3900/kW for the PL9 and Rs 12500/kW for the SL 13. The substantial difference in the cost of avoided power for the two lamps is due to the PL 9 being cheaper and having a separated ballast which is not discarded when the lamp is replaced. On the other hand, the ballast in the SL 13 is integrated with the lamp, leading to a much higher extra cost when the lamp is replaced. Although PL lamps are more cost-effective, in some applications they may require the use of a diffuser and they are longer than SL lamps.

The largest lamp manufacturer in Pakistan is also studying the market for compact fluorescents. There are two main difficulties associated with the introduction of these lamps:
- The wide voltage fluctuations, which can decrease substantially the lamp lifetime.
- The subsidized price of electricity for residential consumers, which makes less attractive the investment in energy saving technologies with a higher initial cost. The residential sector is also a larger potential market for compact fluorescents than the commercial sector.

Figure 3.1 shows the savings associated with the high efficiency lighting technologies, assuming that they replace existing equipment over the period 1988-1996 and that the base case lighting would grow at a rate of 10.6% over the next 20 years.

3.2. Fans and Evaporative Coolers

Fans are used in large numbers in Pakistan during the warmer months to improve comfort. There can be a large difference between the best and worst fans, as much as 34W in the pedestal type and as much as 64 W in the ceiling type (2). Bearing in mind that the fan market estimates range between 1.2 and 1.6 million per year, these differences can have a large impact on electricity consumption. It was not possible to estimate the percentage of fans sold in the non-residential market. Further data collection is needed to determine the penetration of different types of fans and their performance.

As in the motor and ballast industries, there are "garage" industries making poor quality fans which sell at a substantial price discount. Again, they use low quality materials (mild steel instead of silicon steel) and use a lower number of poles in the motors to make them cheaper. Fewer poles in the motors mean higher rotating speed which leads to increased energy consumption.

The use of good quality fans can save over 20% of the total energy used

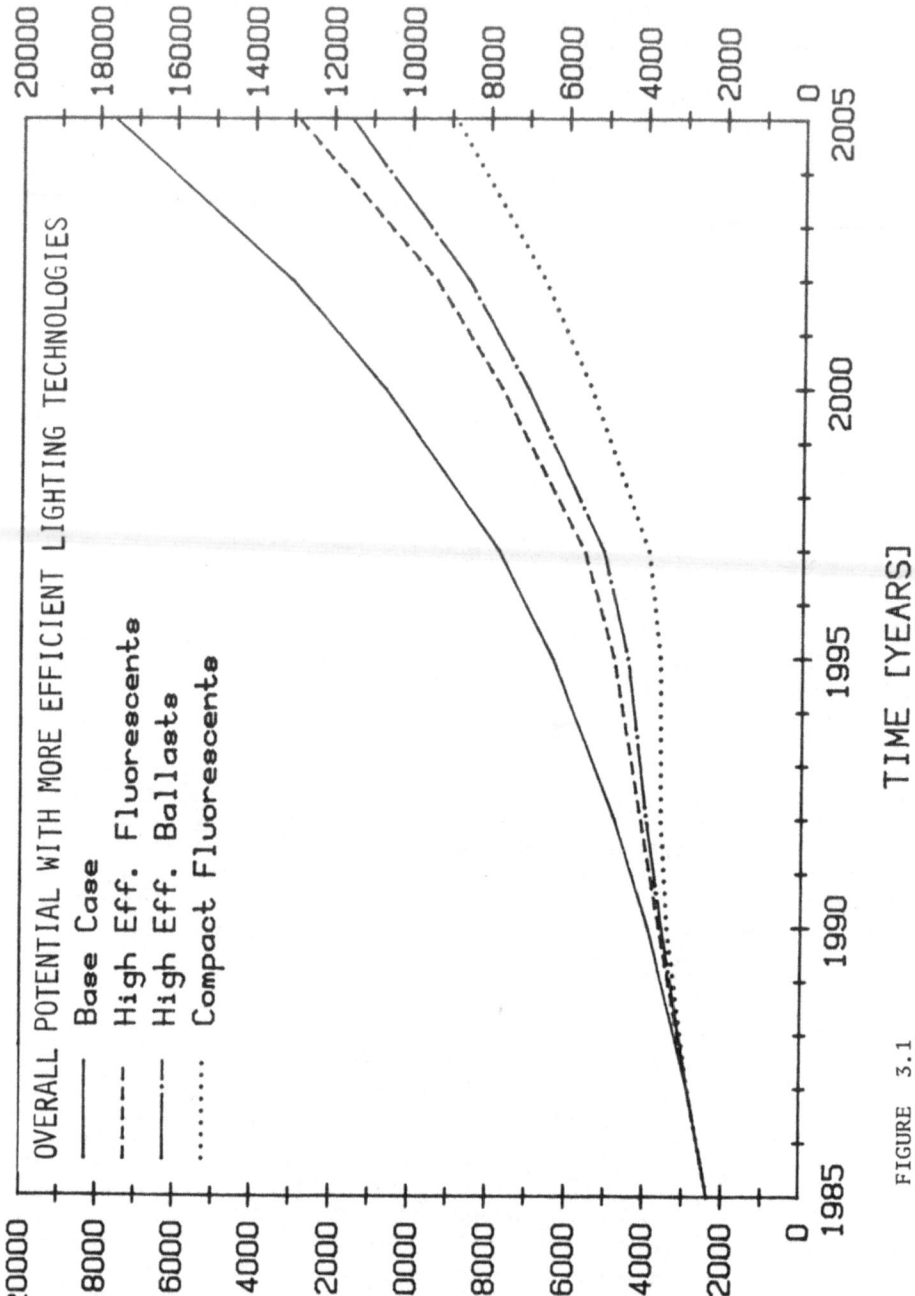

FIGURE 3.1

by fans in buildings. In the commercial sector this means around 1% of the total electricity consumption, which translates to 320 GWh of electricity and 270 MW of intermediate and summer peak generating capacity. Assuming that the price difference between a good and a low quality fan is Rs 200, the typical savings are 35W, and the lifetime 20 years, then the cost of avoided non-diversified power is Rs 5700/kW. The extra investment to buy a good fan gives a rate of return of 22-45% with an annual use of 1000-2000 hours.

Evaporative coolers, called in Pakistan desert coolers, can be very effective during the hot and dry months. However, during the monsoon months, July and August, they cannot provide comfort due to the high relative humidity. During these months the unit is operated as a simple fan. The evaporative cooler market in Pakistan is estimated to be around 0.5 million units per year (this means 0.5 million fans), of which an unknown percentage is applied in non-residential buildings.

3.3. Air Conditioning

Most of Pakistan population lives in areas (Punjab and Sind provinces) with hot weather during most of the year. Air conditioning is generally required to maintain comfort conditions from April to October. In buildings with large heat gains the cooling requirements may be extended to other months.

3.3.1. Room Air Conditioners.

Room air conditioners (RACs) are used to cool medium and small non-residential buildings and in some cases partial areas of big buildings without central air conditioning. The RAC market in Pakistan is around 40000 units per year of which 13000 were manufactured locally in 1984-85 (2). The units made locally use imported compressors, mainly of Japanese origin. The market is dominated by units of unknown performance brought into the country by Pakistani workers coming from abroad. Total room air conditioning stock is estimated at over 200000 units (19), with the non- residential building sector accounting for 40% of that number.

The average energy-efficiency-ratio (EER) of the RACs made in Pakistan is around 8, similar to the average value in the U.S.. However there are units available in the USA market with much higher EER (up to 12). Higher efficiency units have larger heat exchangers, an improved compressor and more efficient fan motors.

The lifetime of RACs is also greatly affected by voltage fluctuations, the typical value being around 5 to 6 years (2). When the units fail, the normal procedure is to have the compressor rebuilt in a rewinding workshop, although it is a lengthy and delicate operation requiring cutting the casing, rebuilding the motor and hermetically welding the casing. Due to the low labor costs, the rebuilding cost for a 1.5 ton RAC is 2000 Rs., which is roughly half of using a new compressor. For these reasons, a room air conditioner ultimately operates for more than 15 years.

The above figures mean that RACs are growing at a rate close to 20%/year. Considering only the commercial sector, the total non-diversified load due to room A/Cs is around 180 MW, with a growth of 36 MW per year. Assuming an average of 1500 hours of annual use, the energy consumed by RACs is now 240 GWh/year.

High efficiency units can reduce considerably the energy and power requirements. These better units carry a cost premium of Rs 1350-1700 per unit of EER in the range of 8-12, if there is mass production (3). An improvement of the EER from 8 to 12 would reduce the demand of a 1.5 tons

RAC from 2.25 kW to 1.5 kW.

Assuming that the lifetime of these improved units is 6 years, prolonged to 18 years by two subsequent rewindings, the rates of return associated with investing in a unit having an EER of 12 are in the range of 13-140% for 1000-8000 hours of annual operation. For typical applications in offices, shops and schools with an annual use of 1000-2000 hours, the rates of return are in the range of 13-36%.

With the projected 16%/year growth rate, loads would grow from the present 180 MW to 3500 MW by the year 2005, if there were no improvements in the equipment used or in the thermal performance of buildings. The use of equipment with an EER of 12 could reduce that demand by one third, that is around 1200 MW. In terms of energy this would correspond to a reduction of 1800 GWh, assuming 1500 hours of annual use.

The cost of avoiding demand through the investment in better RAC units, is Rs 7200-9100/kW of non-diversified demand. Room A/Cs are going to be in the future a major contributor to the summer peak load, and it can be expected that the diversity factor in the commercial sector to be close to one in the afternoons. Therefore the potential reduction in the capacity of intermediate and summer peaking units is around 1500 MW, when system losses of 21% are considered.

If a program for improving the thermal performance of buildings is carried out, the potential savings due to room A/Cs are reduced accordingly. To calculate the aggregate savings in the commercial sector, it is assumed that there will be a 25% reduction in the cooling load by the year 2005, due to the construction of improved buildings and the retrofitting of existing buildings. In these conditions, the potential savings due to RACs are reduced to 1350 GWh of energy and 1100 Mw of intermediate and summer peaking capacity.

Incentives should be considered, both for consumers and manufacturers, to promote the production and purchase of high efficiency RACs. Incentives are called for because the rates of return may not be very attractive to consumers in some applications.

3.3.2. Central Air Conditioners.

Central air conditioners (CACs) are applied in big and medium size buildings. Total CAC capacity in Pakistan is estimated to be around 500000 tons of which 50% are absorption chillers. The central air conditioning load is growing at a 12-15% annual rate (2).

Absorption chillers, although they have higher initial costs, have lower running costs due to the low price of gas in Pakistan. Absorption chillers, which present a coefficient of performance (COP) around 1, are more cost-effective on a life-cycle basis for applications over 200-250 tons.

The costs of CAC equipment, including installation, is in the range of 25000-45000 Rs./ton. The low end includes the lower efficiency packaged type units and the upper end includes the absorption chillers.

CAC equipment is imported, with medium and larger units having EERs of up to 12 and 15 respectively.There is an urgent need to remove from the market units with poor performance, through the application of efficiency standards.Due to the absence of air conditioner standards, there are small packaged type CAC units being sold in Pakistan with an EER as low as 5.5.

There is a decrease in performance of CAC systems with an air-cooled condenser, both in terms of efficiency and capacity, when the condenser inlet temperature rises. The use of wet filters to promote evaporative cooling in the air intake of the condenser can lead to substantial capacity and energy savings (2). Using the ASHRAE design day temperatures for

Lahore it can be seen that those savings can reach 20%.

In buildings with a high demand for fresh air (hospitals, restaurants, food processing), the use of indirect evaporative coolers in the ventilation air intake can also lead to substantial savings in capacity requirements and in energy consumption. The potential benefits of wet filters and indirect evaporative coolers should be assessed based on the temperature and humidities values during the cooling months.

3.3.3. Thermal Insulation. The building cooling load is the sum of the internal heat load (equipment and people) plus the external heat load (heat passed through the building envelope). The use of efficient equipment (such as energy efficient lighting) can minimize the internal heat load leading thus to further savings due to reduced cooling.

There are no building standards in Pakistan which can be applied to the design of the building envelope. The external heat load can be strongly affected factors such as thermal insulation level, orientation of the building, window design and shading.

In buildings with a large window area, the use of window films or window shading should be considered. Thermal insulation is normally the dominant factor to determine the external heat gain. Unfortunately due to the absence of building codes, some of the new buildings are made with no insulation. Some buildings use conventional materials (clay with rice husk, mud) with unknown short and long term characteristics and some even use expanded polyesterene. In weather zones similar to Karachi and Lahore, thermal insulation can be used economically to remove around 30% of the building cooling load (7). The thermal insulation costs can be immediately paid back by the savings in the compressor capacity.

If it is assumed that 25% of the commercial sector cooling load is removed through the improvement in the thermal performance of the buildings (new and existing), the potential electricity savings are 2400GWh by the year 2005. This corresponds to savings of 2000 MW of intermediate and summer peak generating capacity, assuming an average 1500 hours of annual use and system losses of 21%

3.4. Aggregate Savings in the Commercial Sector

Table 3.3 shows the potential aggregate savings in the commercial sector by the year 2005. The total electricity savings, 12850 GWh, corresponds to 40% of the total consumption without conservation. The base generating capacity which can be displaced is 1800 MW. Additionally, 3370 MW of intermediate and summer peak generating capacity can be avoided.

Figure 3.2 shows the electricity consumption in the period 1985-2005, with and without conservation. The lighting improvements are assumed to be made in the period 1988-1996. The other technologies , mainly thermal insulation and air conditioning, are assumed to have a slower penetration, being implemented in the period 1988-2005. The conservation scenario slows the overall growth rate during 1985-2005 to 9.2% per year.

4. CONCLUSIONS

A study of the industrial and commercial sectors of Pakistan found a substantial potential for electricity savings. The two combined sectors have a cost-effective savings potential of 30% of their projected electricity consumption in the year 2005. The conservation potential in the two sectors is larger than the total electricity consumption in Pakistan in 1985. With the projected reduction of the system losses to 21%, the savings

TABLE 3.3

POTENTIAL ELECTRICITY SAVINGS IN THE COMMERCIAL SECTOR IN
THE YEAR 2005

	Power Savings Base/Peak+Int (MW)	Energy Savings at Consumption (GWh)	Energy Savings at Generation (GWh) *
Fluorescent Tubes	1020	5000	6330
High Eff. Ballasts	230	1140	1440
Compact Fluorescents	550	2640	3340
Fans	270 **	320	400
Room Air Conditioners	1100 **	1300	1650
Thermal Insulation	2000 **	2400	3030
TOTAL	1800/3370	12800	16200

Notes: - * Including 21% losses
       - ** Non-diversified summer peak and intermediate
         generating capacity

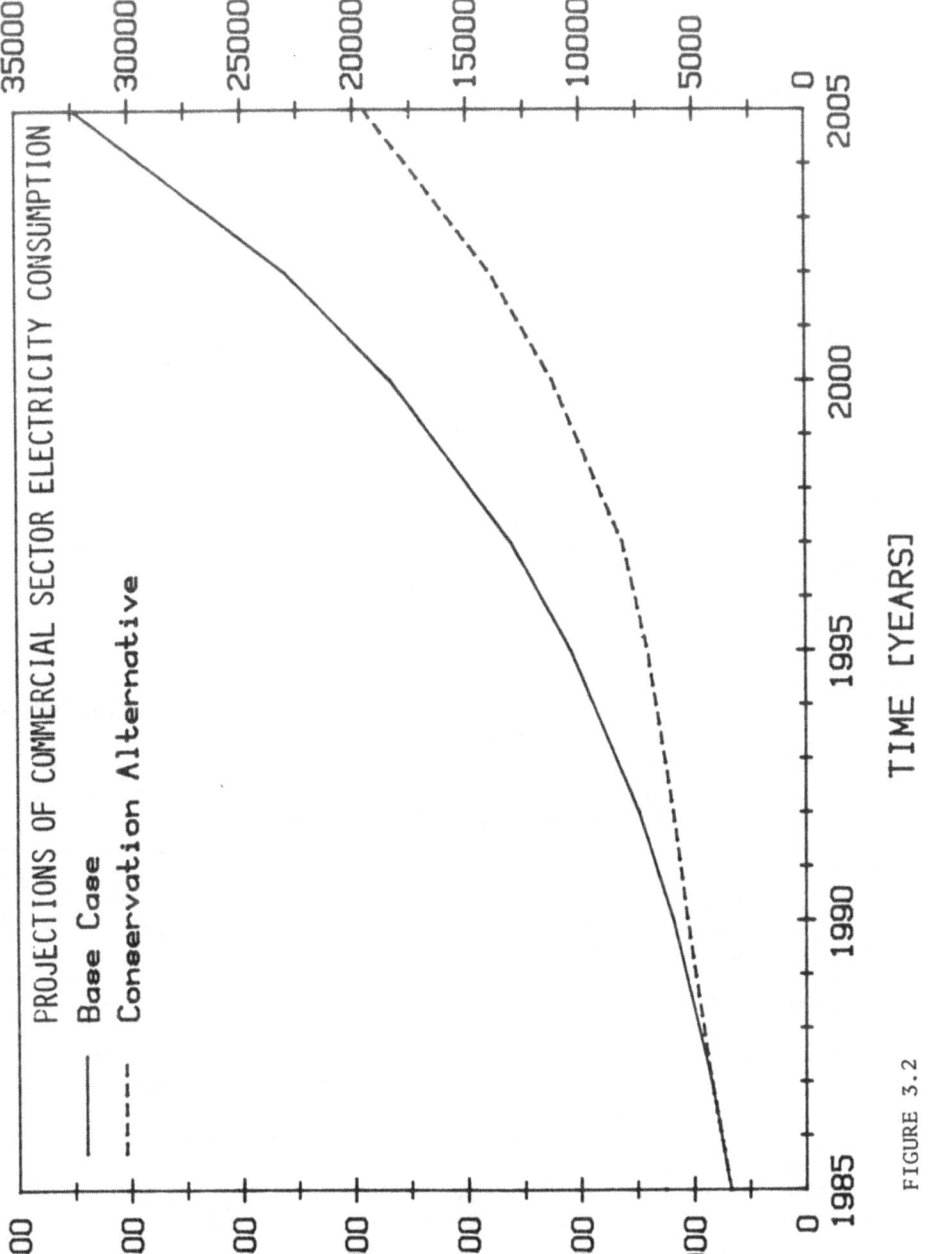

FIGURE 3.2

584

in the electricity generated are over 25000 GWH.

Likewise, there is scope for a large reduction in the expansion of the power supply system. The potential reduction in base load generating capacity is 3200 MW and the reduction in intermediate and peak capacity is close to 4000 MW. The following actions are recommended to tap these energy and demand savings potentials:

-Development and promotion of an information and training program for electricity users about the availability and advantages of energy saving technologies. This program would raise the awareness of consumers about efficient options and could be implemented through courses, brochures and advertisements.

-Establishment and enforcement of standards for main end-use equipment (motors, lamps, ballasts, fans, air conditioners), both for imported and locally made equipment.

-Development of building standards (including requirements for thermal insulation) for different weather regions of the country.

-Creation of incentives, for local manufacturers and consumers, to promote the penetration of high efficiency equipment in the areas with large savings potential. Some industries, namely the small industries, may also need technical assistance.

-Establishment of a tariff structure, for energy and demand in accordance with long term marginal costs. Time of use rates should be implemented for medium and large consumers, to promote load management and alleviate peak demand problems.

-Promotion of the availability of good quality materials, namely silicon steel, at a reasonable cost for the local producers of electricity consuming equipment.

Both the residential and agricultural sectors require similar studies, in order to estimate the global potential savings in Pakistan. In particular the residential sector, accounting for 30% of electricity consumption and with a 16.5% growth rate, presenting a poor load factor and having a large contribution to the peak demand, deserves to be investigated.

A large fraction of the electricity savings potential is already very cost-effective compared to constructing new power supply facilities in Pakistan. Stimulating the wide spread production, sale and use of more efficient electricity end-use equipment is one of the major challenges confronting energy planners during the reminder of this century.

5. REFERENCES

1- A. De Almeida and H. Geller, "The Potential for Electricity Conservation in the Industrial and Commercial Sectors in Pakistan", report for the US Agency for International Development and ENERCON, 1987

2- A. De Almeida, "Data Compilation for Determining Electricity Conservation Potential", ibid

3- H. Geller et al, "Residential Conservation Power Plant Study", American Council for an Energy-Efficient Economy, Washington, 1986

4- M. Munasinghe and J. Warford, "Electricity Pricing-Theory and Case Studies", John Hopkins University Press, 1982

5- Electric Power Research Institute, "Impact of Advanced Power Semiconductor Systems on Utilities and Industry", EPRI-2112, 1981

6- J. Andreas, "Energy Efficient Motors", Marcel Dekker, 1982

7- Owens Corning Fiber Glass, "Thermal Insulation Design Manual for the Middle-East", 1986

POTENTIALS FOR TECHNICAL ELECTRICITY SAVINGS USING KNOWN TECHNOLOGY

JORGEN S. NORGARD and TOM GULDBRANDSEN

Physics Laboratory III, Technical University of Denmark
DK-2800 Lyngby, Denmark.

1. ENERGY SERVICES

Electricity cannot directly be consumed by human beings and hence the electricity consumption is not a proper measure of people's well-being. Electricity can, however, indirectly contribute to the satisfaction of our needs and wants by providing physical <u>energy services</u> like cooking meals, cooling beers, illuminating the desk, storing vegetables and weaving fabrics.

Proposals to save electricity are often interpreted as cutting the energy services. This could, for instance, imply lower indoor temperature, less frequent clothes washing, more manual work, less light in the street, and a generally slower replacement of durable goods. Indeed, many such proposals make good sense and do not necessarily lead to a lower quality of life, often the opposite. However, in this paper we do not consider such reduction in the services. Rather we will look at options for saving electricity by improving the technology used to convert electricity into energy services - <u>the end-use technology</u>. This technology constitutes the last physical part of the total energy chain from primary energy to satisfying needs and wants, and it is also a very inefficient link in the chain (1).

2. POTENTIALS FOR EFFICIENCY IMPROVEMENTS

At Physics Laboratory III we have conducted several analyses of the potentials for saving electricity by improving efficiency of end-use. These have been carried out at a household level (2,3), a local community level (4), a national level for Denmark (5,6), and a regional level for Scandinavia (7).

In the following paragraphs, we will show the overall results for Denmark's electricity savings potential and afterwards illustrate these by examples. Most of the results and examples used here are updated from a study carried out as part of a consulting enterprise for the Danish Government in 1983 and described in reference (5) where more references are also available. Instead of the traditional distribution of electricity consumption, according to the various economic sectors of the country, like industry, households, etc., we have split-up the consumption of electricity into 8 categories according to its technical end-uses, such as pumping, ventilation, etc., as seen in Fig.1. The advantage of this categorization is, that within each category the technological principles are similar which facilitates the efficiency analysis. The disadvantage of the end-use distribution analyses is, that no statistics are

A. T. De Almeida and A. H. Rosenfeld (eds.), Demand-Side Management and Electricity End-Use Efficiency, 585–594.
© 1988 by Kluwer Academic Publishers.

586

readily available. Hence, there is a considerable uncertainty
in the distribution used here and shown as black columns and
percentages in Fig.1. This uncertainty, however, turns out not
to affect very much the total savings potential found in the
analyses, since the savings potentials of the various categories does not differ very much.

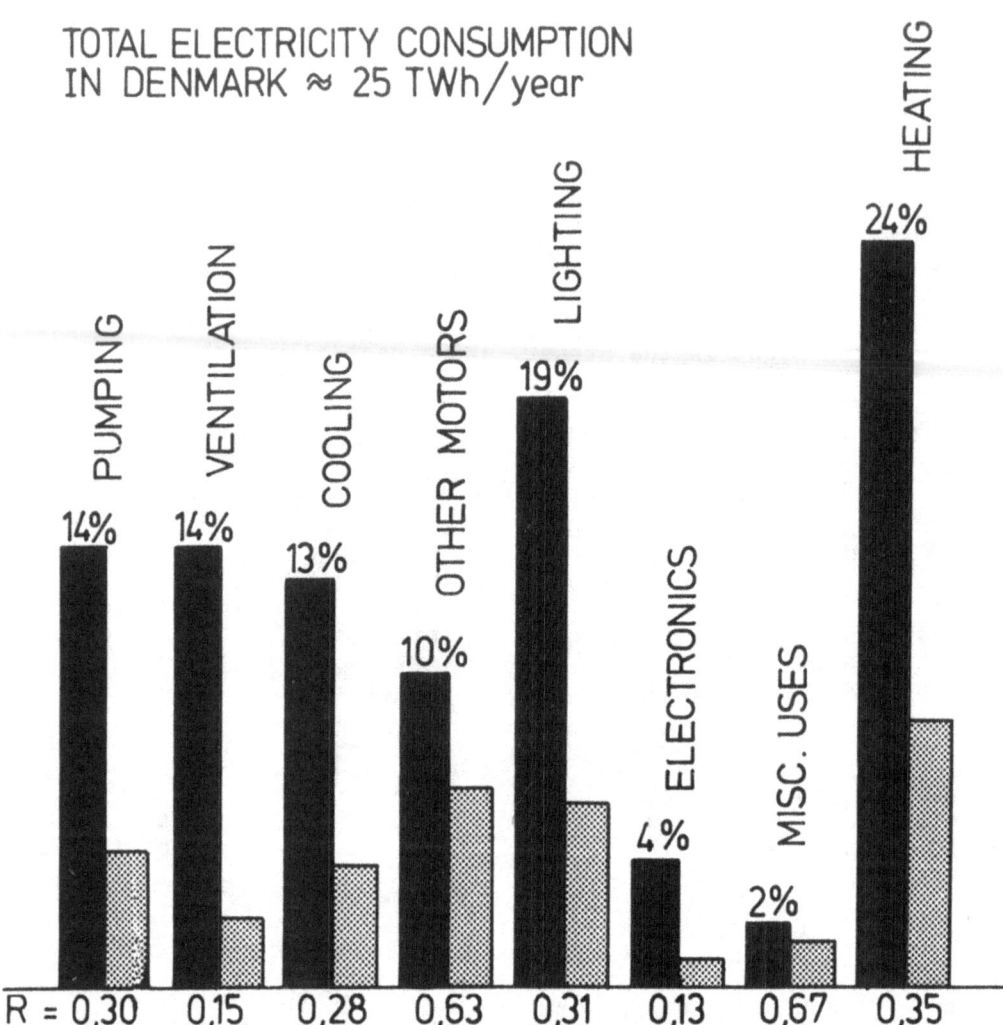

FIGURE 1. Present electricity consumption in Denmark distributed according to 8 main categories of uses, is shown as black
columns together with the possible future demand if efficient
end-use technology is introduced. The percentage indicates for
each category its share of the total with present distribution.
R is the intensity reduction factor, that is, the ratio between
future and present electricity intensity.

The shaded columns indicate to what level each category of electricity consumption can be reduced while still providing the same physical energy services as today. The ratio between the reduced and the present electricity consumption we term the intensity reduction factor, R. As shown at Fig.1, R is found to be 0.30 for pumping, 0.15 for ventilation, etc., if the efficient technology is implemented. In total, there is a potential for reaching R = 0.34, meaning that with better technology we could maintain present Danish standards of comfort, production, etc. with only 34% of the electricity used in Denmark today. In other words, the technical end-use efficiency could for electricity be three times as high as now.

Furthermore, almost all electricity used for heating purposes can relatively easily and thermodynamically advantageously be substituted by other and cheaper forms of energy, such as heat from efficient oil furnaces or from combined heat and power plants. If this path of substituting electricity is pursued as well, we can, in Denmark, be equally well-off with energy services at only 1/4th of our present use of electricity.

3. FUTURE ENERGY SERVICES

It is, of course, unlikely that the general level of energy services and their distribution profiles will remain unchanged in the future.

Often it is anticipated by politicians and economists that there will be an increase in the total level of energy services provided by electricity. This has been the trend for decades and if continued into the future, it will lead to a correspondingly higher consumption than what is indicated by the shaded columns in Fig.1. The ratio between future and present levels of energy services we term the relative energy service demand, D. If, for instance, D is 1.2, this implies a 20% growth in the energy service in question.

For Denmark and a few other countries, however, there are indications from surveys that the national economy is approaching a general saturation in the material standard of living. The majority of people prefer more leisure over more income and consumption (8). In this case, D would average around 1.0.

Finally, increasing concern about environmental, social, and health aspects points towards a future energy service level below the present, that is a value of D less than 1.0.

The relative distribution of electricity services on the different categories will also be an important determinant of future electricity demand. For instance, even a considerable increase in the use of services from electronic equipment might not necessarily lead to a large growth in the need for electricity. The reason is that electronic services have a low and declining electricity intensity; that is, only a small fraction of the cost of these services is for electricity consumption. On the other hand, if a possible future economic growth is extensively turned into electricity services with a high electricity intensity, such as for various electric heating purposes, it will result in a profound growth in the total demand for electricity in Denmark.

4. TECHNOLOGICAL LEVEL

Technologically almost anything can be achieved if one can ignore the cost and anticipate technology to approach the theory of physics. None of this is the case in our analysis.

The savings potential presented in this paper is based on appropriate implementation of basic technology which is well known today. Furthermore, it is limited to the technology which in our analysis we have been able to come across. Concerning the economy of introducing the efficient technology, we have confined ourselves to measures which we estimate could be socio-economically sound within 10-20 years.

No technological breakthrough is anticipated. This implies that research in the area of efficient use of electricity could open-up a savings potential beyond that presented in this paper.

Savings potentials such as those indicated in Fig.1 are often considered unbelievable and characterized as the theoretical maximum of savings. This is, however, not at all the case. We are not even approaching the theoretical limits dictated by the fundamental physical laws of the system. The efficiency of the refrigeration system, for the low energy refrigerator described later, is still only around 20% of the thermodynamical limit, and in this case, insulation also can be improved to a theoretical level approaching zero refrigeration need. It is important to be aware of the very low limits to electricity intensity, in order to encourage research in utilizing this enormous potential for saving electricity.

5. EXAMPLES OF SAVINGS POTENTIALS

To illustrate the methods used in our analysis and to indicate how the large savings are possible we shall describe examples from a few of the categories of electricity use.

5.1 Lighting

More efficient light sources are slowly penetrating the market. Most of them are small flourescent tubes, folded to a compact lamp suitable as replacements for incandescent lamps, but with around 4 times as high efficiency, that is, R = 0.25. Compared to earlier flourescent light tubes, the quality of flourescent light from both normal tubes and the compact types has improved considerably with respect to colour. Still a delay and a flickering at first can be annoying, but these and other difficulties can be overcome in the future with electronic ballast which is already on the market. In addition, these ballasts actually save electricity, but so far they have been rather expensive.

Not only can the light sources be improved; it is important to follow the energy chain from the electric grid all the way to the light reaching the desk. Traditional light fixtures and lamp shades often have very low efficiency, due mainly to poor reflections from shades. One of the most famous and popular lamps in Denmark, the beautifully designed PH-lamp, allows only around 20% of the light to leave the lamp (9). It should be possible to improve reflection of the surface coatings of these and similar lamps without hampering the design. A doubling of the efficiency of shades, for instance, from 20 to 40%, gives an R = 0.50.

Better control of light according to the need can be manual or automatic. Savings of 30% can often be achieved easily, which means R = 0.70.

In summary the efficiency of certain light systems could be increased by 1) replacing incandescent lamps with new flourescent, compact lamps, 2) improving lamp shades as described, and 3) introducing better controls. The total reduction factor for light in this case is:

$$R = 0.25 \cdot 0.50 \cdot 0.70 = 0.09,$$

if the three factors are assumed to be uncorrelated. In other words, the illumination service described above could be the same as now, with less than 10% of present electricity consumption.

The case just described seems more favorable to savings than the average. In offices and institutions like schools etc., efficient flourescent light tubes have been introduced in most places and the average shades do have better efficiences than the 20% mentioned. Nevertheless, we found that by organizing the illumination better, utilizing daylight better, using electronic ballast, automatic control and other features, we could reach R = 0.25 for the whole service sector while improving the illumination comfort.

For Denmark as a whole, the potential for saving electricity for lighting is found to be

$$R = 0.31$$

assuming that we maintain incandescent lamps for some domestic uses.

5.2 Ventilation

There are different purposes of ventilation, of which the most common in Denmark are to provide clean air and to remove excess heat.

The analysis starts by looking at the need for air flow. Standard ventilation systems are usually oversized. This is particularly the case when the need for air replacement is below maximum. Furthermore, more efficient light systems in schools and other institutions as mentioned above, lead to lower need for air flow since there will be less excess heat for most part of the year. On the average, a reduction in the need for air flow to half of the present seems possible without hampering comfort.

With only half the air flow, the required electricity could, in principle, be reduced to 1/8 of the present, since electricity consumption varies with the third power of the flow. In practice, we have found that when improving and adjusting motors, ventilators and control systems, etc., without replacing the duct systems, we could increase the overall efficiency of ventilation systems in Denmark by a factor of 7, which means

R = 0.15

without lowering the energy service or the air quality.

Often the mechanical ventilation systems have replaced natural ventilation systems in order to facilitate control and stabilization of air flow. However, new ventilation systems are being developed, which rely on natural convection flow and wind, but which are automatically controlled by varying the size of the openings for air flow according to needs. Buildings Research Institute in Denmark has developed and tested such systems for ventilating cattle houses, which consume today about 1/3rd of all electricity used in farming (10). In the Danish climate, electricity consumption for that particular purpose could drop close to zero.

Another automatically controlled natural ventilation system, which has been marketed, is a ventilation opening for single rooms. The version on the market regulates the air flow opening according to humidity in the room and is not connected to the electric grid at all.

In the above estimate of savings potentials (R = 0.15) for ventilation, none of these natural ventilation systems have been counted on.

5.3 Refrigeration

In Denmark, mechanical refrigeration is almost exclusively used for cold storage rooms for food, such as domestic refrigerators and freezers. Air conditioning with refrigeration systems is very rarely used and, according to National Building Codes, is now generally banned in new buildings.

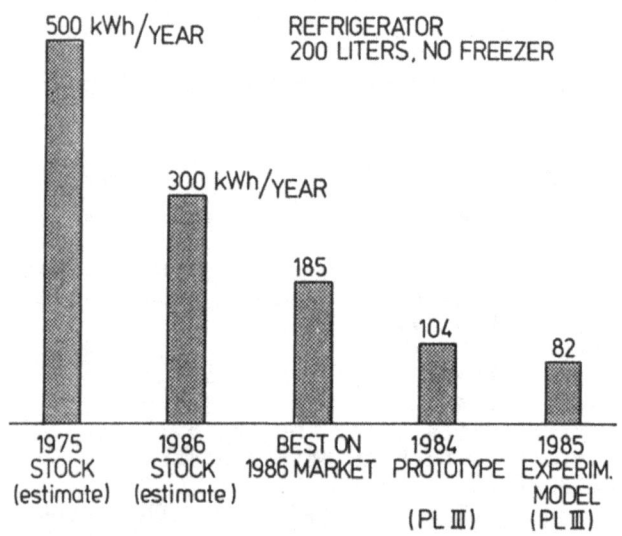

FIGURE 2. Annual electricity consumption according to standard test conditions is shown for different designs of a typical size 200 liter refrigerator with no freezer compartment. In Denmark, most families have a freezer as a separate unit. PL III refers to our models at Physics Laboratory III, Technical University of Denmark.

Refrigeration is an example of a rather long and complicated energy chain of end-use technology. Electricity powers a motor which runs a compressor which is part of a refrigeration system. All this can be made more efficient, but the need for cooling can also be reduced through better insulation of the cabinet, and still provide the same service - a certain cooled storage for food. At our laboratory, we have developed very efficient prototypes of refrigerators. Also, some manufacturers have improved their models considerably since 1973. Fig.2 shows some examples. Our recent prototype consumes only 16% of what was typical in 1973 for a similar size refrigerator (11).

Refrigeration in the commercial and the industrial sector has also been analyzed (12). In all uses of electricity for refrigeration purposes in Denmark, we found a potential for R = 0.28, that is, about three times as high efficiency as today.

A limited number of refrigerators based on the 1984 prototype and the 1985 experimental model shown in Fig.2 will, within six months, be produced in a pilot production. This is the result of a cooperation between Physics Laboratory III and the Danish refrigerator manufacturer Brdr. Gram A/S. The production is sponsored by the Danish Ministry of Energy and three Danish electric utility companies.

6. ECONOMY AND ENVIRONMENT

We will be very brief about the economy of electricity end-use savings.

Many options for more efficient end-use technologies are available essentially free of cost, except for some extra thoughts during the design process. Others require changes in production lines, but require no obvious extra production cost in the long run. Finally, however, many savings will require specific extra investments. In general, the more we have already saved, the more expensive it will be to save another kWh. Fig.3 illustrates this. If the savings we suggest in Fig.1 are achieved in the course of natural replacement of equipment, machinery, and appliances, that is, over a 20-25 years period, we estimate them to be cost-effective up to at least 50% savings, as indicated in Fig.3.

Notice that the extra price for purchasing efficient light bulbs, freezers, etc. does not necessarily reflect the extra production costs. In the early years of introducing such new models, the manufacturers will charge what they can get. They calculate how much the product saves in electricity per year and estimate a payback time which is acceptable to the consumers. That sets the price in the early stage of introducing energy saving technology.

It should also be remembered, that the environmental costs of saving a kWh in most cases are essentially zero, contrary to that of supplying a kWh. Other problems of electricity supply, such as risk of accidents, security of supply, independence of international crisis and expensive import of advanced technology also point in favour of conservation over supply. From a rational standpoint, we should be willing to pay considerably more for a kWh saved than for one produced. However, in many cases today, we don't have to pay more; it is quite the opposite.

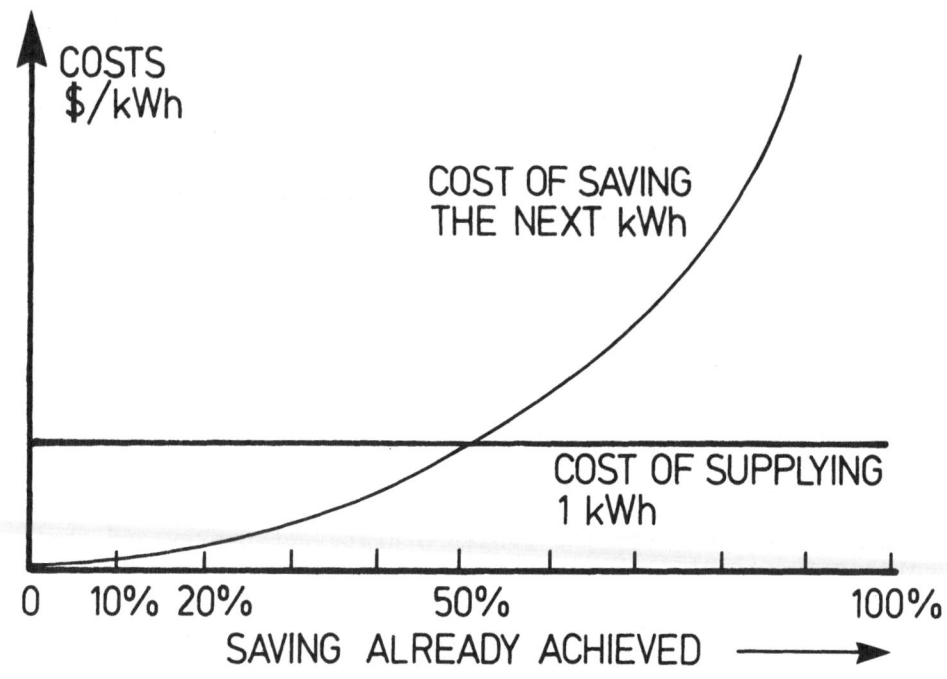

FIGURE 3. Qualitative illustration of the increase in costs of saving another kWh as more and more savings are obtained. The intersection with cost of supplying a kWh indicates an optimum saving if only economic considerations are included.

7. CONCLUDING REMARKS

For the industrialized countries the potential for electricity savings provides an excellent option for reducing the problems of acid rain, nuclear risks, and international dependence, to mention just three of the most obvious problems we are facing today.

Developing countries should not miss the opportunity to choose the efficient end-use technology. It is much easier and cheaper to choose proper equipment from the beginning, rather than replacing it later on. Furthermore, most developing countries (as well as most developed countries) will be capable of producing the conservation technology themselves, contrary to the heavy power plant technology. In general, there is a growing awareness of the benefits of electricity conservation in developing countries where they formerly considered electricity consumption per capita as an indicator of general welfare.

7.1 Savings activities in Scandinavia

This year the Danish Ministry of Energy has initiated a project "Potentials for Saving Electricity" as part of a political compromise. During the last part of 1987, the project will analyse the technical and behavioural potentials for saving electricity. Physics Laboratory III is involved in this project. The savings potentials will be illustrated by concrete examples and will focus on what can be achieved by 1) the best technology already on the market and 2) technology which can be developed and marketed within a 10-15 year period.

The main purpose of the project is to study the means of implementing a conservation policy using one or two local communities as experimental cases. In these areas, various kinds of informative, financial, and other electricity saving incentives will be tried out in order to evaluate their effect on electricity consumption. Also in these local communities, various examples of technical savings might be demonstrated in low electricity buildings.

In Sweden (which has decided some years ago to phase out nuclear power), a project similar to, but larger than that of the Danish Ministry of Energy, is carried out by the dominating electric utility company "Vattenfall".

7.2 Surprising savings potentials

Potentials for saving energy have been investigated in many countries. Most of these studies, however, have been limited to analyzing savings potentials by using the best technological hardware already available on the market. The study presented here goes one step further by including the technological improvements also, which can be achieved by applying present technological knowledge to improve end-use efficiencies. Very few other studies of this kind, if any, have been carried out on electricity savings potentials.

This could explain why the 65% savings potentials found in this study surprise most people, including many energy researchers. However, since we first published studies like this one no critics have claimed the results to be unfeasible in a technical or an economical sense.

Some progress has been made in technological electricity savings. But numerous power plants are still being built all over the world at a much higher economical and environmental cost than that of a corresponding effort on electricity saving end-use technology. An example which illustrates this is a recently developed, American style efficient refrigerator prototype, which consumes only half of the electricity used by the best refrigerators on the market in USA (13). If such models gradually replaced all existing refrigerators in the USA, it would save construction of around 10.000 MW power capacity, for instance, 25 power plant units of 400 MW each. The cost of saving this power would be only a fraction of the cost of generating it.

The obstacles to these electricity savings are discussed in a different paper (1).

REFERENCES
1. Norgard, J.N., and B.L. Christensen: "Sociological and psychological barriers to electricity savings". In this proceedings from NATO Advanced Study Institute in Povoa do Varzim, Portugal, 20-31 July 1987.

594

2. Norgard, J.N., and P. Buhl Pedersen: "Low electricity household for the future". In this proceedings from NATO Advanced Study Institute in Povoa do Varzim, Portugal, 20-31 July 1987.

3. Norgard, J.S.: "Husholdninger og energi" (Households and Energy), Polyteknisk Forlag, Denmark, 1979.

4. Strabo, F., O. Thun, J. Norgard, and S. Prahm: "Elforbrugets udvikling i en landkommune", Nysted Rådhus, DK-4880 Nysted, Denmark, 1981.

5. Norgard, J.S., J. Holck, and K. Mehlsen: "Langsigtede tekniske muligheder for el-besparelser", Physics Lab.III, Technical University of Denmark, DK-2800 Lyngby, Denmark, 1983.

6. Norgard, J.S.: "Same Comfort in Buildings with One Third of Present Electricity Consumption", Vol. J, Proceedings of ACEEE 1984 Conference, Santa Cruz, California, USA, 1984.

7. Meyer, N.I., J.S. Norgard et al.: "Nordiska Energisystem - Möjligheter och begränsninger i ett långsigtigt perspektiv". Lund Universitet, Gerdagatan 13, S-22632 Lund, Sverige, 1981.

8. Norgard, J.N., and B.L. Christensen: "Individual Attitudes in Scandinavia Point Towards a Low-energy Saturated Society. Vol.F, Proceedings from ACEEE 1984 Summer Study on Energy Efficiency in Buildings, Santa Cruz, California, August 1984.

9. Mehlsen, K.: "Bedre lys og luft til servicesektoren", Physics Lab. III, Technical University of Denmark, DK-2800 Lyngby, Denmark, 1986.

10. Strøm, J., and S. Morsing: "Automatically Controlled Natural Ventilation". Proceedings of the second international livestock environment symposium, Iowa, April 1982. Publ. by Am. Society of Agricultural Engineers, 2950 Niles Rd, St. Joseph, Michigan, USA.

11. Guldbrandsen, T., and J.S. Norgard: "Achieving Substantially Reduced Energy Consumption in European Type Refrigerator", presented at 37th Annual International Appliance Technical Conference, Purdue Univ., USA, May 6-7, 1986.

12. Pedersen, P.B.: Electricity Savings in Refrigeration Processes". In this proceedings from NATO Advanced Study Institute in Povoa do Varzim, Portugal, 20-31 July, 1987.

13. Pedersen, P.H. et al.: "Design and Construction of an Efficient US-type Combined Refrigerator/Freezer. Presented at Internationaler Kongress fur Kältetechnik, Wien, 24-29 August, 1987.

ELECTRICITY CONSERVATION IN JAPAN

Haruki Tsuchiya

Research Institute for Systems Technology
1-7-1, Sarugakucho, Chiyodaku, Tokyo, JAPAN

1. INTRODUCTION

Electricity is an extremely expensive energy in Japan which imports 83% of primary energy from outside. The pressure of costly energy forced the development of conservation technologies, improved energy productivity and decreased oil demand drastically. Furthermore, society opened its doors to an "information based society" from a "material and energy intensive society", and the industrial structure began to change itself, though there were factors to increase electricity consumption.

This paper reports on the latest electricity demand situations in the industrial, residential and commercial sectors, and the variety of technologies for end-use electricity conservation in factory and home appliances, including refrigerator, heat pump, motor controller, and so on.

2. END-USE ELECTRICITY

Electricity use in industry was 330 TWH (56.2% of final electricity consumption); 241 TWH (41.1%) was for residential & commercial use and 16 TWH (2.7%) was for railways in the transportation sector in 1985.

Table 1 shows the end-use structure of electricity in industry. Motive power & others (lighting etc) has the largest share. Second largest share is electric heating. Table 2 shows the end-use structure of electricity in the commercial and household sectors. Table 3 shows the change of energy consumption in 1985 relative to that in 1975.

There are energy intensive industries such as iron & steel, paper & pulp, cement & ceramics and chemical industries. They have focussed on energy savings since the oil crisis, and their energy productivity has climbed up. Their electricity demand has not been increasing and is not expected to increase much. In their case, there are opportunities to save energy by introducing efficient processes to recycle thermal energy and generate electricity.

The other industries, which are not energy intensive, have increased their electricity consumption, though fuel demand has been curbed by introduction of better management and efficient processes. Metal products & machine and other industries use machine tools, welders, press machines, robots and induction heating. They have the tendency to be more automated so that electricity demand increases. Furthermore, the basic materials industries are depressed and workers have been on move to service or assembly industries. These factors have pushed electricity demand in these sectors.

The following are energy situations and typical energy conservations in respective industries and sectors focussing on electricity end-use.(1)

A. T. De Almeida and A. H. Rosenfeld (eds.), Demand-Side Management and Electricity End-Use Efficiency, 595–606.
© *1988 by Kluwer Academic Publishers.*

TABLE 1   END-USE STRUCTURE OF ELECTRICITY IN 1984 (2), (5)

	Heating (%)	Electrolysis (%)	Motive Power & others (%)	Electricity demand(TWH)
Manufacturing				319.3
Food	3.5	0.7	95.8	16.4
Textile	1.0	0.0	99.0	15.5
Paper & Pulp	0.8	0.3	98.9	26.9
Chemicals	6.1	18.4	75.5	52.2
Ceramics	14.7	2.9	82.4	24.5
Iron & Steel	29.5	2.9	67.6	76.3
Non ferrous	16.3	41.3	42.4	23.0
Metal products	19.0	3.3	77.7	53.8
Others	9.9	0.6	89.5	30.9
Railways	0.0	0.0	100.0	16.0

TABLE 2   END USE STRUCTURE OF ELECTRICITY IN 1984. (4)
(The numbers in brackets are shares in total end-use energy
including fuels and electricity)

	Heating (%)	Cooling (%)	Hot water (%)	Cocking (%)	Motive power(%)	Electricity demand(TWH)
Commercial Building	4.3 (33.9)	13.3 (4.8)	0.0 (26.7)	0.0 (6.3)	82.4 (28.3)	111.0
Household	5.9 (32.1)	5.2 (1.6)	11.5 (33.3)	3.7 (9.9)	73.7 (23.1)	118.1

TABLE 3   THE CHANGE OF ENERGY CONSUMPTION DURING 1987-1980
AND 1980-85.(%) (2)

	Index of Industrial Production		End use energy Consumption (Fuel+Elect.)		Electricity Consumption	
	'75-80	'80-85	'75-80	'80-85	'75-80	80-85
Manufacturing (total)	+40.1	+23.0	+1.0	-10.5	+18.8	+4.8
Food	+13.4	+2.1	-3.8	-7.7	+33.4	+22.9
Textile	+6.8	-2.0	-25.7	-12.3	+18.8	+11.9
Paper & Pulp	+34.2	+12.8	+0.7	-20.4	+21.5	+2.4
Chemicals	+41.8	+20.4	+5.5	-6.6	+1.4	+1.0
Ceramics & Cement	+32.5	-4.5	+19.0	-31.5	+37.2	+4.1
Iron & Steel	+25.1	+0.4	-10.0	-15.7	+13.3	-4.7
Non ferrous metal	+40.6	+0.1	+15.3	-35.8	-8.3	-31.1
Metal products	+61.8	+46.5	+11.1	+12.5	+41.9	+38.2
Others	+23.1	-0.5	+35.1	+26.9	+39.4	+23.0
Railways	----	----	+9.9	-15.7	+10.5	+5.0
Household & Commercial	----	----	+14.9	+13.3	+38.6	+29.1

2.1 Iron & steel industry

The demand for crude steel has not increased and has stayed around 100 Million tons (of which 30% is for export) since 1973. And the iron & steel industry are looking for other business opportunities such as electronics, information, new material and bio-chemistry, as they cannot expect steel demand growth any more. The trend of energy demand is that oil demand has been decreasing while coal and electricity demand has increased relatively. The energy efficiency has been drastically improved as shown in Table 3.

The largest electricity demand is for arc furnaces which recycle scrap steels. Recently, increasing demand for special steels has caused demand increases in electricity slightly.

The blast furnace gasses are recycled by top gas recovery turbine to generate 3% of total electricity. The coke dry quenching systems also recycle thermal energy to generate electricity. The scrap steels, nearly 30 million tons per year, are recycled by electric arc furnaces, a process in which the scrap steels are preheated to save electricity and to shorten melting time by recovery of exhaust gas from the furnace.(3)

2.2 Paper & pulp industry

The effort of energy conservation has been focussed on fuels rather than electricity. The private electric generation with steam occupies 50% of electricity consumption. Electricity end-uses are mainly motive power - fan, pump and axle power in which rotary control system has been introduced to improve efficiency.

It is hoped to improve the production process drastically by direct sodium method without process of burning sodium carbonate, and iBio pulping i which uses special bacteria to separate cellulose from lignine at room temperature and pressure and saves most of energy, half of the cost and eliminates addition of chemical materials.

2.3 Cement industry

Cement industry succeeded in changing its main fuels from heavy oil to coal completely within 4 years only (1978-1981). Already wet kiln were not seen. New Suspension Pre-heater Kiln has been introduced which has led to efficiency improvements. The chemical reaction of cement production is at a temperature of 1500°C maximum. The exhaust gas is at 900-1100°C and thermal energy is absorbed by input materials by heat exchange. There is still excess heat at 250-380°C at the outlet. This hot gas can be recycled to generate electricity, an area for which technical development is under way. The electricity use is mainly for motive power -- dryer, crusher and particle separator. Power and speed control of motor by micro computer has been successfully introduced.

2.4 Chemical industry

Electricity use in the chemical industry is for motive power (pump, fan, blower), electric chemical process, and heating. The biggest end-use is pump, for which energy conservation methods are to introduce efficient compressor, rotation control, lower pressure operation and discontinuous operation of pump. Furthermore, waste heat is recovered to generate electricity. For these reasons, electricity use did not increase during 1975-85, while index of chemical industrial production showed 70.8% increase.

In sodium production industries an innovative, energy efficient, ion exchange membrane method is introduced because the mercury method was banned completely after 1986.

TABLE 4   EFFICIENCY OF SODIUM PRODUCTION METHODS (3)

	electricity consumption(kWh/ton)
Mercury method	3000 – 3200
As best membrane method	2800
Ion exchange membrane method	2600

The ion exchange membrane method is the most energy efficient, has the least impact on environment and still provides room for technical improvement.

2.5 Non ferrous metal industry

The electricity demand for non ferrous metal industry has drastically decreased in the last ten years because of the banning of aluminum electrolysis plants. At peak period the plant capacity was 1100 thousand tons/year, but now it is less than 100 thousand tons/year. The aluminum industry switched their strategy from domestic smelting to ingot import. In the international market place, aluminum is produced at electricity cost of around 2 cent/kWh, which is ten times less than that in Japan. So only the smelters that could use private old hydro electricity plants survived.

This case showed that market economy could automatically expel the uneconomical production processes completely. The aluminum related industry survived cleverly to produce and sell final aluminum products, for which demand at present is more than 1600 thousand tons/year.

2.6 Commercial and public buildings

Energy demand for commercial and public buildings has increased 3.3 times from 1965 to 1985 at the rate 6.2% per year. But from 1975 to 1985, the growth rate was as low as only 1.6% per year. This rate is lower than that of the household sector. The energy efficiency was much improved.

The energy efficiency was at bottom 374.8 Mcal/square meter/year in 1973. Since then it was improved to 265.1 Mcal/square meter/year in 1985. The electricity use is mainly for motive power and lighting.

Electricity use is increasing together with gas, in spite of the decrease in oil, coal and other sources. The electricity demand for motive power and others has increased from 55.8 kWh/square meter/year in 1973 to 87.2 kWh/square meter/year in 1975. This is because office environment has changed and equipment of office automation has increased.

2.7 Household sectors

Energy demand for the household sector has increased 3 times at a growth rate of 5.8% per year from 1965 to 1985. The growth rate is not so affected even after 1973. End-use share of electricity increased from 19.0%(1965) to 31.4%(1985). Also, the growth rate of electricity demand per household was 5.3% per year (from 849 kWh per household in 1965 to 2398 kWh in 1985). The reason is the penetration of TV, refrigerators, washing machines and other electric appliances. These markets are now almost saturated and the followers are air conditioners (heat pumps), video recorders, audio visual devices, cooking devices and personal computers, which are relatively less energy consuming, and have relatively less working time throughout the year.

TABLE 5 PENETRATION OF ELECTRIC APPLIANCES IN HOUSEHOLD IN 1985.(6)

	penetration rate to household(%)	
refrigerator	98.4	(*)
washing machine	98.1	Kotatsu is a spot
vacuum cleaner	97.4	heating device with
fan	95.3(1982)	table, infrared heater
kotatsu(*)	91.9	and kilt cover. People
air conditioner	52.3	gather there and warm
microwave oven	42.8	up lower half of body.

The largest electricity consumer in household is refrigerator. It is the only machine that is switched on all round the year and consumes nearly 30% of electricity demand in the household. But its efficiency has been greatly improved in the last ten years.

3. TECHNOLOGIES FOR ELECTRICITY CONSERVATION

There are many ways to save electricity in factories, commercial buildings and households. The most cost effective way is merely a good management of demand and devices, decreasing the loss of energy, which sometimes cost nothing in practice. The next is to introduce new efficient devices developed and commercialized since 1973.

The methods to save electricity are gathered and summarized here. They are already used widely in Japan, including lighting, heating, refrigerator, motors and other industrial electricity uses.

3.1 Lighting

Energy efficiency of incandescent bulb is nearly 1/3 of a fluorescent bulb, and has 1/5 - 1/10 of its life time. Furthermore, in summer time the incandescent bulb is a heat source which increases cooling demand.

TABLE 6 CHARACTERISTICS OF LIGHTING BULBS

	Incandescent bulb			Fluorescent bulb	
Catalogue power(W)	40	60	100	20	40
Lumen/W	12	13	15	50	65
Life time(hour)	1000	1000	1000	7500	10000

In the commercial buildings, incandescent bulbs are substituted with fluorescent bulbs increasingly. And furthermore, fluorescent bulb with plug compatible to incandescent bulb has been developed and is widely used in commercial buildings. It is a ball type bulb having U-shaped slender fluorescent tube inside. It consumes only 13 Watts but is equivalent to a 60 W incandescent bulb.

Several manufacturers have released a fluorescent bulb equivalent to a 100 W incandescent bulb. It has a shape of combining 4 straight fluorescent bulbs together; its size is smaller than conventional U-shaped fluorescent bulb. The point of compact, high power bulb design was to keep fluorescent materials cool and to maintain higher density of ultra violet rays which result in short life time. The new products solved these problems by introducing rare earth fluorescent materials and low temperature design.

The latest technology of fluorescent lamps is enhancement of 3 wave lengths (450 nm blue, 540 nm green and 610 nm red) to fit to human sensitivity. This method is successfully introduced and now occupies nearly 20-30% of the market.

TABLE 7   IMPROVEMENT OF 3 WAVE ENHANCING FLUORESCENT LAMP
(Rated Power 28W)

Year	1980	1981	1982	1983	1984	1985
Lumen	1670(*)	1720	1720	1950	2000	2200

(*) Conventional bulb still in the market.

The conventional straight fluorescent bulb has relatively large surface area and is often affected by dust. It is known that if a fluorescent bulb is used without cleaning during a year in dusty places such as forging factory, the light from the bulb reduces to nearly half of the original output. So cleaning the bulb is very important, especially in severe environments.

The efficiency of the high pressure sodium bulb is 4.65 times better than that of the incandescent bulb, and there has also been improvement in its light performance. These are mainly used for light in highways, city streets and parking places.

The reflecting covers for bulbs are usually designed without vertical air penetration, so dust is accumulated on the inner surface of reflecting covers. The substitution of reflecting covers to the ones having slits for vertical air penetration decrease the dust accumulation and cleaning cost with a 2 year pay-back time.

The materials for ceiling, wall and floor also affect the lighting efficiency as follows:

TABLE 8   LIGHT REFLECTIONS OF WALL MATERIALS

MATERIALS	LIGHT REFLECTIONS (%)
paints (white)	60 – 80
(black)	5
(red)	20
concrete	25
brown wall	5 – 10
white wall	60
paper	30 – 50
white tile	65 – 80

If high reflection materials are used in buildings, the electricity demand can be reduced, and less bulbs are necessary.

The switching circuits for lighting big rooms can be improved by dividing the circuits into several parts so as to be controllable depending on the local lighting demand. Also sun light sensor can be connected to automatic lighting control.

3.2 Refrigerators

Japanese refrigerators have become very energy efficient, as shown in Table 9. The methods of efficiency improvement are the design of efficient compressors and motors, thicker insulation with micro babble poly-urethane materials, elimination of heaters (super cool preventing heater, dew heater and drain heater), fan inside while fan motor outside the room, door sealing materials with multiple fin and so on. Furthermore, due to modernization of living standards, the capacity of refrigerators has been increasing from the average level 150 -170 liters ten years ago to the level of 300 liters these days, as in Table 10.(6)

TABLE 9   MONTHLY ELECTRICITY CONSUMPTION OF
           REFRIGERATORS   (kWh/month) (7)

Volume	1973	1974	1975	1976	1977	1978	1979	1980	1981	1982	1983	1984	1985
170 L	80	73	69	60	55	48	43	36	33	31	30	27	26
230 L	--	91	79	66	58	52	46	37	31	26	24	24	24
260 L	--	--	--	--	--	53	49	42	36	29	27	25	24

Now the best model of 300 liter capacity with direct flow fan and three doors (freezer/refrigerator/vegetable box) type consumes an average of 22-24 kWh/month.

The purpose of purchase is nearly 10% for new, 20% for add-on and 70% for substitution. The production is 4 - 4.8 million annually and around 800 thousand are exported mainly to Asia and the Middle East.

TABLE 10 DISTRIBUTION OF REFRIGERATOR SIZE (%)(6)

Volume	1978	1979	1980	1981	1982	1983	1984	1985
< 120	22.0	23.9	23.5	22.6	24.3	25.9	26.4	28.0
121 - 170	31.4	20.9	17.2	15.1	13.4	11.3	9.4	8.7
171 - 220	24.3	22.5	19.0	17.2	15.5	13.9	12.9	11.7
221 - 250	10.1	15.8	20.0	25.0	23.9	22.7	15.6	12.8
251 - 300	7.0	10.7	13.9	14.1	17.1	18.5	19.8	13.3
301 - 400	0.6	1.5	1.8	1.5	2.6	5.1	13.7	23.7
400 >	4.3	4.7	4.6	4.5	3.2	2.7	2.3	1.9
(liter)								

3.3 Television

Efficiency of T.V. is also improved from 140 watt in 1973 to 83 watt in 1985 for the 19-20 inch class, as shown in Table 11. The methods of improvement are transition of circuit elements such as vacuum tube -> transistor -> integrated circuit -> LSI, CRT with less energy consumption in intensity power and deflection power, substitute of preheater to quick heater (preheater consumed 5-10 W while unused).

TABLE 11   EFFICIENCY IMPROVEMENT OF TELEVISION (7)
            (Electricity Consumption: Watts)

class	1967	68	69	70	71	72	73	74	75	76	77	78	79	80	81	82	83	84	85
13-14i	--	123	107	95	88	83	83	77	71	64	66	65	60	60	57	55	55	55	55
19-20i	325	300	233	183	138	150	140	138	102	95	95	92	90	90	87	85	85	84	83

3.4 Motors

The efficiency of motors, to convert electricity to mechanical works, is believed to be more than 80% and there seems no more room for improvement.

However, in the end-use, there are various ways to decrease loss of energy in motor use. The methods are as follows:

(1) *Keep the rated voltage*. If the voltage is high, torque and power is proportional to the voltage. Excess power is unnecessary.

(2) *Decrease idling use.* Usually motor is connected to some rotary component of the machineries, the energy loss of idling is two or three times more than the idling loss in the motor itself. The following is a sample of idling.

TABLE 12   ELECTRICITY LOSS IN MOTOR USE

	3.7 kW	5.5 kW
rated power		
idling loss in motor only	0.29	0.3
idling loss in practice	0.53	0.98
connected to some component		
Annual loss (kWh)	159	294

The sample shows one hour idling use per day for 300 days result in 159 kWh and 294 kWh loss for 3.7 kW and 5.5 kW motors respectively.

To prevent idling loss, it is recommended to

a)   install switch at the place easily accessible.

b)   install an automatic control system to stop the motor when it begins idling.

c)   install a warning bell to inform of idling.

d)   reform the production process into continuous line so as to avoid waiting for next stage.

e)   to simplify and save time by improving the preparation work by changing layout and tools.

(3) *Optimum load.* Motor can work most effectively at the 80 - 100% load of rated output. In a factory, the press machines and others used 18 motors, a total of 170 kW. But the load was found to be distributed from 20% to 95%, an average 68% of rated output. They substituted motors to get optimum load, then 1700 kWh could be saved monthly, and the load change to 81% of the rated output. The motor load checker is commercially available and is useful in finding inefficient motor use.

(4) *Improve motor drive.* The efficiency of transmission of mechanical work is 100% for direct coupling, 96 - 97% for belt, 93 - 96% for gear, 85 -90% for worm gear. In the case of V-shape belt, the contact angle should be more than 140 degrees and the speed ratio should be less than 1 to 10 for each transmission. The efficient motor drive should include better selection of devices and good maintenance.

Eddy current couplings are also used to control speed. In an electric wire production factory, the idling time of eddy current coupling was found to be nearly 50%; 3.5 hours per day while it is activated 7 hours per day. So a sequence control system which can work as programmed operation and preparation work, was introduced and resulted in savings of 12,800 kWh/month for 90 motors and eddy current couplings, total capacity 1593 KW. The cost of controller was $4,000.00, which was paid back within 10 months.

(5) *Power factor control in motor use.* The power factor in motor use increases according to the load ratio as follows:

TABLE 13 RELATIONSHIP BETWEEN LOAD RATIO AND POWER FACTOR
(2 KW motor 200 V,50 HZ)

load ratio (%)	power factor (%)
25	40
50	58
75	72
100	85

Japanese electricity fee system depends on the power factor. Usually, it increases more when power factor becomes less than 85%. To improve power factor in motor use, advanced phase condenser is recommended. The appropriate capacity of condenser is as follows:

TABLE 14 CONDENSERS TO IMPROVE POWER FACTOR

Rated power (Kw)	0.2	0.4	0.75	1.0	2.0	3.0	5.0	7.5	10	20	30	40	50
Capacity of Condenser (F)	15	20	30	30	50	50	100	150	200	400	500	600	900

Usually the introduction of advanced phase condenser is cost effective and the pay-back time is less than a year, and furthermore it can decrease power loss in the distribution line, and give swing capacity to transformer and distribution equipment.

Power factor controller is also commercially available. It controls electric current according to the change of load, but this new technology is not widely used.

(6) *Speed control in motor use.* Speed control or rotary control is effective in the case of a blower, fan and pump used for variable load. The conventional method used to control fluid flow is to install a damper to adjust the area of flow to load while running a motor at a rated output. In such cases, speed control of the motor is designed by means of frequency control, fluid coupling and so on.

A coke production factory used 30 kW x 2 unit motor with damper control to drive blowers of dust collectors and consumed 6000 MWh annually. A fluid coupling was introduced as speed controller to adjust the load fluctuating from 10% - 20% load during 80% time and 100% load during 20 % time. The cost of fluid coupling was $150,000.00 and was paid back in a year by reducing electricity demand to 2750 MWh.

3.5 Air Conditioners

The air conditioner is gradually penetrating in factory, office and household. The heat pump is used as a cooler in summer and as a heater in winter in the area except Hokkaido, where cooling demand is extremely low. The conventional device for heating is a kerosene heater in ordinary households. It is the cheapest way of heating. However, the manufacturers of the latest heat pumps with COP of 2 - 2.8 claim that they are cheaper than gas heating and are cost equivalent to kerosene heating.

The Japanese housing condition is affected by scarcity of land and space, and the heat pump performs as both cooler and heater to save space requirements. While 3.3 million air conditioners were produced, 2/3rds of them were heat pumps in 1985.

The ordinary size of heat pump for a room of individual house is around 1800 -4000 kcal/h for heating and 1500 - 3000 kcal/h for cooling. The heat pumps are expected to be cheaper than electric heaters or gas and some equivalent to kerosene heating, if the outdoor temperature is not less than 7°C. When outdoor temperature goes down to 0°C, then the capacity of heat pump decreases. However, heat pumps run by sensoring room temperature automatically and an inverter control circuit controls the power at the optimum. This mechanism greatly contributes to energy savings also.

TABLE 15 EFFICIENCY IMPROVEMENT OF COLLING MACHINES (7)
(Electric Consumption: Watt)

class (kcal/h)	1973	1974	1975	1976	1977	1978	1979	1980	1981	1982	1983	1984	1985
1600	847	838	755	711	687	650	606	567	530	511	504	495	493
2000	1031	1005	943	905	881	844	797	766	741	714	706	699	682

When outdoor temperature is low, the heat pump does not seem to be enough, and some backup heating is necessary for the less insulated household. Twenty-eight% of newly built Japanese houses were insulated in 1975, and 70% in 1985. Insulated houses form an estimated 20% of the housing stock today.(8)

3.6 Electric Welding Machine

Welding machines use low voltage and large current electricity. Careful use of secondary cable is important. The cable should be as short as possible, and never wound. If it is wound, the loss is 3 - 4 times bigger than straight cable, and voltage drop increases to 20 volts.

The electric current in electric welding is also important. In the case of spot welding, points of welding must be carefully selected not to generate ineffective separate electric currents.

Other importants factors for energy saving are (1) to keep the surface clean, (2) to make the parts in high accuracy for homogeneous surface contacts, (3) to standardize the welding conditions (electric current, welding time and voltage) for respective materials, (4) to keep the shape of welding tip by careful dressing.

In cases using many alternative current arc welding machines, it is recommended to use welding machines including condensers to improve power factors. This enables a reduction in contract electricity capacity, and size of main transformers. The welding machine including condensers reduces nearly 30% of input kVA and contributes to less voltage drop in primary cable, and to the stability of voltage and arc.

Welding machines seem to have no room for improvement in energy efficiency, but there are many factors that can save electricity at a small cost, as mentioned above.

3.7 Industrial Electric Furnace

Industrial furnaces are widely used in mechanical and electrical industries and have large potentials to save energy and cost.

One of the most simple methods to save energy in industrial furnaces is to reduce the weight of the products. Reducing weight means (1) less thermal capacity, (2) less heating time, (3) less energy consumption. In a factory producing automobile parts, the weight of parts is reduced to 80% from the initial design. This achieves the same reduction in cost and electricity. Less heating time is a key factor in increasing production speeds. Not only the products, sometimes the weight of containers carried into furnaces can be reduced and this would contribute to energy savings.

Small industrial furnaces are often used in daytime only and not used at night. In the morning they must wait until the furnace is warmed up. In case of metal vapor deposit continuous furnace, cooling effects at night can be prevented by building insulation doors at inlet and outlet. This simple method reduces heating-up in the morning to 40% of energy and warm-up time with a pay-back of 10 months.

Also, a metal curtain of stainless steel at the gate of continuous furnace can improve efficiency by 10% during operation with a pay-back time of 8 months.

In case of plastics melting furnace, it is controlled to keep at a temperature more than the melting point in 24 hours. It is necessary to keep such high temperature while melting because the excess temperature can drive melting. But once melted, it is unnecessary to keep the plastic at temperatures more than the melting point. This problem can be solved easily by installing a timer which can control the temperature in two ways. For example, 90°C degree for melting in 10 hours, and 70°C degree in 14 hours. This method can save 20% of electricity demand with a pay-back time of 3 months.

Insulation of furnace wall is needless to mention, though heat recovery is important. In case of high temperature coating, chain conveyors are heated and used to carry a product in and out of the furnace. The heated conveyor could preheat the products by changing layout and save 30% of electricity demand with a pay-back time of 6 months in a can making factory.

Energy conservation in industrial furnaces requires only good management of heat, temperature and products. These points are sometimes forgotten because they are too simple and not attractive. But the results are astonishing.

Laser beam is still expensive but has a great potential to save energy for spot heating demand. Laser cutting is expected to reduce material requirements because it can cut flexibly and produce less residue. Microwave is also useful for heating directly inside of the materials. These technologies are not reported to have contributed to energy savings as yet.

3.8 Co-generation systems

CGS (Co-generation system) is an energy efficient system to supply electricity and low temperature heat. CGS applies gas engine, diesel engine, gas turbine or fuel cell as power generator and gas, oil, or solar heat as energy sources. Though conventional power plants have conversion efficiency of 35% at maximum, the efficiency of CGS is 20-30% for electricity generation and 40-50% for heat recovery from exhaust gas; the integrated efficiency is expected to be as much as 70-80%.

In industrial processes such as paper and chemical industry, CGS was already used to generate electricity and steam and the total capacity was 15GW, 9% of national power plant capacity in Japan in 1985. CGS recently focussed on are those for commercial and public buildings in urban areas, which have demand of electricity, hot water, heating and cooling.

Gas supplying companies and engine manufacturers are dashed into the CGS market and 83 units worth more than 40 thousand kW, were already introduced until 1986. CGS units range from 16 kW to 14,000 kW. Their energy sources are town gas(60%), heavy oil(20%), kerosene(10%) and LPG(10%). Electric utility companies can decrease their peak demand at summer as CGS can substitute cooling demand. To introduce CGS efficiently, it is important that CGS must work as base load and operate following electricity demand (because heat demand fluctuates much). Now, the user can sell excess heat to others but cannot sell electricity as regulated by law. Economics of CGS were widely studied and pay-back time depends on demand patterns. Expected pay-back time is 3 years for hotels, 6 years for hospitals and 11 years for shops in typical cases, supplying 70-85% of the total demand.

4. CONCLUSION

The trend of electricity demand shows that though the growth rate has been winding down, demand is still growing due to the shift to a modernized life style, using convenient equipment and favoring artificial comfortable environment with affluent industrial materials. However, there are some signs that show that this shift has become saturated and consumption based on hardware materials is growing less. Big smoke stack industries are depressed and are looking for business opportunities in software based businesses.

Energy efficiency has improved drastically, where do we go from here? At this point it seems very difficult to foresee future energy demand and price. From the view point of technical development, we can expect more efficient energy use and the introduction of renewable sources in the near future. Photovoltaics are massively produced and are expanding their applications. Their cost reduction is expected feasible to be competitive with conventional electricity supply systems in the 1990s. If energy efficiency should have improved by then, photovoltaics could be easily connected to efficient end use in factory and household.

Efficiency improvement is a key resource when we discuss long term energy supply and demand on a global scale, to reduce the environmental effect of massive uses of fossil fuels, and to construct sustainable energy systems.

References

1. Energy Conservation in Japan, 1984. The Energy Conservation Center, Japan.

2. Energy Balances in Japan(1985). The Energy Data and Modeling Center, The Institute for Energy Economics, Japan. 1986.

3. Study on End Use Structure of Industrial Energy Demand. The Institute for Energy Economics, Japan. 1986.

4. Enquete and Hearing Survey on Energy Consumption in Commercial Sector. New Energy Development Organization, Japan. 1987.

5. Econometric Analysis on Oil-Substituting Energy. New Energy Development Organization. 1987.

6. Japanese Electrical Manufacturing, 1985. The Japan Electrical Manufacturers Association.

7. Handbook of Electric Appliance Industry, 1985. Association of Japan Electric Appliances.

8. Energy Conservation Handbook, 1985. Institute of Building Energy conservation, Japan.

THE POTENTIAL FOR RESIDENTIAL ELECTRICITY CONSERVATION IN THE U.S.: THE PG&E CASE STUDY

HOWARD S. GELLER

AMERICAN COUNCIL FOR AN ENERGY-EFFICIENT ECONOMY
WASHINGTON, DC

1. BACKGROUND

This paper examines the potential for cost-effective electricity savings in the residential sector of one large utility in the U.S. -- the Pacific Gas and Electric Company (PG&E). PG&E serves about 3.5 million residential customers in northern and central California. The paper is a summary of a detailed study conducted for the utility (1). In this study, the potential savings are termed a "conservation power plant" to signify that improved end-use efficiency is one of the resource options available to the utility. This conceptual approach is in accordance with other studies conducted by PG&E.

The first step in assessing the feasibility of a residential conservation power plant is to determine its potential size and composition -- How much cost-effective conservation is potentially available? What technologies and options look most promising in terms of savings, cost-effectiveness, and commercial availability?

To answer these questions, we analyzed the potential for electricity conservation in seven major residential end-uses, namely:
o refrigerators;
o freezers;
o water heating;
o lighting;
o central air conditioning;
o cooking;
o clothes drying.
Together, these end-uses account for about 70% of the electricity consumed in PG&E's residential sector.

Technology assessments in these seven end-use areas cover currently available and advanced conservation technologies. Each option is evaluated on the basis of cost, electricity and peak power savings, cost-effectiveness and status. The electricity-conserving options do not involve any reduction in lifestyle or comfort level.

Following the technology assessments, the study develops three scenarios for electricity use in the seven end-use areas over the next 20 years. One is a base scenario that is close to PG&E's 1985 end-use forecast; the second is a current technology scenario assuming a higher penetration of cost-effective, energy-efficient technologies now available but not yet widely used; and the third is a technical potential scenario assuming a high penetration of

A. T. De Almeida and A. H. Rosenfeld (eds.), Demand-Side Management and Electricity End-Use Efficiency, 607–620.
© 1988 by Kluwer Academic Publishers.

both energy-efficient products now available and advanced technologies not yet commercially available. The same overall equipment stocks and replacement rates are used in all scenarios.

The scenarios analysis shows that relative to the base case, the current technology scenario leads to a 25% reduction in electricity consumption in 2005. In the technical potential scenario, electricity consumption in 2005 is 44% lower than in the base scenario. The end-uses presenting the greatest electricity savings potential are refrigerators and lighting, while CAC offers the majority of the potential savings in peak summer demand.

2. TECHNOLOGY ASSESSMENTS

In each technology assessment, a "baseline technology" and various electricity-conserving options are considered. The energy performance of the baseline technology is close to that of the typical appliance model sold in 1985.

The cost-effectiveness of the conservation measures is evaluated from the perspective of utility ownership, considering only the extra first cost for the efficiency measures. We calculate both the cost of saved energy (CSE) and the cost of conserved peak power over 20 years (CCPP20) for each conservation option. The CSE is given by the annualized extra cost for an option divided by the annual energy savings (2). CCPP20 is the net present value of investments required to save a kW of peak demand with a particular conservation measure for a 20 year period.

Cost-effectiveness is determined by comparing the costs of electricity and peak demand savings to PG&E's marginal electricity supply cost -- $0.06-0.10/kWh during 1986-2005 (1). Important assumptions that are used in the economic evaluation include a 7% real discount rate and amortization periods equal to the estimated product lifetime. Programs to promote the purchase of more efficient equipment are not included in the analysis, nor is the potential increase in equipment usage as a consequence of lower utility bills accounted for.

2.1. Refrigerators and freezers

The typical refrigerator-freezer produced in the U.S. in 1985, with a total volume of about 500 liters, consumes about 1100 kWh/yr according to the standard test procedure in the U.S. Although considerable progress has been made in improving the efficiency of new refrigerators and freezers in recent years (see Figure 1), there is large potential for further cost-effective energy savings.

Table 1 shows the analysis of conservation options for refrigerator-freezers with a top-mount freezer. By combining a variety of design options such as more efficient motor-compressors, improved insulation, and better refrigeration system design, it is possible to reduce energy consumption by as much as 85% relative to the electricity use typical of models produced in 1985. Furthermore, realizing the full savings potential from refrigerators and freezers has an average cost of saved energy of only $0.03/kWh, much less than PG&E's marginal electricity supply cost. Similar savings potential exists for other types of refrigerators and for freezers.

Most of the potential savings is not yet available in mass-produced refrigerators and freezers in the U.S. However, commercial models exhibiting very low energy consumption could become widely available by the early-1990's because of the adoption of minimum efficiency standards and expected revisions to the initial standards levels (3), and the broad interest in electricity conservation in the U.S. It should be noted that a prototype 510 liter refrigerator-freezer consuming about 480 kWh/yr has been built in Denmark (4). Furthermore, a very efficient custom-made refrigerator is already produced in the U.S. (5).

2.2. Water heating

Only about 9% of the households served by PG&E heat water with electricity. For those households with electric water heating, electricity use is approximately 4400 kWh/yr in single family homes and about 3000 kWh/yr in apartments and mobile homes. Despite the low saturation of electric water heaters, there is substantial potential for conserving electricity and peak power demand in this end use.

Table 2 shows the water heater conservation options for single family homes with electric water heating. It is assumed that such households typically consume 50 gallons (190 liters) of hot water per day. The energy factor values shown in Table 2 are overall efficiencies taking into account heat losses from the storage tank and distribution pipes. In addition to the water heater options listed in Table 2, there are conservation options such as low-flow showerheads that reduce the amount of hot water usage.

Electricity consumption for water heating can be reduced by 50-75% using heat pump water heaters rather than ordinary electric resistance heating (6). Domestic heat pump water heaters are now commercially available in the U.S. and other industrialized countries. As long as hot water consumption is sufficiently high (150-190 l/day), heat pump water heaters are generally cost-effective compared to PG&E's marginal electricity supply costs.

Reducing hot water demand for uses such as showers, clothes washing and dishwashing can provide cost-effective energy savings as well. Front-loading clothes washers are a water-conserving option currently available; technologies allowing low-temperature dishwashing are expected to become available in the near future (1).

2.3. Lighting

Compact fluorescent light bulbs are an important conservation technology currently available in the U.S. and other industrialized countries. Compact fluorescent bulbs consume 60-75% less power than incandescent bulbs for the same amount of light output and last 5-10 times as long. Although compact fluorescent bulbs cost $15-20 in the U.S., they are cost-effective in commonly used lamps and fixtures (usage at least 1.7 hours per day) (1).

Other conservation options for residential lighting include Krypton-filled incandescents that are available at wattages 5-10% lower than ordinary incandescent bulbs. These bulbs are widely available and are economical in low-use applications. In addition, a coated incandescent bulb that consumes 50% less electricity than a conventional incandescent bulb is available on a limited basis in the U.S.

2.4. Central air conditioners

Central air conditioning (CAC) systems (including heat pumps) are used in about 21% of the households served by PG&E. However, residential CAC systems accounted for about 12% of PG&E's entire peak demand in 1982.

The efficiency rating for new CAC systems in the U.S. ranges from under 7.0 to as high as 15.0, while the average new model has an efficiency rating of about 9.0 (7). (The ratings are expressed in terms of Btu/hr of cooling output per watt of power input.) More efficient CAC systems providing about 30% savings compared to ordinary systems appear to be cost-effective on the basis of annual electricity savings in high-use applications (3000 hours/yr or more) if expected equipment price reductions are realized (1).

Indirect evaporative cooling is an emerging technology that shows great promise for providing on the order of 75% energy and peak power savings in residences in a cost-effective manner (8). Indirect evaporative cooling does not create the high indoor humidity levels experienced with ordinary evaporative coolers. It is estimated that the cost of saved energy is $0.025-0.05/kWh as long as air conditioning is needed at least 1000 hours/yr.

Thermal storage of "coolth" is another potentially attractive technique for reducing the peak power demand for air conditioning. Various systems are currently being developed and commercialized for the residential market.

2.5. Cooking ranges

Using simpler technologies such as increased insulation, better oven door seals, reduced thermal mass, and burner elements with less contact resistance, it should be possible to lower the electricity consumption of electric ranges by about 20% in a very cost-effective manner. These improvements present no significant technical challenges and it is believed that they are used to some degree in newer cooking ranges.

Induction cooktops, now commercially available, are estimated to consume about 15-25% less electricity than conventional electric cooktops besides providing other important benefits (1). Induction cooktops are not cost-effective, however, when evaluated strictly on the basis of energy performance. Thus, the other benefits must be taken into account to justify their use.

An innovative oven design, known as the bi-radiant oven, consumes about 70% less electricity than conventional ovens and is cost-effective on the basis of its cost of saved energy (9). Unfortunately, manufacturers do not appear to be interested in producing this oven design at the present time because of uncertain consumer response and other factors.

2.6. Clothes dryers

Moisture sensor and automatic termination controls are available with clothes dryers. This cost-effective feature typically results in 10-15% electricity savings. A number of innovative clothes dryer technologies are under development and some are already commercially available overseas. These include dryers with exhaust heat recovery, heat pump dehumidification dryers, and microwave dryers. The estimated electricity savings are 40% with the microwave dryer and 50-70% with the heat pump (1).

Both of these advanced technologies appear to cost-effective relative to PG&E's marginal electricity costs. One company is planning to market a heat pump clothes dryer in the U.S. in the near future (4).

3. SCENARIOS ANALYSIS

The three scenarios mentioned previously are developed by modeling equipment stocks, new purchases, and retirements along with average energy consumption year-by-year for each end-use (1). Most of the demographic assumptions as well as the unit energy consumption values in the base scenario are taken from PG&E's residential end-use model. The technology assessments serve as the basis for the assumptions regarding the energy consumption of new models in the current technology and technical potential scenarios. Those conservation options deemed to be cost-effective are included in the scenarios when they are expected to be available. No attempt is made to limit the savings due to implementation problems or other constraints.

Table 3 shows the principal results of the scenarios analysis. In the base case, overall electricity consumption for the seven end-uses increases 37% between 1985 and 2005. Electricity consumption per household declines about 8% over this period. In the current technology scenario, total electricity consumption for the seven end-uses remains nearly constant during the next 20 years, and consumption per household declines 31%. In the technical potential scenario, absolute electricity consumption drops 24% between 1985 and 2005, with consumption per household falling by nearly 50%.

Figures 2 and 3 show the savings in electricity consumption and peak power demand over time in the current technology and technical potential scenarios. The savings are determined relative to the base case scenario. As indicated in Table 3 and Figures 2 and 3, the savings potentials in the current technology scenario are 5180 GWh/yr and 1790 MW of peak demand by 2005. These are 25% and 31% reductions from the base case, respectively. A reduction in electricity consumption of 5180 GWh/yr is equivalent to the output from approximately 1000 MW of baseload generating capacity.

In the technical potential scenario, the savings potentials by the year 2005 are 9260 GWh/yr and 3240 MW of peak demand, 44% and 56% reductions from the base case. A reduction in consumption of 9260 GWh/yr is equivalent to the output from about 1800 MW of baseload generating capacity.

Figures 4 and 5 show the estimated electricity consumption and peak power demand in the year 2005 by end-use and scenario. Refrigerators and lighting are the end-uses offering the greatest electricity savings potential. Lighting provides about 40% and refrigerators about 25% of the total savings in the technical potential scenario. In terms of peak power demand, air conditioning stands out, providing about two-thirds of the total savings in the technical potential scenario.

4. QUALITATIVE ISSUES AND CONCLUSION

End-use efficiency improvements, the so-called "conservation power plant", have a number of qualitative advantages in comparison to traditional power plants. The advantages include shorter lead time, low risk of environmental degradation, no interest charges or capital exposure during implementation, maximum flexibility, and improved control over load shape. Efficiency improvements would be more likely to receive regulatory approval and enhance customer relations.

On the other hand, there are significant uncertainties related to the effectiveness of conservation technologies, implementation, customer response, and changes in the regulatory and political environment. But through technical R&D as well as program experimentation and evaluation, it should be possible to limit these uncertainties to manageable levels.

In conclusion, the PG&E case study has demonstrated that there is substantial potential for cost-effective electricity and peak demand savings in the residential sector in the U.S. Building "conservation power plants" through end-use efficiency improvements is a way to avoid costly, unnecessary power plants and to move towards least-cost energy services.

REFERENCES

1. H.S. Geller, et al., "Residential Conservation Power Plant Study Phase I - Technical Potential", report prepared for Pacific Gas and Electric Co. by the American Council for an Energy-Efficient Economy, Washington, DC, Feb. 1986.
2. A. Meier, et al., Supplying Energy Through Greater Efficiency, University of California Press, Berkeley, CA, 1983.
3. H.S. Geller, "Energy and Economic Savings from National Appliance Efficiency Standards in the U.S.", (this volume).
4. H.S. Geller, "Energy Efficient Appliances: 1986 Update", Proceedings of the ACEEE 1986 Summer Study on Energy Efficiency in Buildings, American Council for an Energy-Efficient Economy, Washington, DC, Aug. 1986.
5. H.S. Geller, "Progress in the Energy Efficiency of Residential Appliances and Space Conditioning Equipment", in D. Hafemeister, H. Kelly, and B. Levi, eds., Energy Sources: Conservation and Renewables, American Institute of Physics, New York, 1985.
6. J.E. Dobyns and M.H. Blatt, "Heat Pump Water Heaters, EPRI EM-3582, Electric Power Research Institute, Palo Alto, CA, May 1984.
7. Date provided by the Air-Conditioning and Refrigeration Institute, Arlington, VA, 1987.
8. N. Eskra, "Indirect-Direct Evaporative Cooling Systems", ASHRAE Journal, May 1980.
9. D. DeWitt and V. Peart, "Bi-Radiant Oven: A Low-Energy Oven System", ORNL/Sub-80/0082/1, Oak Ridge National Laboratory, Oak Ridge, TN, April 1980.

TABLE 1 - OVERVIEW OF REFRIGERATOR-FREEZER CONSERVATION OPTIONS (a)

Option	Electricity use (kWh/yr)	Peak demand (kW)	First cost (1985$)	Avg. CSE ($/kWh)	Marginal CSE ($/kWh)	CCPP(20) ($/kW)	Est. year avail.
Baseline	1165	155	671	--	--	--	1985
1992 Standards Model (b)	610	81	731	0.010	0.010	810	1989
Low Technology Measures (c)	460	62	807	0.018	0.050	3920	1991
Intermediate Technologies (d)	385	51	880	0.025	0.089	6740	1993
Advanced Technologies (e)	175	23.5	985	0.030	0.047	3770	1995

(a) Based on a 450-510 liter (16-18 cubic foot) top-mount refrigerator-freezer with automatic defrost.

(b) Includes a moderately improved compressor, more insulation, a more efficient fan motor, and a double freezer gasket.

(c) Includes a 4.5 EER compressor and a double refrigerator gasket in addition to previous measures.

(d) Includes an external fan motor, 5.0 EER compressor, and dual evaporator in addition to pervious measures.

(e) Includes evacuated panel insulation and bottom-mounted condenser in addition to perivous measures.

TABLE 2 – OVERVIEW OF WATER HEATER CONSERVATION OPTIONS (a)

Option	Energy Factor	Electricity use (kWh/yr)	Peak demand (kW)	First cost (1985$)	Avg. CSE ($/kWh)	Marginal CSE ($/kWh)	CCPP(20) ($/kW)	Est. year avail.
Baseline	0.81	4400	542	300	--	--	--	1985
Thermal traps & insulation blanket	0.9	3960	488	335	0.010	0.010	840	1985
Avg. heat pump water heater	1.6	2230	275	1050	0.041	0.049	4350	1985
Improved heat pump water heater	2.2	1620	200	1350	0.045	0.059	5180	1985
Advanced heat pump water heater	2.6	1370	169	1500	0.047	0.072	6270	1987
Exhaust heat recovery HPWH	2.0	1280	137(b)	1750	0.056	0.332	2360	1987

(a) Based on a hot water demand of 190 l/day.

(b) Some proposed designs reverse air flow during the summer and therefore cool incoming air. This cooling benefit has not been included, even though it could significantly increase peak demand savings.

(c) Estimate assumes that the de-superheater contributes three months of hot water and that the original unit had an EF = 0.90. The cost for the de-superheater does not include the cost for the hot water heater itself.

TABLE 3 - ENERGY CONSUMPTION AND PEAK POWER DEMAND
IN THE THREE SCENARIOS

	Base Scenario	Current Technology Scenario	Technical Potential Scenario
Electricity consumption in 2005 (GWh/yr)	20,800	15,600	11,600
Peak power demand in 2005 (MW)	5,750	3,960	2,510
Change in electricity consumption (1985-2005)	+37%	+3%	-24%
Change in el. consumption per household (1985-2005)	-8%	-31%	-50%
Change in peak power demand (1985-2005)	+57%	+8%	-32%
Change in peak demand per household (1985-2005)	+5%	-28%	-55%
Change in el. consumption in 2005 relative to base scenario	--	-25%	-44%
Change in peak demand in 2005 relative to base scenario	--	-31%	-56%

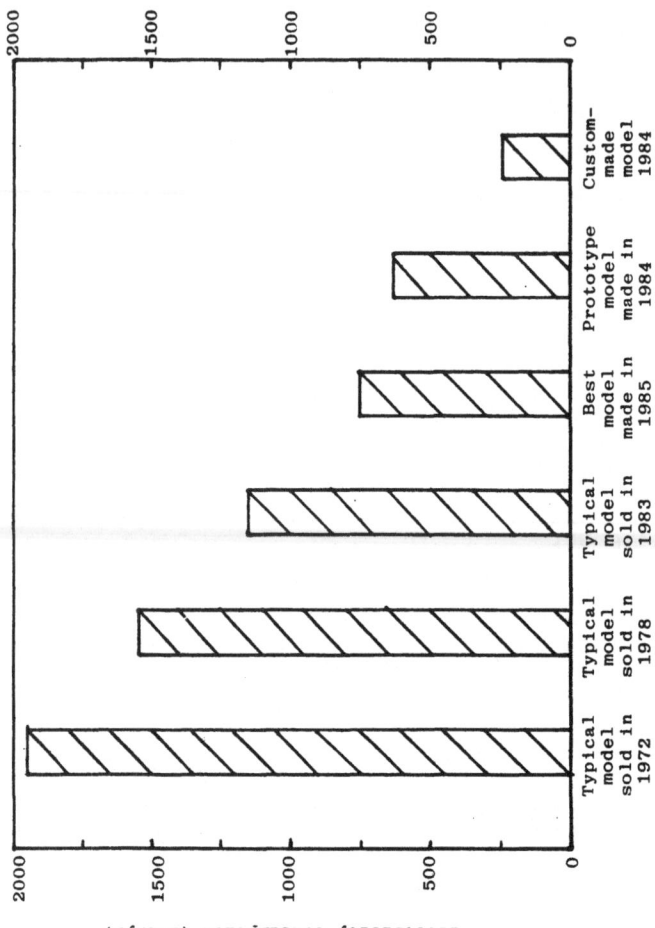

Figure 1 - Progress in the electricity consumption of
top mount freezer, automatic defrosting refrigerator-
freezers, 16-18 cubic feet manufactured in the U.S.

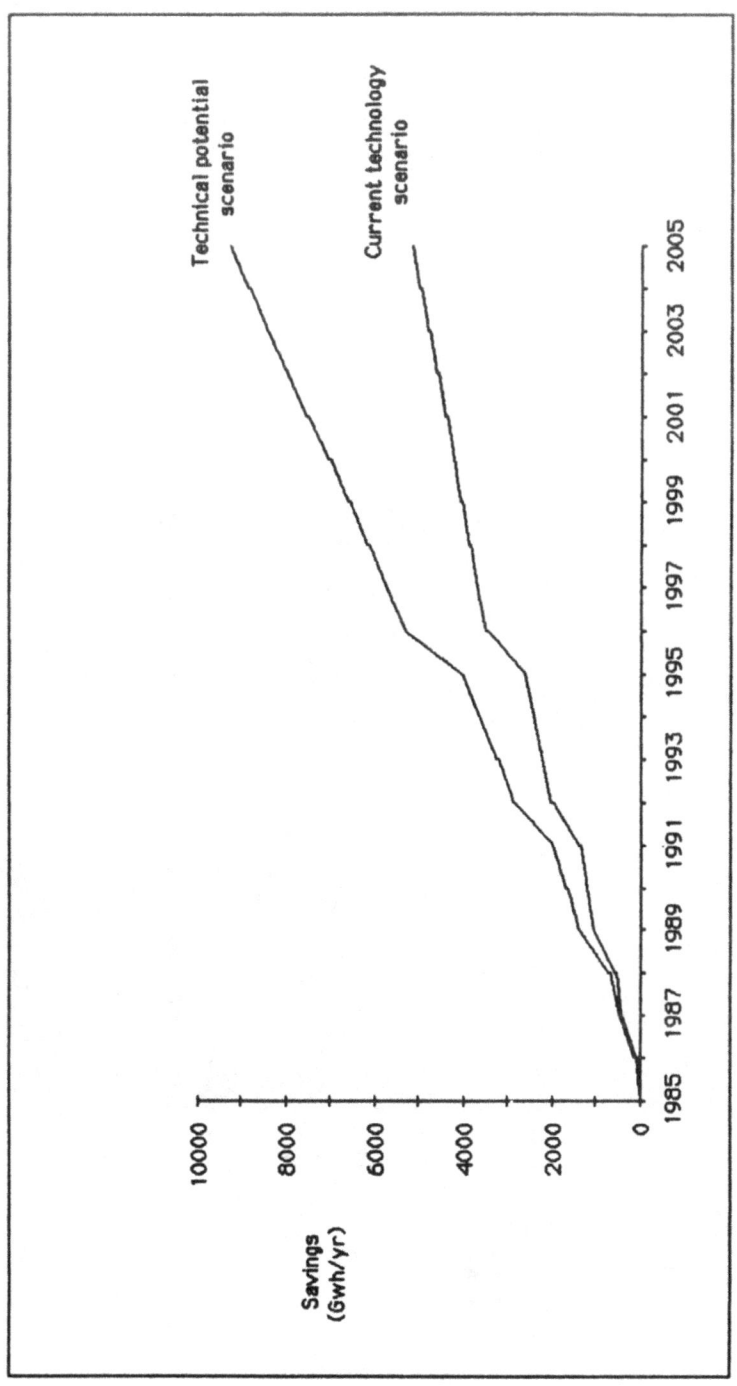

Figure 2 – Total Electricity Savings
(relative to base case scenario)

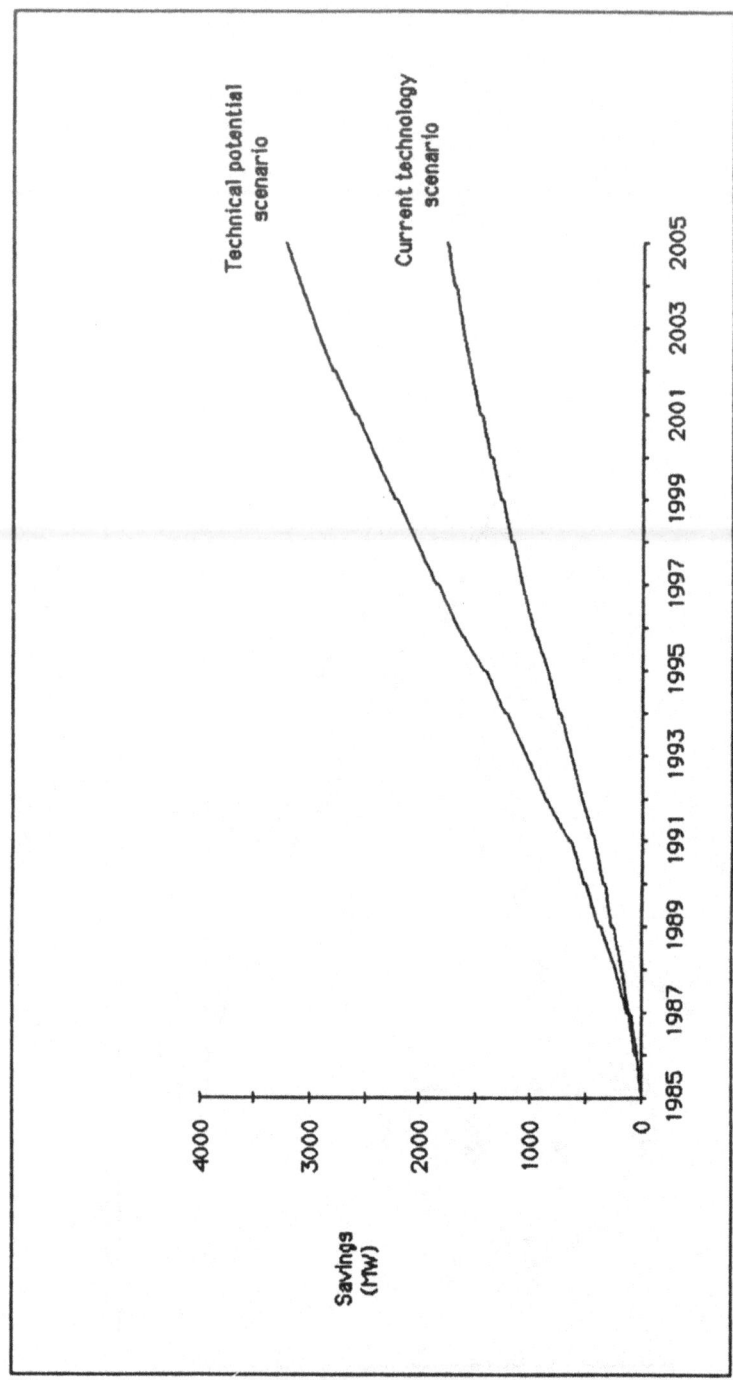

Figure 3 - Total Peak Demand Savings
(relative to base case scenario)

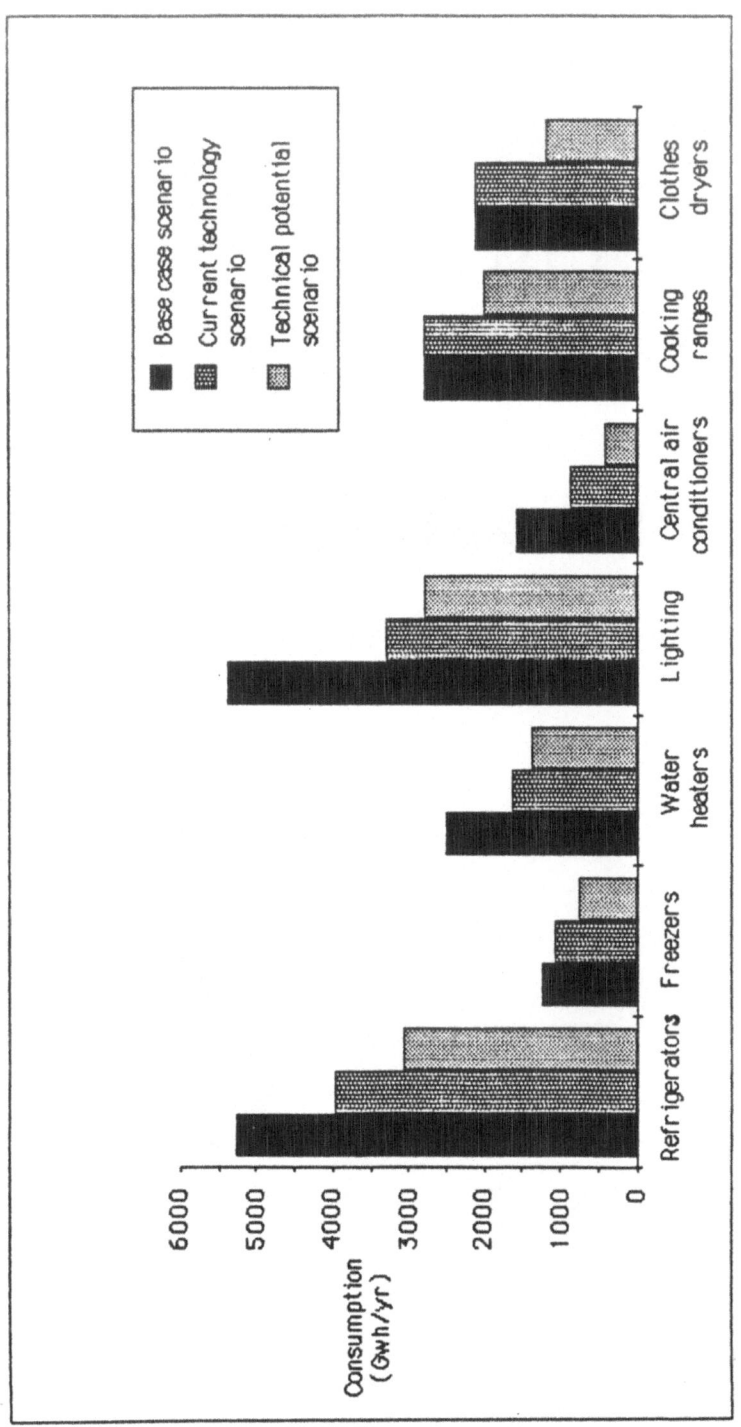

Figure 4 - Electricity Consumption in 2005

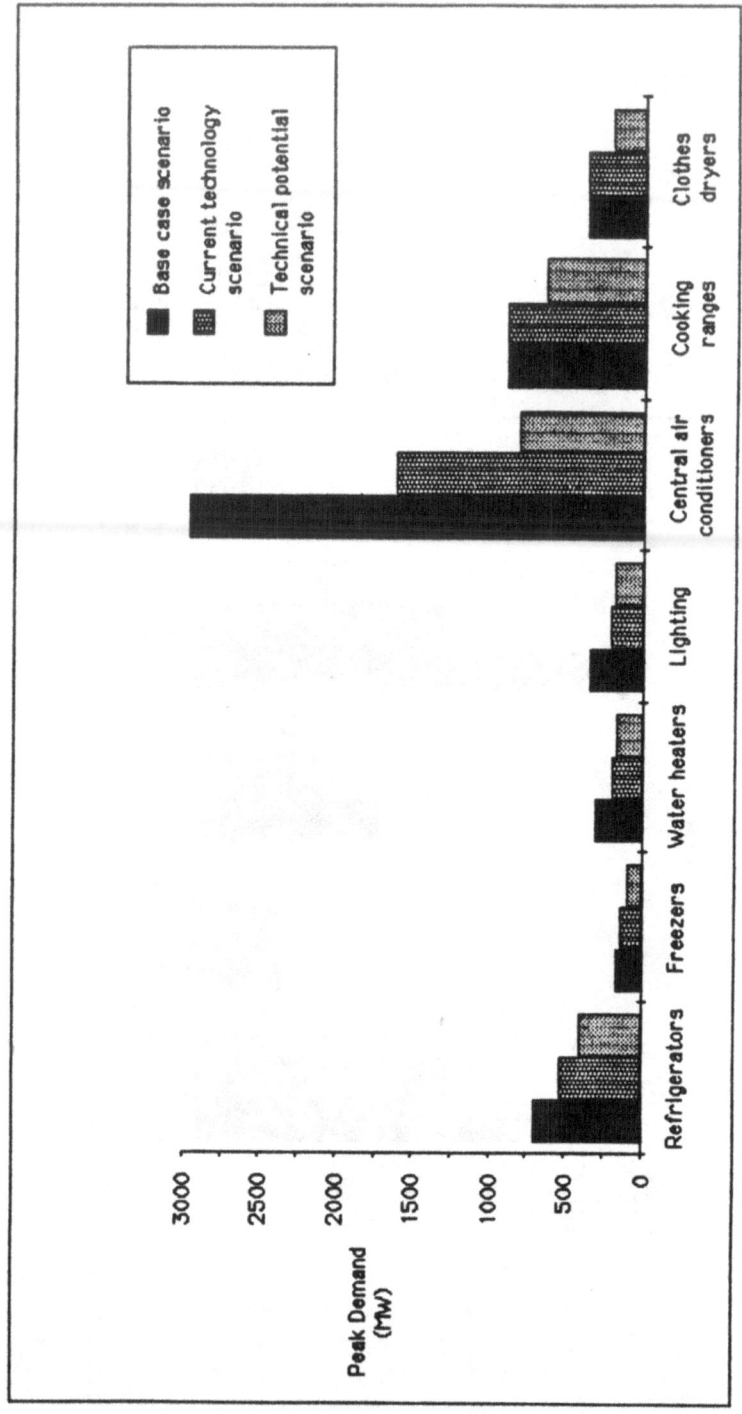

Figure 5 - Peak Power Demand in 2005

THE POTENTIAL FOR ELECTRICAL EFFICIENCY IN THE RESIDENTIAL SECTOR CASE STUDY MICHIGAN, USA

Florentin Krause, William G. Colborne, Arthur Rosenfeld

Lawrence Berkeley Laboratory, University of California, Berkeley, CA 94720

1. INTRODUCTION.

Over the last few years, regulatory agencies and electric utilities have increasingly adopted a new planning approach known as least-cost utility planning or integrated resource planning. This approach involves the consideration of a broadened set of supply-side options (including refurbishment, cogeneration and small-scale renewables-based generating plants, power purchases), along with a large number of demand-side options (various efficiency improvements and load management programs) on a "level playing field." The analytic challenge posed by this new approach is to (1) quantify demand-side resources in a way that makes them directly comparable to conventional supply-side resources, and (2) to construct utility resource plans using a year-by-year, side-by-side integration of both kinds of resources over an extended planning horizon.

This study was part of an attempt to perform such an integrated resource planning exercise: The Michigan Electricity Options Study (MEOS). The Michigan project was prompted by the enormous cost of two nuclear power plants that had been recently constructed in the state and the rate shock that was anticipated as these investments would enter the rate base. To analyze the state's future options, MEOS commissioned several research organizations to develop the cost and size of supply-side and demand-side resources. Forecasts of the future demand for energy services were developed by a special work group, and formed the basis for the resource assessments. At this writing, the various supply and demand-side analyses are being combined into several alternative integrated resource plans. This integration is based on an hourly valuation of both demand-side and supply-side resources.

Lawrence Berkeley Laboratory's (LBL) contribution to the MEOS effort was an analysis of the technical and achievable potential for increased electrical efficiency, load management, and fuel switching in the major residential end-uses. The basic method used was a supply-curve analysis of presently available demand-side technologies.

The present study differs from early conservation potential/supply curve analyses in a number of respects:

- Unlike previous analyses, which only provided technical potentials of demand-side resources, the study estimates an achievable potential specifying what amount of that resource could be deployed through specific state- or utility-sponsored incentives programs, and at what program-based costs.

- The study estimates hourly end-use based load profiles for different day types and seasons. Based on these, the approximate impact of efficiency investments on system peak demand and the cost of conserved peak power are calculated.

A. T. De Almeida and A. H. Rosenfeld (eds.), Demand-Side Management and Electricity End-Use Efficiency, 621–642.
© *1988 by Kluwer Academic Publishers.*

2. BASIC APPROACH.

2.1. Preparation of analysis inputs

    2.1.1. <u>Inputs.</u> The principal inputs to the analysis are:

- Baseline data concerning equipment and building stocks, turnover rates, and unit energy and peak power consumptions.
- A projection by end-use of business-as-usual fuel shares, equipment saturations by type, and of changes in demographic and behavioral factors that influence the consumption of *energy services*.
- Technology cost and performance data, characterized in terms of costs of conserved energy or peak power.
- Quantitative and qualitative results and design experience from past and on-going demand-side programs by utilities and states.

    2.1.2. <u>Forecast and scenario development.</u> These inputs are combined to develop a set of two forecasts and two scenarios:

- A frozen efficiency forecast,
- a business-as-usual (base case) forecast,
- a technical potential/best available technology scenario, and
- a program-based scenario.

All four are disaggregated, end-use based accounts of electricity use covering the years 1985-2005.

The *frozen efficiency forecast* is based on the assumption that all existing equipment will remain at its 1985 stock-weighted energy efficiency until replaced (no efficiency retrofits). Further, all new equipment and buildings will be no more efficient than 1985 sales-weighted averages (no efficiency improvements in available equipment). Behavior functions, population and household growth, turnover of old capital stocks, and saturation changes occurs as in the other scenarios. In so far as the unit energy consumptions of buildings and equipment *sold* in 1985 were lower than the stock-weighted averages, the frozen efficiency forecast does itself lead to reductions in stock-average unit energy consumptions as old equipment is replaced by new stocks over time.

The *business-as-usual forecast* combines the same demographic, saturation, and behavioral data with business-as-usual trends in appliance, lighting, and building efficiency. These trends are based, in part, on data supplied by the Association of Home Appliance Manufacturers (AHAM). This forecast is henceforth called MEOS/AHAM or, abbreviated, MEOS baseline forecast.

The *technical potential scenario* is the hypothetical upper-limit case. This scenario estimates the electricity savings and load shifts that could be achieved if the best presently available efficiency and load management technologies were deployed in all households where they can be physically applied.

The *program-based scenario*, or *achievable* potential, investigates the extent to which incentives programs and efficiency standards could be used to motivate customers to invest in more energy-efficient buildings, lighting, and appliances, and to participate in load management programs.

MEOS revised its forecast during the preparation of this study. All numbers quoted here reflect the original residential forecast as developed by MEOS Work Group 5 in Sept. 1986.

2.1.3. Supply curve representation. Where applicable, the demand-side resource is summarized in the form of supply curves. The supply curves specify for a given year the relative cost and size of the available demand-side resource from each technology.

Figure 1 shows how the four scenarios for total electricity use and the supply-curve of efficiency improvements relate to each other. In effect, the supply curve of program-based savings connects the frozen efficiency forecast with the projections of achievable savings, while the supply curve of technically available savings does the same for the projection of the technical potential scenario. Depending on the marginal cost of supply, a greater or lesser amount of savings will be economically efficient, and the cost-effective portions of the supply curve contract or expand. This, in turn, moves the end-points of the scenario projections upward or downward.

2.1.4. Preliminary evaluation of demand-side resource cost-effectiveness. In this evaluation, a distinction is made between the consumer perspective, the utility perspective, and the all-ratepayer perspective. We aggregate the demand-side resources into cost bins related to the short-run marginal costs and long-run marginal costs of generating electricity and peak power from conventional supply sources. This allows a preliminary cost-effectiveness evaluation from the utility perspective. In this assessment, dynamic interactions between the implementation of demand-side resources and supply-side marginal costs are not captured.

3. BASELINE DATA AND METHODOLOGY

3.1. Scope and baseline data

Our study investigates electricity use in the service territories of the two major Michigan utilities, Consumers Power (CP) and Detroit Edison (DE). These two territories account for about 85 percent of all electricity use in Michigan.

3.1.1. End-uses and technologies. We investigated the following demand-side measures:

Demand-side measures with impact on energy use and peak demand (efficiency improvements):

> More efficient refrigerators
> More efficient freezers
> More efficient air conditioners
> More efficient building shells
> More efficient electric space heating equipment (heat pumps)
> More efficient hot water use
> More efficient electric water heaters
> Solar water heaters

Fuel-switching measures:

> Fuel switching from electricity to gas in water heaters, clothes dryers, and ranges

Load control measures:

> Air conditioner cycling
> Air conditioner load shedding
> Water heater cycling
> Thermal storage
> Demand subscription

The total energy conservation resource reported in this study refers to these end-uses and measures only, and only to the resource for the two major utilities. We have not extrapolated to all residential end-uses combined, nor do we develop state-wide potentials.

3.1.2. Baseline Data. The baseline year for our study was 1985. Key data on the two utilities are summarized in Table 1.

Table 1. Michigan Utility Profiles.				
	Detroit Edison		Consumers Power	
	1984	1985	1984	1985
Total Sales (10^9 kWh)	35.887	36.695	25.230	25.483
Residential Sales (10^9 kWh)	10.150	10.077	8.181	8.178
% of Total Sales	28.28	27.46	32.43	32.09
Residential Customers	1,630,000	1,643,000	1,210,000	1,220,000
Avg Ann Use Per Customer (kWh)	6253	6165	6789	6720
System Peak Demand (summer MW)	7350	7171	4840	4700
Heating Degree Days (base 65°F) *	6869	6846	6869	6846
Cooling Degree Days (base 65°F) *	603	507	603	507
Normal HDD (30yrs) (65°F) *	6802	6802	6802	6802
Normal CDD (30yrs) (65°F) *	604	604	604	604

* Degree-day data reflect average Michigan temperatures. All our findings are reported for both companies combined.

The MEOS frozen efficiency and business-as-usual forecasts relied on detailed historical saturation, efficiency, and consumption data that reached back to 1967. These data were used to calculate the energy impacts of replacements of old equipment, based on fixed lifetimes for each category of equipment. These MEOS household numbers, saturations, and behavior functions were also used for the other three scenarios and forecasts. The size and availability of demand-side potentials over time is thus mapped in close correlation with the actual vintage and efficiency composition of Michigan's equipment stocks. In addition, hourly loadshapes by end-use and by 8 day-types were developed from utility submetering studies.

The end-uses studied by LBL (excluding fuel switching) cover 67 percent of combined 1985 sales. Figure 2 shows graphically how the major end-uses contribute to total combined 1985 sales. These end-uses consumed the output of two and a half large (1000) MW central stations.*

Using measured end-use load shape data, we also estimated the baseline and system peak demand contributions of the major end-uses. On average, the end-uses studied by LBL make up 21 percent of the total system peak. Both utilities are summer peaking. The end-uses we studied are collectively summer peaking only for Detroit Edison, which has much larger air conditioning loads than Consumers Power.

3.2. Characterization of demand-side resource costs

3.2.1. Definitions of cost of conserved energy and peak power. The *cost of conserved energy (CCE)* is the annual cost of implementing an efficiency or peak demand reduction

* To convert savings into equivalent baseload capacity, we used a 60.4 percent capacity factor (5300 full-load hours per year) and a six percent transmission and distribution loss.

measure, divided by the annual energy savings. It is defined by the following formula:

$$cost \ of \ conserved \ energy = \frac{investment \ rate \times capital \ recovery + O/M \ incremental \ cost}{annual \ energy \ saved} \quad (1)$$

The capital recovery rate (CRR) annualizes the investment. In terms of the real annual discount rate d and the lifetime n, it is given by the expression:

$$f = \frac{d}{1 - (1+d)^{-n}} \quad (2)$$

The Cost of conserved peak power (CCPP$_{20}$). While the CCE is annualized over the life of the hardware (e.g. ten years for a room air conditioner), the CCPP is present-valued over the life of the avoided peak power plant, which is taken to be 20 years. The formula is:

$$cost \ of \ conserved \ peak \ power = \frac{net \ present \ value \ of \ (investments + O/M \ incremental \ cost)}{diversified \ peak \ demand \ saved} \quad (3)$$

For hardware that is replaced sooner than 20 years, we add the present value of all replacement costs.

3.2.2. Selection of technology levels. With the exception of refrigerators, technologies are limited to measures that are commercially available in 1987 in the U.S. This excludes many advanced technologies that are currently in the stage of advanced prototype development and testing. In the case of refrigerators, the 1992 California appliance efficiency standard, which exceeds the performance of the best 1987 models on the market, was incorporated.

The limitation to currently commercialized technologies leads to a truncation of efficiency improvements in the later part of the 20-year scenario period. In practice, significant further efficiency improvements could be expected.

3.2.3. Cost data. Existing data bases on the costs and performance characteristics of a range of demand-side technologies were expanded, updated, and adjusted to Michigan climate and economic conditions. The cost data reflected in our supply curves include materials, labor, and maintenance costs other than replacement of the measure. The sources for our cost data include manufacturer's retail prices, monitored construction experience, price lists from audit and conservation programs, and engineering-economic calculations. For building shell measures, we distinguish between retrofit costs and new construction costs, since these can be very different.

3.2.4 Lifetimes and discount rates. Amortization of investments is done over the useful life of the measure, i.e., the period during which it will provide energy or peak power savings. Two discount rates for the the analysis were uniformly set by MEOS, i.e., 3 percent and 7 percent in constant dollars.

3.2.5. Program cost data. Data on the program administration costs, incentives requirements, participation rates, and market penetration fractions were gathered from utilities in various parts of the U.S., and complemented by interviews with program managers.

3.3. Incentive Program Analysis

It is widely recognized that the implicit economic decision-making in the purchase of energy-consuming devices by energy consumers does not follow lines of least-cost rationality. One objective of the MEOS demand-side analyses was to define the amount of energy and peak demand savings that could be realistically obtained from the technologies considered, if utilities were to *aggressively* pursue specific incentive programs to shift the purchasing behavior of their customers. This requires incorporation of achievable annual participation rates and cumulative penetration fractions into the analysis. Also, the administration and incentives cost of such programs must be ascertained. Such an analysis is presently difficult because only a limited number of programs have been carefully monitored over an extended period of time, and even fewer programs have been aggressively pursued.

3.3.1. Program participation rates. Our method was to assume that aggressive programs would provide incentives in such a way as to eliminate all extra first costs for participants (full incentives). High incentives alone do not guarantee high participation rates, but they do make participation less sensitive to non-financial program features. In addition, we assumed that all programs would observe a number of lessons that have emerged from past experience with residential programs, such as:

- Large-scale programs should be preceded by well-designed and thoroughly evaluated pilot and demonstration projects.

- Monitoring, feedback, and quality control functions should be built into all aspects of the implementation process.

- Programs should make use of market segmentation techniques and other methods to flexibly target different consumer groups and local conditions.

- Community groups and trade allies can be one of the most effective agents in the implementation process.

- Promotion of demand-side measures should emphasize how such measures contribute to the broad values sought by customers, such as increased comfort, safety, reliability, environmental health, and productivity.

- Information and incentives strategies should build upon market forces wherever possible, and reward savings rather than expenditures.

- Efficiency standards can be one of the most effective complements to incentives-based programs.

The program-based scenario in this study assumes that full incentives are offered until most or all existing stocks have been turned over. The scenario also foresees for each end-use a two- to five-year pilot project phase in which program designs are optimized before full-scale implementation begins. The onset of major savings is correspondingly delayed. Figure 3 shows schematically how various program phases and the customer response might evolve over time.

The impact of increasing incentives on penetration fractions is schematically illustrated in Fig. 4, which shows the cost of conserved energy or peak power based on the sum of technology costs and costs for program administration (processing of rebates, advertising, etc.), for a stylized two-measure supply curve.

3.3.2 Program incentive and administration costs. Two issues in calculating the cost of program-based demands-side resources are the free rider and spill-over effects of the program. In most cases, variable incentives based on careful analysis of existing purchasing patterns can minimize the free-rider problem. Spill-over effects will also counteract the free-rider problem. Both together may eliminate the problem entirely while leaving a significant net spill-over effect that provides program-induced savings from non-participants at zero program cost.

Depending on the technology and end-use, we take into account free-rider fractions of up to 20 percent. At the same time, we neglect the spill-over effect of programs on consumer behavior and retail markets.

Another important variable in the utility cost of demand-side resources is the level of incentive that is needed to achieve desired levels of customer participation. Utility experience shows that substantial participation can be achieved at significantly less than full incentives. Utilities have also found that as programs mature, the level of incentives can be lowered and in some cases entirely substituted by effective information and promotion. This is partly the result of the programs' spill-over effects on the retail markets, such as better informed salespeople and greater stocking of high efficiency merchandise. Another reason is that in the larger utility incentives programs, the prices of demand-side technologies tend to drop as the program moves them out of their small, specialty or high-income market niche, where dealer mark-ups are high.

Since in our program scenario full incentives are provided over extended periods of time (eight to 15 years), the program costs developed in this study define an upper bound for the likely cost of buying efficiency from customers.

3.3.3. Standards vs incentives programs. Our program-based scenario relies on comparatively expensive rebate programs rather than on state-promulgated standards or combinations of both. Only the 1987 federal US appliance efficiency standards are incorporated. They are also taken into account in the baseline forecast and thus do not contribute to the savings as defined here. We do not assume tighter standards though more stringent requirements for the mid-1990s would be feasible and economically justified for a number of end-uses.

To the utility, efficiency standards are one to two orders of magnitude cheaper than incentives programs. An optimal program mix would combine standards to raise the efficiency floor of the market and use incentives to create a market pull on the high efficiency end. Such an optimization could reduce the utility costs of demand-side resources significantly below the level estimated in this study.

3.3.4. Analysis of dispatchable load management options. In analyzing load management options, we limited ourselves to an assessment of their technical potential. The technical potentials we calculate are maximum potentials based on 1985 end-use efficiencies and diversified loads. We did not develop program-based scenarios for implementing the technically feasible load shifts, because the economic priority between load control and conservation options can only be evaluated in an integrated hourly analysis, and because several of the options studied overlap in complex ways.

We calculate the maximum load shift that could be achieved with load management techniques, based on 1985 end-use efficiencies and diversified loads. The maximum participation fractions of residential customers are based on an analysis of the system load curve at system peak, and an optimization between peak savings at the peak hour and the subsequent peak when load control is ended ("payback spike").

3.4. Least-Cost Perspectives in Evaluating Demand-Side Resources

3.4.1. Least-cost perspectives for integrated planning. In the context of a least-cost planning exercise, the cost-effectiveness of demand-side resources can be evaluated from a number of perspectives. The delineation among some of these perspectives is a function of regulatory practice. We define three: the customer perspective, the utility perspective, and the all-ratepayer perspective.

- In the *customer perspective*, the capital and recurring costs of the demand-side measure to the customer are compared with the electricity bill reductions, based on projected average rates over the lifetime of the measure (life cycle costs). Tax savings may also apply. We use the term *technology cost* to denote the the the customer perspective.

• In the *utility perspective*, the utility buys from a subgroup of customers a certain increment of energy or peak power savings. Only program administration and incentives costs are counted. These may in sum be higher or lower than technology costs. Program costs are then compared to the change in electricity production costs. This change consists of avoided energy costs and avoided or deferred capital costs for capacity expansion. Revenue losses are not counted, because it is assumed that the utility will be recovering these through rate increases from the ratepayers.

• In the *all-ratepayer perspective*, the program administration costs of the utility and the full technology cost are counted, Again, these costs are compared with the benefits of reduced generating costs.

3.4.2. Relationship of technology costs and program costs in the utility perspective. When programs are based on full incentives, the program cost from the utility perspective is the technology cost plus the cost of program administration. In terms of the cost of conserved energy (CCE) we can write:

$$Program\ based\ CCE = technology\ CCE\ (1 + \frac{administration\ costs}{incentive\ costs})$$

With full incentives, the administration-to-incentives ratio in well-managed programs is typically of the order of ten percent. Where applicable a free rider or spill-over term can be added to the equation.

3.4.3. Methods for preliminary cost-effectiveness evaluation. At the time of this writing, results from the MEOS integrated analysis are not available. An integrated supply-side and demand-side analysis is important from the utility and all-ratepayer perspective. Both incentives payments and revenue losses to the utility are treated as transfer payments. An informative preliminary assessment of cost-effectiveness can be made by assigning the demand-side resources to different cost bins or blocks that correspond with the marginal energy and capacity cost structure of the utility. For electricity savings, the following three cost blocks are used:

1. *Electricity resources with costs lower than short-run marginal costs.* These savings would generally be cost-effective, since few utilities if any have access to new supplies at costs below their own short-run marginal costs. It is then cheaper to buy this demand-side resource than to operate existing capacities. The resource moves ahead of existing plants in the dispatch order.

2. *Electricity resources with costs comparable to short-run marginal costs* are cost-competitive with existing capacities but their dispatch priority needs to be evaluated on the basis of additional analyses. Conservation resources will tend to be cost-effective if the lifetime of the conservation measure extends into the period where new capacities or more expensive fuels will be needed.

3. *Electricity resources with costs higher than current short-run marginal costs but lower than the cost of power from new power plants.* These may or may not be cost-effective on a life-cycle basis. Theoretically, one would defer such resources until additional capacities are needed, and then dispatch them. In practice, conservation resources cannot be switched on and off like a power plant. Suppose an appliance has a 20 year life and efficiency improvements fall into this cost block. Not investing in a more efficient appliance now foregoes savings in that application for 20 years. Extra costs incurred in

early years when marginal costs are low must therefore be balanced on a net present value basis with benefits in later years when new capacities are needed and marginal costs are high.

For load management options the evaluation is somewhat different. The common reference point is the peaking turbine on the supply-side. All load management options have approximately zero capacity value so long as existing capacities are sufficient to keep loss-of-load probabilities low. However, load management may also have some energy cost benefits to the utility in so far as some load is moved to power plants with fuel costs that are lower than those of peaking plants. Neglecting this second-order effect, the cost bins for load management options can be interpreted as follows:

1. *Load management and conservation resources with costs of conserved peak power ($CCPP_{20}$) less than, or comparable to, that of a peaking turbine.* These options may or may not be cost-effective. A more detailed investigation is needed to determine the point in time when such peak load savings begin to have capacity value in the utility system. Again the degree of dispatchability is an issue. Direct control type programs usually can be simply deferred until such time when they become cost-effective.

2. *Load management and conservation resources with $CCPP_{20}$s greater than that of a peaking turbine.* As load control programs, these are clearly not economical. On the other hand, many conservation resources with comparatively high $CCPP_{20}$s would still be cost-effective on energy grounds alone, and would thus be dispatched irrespective of their capacity value.

Table 2 shows the estimated short-run marginal costs in 1986 mills/kWh from existing Michigan power plants.

Table 2. Short-run marginal electricity costs from existing capacities (1986 mills/kWh)

Load Segment	Segment Number	Annual-Average Marginal Cost
peak	1 oil-fired peaking	60.4
peak	2	35.7
peak	3	33.7
peak	4	33.3
peak	5	33.0
peak	6	32.7
peak	7	32.3
mid-peak	8	31.5
off-peak	9	29.5
off-peak	10	28.1
base-load	11 coal, nuclear	27.4

The table distinguishes 11 load segments and three seasons. On average, baseload power costs 27.4 mills/kWh, and power in the mid-range of the load duration curve costs up to 35.7 mills/kWh. The last load segment, corresponding to peak load power production from small oil-fired peaking plants, costs 60.4 mills/kWh. These figures need to be corrected to account for transmission and distribution losses, which we take to be 6 percent. Figures for long-run marginal costs from conventional power plants were not provided, but are generally expected, under most favorable assumptions, to be at least as high or higher than current average electricity rates, i.e., at least 8 cents/kWh or more. For peaking turbines, capital costs are commonly estimated as $500-700/kW.

4. RESULTS: THE SIZE OF MICHIGAN'S RESIDENTIAL DEMAND-SIDE RESOURCE

4.1. Cost of conserved electricity

Table 3 shows the costs of conserved energy for the efficiency and fuel switching measures examined. Costs of conserved peak power were almost always significantly higher than the cost of peaking turbines and are not shown here. Among the efficiency measures, several high-cost items stand out: evaporative coolers, high efficiency air conditioners, solar water heaters, and heat pumps in buildings that do not use air conditioning. With the exception of solar water heaters, the high cost of these technologies is mainly climate-related. These options were not considered in our program-based scenario.

Table 3. Cost of conserved electricity
from demand-side efficiency
and fuel switching options under Michigan climate conditions.

Measures	CCE ¢/kWh
More efficient refrigerators	
Standard	1.6-5.0
Auto defrost, top freezer	0.9-2.2
Auto defrost, side-by side	1.0-1.9
More efficient freezers	
Manual defrost	0.8-2.7
Auto defrost	0.7-1.7
Lighting	1.2
Water heating	
Water temp. setback	0.0
More eff. hot water use	0.3-2.8
More efficient water heaters	1.2-1.6
Heat pump water heaters	4.6
Solar water heaters	11.4-20.5
Air conditioning	
Central	10-21
Room	5.4-15
Evaporative coolers	25-32
Shell improvements, SFH	
New homes	0.5-4.3
Existing homes	0.5-6.2
Heat pump systems	
Home without AC	5.2-24.1
Home with AC	2.2-13.5
New homes with AC	0.5-13.8
Fuel switching	
Water heater	5.0
Clothes dryer	3.7
Range	6.8

4.2. Demand-side resources from efficiency improvements

The results for the four electrical efficiency projections are shown graphically in Fig. 5. Table 4 summarizes the efficiency savings for 1995 and 2005.

Table 4. Comparison of Projected Residential Sector
Baseload Equivalent Demand (GW), 1995 and 2005:
Program Achievable Potential and Technical Potential

	MEOS Baseline	Program Achievable	Technical Potential
1995	2.41	1.91	1.39
2005	2.36	1.68	1.04

4.2.1. Technical potential scenario.

Electricity savings:

Technical potential electricity savings are 42 percent (5110 GWh) in 1995 and 56 percent (6590 GWh) in 2005 compared to the MEOS forecast. Total savings are equivalent to 1020 and 1320 MW baseload capacity, respectively. These figures do not include fuel switching.

Peak demand savings:

1995 combined peak load savings are 1100 MW in the winter and 800 MW in the summer, or 46 percent and 35 percent. The corresponding figures in 2005 are 1380 MW and 1110 MW, or 56 and 49 percent. These savings do not include those available from direct load control strategies.

4.2.2. Program-based scenario

Electricity savings:

Table 4 shows the relationship of program-based savings to the MEOS forecasts. Compared to the MEOS business-as-usual forecast, conservation programs achieve a 21 percent saving by 1995 (2500 GWh), and a 29 percent saving by 2005 (3410 GWh). The baseload equivalent MW savings for the two companies combined are 500 MW in 1995 and 680 MW in 2005.

Figure 6 shows a pie chart of the savings by end-use for 2005. The largest contributor to the total savings is improvement in lighting efficiency (45 percent in 1995 and 33 percent in 2005), followed by hot water and water heating savings (24 percent and 25 percent), refrigerator and freezer savings (18 and 25 percent), space heating savings (10 percent and 14 percent), and finally savings in air conditioning from better building shells (3 percent and 3 percent).

Peak demand savings:

In 2005, the demand reduction for the combined territories at summer system peak is 20 percent (450 MW), and 34 percent during winter peak (835 MW). The corresponding figures for 1995 are 14 percent (320 MW) and 26 percent (620 MW). While the MEOS forecast would invert the winter to summer peaking situation in the *combined* territories for the end-uses studied, the program-scenario maintains the summer peaking. Within each company's territory, the qualitative winter/summer peak relationship remains unchanged.

4.2.3. Comparison of Technical Potential and Program-Based Efficiency Resources. In the program-based scenario demand-side efficiency improvements achieve 50 percent of the technical potential savings in 1995, and 52 percent of the savings in 2005 (see Fig. 5). This penetration into the technical potential is an average over all end-uses and varies somewhat from end-use to end-use. Space heating and lighting approach the technical potential most closely, while the program-based refrigerator, freezer and air conditioning savings lag furthest behind.

4.3. Supply curve of demand-side resources from efficiency improvements

Table 5 shows how the 2005 savings are distributed over three marginal cost ranges, for 3 percent and 7 percent real discount rates, respectively. The corresponding supply curve is shown in Fig. 7.

With a 3% discount rate, 79 percent of the total program-based savings can be bought for less than the short-run marginal cost of electricity production, and another 14 percent of the resource is cost-competitive with existing supplies. For reference, only 7 percent of the achievable savings cost more than the operation of current capacity, but even these savings cost less than the marginal cost of power from adding new capacity to the ratebase.

Figure 7 and Table 5 show that for a 7 percent discount rate, results do not change substantially. 77 percent of the efficiency resource are still cheaper than short run marginal costs. 2 percent are competitive, and 21 percent are higher.

4.4. Demand-side resources from fuel switching.

The potential electricity and peak power savings from fuel switching for three appliances are shown in Table 6.

Table 6. Residential Fuel Switching Potential in Michigan				
	Units	Consumers Power	Detroit Edison	Both Utilities
Grand total switchable electricity for all appliances	GWh/year	1860	1620	3480
% of 1985 residential usage	%	18	20	19
Equivalent baseload (@ 57% plant factor)	MW	370	320	690
Grand total switchable demand at system peak				
summer	MW	201	228	429
winter	MW	130	147	277
Fraction of 1985 system peak (summer)*	%	4.3	3.2	3.6
* This is measured at the utility's 3PM system peak; the residential peak occurs at 7PM.				

Table 5   Macro Supply Curve of Annual
Electricity Savings by Block, Consumer's
Power and Detroit Edison
Territories, Year 2005: 7% Discount Rate

End-Use/ DS-Measure	Annual Savings over MEOS (GWh)			Utility Cost of Conserved Energy ¢/kWh	Rank
	Block 1 0-3¢	Block 2 3-4¢	Block 3 >4¢		
1. Refrigerators					
a. Std. 1992/low tech.			51	9.3	18
b. Frost-free 1992 Std.	189			2.5	7
c. Frost-free low tech.			149	4.1	13
d. 2nd units	226			1.7	5
e. Low-income prog.			38	11.3	19
2.Freezers					
a. Manual low tech.			162	4.9	15
b. Auto-defrost low tech.		20		3.1	10
3. Air Cond.	114			0.0	1
4. Lighting	1123			1.9	6
5. Space Heating					
a. Exist. EHH, Block 1	86			2.9	9
b. Exist. EHH, Block 2		15		4.0	12
c. New Houses, Block 1	140			2.7	8
d. New Houses, Block 2		26		3.6	11
e. New EHH, AAHEX			41	5.0	16
f. Furnace fans	173			0.0	2
6. Water Heating					
a. Temp. setback	191			0.0	3
b. Hi-eff. showers & faucets	385			0.3	4
c. Clothes washers			112	4.5	14
d. Eff. water heaters			168	8.0	17
Total	2627 77%	61 2%	721 21%		
Average CCE					
Block 1				1.4	
Block 2				3.5	
Block 3				6.0	
All Blocks				2.4	

Switching electric water heaters, ranges, and clothes dryers to gas would save 3500 GWh of electricity. Summer peak savings are 430 MW. Fuel switching in water heating is the largest contributor with 1480 GWh and 310 MW of peak demand savings.

As shown in Tab. 3, the costs of conserved electricity for these measures range from 3.7 to 6.8 cents/kWh. Most of this cost is, however, the cost of buying gas, which is included in the CCE as an operating cost. The extra cost of buying the gas appliance upon replacement of the electric appliance and of hooking it up are only about 10 to 33 percent of the total CCE. A utility program could likely buy the fuel switching resource by simply offering to pay for these conversion costs. Including administration, such a fuel switching program would cost an estimated 0.5 cents/kWh for water heating, 1.4 cents/kWh for clothes dryers, and 3.0 cents/kWh for ranges. Like many of the efficiency improvements, fuel switching would cost less than operating existing generating capacities.

4.5 Peak demand savings from load management options: Technical potential and costs

Table 7 shows the potential peak demand savings and associated costs per kW peak savings for the load-control measures.

Demand subscription appears to be the most cost-effective program, with a maximum potential of about 600 MW savings during summer peak. Costs are well below those of a peaking turbine. Air conditioner cycling, water heater interruption, and space heating thermal storage are more expensive than a reference peaking turbine. Extending the air conditioner cycling period leads predictably to greater cost-effectiveness of that option, but makes that measure simultaneously less distinguishable from load shedding.

It is important to note that if the energy savings of the program scenario are implemented, the size of shiftable loads, and therefore the number of people with sufficiently large loads to be eligible for the program, will also decrease. For example, the central air conditioning loads will have decreased by 25 percent on account of improved building shells in gas-heated homes, and by a further (multiplicative) 18 percent on account of air conditioner standards. The combined 38.5 percent reduction in peak loads will increase the average $CCPP_{20}$ of air conditioner load shedding from \$219/kW (diversified) in the case of DE to \$356/kW. The peak power cost of demand subscription, now estimated to be \$266 for high use customers, would rise to \$433/kW.

5. CONCLUSIONS

Providing residential electrical services at least economic cost would mean a major increase of utility investments on the demand-side. Most of Michigan's residential demand-side resource is cost-effective against the cost of power from existing coal and nuclear baseload capacities, which represent the least expensive portion of the generation mix. In the case of Michigan, where business-as-usual forecasts already show stagnating residential electricity use, this strategy would result in negative residential electricity growth. An integrated analysis of supply- and demand-side options needs to be performed to assess the costs of such a strategy from the all-ratepayer perspective. The environmental benefits of demand-side investments also need to be considered. Finally, more detailed program implementation studies are needed to reduce uncertainties in the size and deployment time of demand-side resources.

REFERENCE

Krause, Florentin, et al., "Analysis of Michigan's Demand-Side Electricity Resources in the Residential Sector," Lawrence Berkeley Laboratory Report, LBL-23025. February 1987.

ACKNOWLEDGEMENT

The work described in this report was funded by the Assistant Secratary for Conservation and Renewable Energy, Office of Building and Community Systems, Building Systems Division of the U.S. Department of Energy under Contract No. DE-AC03-76SF00098.

Table 7. Summary of Dispatchable Demand-Side Options:
Technical Performance and Cost Effectiveness

STRATEGY	PARTICIPANTS	LOAD SHIFT (MW)	CAPITAL COST ($1985/kW)	$CCPP_{20}$,3% ($1985/kW)	$CCPP_{20}$,7% ($1985/kW)
Demand Subscription					
Consumers Power (base case)	78,261	216	51	266	203
Detroit Edison (base case)	153,541	423	51	266	203
Thermal Storage—SF homes only, 53%					
Consumers Power (base case)	27,030	134	815	981	933
Detroit Edison (base case)	13,780	68	815	981	933
Water Heater Interruption					
Consumers Power (base case)	59,619	34	151	928	704
Detroit Edison (base case)	89,897	52	151	928	704
Air Conditioner Load Shedding					
Consumers Power (base case)	64,799	128	44	270	203
Detroit Edison (base case)	73,943	180	36	219	164
Case I: 20 minute cycling periods					
Consumers Power	111,000	73	132	809	614
Detroit Edison	221,000	180	107	656	498
Case II: 40 minute cycling periods					
Consumers Power	97,198	128	66	404	307
Detroit Edison	110,970	180	54	328	249

Figure 1.

Forecasts, technical potentials, and conservation supply curve.

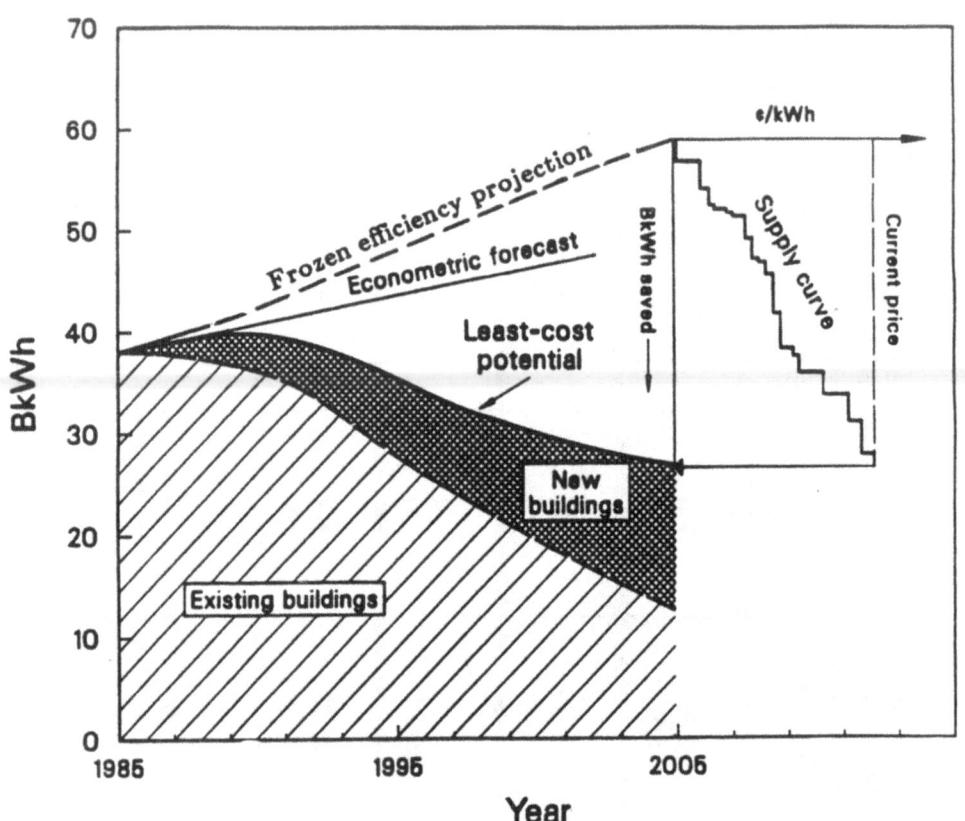

Figure 2.

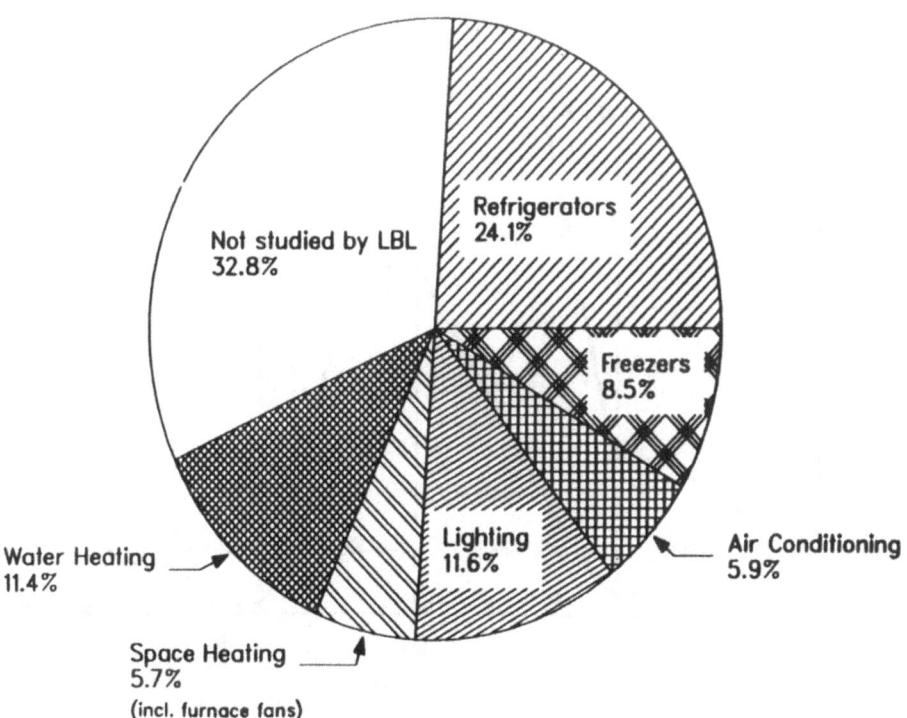

1985 Residential Electricity Sales
Combined CP and DE Territories,
Total Sales = 18700 GWh

638

Figure 3.

Program Phases and Timing

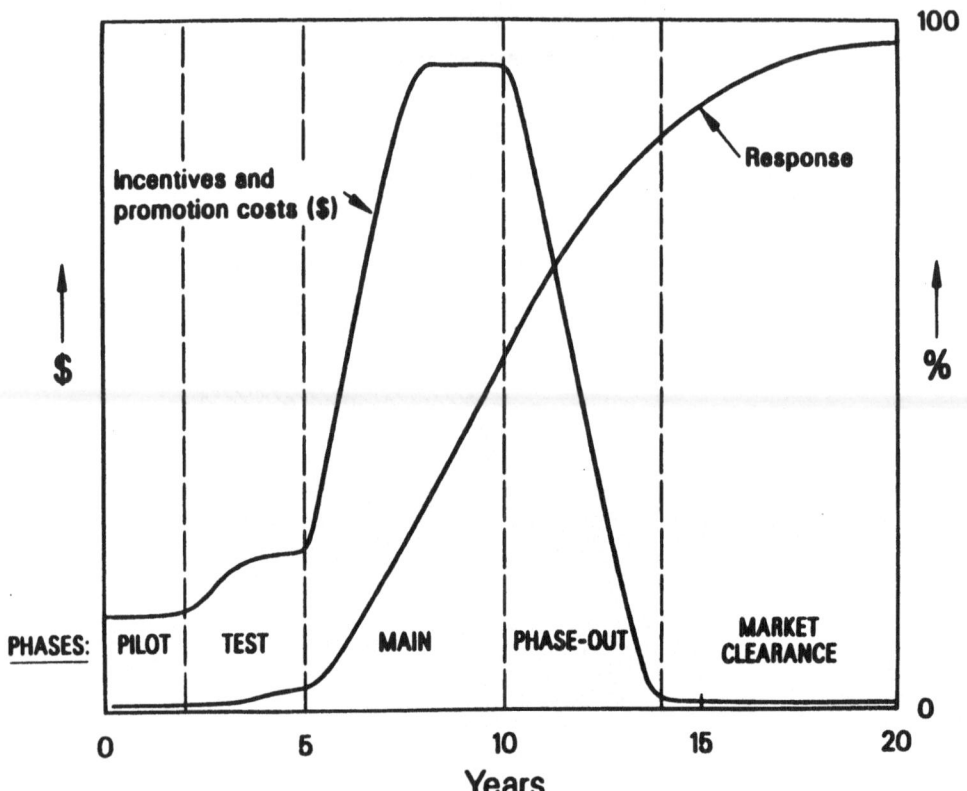

Figure 4.

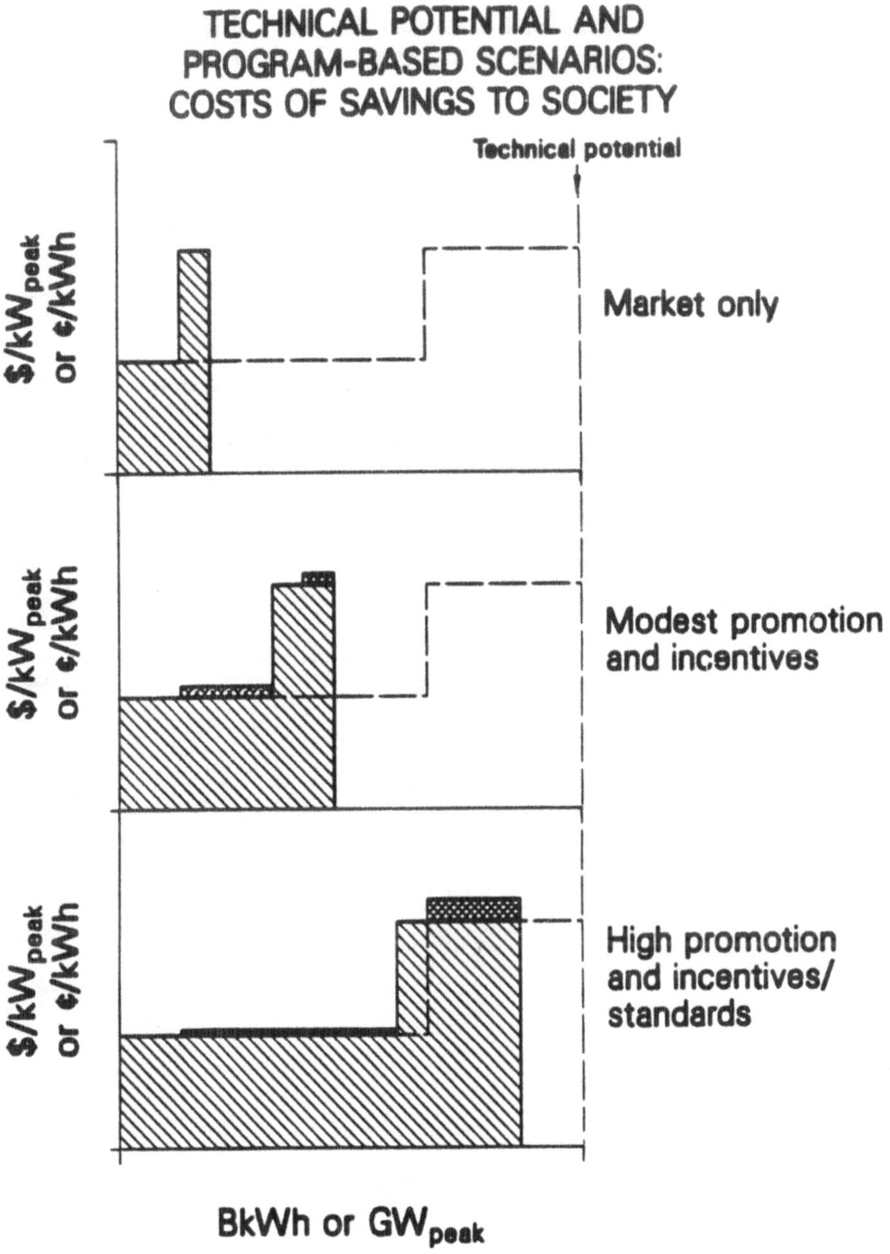

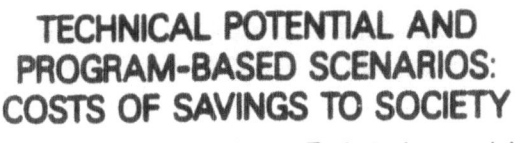

TECHNICAL POTENTIAL AND
PROGRAM-BASED SCENARIOS:
COSTS OF SAVINGS TO SOCIETY
Technical potential
$/kW_peak or ¢/kWh
Market only
$/kW_peak or ¢/kWh
Modest promotion and incentives
$/kW_peak or ¢/kWh
High promotion and incentives/ standards
BkWh or GW_peak

Figure 5.

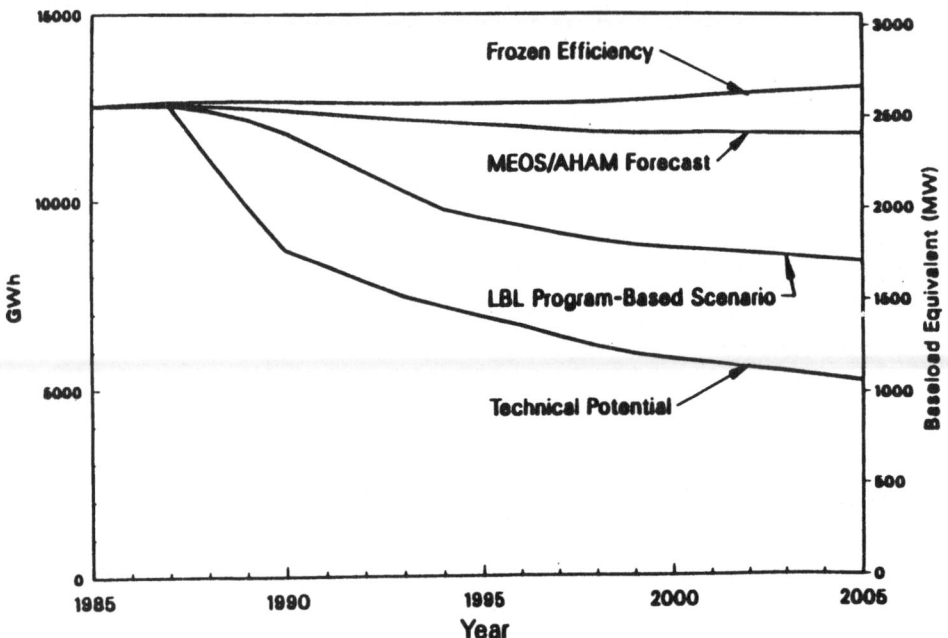

Change in Residential Electricity Use, 1985-2005

End-uses Studied by LBL, CP and DE Territories, no fuel switching

Figure 6.

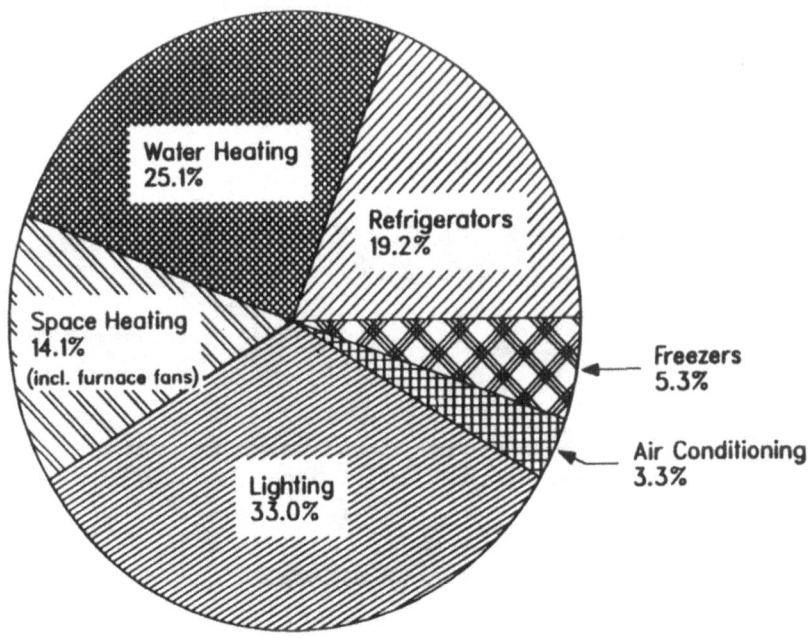

Annual Program-Based Electricity Savings
Over MEOS Forecast in Year 2005
Breakdown by End-use, CP and DE Territories
Total Savings = 3408 GWh

XCG 8612-12325

Figure 7.

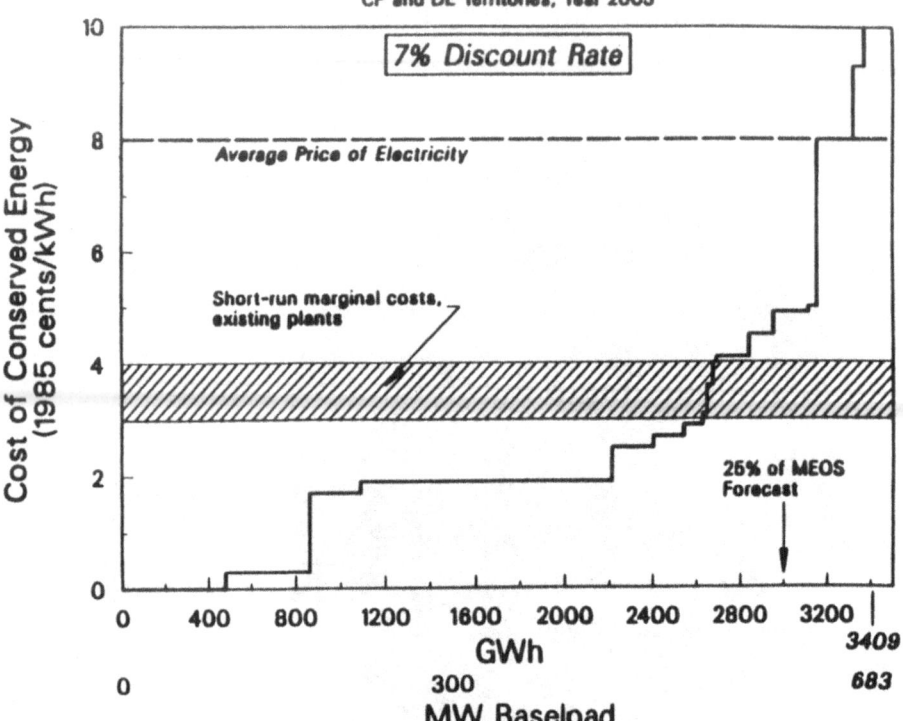

Macro Supply Curve of Electricity Savings
Program-Based Scenario
CP and DE Territories, Year 2005

Index

absorption heat pumps 30, 403, 406, 408
AC drives 298, 299
AC motor 299
AC-induction motors 393
ACCET 173
Accet's 181
accuracy 177
acid rain 250
actuators 127
Adjustable Speed Drives (ASDs) 36, 273, 278, 321, 567
advanced technologies 613
AFUE 493
aggregate savings 572
agricultural sector 561
air conditioner standards 40
air conditioners 19, 39, 49, 486, 603,
air conditioning 102, 132, 423, 485, 542, 572, 579, 630
air knives 383
air quality 424
alarm reporting 131
Alaska pipeline 34
Algeria 253
analog control 130
annual cooling load 473
annual energy intensity 52
appliance standards 37, 51, 90, 485
appliances 49, 435, 449, 450, 591, 598, 622
arc furnaces 134, 201, 597
arc lamp 337
ASDs 36, 273, 278, 321, 323, 567
ASHRAE 51, 323, 430, 469, 477, 580
asynchronous machines 281
asynchronous motors 304, 307
audit 142, 188
Australasia 18
Austria 408, 409, 418, 524, 526
automatic adjustment 132
automatic control 542, 589
automation 126
automobile energy-use 48
automobiles 48

balance of payments 26
ballast 352, 353, 574, 576

base case scenario 617
base scenario 607
baseline data 622
baseload 97, 113, 488
baseload capacity 496
baseload plants 22
batteries 529
battery back-up 127
battery chargers 134
behavior 74, 460, 463, 497
benefit-cost ratio 488
BEPS 477
beverage industry 375
bi-radiant oven 610
billing 217
biomass conditioning 531
biotechnologies 375
blast furnace 380, 597
BRITE 35
BTUs 17
building codes 522, 529
building cost 482
building energy R&D 30
building fabric 51
building sector 32
building standards 26, 469, 477, 581
buildings 19, 519, 522
buildings RD&D 527
business-as-usual forecast 42, 622

CACs 49, 580, 610
California 26, 486
California energy commission 487
California peak power 41
CALMS 174
CALMU 174
Canada 490
canning industry 375
capacity factor 570
capital cost 90, 97
capital expenditure 23
carnot 393, 401
carrier signal 140
CCE 624
cement industry 597
central AC 493